Principles of Environmental Thermodynamics and Kinetics
Fourth Edition

T0138810

Principles of Environmental Thermodynamics and Kinetics
Fourth Edition

Kalliat T. Valsaraj and Elizabeth M. Melvin

CRC Press
Taylor & Francis Group
Boca Raton London New York

CRC Press is an imprint of the
Taylor & Francis Group, an **informa** business

CRC Press
Taylor & Francis Group
6000 Broken Sound Parkway NW, Suite 300
Boca Raton, FL 33487-2742

First issued in paperback 2020

ISBN-13: 978-0-367-57205-1 (pbk)
ISBN-13: 978-1-4987-3363-2 (hbk)

Library of Congress Cataloging-in-Publication Data

Names: Valsaraj, K. T. (Kalliat T.) author. | Melvin, Elizabeth M., author.
Title: Principles of environmental thermodynamics and kinetics / Kalliat T. Valsaraj and Elizabeth M. Melvin.
Other titles: Elements of environmental engineering
Description: Fourth edition. | Boca Raton : Taylor & Francis, CRC Press, 2018. | Revised edition of: Elements of environmental engineering / Kalliat T. Valsaraj. 2009. | Includes bibliographical references and index.
Identifiers: LCCN 2017040508 | ISBN 9781498733632 (hardback : alk. paper)
Subjects: LCSH: Environmental engineering. | Thermodynamics. | Chemical reactions.
Classification: LCC TD153 .V35 2018 | DDC 628--dc23
LC record available at https://lccn.loc.gov/2017040508

Visit the Taylor & Francis Web site at
http://www.taylorandfrancis.com

and the CRC Press Web site at
http://www.crcpress.com

Contents

Preface

During the twentieth century and the beginning of the twenty-first century, a number of issues relevant to both energy utilization and environmental concerns confronted the present generation. Resource challenges, solutions to environmental changes, and efforts to tackle them have been mired in debates, both scientific and nonscientific. Global climate change, water quality, air quality, and plastics and solid waste disposal are a few example topics of present-day debates.

Our prescription for alleviating environmental problems requires that we train students in the interdisciplinary aspects of environmental issues. A well-trained cadre of professionals can ensure that sound science is involved in the formulation of public policy. Environmental engineering is, by its very nature, interdisciplinary. It is a challenge to develop courses that will provide students a through, broad-based curriculum that includes every aspect of the environmental engineering profession. Traditionally, environmental engineering was a subdiscipline of civil engineering with primary emphasis on municipal wastewater treatment, sewage treatment, and landfill management practices. During the second half of the twentieth century, emphasis was placed on end-of-pipe treatment in manufacturing operations to control release of water, air, and solid wastes. With the realization that environmental problems are not confined to end-of-pipe treatment, attention turned toward pollutant fate and transport in the general environment, waste minimization, pollution prevention, and green engineering. Thus, stand-alone programs in environmental engineering began to appear in many universities.

Environmental engineers perform a variety of functions, most critical of which are process design for waste treatment or pollution prevention, fate and transport modeling, green engineering, and risk assessment. Chemical thermodynamics and chemical kinetics, the two main pillars of physical chemistry, are also crucial to environmental engineering. Unfortunately, these topics are not covered in the environmental engineering curricula in most universities. Chemical engineers take separate courses in thermodynamics and kinetics. They also take several prerequisites in chemistry, such as physical chemistry, organic chemistry, inorganic chemistry, and analytical chemistry. Most environmental engineering programs do not require such a broad spectrum of chemistry courses but rely on an introductory chemistry course for environmental engineers. However, such a course lacks proper depth in the two important aspects of physical chemistry. We believe that an additional course has to be taught that introduces these subjects and lays the foundation for more advanced courses in environmental process design, environmental transport modeling, green engineering, and risk assessment. It was to accomplish this objective that the first edition of this textbook was written in 1995. Based on its use in various universities, numerous suggestions for improvement were incorporated into the second and third revisions of this book in 2000 and 2009, respectively.

In writing this fourth edition, we have refocused the chapters so that the introductory aspects of chemical thermodynamics and kinetics are explored in one chapter with a subsequent chapter containing all applications. The examples are chosen to represent important applications but, since the choice is subjective, we do admit that some may be more relevant than others. The problems are of varying level of difficulty and they are ranked 1, 2, and 3, with 1 indicating the "least difficult" and 3 indicating the "most difficult" or "advanced." These are represented by subscripts beside the problem number. A separate Solutions Manual is available from the publisher for all the problems in the various chapters. We have provided several case studies and placed them in the Solutions Manual as a separate chapter.

Acknowledgments

This work would not have been possible without the unconditional love and support from our families, and we thank them. Valsaraj thanks his wife, Nisha, and his two children, Viveca and Vinay, for their support and encouragement throughout the completion of this project. Melvin thanks her husband, Adam, and her three children, Abigail, Charles, and Zachary, for their patience, sincere input, and support while working on this fourth edition.

Authors

Professor Kalliat T. Valsaraj is the vice president for Research & Economic Development at Louisiana State University (LSU). He holds the titles of Charles and Hilda Roddey Distinguished Professor and Ike East Professorship in chemical engineering. He is a fellow of the National Academy of Inventors (NAI), American Association for the Advancement of Science (AAAS), and the American Institute of Chemical Engineers (AIChE). In 2010, he was awarded the LSU Rainmaker Award in the Senior STEM category and in 2011 he was awarded the Distinguished Research Master award by LSU. The professional societies of AIChE and ACS (American Chemical Society) awarded him the Charles E. Coates award in 2012.

He received an MSc in chemistry from the Indian Institute of Technology, Madras, in 1980 and a PhD in chemistry (with chemical engineering as minor) from Vanderbilt University in 1983. He has mentored 15 PhD, 22 MS, and several postdoctoral students in addition to hosting a number of visiting professors in his laboratory.

Dr. Valsaraj's research area is in environmental chemical engineering. He has broad research experience in wastewater treatment, atmospheric chemistry, and modeling the fate and transport of contaminants in all three environmental media (air, water, and soil/sediment). His present research is concerned with the transformations of pollutants on atmospheric aerosols (fog, rain, ice, and snow), mercury sequestration in sediments, and studies on chemical dispersant design for sub-sea oil/gas spill. He is the author of 1 textbook (with three editions), 203 peer-reviewed journal articles, 28 book chapters, and 2 U.S. patents. He has given more than 250 national and international presentations and 27 invited seminars and plenary lectures on his research. His research has been supported by the NSF, EPA, DOE, DOD, USGS, and several private industries. Dr. Valsaraj was very active in one of the longest lasting (1982–2002) Centers of Excellence at LSU in the college of engineering, viz., the USEPA Hazardous Substances Research Center, which brought in about $30 million in research funds during its lifetime. He codirected another $10.34 million Consortium on Molecular Engineering of Dispersants (CMEDS) funded by the Gulf of Mexico Research Initiative. He has consulted for various private industries and also provided service to several review panels and state and federal agencies.

Elizabeth M. Melvin is currently a director of Academic Affairs in the College of Engineering at Louisiana State University (LSU). Dr. Melvin was formally a Professional in Residence and the Undergraduate Coordinator in the Cain Department of Chemical Engineering at LSU in Baton Rouge. She earned her BS in chemical engineering from The Ohio State University in Columbus, Ohio, in 2002 and her MS and PhD in chemical engineering from North Carolina State University in Raleigh, North Carolina, in 2008 and 2011, respectively. While in North Carolina State, the focus of her research was to design microfluidic devices for the detection and manipulation of various cell types. One application in particular was to design an electrically driven cell-focusing microfluidic device to be used in conjunction with an optical waveguide for environmental-based applications. Although she has held a number of positions in industry with companies such as Dow Corning, Johns Manville, and Hospira, Dr. Melvin's passion lies in teaching, mentoring undergraduate students, and promoting excellence in education for engineers.

1 Introduction

Pollution is an inevitable consequence of advances in human endeavors to improve our quality of life on this planet. As civilized societies learned to organize, humans realized methods to purify water for drinking purposes, dispose of excrement, build sanitary sewers, and perform municipal wastewater treatment to prevent communicable diseases. They also realized how air pollution can adversely affect human health and the necessity to control the same. Thus, the history of pollution is as old as the human species itself.

The twentieth and twenty-first centuries have been a period of rapid technological advances, which have helped us to harness the natural resources available. Along with these advancements, we have also created myriad environmental pollution problems. Pollution is undesirable and expensive, but it is an inevitable consequence of modern life. The reality is that we cannot eliminate pollution altogether, but we can certainly mitigate it through recycle, reuse, and reclamation.

1.1 ENERGY, POPULATION, AND POLLUTION

Energy consumption has increased dramatically as the population has exploded. Increased utilization of natural resources is necessary to sustain the various industries that drive the economy of industrialized nations. Unfortunately, increased industrial activity has also produced anthropogenic pollutants. The harnessing of nuclear power has left us the legacy of radioactive waste. Increased agricultural activity in both developed and developing nations has been necessitated in order to sustain the bourgeoning world population. Intensive agricultural uses of pesticides and herbicides have contributed to the pollution of our environment. Some of the major environmental problems that one can cite as examples of anthropogenic origin are (1) increased carbon dioxide and other greenhouse gases in the atmosphere, (2) depletion of the earth's protective ozone layer due to man-made chlorofluorocarbons, (3) acid rain due to increased sulfur dioxide as a result of fossil fuel utilization, (4) atmospheric haze and smog, polluted lakes, waterways, rivers, and coastal sediments, (5) contaminated groundwater, and (6) industrial and municipal wastes.

The environmental stress (impact) due to the needs of the population and the better standard of living is given by the master equation (Graedel and Allenby, 1996)

$$EI = P \times \frac{GDP}{Person} \times \frac{EI}{GDP_{unit\ per\ capita}} \qquad (1.1)$$

where
   EI is the environmental impact
   P is the population
   GDP is the gross domestic product

Population, P, in this equation is unarguably increasing with time and that too in geometric progression. The second term on the right-hand side, (GDP/person), denotes the general aspiration of humans for a better life, and is generally increasing as well. The third term on the right-hand side (EI/GDP$_{unit\ per\ capita}$) denotes the extent to which technological advances can be sustained without serious environmental consequences. The third term can be minimized to limit the overall environmental impact and enable the transition to a sustainable environment. Both societal and economic issues are pivotal in determining whether we can sustain the quality of life while at the same time mitigating the environmental consequences of the technologies that we adapt.

The quality of our life is inextricably linked to industrial growth and improvements in agricultural practices. Chemicals are used in both sectors, to sustain innovations in the industrial sector and to improve agricultural efforts at maintaining a high rate of crop production. Thus, the chemical manufacturing industry has been at the forefront of both productivity and growth, especially through the twentieth and twenty-first centuries. Overwhelming arrays of chemicals have been produced each year. Over the whole period of human history, one estimate suggests that approximately six million chemical compounds have been created, of which only 1% are in commercial use today. Many of the compounds are important for sustaining and improving our health and well-being, since they are starting products for various products we utilize every day.

1.2   ENVIRONMENTAL STANDARDS AND CRITERIA

There exist several Congressional statutes in the United States that are instrumental in setting standards for drinking water, ambient water quality, and ambient air quality.

The Environmental Protection Agency (EPA) takes several steps before a standard is set for a specific compound in a specific environment. These include factors such as occurrence in the environment, human exposure and risks of adverse health effects in the general population and sensitive subpopulations, analytical methods of detection, technical feasibility, impacts of regulation of water systems, the economy, and public health.

Environmental quality standards refer to maximum contaminant concentrations allowed for compounds in different environmental media. These concentration limits are expected to be protective of human health and useful for ecological risk management. It has long been recognized that a realistic assessment of the effects of chemicals on humans and ecosystems is mandatory for setting environmental quality standards. The implication is that our concerns for the environment should be driven by sound science. A prudent policy with regard to environmental regulations should consider the weight of evidence in favor of acceptable risk against potential benefits. In fact, risk assessment should be the paradigm that is the basis for current and future environmental legislation.

The United States is not alone in setting such standards. Other nations' governing bodies have created agencies or departments like the U.S. EPA to set environmental quality standards to promote the health of its citizens and the protection of the environment. Such organizations include Procuraduría Federal de Protección al Ambiente (PROFEPA, Mexico), Environment Canada (Canada), European Environmental Agency (Europe), and Ministry of the Environment (Japan).

1.3 THE DISCIPLINE OF ENVIRONMENTAL ENGINEERING

Environmental awareness is the first step in understanding pollution problems. On one hand, we should better understand the potential impacts, so that we can focus our resources on the most serious problems. On the other hand, we should also better understand existing and emerging technologies, so that we can improve their effectiveness and reduce their costs. These activities are within the discipline of environmental engineering.

"Environmental engineering" is the study of the fate, transport, and effects of chemicals in the natural and engineered environments and includes the formulation of options for treatment, mitigation, and prevention of pollution in both natural and engineered systems.

It has only been in the latter half of the twentieth century that environmental engineering grew to a mature field with depth and focus. Environmental engineering is an interdisciplinary field. It involves the applications of fundamental sciences, that is, chemistry, physics, mathematics, and biology, to waste treatment, environmental fate and transport of chemicals, and pollution prevention. The main pillars that support an environmental engineering curriculum are physics (statics and fluids), chemistry (organic, inorganic, kinetics, thermodynamics, material and energy balances), mathematics (calculus, algebra), and biology (toxicology, microbiology, biochemistry) (Figure 1.1). Statics, fluids, and energy and material balances are prerequisites for several engineering disciplines (chemical, civil, petroleum, and biological).

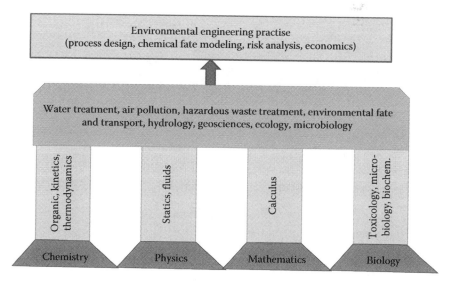

FIGURE 1.1   The pillars of environmental engineering.

For environmental engineering, two main pillars are from chemistry (chemical thermodynamics and kinetics). Chemical thermodynamics is that branch of chemistry that deals with the study of the physicochemical properties of a compound in the three states of matter (solid, liquid, and gas). It also describes the potential for a compound to move between the various phases and the final equilibrium distribution in the different phases. Chemical kinetics, on the other hand, describes the rate of movement between phases and also the rate at which a compound reacts within a phase.

1.4 CHEMICAL THERMODYNAMICS AND KINETICS IN ENVIRONMENTAL ENGINEERING

The environment may be conceptualized as consisting of various compartments. Table 1.1 summarizes the total mass and surface areas of various compartments in the natural environment. The transport of materials between the compartments on a global scale depends only on forces of global nature. However, on a local scale, the partitioning and transport depend on the composition, pressure, temperature, and other variables in each compartment.

There are four distinct environmental compartments—hydrosphere, atmosphere, lithosphere, and biosphere (Figure 1.2). These may be in continuous contact with a sharp boundary between them (air–water) or may be discontinuous (e.g., soil–water). In some cases, one phase will be dispersed in another (e.g., air bubbles in water, fog droplets in air, aerosols and dust particles in air, colloids suspended in water, oil droplets in water, and soap bubbles in water). Some compartments may have the same chemical composition throughout but differ significantly in their spatial characteristics (e.g., the lower troposphere versus the upper stratosphere, a stratified deep water body or a highly stratified atmosphere). The biosphere, which includes all plant and animal species, is in contact with the three other compartments in the overall scheme. Reactions and transformations occur in each phase, and the rate of exchange of mass and energy between the compartments is a function of the extent to which the respective compartments are in nonequilibrium.

TABLE 1.1
Composition of Natural Environment

Compartment	Value
Air (atmosphere) (mass)	5.1×10^{18} kg
Water (hydrosphere) (mass)	1.7×10^{21} kg
Land (lithosphere) (mass)	1.8×10^{21} kg
Land on Earth (area)	1.5×10^{14} m^2
Water on Earth (area)	3.6×10^{14} m^2

Sources: Weast, R.C. and Astle, M.J. (eds.), *CRC Handbook of Chemistry and Physics*, 62nd edn., CRC Press, Inc., Boca Raton, FL, 1981; Stumm, W. and Morgan, J.J., *Aquatic Chemistry*, 3rd edn., John Wiley & Sons, New York, 1996.

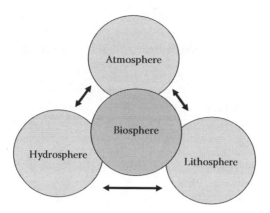

FIGURE 1.2   Equilibrium between various environmental compartments, reactions within compartments, and material exchange between the compartments.

1.4.1 Applications of Thermodynamics and Kinetics

1.4.1.1 Equilibrium Partitioning

Chemical thermodynamics is central to the application of the equilibrium partitioning concept to estimate pollutant levels in environmental compartments. It assumes that the environmental compartments are in a state where they have reached constant chemical composition, temperature, and pressure and have no tendency to change their state. Equilibrium models only give us the chemical composition within the individual compartments and not how fast the system reached the given equilibrium state. Although, admittedly, a time-variant (kinetic) model may have significant advantages, there is general agreement that equilibrium partitioning is a starting point in this exercise. True equilibrium does not exist in the environment. In fact, much of what we observe in the environment occurs as a result of the lack of equilibrium between compartments. Every system in the environment strives toward equilibrium as its ultimate state. Hence, the study of equilibrium is a first approximation toward the final state of any environmental system.

1.4.1.2 Fate and Transport Modeling

The environment is a continuum in that as pollutants interact with various phases, they undergo both physical and chemical changes and are finally incorporated into the environment. Fate models, based on the mass balance principle, are necessary to simulate the transport between and transformations within various environmental media. This is called the *multimedia approach*. The mass balance principle can be applied through the use of a multimedia fate and transport model to obtain the rates of emissions and the relative concentrations in each compartment. A number of such models already exist, some of which are described in Table 1.2.

Consider a chemical that is released from a source to one of the environmental compartments (air, water, or soil). In order to assess the effect of the pollutant to the ecosystem, we need to first identify the various pathways of exposure.

TABLE 1.2

Descriptions of a Few Multimedia Fate and Transport Models That Are Currently in Use

Model Acronym	Description
CalTOX	A fugacity-based model to assist the California Environmental Protection Agency to estimate chemical fate and human exposure in the vicinity of hazardous waste sites.
ChemCAN	A steady-state fugacity-based model developed for Health Canada to predict a chemical's fate in any of the 24 regions of Canada.
HAZCHEM	A fugacity-based model developed as a regional scale model for the European Union member states.
SimpleBOX	Developed by RIVM, the Netherlands, it uses the classical concentration concept to compute mass balances.

Source: SETAC, *The Multi-Media Fate Model: A Vital Tool for Predicting the Fate of Chemicals*, SETAC Press, Pensacola, FL, 1995.

Figure 1.3 illustrates the three primary pathways that are responsible for exposure from an accidental release. Direct exposure routes are through inhalation from air and drinking water from the groundwater aquifer. Indirect exposure results, for example, from ingestion of contaminated fish from a lake that has received a pollutant discharge via the soil pathway. Coupling the toxicology with a multimedia fate and transport model thus provides a powerful tool to estimate the risk potential for both humans and the biota. Risk assessment is typically divided into the assessment of cancer and noncancer health effects. Animal toxicology studies or human epidemiological data are used to establish a unit risk level for cancer. A dose-response model then relates the response to the dosage to which the subjects are exposed. Most cancer dose-response models conservatively assume a no-threshold model, so that the risk is extrapolated from high to low dosages assuming that there is some response at any dosage above zero.

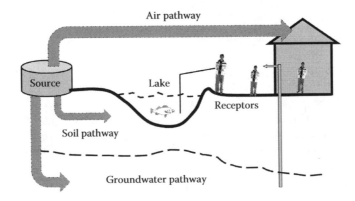

FIGURE 1.3   Environmental risks and exposure pathways.

The risk alluded to is the incremental lifetime cancer risk (ILCR), and it is given by

$$ILCR = SF \times CDI \qquad (1.2)$$

where
   SF is the slope factor for the chemical
   ILCR is the incremental lifetime cancer risk
   CDI is the chronic daily intake of the chemical by the defined exposure route

A listing of the slope factors can be found in Appendix I. The chronic daily intake requires the use of a multimedia fate model to obtain the concentration value that goes into its determination. Thus, knowledge of the slope factor and the CDI allows the determination of the lifetime cancer risk.

If the excess cancer risk from the inhalation route is to be estimated, we can also write the following equation:

$$Risk_{air} = IUR \times C_{air} \qquad (1.3)$$

where
   IUR is the inhalation unit risk (in per $\mu g\,m^{-3}$) (see Appendix I)
   C_{air} is the average exposure concentration in air

For example, for formaldehyde, the unit risk is $1.3 \times 10^{-5}/\mu g\,m^{-3}$ and if someone is exposed to a concentration of $0.77\ \mu g\,m^{-3}$ over a lifetime (70 years), the lifetime risk is 1 in 10^5. For noncancer risk assessment in air, we use a hazard quotient (HQ):

$$HQ = \frac{C_{air}}{RfC} \qquad (1.4)$$

where RfC is the reference concentration (see Appendix I for definitions). If HQ is less than 1, the risk is acceptable.

Within the regulatory framework, it is now mandatory to assess the potential harmful effects on humans and the environment from the use of new chemicals and from the continued use of existing ones. Examples are the Toxic Substances Control Act (TSCA) in the United States, Canadian Environmental Protection Act (CEPA) in Canada, and the Seventh amendment in the European Union (EU). Once risk is established, the next step will be to isolate the pollutant from the ecosystem to minimize the risk.

In this context, the following specific questions will need to be addressed:

1. What is the final equilibrium state of the pollutant in the environment, i.e., which of the environmental compartments is the most favorable, how much resides in each compartment at equilibrium, and what chemical properties are important in determining the distribution?
2. How fast does the pollutant move from one compartment to another, what is the residence time in each compartment, and how fast does it react within each compartment or at the boundary between compartments?

The answer to the first question employs the tools of chemical thermodynamics and the second question requires the applications of concepts from chemical kinetics.

Multimedia fate and transport (F & T) models are recognized as a necessary component for risk assessment, chemical ranking, management of hazardous waste sites, optimization of testing and monitoring strategies, and determination of global dispersion and recovery times. In order to construct F & T models, one needs to obtain a range of physicochemical parameters—thermodynamic, kinetic, and toxicological.

1.4.1.3 Design of Separation Processes

Environmental engineers design separation processes for isolation of contaminants from waste streams before they are discharged into the environment. Both physical and chemical separation processes are used in environmental engineering. For example, particulate separation from air and water involves physical separation techniques that use mechanisms such as aggregation, coagulation, impaction, centrifugal force, and electromotive force, to name a few. The removal of dissolved gases and vapors from air and water involves, on the other hand, chemical separation methods. While mixing of chemicals to form a mixture is a spontaneous process and, as we will see in Chapter 2, a thermodynamically favorable process, the reverse, namely, separation into the component species, requires the expenditure of work. The overall objective of an environmental separation process is not only to isolate the pollutant but also to recycle and reuse where possible the materials separated and separation agents that were used during the operation.

Invariably, environmental separation processes involve contact between two or more phases and the exchange of material and energy between them. As an example, consider the removal of organic contaminants from water by contacting with a solid phase such as powdered activated carbon. The process has two distinct stages. In stage I, those contaminants in water having a greater affinity for the carbon are concentrated on the carbon by a process called adsorption. The removal depends not only on the rate of transport of compounds from water to carbon (the realm of kinetics) but also on the ultimate capacity of the carbon bed (the realm of thermodynamics). In stage II, we have to regenerate the medium within the reactor so that it can be reused. In stage II, we also need information on how fast the pollutant can be recovered from the activated carbon, and also the fate and transformation of the adsorbed pollutant (namely, the kinetic aspects).

Separation techniques can be classified broadly into five categories as shown in Figure 1.4 (Seader and Henley, 1998). In every one of these processes, the rate of separation is dependent on selectively increasing the rate of diffusion of the contaminant species relative to the transfer of all other species by advection (bulk movement) within the contaminated feed. Equilibrium limits the ultimate compositions of the effluent streams (pollutant-rich and pollutant-free). The rate of separation within the reactor is determined by the mass transfer limitations ("driving force"), while the extent of separation is determined by equilibrium thermodynamic

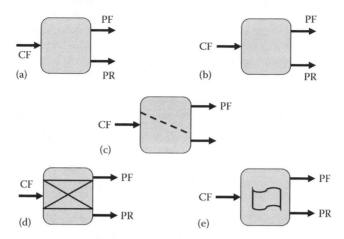

FIGURE 1.4 General separation techniques. CF, contaminated feed; PF, pollutant-free stream; PR, pollutant-rich stream. (a) Separation by phase creation; (b) separation by phase addition; (c) separation by barrier; (d) separation by solid agent; (e) separation by field gradient. (Modified from Seader, E.D. and Henley, E.J., *Separation Process Principles*, John Wiley & Sons, Inc., New York, 1998.)

factors. Both thermodynamic and kinetic (transport) properties are thus instrumental in environmental separations.

The discussions in this chapter of applications in environmental engineering show that both for natural and engineered systems, two types of issues are paramount—an equilibrium study to describe the final distribution of pollutants within different compartments and a kinetic study to describe the rate of transformations of chemicals within each compartment and the rate of movement of chemicals between compartments. A list of processes common to both natural and engineered systems is given in Table 1.3. It summarizes the associated thermodynamic and kinetic properties that are necessary for understanding each process.

In summary, we can look at chemical thermodynamics in environmental engineering as the essential element by which we connect observations on simplified systems to the segment of the environment (natural or engineered) that we are interested in. To put it in context, Figure 1.5 appropriately places the role of thermodynamics in our endeavor to understand and design processes in the natural and engineered systems.

1.5 UNITS AND DIMENSIONS

The SI (International System of Units) is the standard set of units for engineering and science. The International Union of Pure and Applied Chemistry (IUPAC) produced a document (Cohen et al., 2007) entitled *Quantities, Units and Symbols in Physical Chemistry,* which recommended a uniform set of units for all measurements in physical chemistry (Table 1.4). This document should be consulted for a more elaborate discussion of the various units. However, in a number of cases, environmental engineers still prefer to use the CGS (centimeter-gram-second) system of units.

TABLE 1.3
Applications of Chemical Thermodynamics and Kinetics in Environmental Processes

Process	Typical Equilibrium Representation	Thermodynamic Property	Kinetic Property
Solubility in water	A (pure) ⇌ A (water)	Saturation solubility, C_i^*	Dissolution rate
Absorption in water	A (air) ⇌ A (water)	Henry's law constant, K_H	Absorption rate
Precipitation from water	A (water) ⇌ A (crystal)	Solubility product, K_{sp}	Precipitation rate
Volatilization from water	A (water) ⇌ A (air)	Henry's law constant, K_H	Volatilization rate
Evaporation from pure liquid	A (pure) ⇌ A (vapor)	Vapor pressure, P_i^*	Evaporation rate
Acid/base dissociation	$A \rightleftharpoons A^- + H^+$	Acidity or basicity constant, K_a or K_b	Acidification rate
Ion exchange	$A^+ + BX \rightleftharpoons BA + X^+$	Ion-exchange partition constant, K_{exc}	Ion exchange rate
Oxidation/reduction	$A_{ox} + B_{red} \rightleftharpoons A_{red} + B_{ox}$	Equilibrium constant, K_i	Redox reaction rate
Adsorption from water	A (water) ⇌ A (surface)	Soil/water partition constant, K_{sw}	Adsorption rate
Adsorption from air	A (air) ⇌ A (surface)	Particle/air partition constant, K_{AP}	Adsorption rate
Uptake by biota	A (water) ⇌ A (biota)	Bioconcentration factor, K_{BW}	Rate of uptake
Uptake by plants	A (air) ⇌ A (plant)	Plant/air partition constant, K_{PA}	Rate of uptake
Chemical reaction	$A + B \rightleftharpoons$ Products	Equilibrium constant, K_{eq}	Rate of chemical reaction
Photochemical reaction	$A + (h\nu) \rightleftharpoons$ Products	Equilibrium constant, K_{eq}	Rate of photolysis
Biodegradation reaction	$A + (enzymes) \rightleftharpoons$ Products	Equilibrium constant, K_{eq}	Michaelis–Menten and Monod kinetics constants

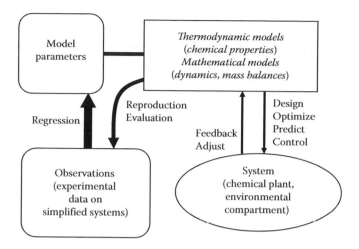

FIGURE 1.5 The role of chemical thermodynamics in understanding natural and engineered systems. (Modified from CODATA: Committee on Data of the International Council for Science, http://www.codata.org/codata02/04physci/rarely.pdf.)

TABLE 1.4

Base Quantities, Units, and Symbols in SI Units

Physical Quantity	Name	Symbol
Length	meter	m
Mass	kilogram	kg
Time	second	s
Electric current	ampere	A
Thermodynamic temperature	Kelvin	K
Amount of substance	mole	mol
Luminous intensity	candela	cd

Source: IUPAC Physical Chemistry Division, in *Quantities, Units and Symbols in Physical Chemistry*, Mills, I., Cvitas, T., Homann, K., Kallay, N., and Kuchitsu, K. (eds.), Blackwell Scientific Publications, Oxford, U.K., 1988.

Table 1.5 gives the seven *base quantities* and their symbols that SI units are based on. All other physical quantities are called *derived quantities* and can be algebraically derived from the seven base quantities by multiplication or division.

The physical quantity *amount of substance* is of paramount importance to environmental engineers. The SI unit for this quantity is the mole defined as the amount of substance of a system that contains as many elementary entities as there are atoms in 0.012 kg of carbon-12. IUPAC recommends that we should refrain from calling it the "number of moles." Much of the published literature is based on the more familiar CGS units. The relations between the common CGS and SI units for some important derived quantities of interest in environmental engineering are given in Table 1.5.

TABLE 1.5
Relations between SI and CGS Units for Some Derived Quantities

Derived Quantity	Unit	CGS Symbol	Equivalent SI Unit
Force	dyne	dyn	10^5 Newtons (N)
Pressure	bar	bar	10^5 Pascals (Pa)
	atmosphere	atm	101,325 Pa
	torr	torr	133.32 Pa
	millimeter of mercury	mm Hg	133.32 Pa
	pounds per square inch	psi	6.89×10^3 Pa
Energy, work, heat	ergs	erg	10^{-7} Joules (J)
	calories	cal	4.184 J
	liter atmospheres	L atm	101.325 J
Concentration	molar (mol/L)	M	10^3 mol/m^3
			1 mol/dm^3
Viscosity	centipoise	cP	10^{-3} kg/m s

A description of the most common units in environmental engineering is given in Appendix D and the reader should familiarize himself or herself with these before proceeding further.

1.6   STRUCTURE OF THIS BOOK

This book is divided into four chapters. Chapter 2 introduces the fundamental principles of thermodynamics, transport, and kinetics that will be used in the applications chapters (Chapters 3 and 4). Chapter 2 begins with an introduction to the thermodynamics of homogeneous phases composed of single or multiple species. It also introduces the important concepts of free energy and chemical potential that are of paramount importance in dealing with equilibrium systems in environmental engineering. A concise description of surface thermodynamics is also included in Chapter 2. Section 2.2 is an extension of the thermodynamics of homogeneous systems to heterogeneous and multicomponent systems. The important concepts of activity and fugacity and nonideal solutions and gases are dealt within this chapter. Section 2.3 gives a short summary of the essential aspects of chemical reaction kinetics. Concepts such as reaction rates and activation energies are introduced and discussed. In Section 2.4, principles of reactor design as they pertain to environmental systems are introduced. Chapter 3 deals with the applications of the concepts developed in the thermodynamics portion of Chapter 2 on air–water, soil–water, and air–soil equilibria to illustrate the concept of equilibrium partitioning between compartments in environmental engineering. Applications of equilibrium thermodynamics in waste treatment operations are also described. The concepts developed in Sections 2.3 and 2.4 are used to illustrate the applications of chemical kinetics in environmental waste treatment processes and biological systems in Chapter 4.

REFERENCES

Cohen, E.R., Cvitas, T., Frey, J.G., Holmstrom, B., Kuchitsu, K., Marquardt, R., Mills, I.,
    Pavese, F., Quack, M., Stohner, J., Strass, H.L., Takami, M., and Thor, A.J. (2007)
    Quantities, Units and Symbols in Physical Chemistry, 3rd edn. Cambridge, U.K.: The
    Royal Society of Chemistry (for IUPAC).
Graedel, T.E. and Allenby, B.R. (1996) *Design for the Environment*. Upper Saddle River, NJ:
    Prentice Hall.
IUPAC Physical Chemistry Division. (1988) In: Mills, I., Cvitas, T., Homann, K., Kallay, N.,
    Kuchitsu, K. (eds.), *Quantities, Units and Symbols in Physical Chemistry*. Oxford, U.K.:
    Blackwell Scientific Publications.
Seader, J.D., and Henley, E.J. (1998) *Separation Process Principles*. New York: John Wiley
    & Sons, Inc.
SETAC. (1995) *The Multi-Media Fate Model: A Vital Tool for Predicting the Fate of Chemicals*.
    Pensacola, FL: SETAC Press.
Stumm, W. and Morgan, J.J. (1996) *Aquatic Chemistry*, 3rd edn. New York: John Wiley &
    Sons.
Weast, R.C. and Astle, M.J. (eds.) (1981) *CRC Handbook of Chemistry and Physics*, 62nd edn.
    Boca Raton, FL: CRC Press, Inc.

2 Basic Chemical Thermodynamics and Kinetics

As we discussed in Chapter 1, cognizance of both physical and chemical equilibrium between different compartments is important in environmental engineering. Thermodynamics was initially a subject that only involved the relationship between energy and mechanical work, but it has come to involve all forms of relationships between macroscopic variables within a system (e.g., pressure, temperature, volume, composition). The basic principles of thermodynamics are germane to the understanding of equilibrium in environmental systems. There are four fundamental laws of thermodynamics that epitomize the entire subject.

In the following sections, our discussion of thermodynamics is limited to its basic laws, since a number of excellent references are available elsewhere (e.g., Lewis and Randall, 1961; Denbigh, 1981). Phase equilibrium, as it relates to bulk phases (single component, homogeneous), followed by an expanded discussion of multicomponent and heterogeneous systems, will be discussed in this chapter. Since most of the applications of thermodynamics in environmental engineering are confined to a narrow range of temperature (approx. $-40°C$ to $+40°C$) and atmospheric pressure, we need not focus on extreme temperatures or pressures. A brief review of the concept of equilibrium and the basic laws of thermodynamics is presented, followed by the introduction of the concepts of free energy and chemical potential. The effect of surface area on the total thermodynamic property of a system is negligible, except when the subdivision in phases is exceedingly small, which occurs in many environmental systems. Therefore, an introductory discussion of surface thermodynamics is also included.

2.1 SINGLE-COMPONENT THERMODYNAMIC EQUILIBRIUM

2.1.1 EQUILIBRIUM

The natural environment is a complex system. However, first assuming that equilibrium exists between the different environmental phases and that the properties are time invariant, we can assess an environmental system. Equilibrium models are easy to apply since they need only few inputs. They are sometimes grossly inappropriate and have to be replaced with kinetic models, which assume time-variant properties for the phases; we discuss this later in this chapter. Kinetic models are complex and require a number of input parameters that are poorly understood and sometimes unavailable for environmental situations.

Before beginning a discussion of the laws of thermodynamics, we need to define some terms. To quote Lewis and Randall (1961): "whatever part of the objective world is the subject of thermodynamic discourse is customarily called a *system*." The system may be separated from the surroundings by a physical boundary, like the walls of a container or maybe less concrete. A system may be *closed* if there is no mass transfer either to or from the surroundings or *open* if either matter or energy can transfer to and from the surroundings. For example, the earth is an open system in contact with the atmosphere around it with which it exchanges both matter and energy. A system is said to be "isolated" if it cannot exchange either mass or energy with its surroundings. The universe is considered an isolated system. A system is said to be "adiabatic" if it cannot exchange heat with the surroundings. For example, an insulated thermos is such a system. A system is composed of spatially uniform entities called "phases." Air, water, and soil are three environmental phases. The "properties" that characterize a system are defined by temperature, pressure, volume, density, composition, surface tension, viscosity, etc. There exist two types of properties—"intensive" and "extensive." Intensive properties are those that do not require any reference to the mass of the system and are nonadditive. Extensive properties are dependent upon the mass in the system and they are additive. Any change in system properties is termed a "process."

Despite its name thermodynamics does not deal with the dynamics of a system, but its "equilibrium state." The word equilibrium implies a state of balance. When a system is in such a state that upon slight disturbances it reverts to its original state, it is said to be in a "state of equilibrium." This state of equilibrium is one in which the system is time invariant in properties. A thermodynamic system can experience thermal, mechanical, and chemical equilibrium.

If a system undergoes no apparent changes, it is said to be "stable." However, there are systems that are apparently stable because the rates of change within them are imperceptible. Such systems are called "inert." The degree of stability of a system can vary. For example, a book placed flat on a table is at its state of rest and is stable. However, when pushed over the table to the ground it reaches another state of rest, which is also stable.

The fundamental principles of thermodynamics can be stated in terms of what are called "system variables," "modes of energy transfer," and characteristic "state functions" (Stumm and Morgan, 1981). The system variables are five in number, namely, temperature T, pressure P, volume V, moles n, and entropy S. There are only two modes of energy transfer—heat, q, transferred to a system from the surroundings and work, w, done on the system by its surroundings. A state function is one that depends only on the initial and final states of the system and not upon the path by which the system may pass between states. Only one such characteristic state function, internal energy U, is necessary to define the system. However, three other characteristic state functions are also used in thermodynamics. They are Helmholz free energy A, Gibbs free energy G, and enthalpy H.

2.1.2 FUNDAMENTAL LAWS OF THERMODYNAMICS

There are four fundamental laws on which the edifice of thermodynamics is built. These are called zeroth, first, second, and third laws. These laws, which dominate

TABLE 2.1
Major Developments in Thermodynamics over Two Centuries

Year	Person	Contribution
1760–1766	Joseph Black	Calorimetry
1842	Julius Meyer	Conversion of heat to work and vice versa
1843–1852	James Joule	Heat, work, and first law
1848–1851	William Thompson	Absolute temperature scale, second law
1854–1865	Rudolph Clausius	Concept of entropy
1873–1878	Josiah Willard Gibbs	Chemical thermodynamics, chemical potential, phase rule
1882	H. von Helmholtz	Equilibrium, free energy
1884–1887	Jacobus van't Hoff	Equilibrium constant, solution theory
1906	Walter Nernst	Heat theorem (third law)
1927	Franz Simon	Third law
1900–1907	Gilbert N. Lewis	Fugacity, nonideality
1929	W. F. Giauque	Third law verification
1931	Lars Onsager	Nonequilibrium thermodynamics, reciprocity relations
1949	Ilya Prigogine	Irreversible processes, dissipative structures

Source: Data from Laidler, K.J., *The World of Physical Chemistry*, Oxford University Press, New York, 1993.

modern thermodynamics, are the work of several scientists over a considerable time span (see Table 2.1).

The unique feature of these laws is that although derived through generalizations resulting from numerous experimental data, they have not yet been disproved. Einstein (1949) remarked that "…classical thermodynamics….It is the only physical theory of universal content concerning which I am convinced that, within the framework of the applicability of its basic concepts, it will never be overthrown…." Thermodynamics is, as Einstein called it, "a theory of principle" based on "empirically observed general properties of phenomena" that does not rely on any assumptions on "hypothetical constituents." Thus, the laws of thermodynamics have stood the test of time and are probably inviolable.

2.1.2.1 Zeroth Law of Thermodynamics

One of the fundamental system variables in thermodynamics is temperature. The zeroth law states that "if two systems are in thermal equilibrium with a third, then they are also in thermal equilibrium with each other." This allows us to create a "thermometer." We can calibrate the change in a property, such as the length of a column of mercury, by placing the thermometer in thermal equilibrium with a known physical system at several reference points. Celsius thermometers use the reference points fixed as freezing and boiling points of pure water. If we then bring the thermometer in thermal equilibrium with a human body, for example, we can determine the body temperature by noting the change in the thermal property. This law also establishes a temperature scale, that is, only a system at a higher temperature can lose its thermal energy to one with a lower temperature and not vice versa. The International Union of Pure and Applied Chemistry (IUPAC) has adopted as the unit of temperature the Kelvin, which is the fraction 1/273.16 of the thermodynamic temperature of the triple point of water (IUPAC, 1988).

2.1.2.2 First Law of Thermodynamics

The universal principle of conservation of energy is the first law of thermodynamics. The term "energy," although a subtle concept and introduced first by William Thompson in 1852, can be composed of three types, viz., potential energy, kinetic energy, and internal energy (molecular energy). According to the first law, "for any closed system the change in total energy (kinetic energy + potential energy + internal energy) is the sum of the heat absorbed from the surroundings and the work done by the system." For macroscopic changes in a closed system with kinetic energy E_K, potential energy E_P, and internal energy U, and that which absorbs heat q, and does work equivalent to w, the first law can be written in the following form:

$$\Delta E_P + \Delta E_K + \Delta U = q - w \tag{2.1}$$

The potential energy term E_P is generally given by mgz, where m is the mass, g is the gravitation acceleration, and z is the height. The kinetic energy term E_K is generally of the form $\frac{1}{2}mu^2$, where u is the velocity. For a macroscopic system at rest, we have both ΔE_P and ΔE_K equal to zero. If δq represents the infinitesimal heat absorbed from the surroundings and $-\delta w$ represents the infinitesimal work done *by* the system (by definition $+\delta w$ is the infinitesimal work done *on* the system), then the infinitesimal change in energy, dU, is given by

$$dU = \delta q - \delta w \tag{2.2}$$

Notice that in this equation we have used δ instead of d to remind us that q and w are defined only for a given path of change, that is, the values of q and w are dependent on how we reached the particular state of the system. dU, however, is independent of the path taken by the system and is determined only by the initial and final states of the system. One measures only changes in energies as a result of a change in state of the system and not the absolute energies. Thus, for a closed system at rest and for finite macroscopic changes in q and w, we have

$$\Delta U = q - w \tag{2.3}$$

The work term in this equation can involve any of the following: pressure-volume work (w_{pv}) and any of the other forms of shaft work, w_{shaft} (e.g., push-pull, electrical, elastic, magnetic, surface). When the work done by a system is only work of expansion against an external pressure P_{ext}, then w is given by[*]

$$w = \int_{V_1}^{V_2} P_{ext}dV = P_{ext}\Delta V \tag{2.4}$$

For an adiabatic process, q = 0 and hence $\Delta U = -w$, whereas for a process such as a chemical reaction occurring in a constant volume container, $\Delta V = 0$ and hence $\Delta U = q$.

---

[*] Note that we will henceforth drop the subscript on P.

Most environmental processes are constant pressure processes and hence they invariably involve pressure-volume (PV) work. An example of an adiabatic process in the environment is the expansion of an air parcel and the attendant decrease in temperature as it rises through the atmosphere leading to what are called "dry and moist adiabatic lapse rates" (see Example 2.1).

For open systems, there is an additional contribution due to the flow of matter dU_{matter}, and therefore we have

$$dU = \delta q - \delta w + dU_{matter} \qquad (2.5)$$

For open systems (environmental systems), Equation 2.3 for macroscopic changes can be modified as follows:

$$\Delta U = q - w + m_{in}\left(U + PV\right)_{in} - m_{out}\left(U + PV\right)_{out} \qquad (2.6)$$

where m_{in} and m_{out} are the masses of the entering and exiting streams.

For an open system with time-varying properties such that the following equation applies:

$$\frac{dU}{dt} = \dot{q} - \dot{w} + \dot{m}_{in}\left(U + PV\right)_{in} - \dot{m}_{out}\left(U + PV\right)_{out} \qquad (2.7)$$

where the "·" above the terms represents rates of flow.

The most general system will be one in which all forms of energy are included such that Equation 2.1 is used, and for which we consider the flow of material between different heights at inlet and outlet (z_{in} to a height z_{out}), and where the velocities of flow are u_{in} and u_{out}, respectively, at the two points. Then we can write the following general equation for the first law:

$$\frac{dE_T}{dt} = \dot{q} - \dot{w} + \dot{m}_{in}\left(U + PV + \frac{u^2}{2} + gz\right)_{in} - \dot{m}_{out}\left(U + PV + \frac{u^2}{2} + gz\right)_{out} \qquad (2.8)$$

2.1.2.3 Second Law of Thermodynamics

The French engineer Sadi Carnot derived the second law of thermodynamics from the principles of heat engines that he studied. This can be best explained with reference to Figure 2.1, in which a heat engine converts heat to work. The heat engine

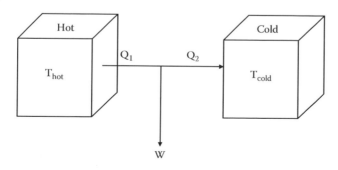

FIGURE 2.1 Carnot's heat engine.

absorbs heat Q_1 from the hot reservoir (heat source) at temperature T_{hot}, converts part of it to work W, and discards heat Q_2 to the cold reservoir (sink) at temperature T_{cold}. The efficiency of a heat engine is defined as the ratio of the work it produces to the heat that is absorbs: $\eta_{eff} = W/Q_1$. Thus, if all heat is converted to work, the efficiency will be unity. By first law, $W = Q_1 - Q_2$. Further, $Q \propto T$. Therefore, the efficiency of a Carnot's heat engine is given by

$$\eta_{eff} = \frac{Q_1 - Q_2}{Q_1} = \frac{T_{hot} - T_{cold}}{T_{hot}} \tag{2.9}$$

From the generalizations of the concept of efficiency of heat engines for an arbitrary cycle, Carnot defined the following for a closed cycle:

$$\oint \frac{dQ}{T} = 0 \tag{2.10}$$

where the integral is over a path representing a reversible process from state A to state B. Thus, Carnot realized that the function dQ/T is independent of the path taken by the system and depends only on the initial and final states. Such a function was given the name "entropy," S.

$$dS = \frac{dQ}{T} \tag{2.11}$$

In other words, the change in entropy (dS) is obtained from the ratio of energy transferred as heat (dQ) to the absolute temperature (T) at which the transfer took place.

The second law of thermodynamics is a general statement regarding the spontaneous changes that are possible for a system. When a change occurs in an isolated system (i.e., the universe) the total energy remains constant, however, the energy may be distributed within the system in any possible manner. For all natural processes occurring within the universe, any spontaneous change leads to a chaotic dispersal of the total energy. It is unlikely that a chaotically distributed energy will in time reorganize itself into the original ordered state. A large number of practical examples can be given here and is well described in various texts (e.g., Atkins and de Paula, 2006). In order to quantify this inexorable transition toward a chaotic state for the universe, the second law introduces the term called "entropy" denoted by S. It is also a state function. After a spontaneous change for a system, the total entropy of the system plus surroundings is greater than the initial state. Clausius enunciated the second law thus: "the entropy of the universe tends to a maximum." A more precise statement of the second law is: "as a result of any spontaneous change within an isolated system the entropy increases."

The entropy transferred to a system by the surroundings is defined by

$$dS_{surr} = \frac{\delta q}{T} \tag{2.12}$$

The total entropy change of the isolated system, viz., the universe is the sum of the entropy changes within the system, dS_{sys}, and the entropy change transferred from the system to the surroundings, $-dS_{surr}$. Thus,

$$dS_{univ} = dS_{sys} - dS_{surr} \tag{2.13}$$

Since for any permissible process within an isolated system, $dS_{univ} \geq 0$, we have $dS_{sys} \geq \delta q/T$. This is called the "Clausius inequality." Entropy is a measure of the "quality" of the stored energy; it can be appropriately identified with disorder.

2.1.2.4 Third Law of Thermodynamics

The third law defines the value of entropy at absolute zero. It states that "the entropy of all substances is positive and becomes zero at $T = 0$ and does so for a perfectly crystalline substance." Thus, $S \rightarrow 0$ as $T \rightarrow 0$. Although there have been several attempts, no one has ever succeeded in achieving absolute zero of temperature. Thus, third law explicitly recognizes the unattainability of absolute zero in temperature.

To summarize: according to the first law, energy is always conserved and if work has to be done by a system it has to absorb an equal amount of energy in the form of heat from the surroundings. Therefore, a perpetual motion machine capable of creating energy without expending work is a utopia. The permissible changes are only those for which the total energy of the isolated system, that is, the universe, is maintained constant. First law of thermodynamics is thus a statement regarding the "permissible" energy changes for a system. The second law gives the "spontaneous" changes among these "permissible" changes. The third law recognizes the difficulty in attaining the absolute zero of temperature.

Example 2.1 Pressure and Temperature Profiles in the Atmosphere

Problem statement: If atmospheric air can be considered an ideal gas, it obeys the ideal gas law in the form $PV = R$, where V is the molar volume of air $V = M/\rho$. ρ is the density of air ($kg\,m^{-3}$) and M is the average molecular weight of air (≈ 29). Consider a parcel of air rising from the ground (sea) level to the lower atmosphere. It experiences a change in pressure with altitude given by the well-known relation, $dP(h)/dh = -\rho g$. Use the ideal gas law expression to obtain an expression for $P(h)$ given P_0 is the atmospheric pressure at sea level.

Assume that as the parcel rises it undergoes a change in volume in relation to a decreasing pressure, but that there is no net heat exchange between it and the surroundings, that is, it undergoes an "adiabatic expansion." This is a reasonable assumption since the size of the reservoir (ambient atmosphere) is much larger than the size of the air parcel and any changes in the ambient temperature are imperceptible whereas those within the air parcel will be substantial. Therefore, the volume expansion of the parcel leads to a decrease in temperature. This variation in temperature of dry air with height is called the "dry adiabatic lapse rate." Apply the first law relation and use the expression $dU = C_v dT$ (as in Section 2.1.2.2) to obtain an equation for the variation in temperature of the air parcel with height.

Solution: Considering the air to be an ideal gas at any point in the atmosphere we have the following:

$$P = \frac{\rho RT}{M} \tag{2.14}$$

Since the pressure at any point is due to the weight of air above, we have $dP(h)/dh = -\rho g$, and hence,

$$\frac{dP(h)}{dh} = -\frac{P(h)Mg}{RT} \tag{2.15}$$

Integrating this equation with $P(0) = P_0$, we get

$$P(h) = P_0 e^{-\left(\frac{Mhg}{RT}\right)} \tag{2.16}$$

This equation gives the pressure variation with height in the atmosphere.

Now we shall obtain the temperature profile in the atmosphere using the concepts from first law. For an adiabatic process, we know that $\delta q = 0$. Hence, $dU = \delta w$. As will be seen in Example 2.1, $dU = C_v dT$. Since it is more convenient to work with P and T as the variables rather than with P and V, we shall convert PdV term to a form involving P and T using ideal gas law. Thus,

$$d(PV) = PdV + VdP = \frac{m}{M}RdT \tag{2.17}$$

Hence, we have for the first law expression:

$$C_v dT = \left(\frac{m}{M}RT\right)\frac{dP}{P} - \left(\frac{m}{M}R\right)dT \tag{2.18}$$

Rearranging, we obtain

$$\frac{dT}{dP} = \frac{\frac{m}{M} \cdot \frac{RT}{P}}{C_v + \frac{m}{M} \cdot R} \tag{2.19}$$

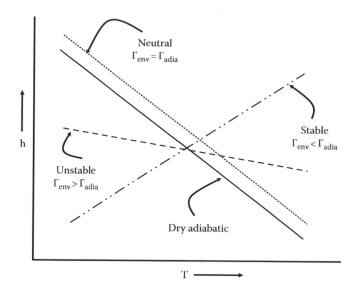

FIGURE 2.2 Dry adiabatic and environmental lapse rates related to atmospheric stability. Neutral, stable, and unstable atmospheric conditions are shown.

Combining the equation for dP/dh with Equation 2.19, we get

$$\frac{dT}{dh} = -\frac{g}{C_{p,m}}$$
(2.20)

where $C_{p,m}$ is the heat capacity per unit mass of air at constant pressure. It is defined as $C_{v,m} + R/M$. The term $\Gamma_{adia} = g/C_{p,m}$ is called the "dry adiabatic lapse rate."

The student is urged to work out Problem 2.5 to get a feel for the magnitude of temperature changes in the lower atmosphere. In order to quantify the atmospheric stability (the capacity of air to disperse pollutants is related to this property), one compares the prevailing (environmental) lapse rate $\Gamma_{env} = -dT/dh$ to the dry adiabatic lapse rate. Figure 2.2 shows the characteristic profiles for the unstable, stable, and neutral atmospheres. It can be seen that for an unstable atmosphere, $\Gamma_{env} > \Gamma_{adia}$, whereas for a stable atmosphere, $\Gamma_{env} < \Gamma_{adia}$. For an unstable atmosphere, the less dense air parcel at the higher altitude continues to rise, and that at the lower altitude is denser and continues to sink. The condition is unstable since any perturbation in the vertical direction is enhanced. Pollutant dispersal is rapid in an unstable atmosphere.

Example 2.2 Hurricane as a Heat Engine

Emanuel (2005) described the hurricane as Nature's steam engine that obeys the Carnot's energy cycle, which is the basis for the second law of thermodynamics. The hurricane can be considered to obey a four-step cycle described in Figure 2.3, where a cross section through a hurricane is shown. The axis on the left side is the central axis of the storm (eye). At point A near the sea surface, the air is at least

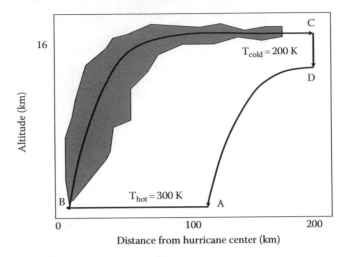

FIGURE 2.3   Hurricane heat engine.

hundreds of kilometers from the storm center. The air slowly spirals inward toward the eye wall (point B). From point A to point B, there is a significant decrease in pressure. The sea surface is an infinite heat reservoir as the relative size of the storm is significantly smaller than the ocean itself; therefore, its temperature is assumed to be constant. Unlimited heat is therefore added to the air parcel as it moves from A to B. This amounts to the increased humidity in the air as seawater evaporates into the incoming air. At point B, the air turns and moves upward, making the hurricane eye wall. Here the latent heat is converted into sensible heat as water vapor condenses and pressure decreases rapidly. This is an example of adiabatic expansion of air. At point C (12–18 km from the surface), the adiabatic expansion decreases the air temperature to 200 K. As the air remains at this temperature, it sinks to point D toward the tropopause, which is near isothermal conditions. Finally, the air from point D returns to point A in a near adiabatic compression. The thermodynamic cycle is thus complete and a perfect steam engine is created, except that the energy produced by the hurricane shows up as the energy of the wind. Thus, we can apply Carnot's theorem to obtain the maximum wind for an idealized hurricane.

The kinetic energy dissipated by the atmosphere near the surface is given by

$$D \approx C_D \rho V^3 \qquad (2.21)$$

where
     D is the rate of heat energy dissipation per unit surface area
     ρ is the air density
     V is the wind speed
     C_D is the drag coefficient (remember fluid mechanics)
     The total mechanical work done by a Carnot engine is given by

$$W = Q\left(\frac{T_{hot} - T_{cold}}{T_{hot}}\right) \qquad (2.22)$$

where Q is the heat input rate. The total heat input rate per unit surface area is given by

$$Q = C_K \rho VE^2 + C_D \rho V^2 \tag{2.23}$$

where
 C_K is the enthalpy exchange coefficient
 E is the evaporative potential of the sea surface

E is a measure of the air-sea thermodynamic equilibrium and is related to the Greenhouse effect (see Chapter 4). The larger the greenhouse effect of gases emitted to the atmosphere, the larger the E value. We now have

$$W = \left(C_K \rho VE^2 + C_D \rho V^2\right) \cdot \left(\frac{T_{hot} - T_{cold}}{T_{hot}}\right) \tag{2.24}$$

Equating W to D and using algebraic manipulations, we have the final equation for the hurricane's maximum wind velocity (Emanuel, 2003):

$$V_{max} = \sqrt{\left(\frac{T_{hot} - T_{cold}}{T_{cold}}\right) \cdot E} \tag{2.25}$$

Note that T_{cold} and not T_{hot} appear in the denominator of this equation. This is a result of the feedback mechanisms of dissipative heating. The evaporative potential of the sea surface is approximated by the following equation:

$$E \approx \frac{C_K}{C_D} \cdot \left(h^* - h\right) \tag{2.26}$$

where h* and h are the specific enthalpy (enthalpy per unit mass) of air near the ocean boundary surface and of the inflowing dry air in the ambient boundary layer, respectively. The ratio C_K/C_D has been observed to be about 1 (Emanuel, 2005). The student is encouraged to work out Problem 2.33 to learn more about the applicability of these equations.

Example 2.3   Application of the First Law to a Pump

A pump problem is a common one in environmental engineering. For example, moving water or other fluids from one point to another and energy required to pump such fluids is a frequent issue in design of wastewater and other treatment facilities. We can apply Equation 2.8 in this case if we consider two points 1 and 2, viz., the pump intake and the discharge point.

$$\frac{dE_T}{dt} = \dot{q} - \dot{w} + \dot{m}_1\left(U_1 + P_1V_1 + \frac{u_1^2}{2} + gz_1\right) - \dot{m}_2\left(U_2 + P_2V_2 + \frac{u_2^2}{2} + gz_2\right) \quad (2.27)$$

We can assume that the total energy of the pump does not change with time; in other words, it is at steady state and $dE_T/dt = 0$. We can also assume that there is no heat transfer to the pump; hence, $q = 0$. If the cross-sectional areas at 1 and 2 are the same, then $u_1 = u_2$. We shall assume constant mass rates. Further, if the height differential is small, then $z_1 = z_2$. Hence *work done on the pump by the system* is $\dot{w} = \dot{m}_1\left[(U_1 + P_1V_1) - (U_2 + P_2V_2)\right]$. The *work done by the pump* is the negative of this work $w_{pump} = \dot{m}_1\left[(U_2 + P_2V_2) - (U_1 + P_1V_1)\right]$. Further, in Section 2.1.2.2, we shall define $U + PV = H$, the enthalpy of the process. Hence $w_{pump} = H_2 - H_1 = \Delta H$, which means that the work done by the pump is simply the enthalpy of the process.

2.1.2.5 Combination of First and Second Laws

Combining the first and second laws, we can obtain many useful thermodynamic relationships. Consider a closed system, which does only PV work of expansion and for which δq is given by TdS:

$$dU = TdS - PdV \quad (2.28)$$

From the definition of enthalpy H, we can also derive the following:

$$dH = TdS + VdP \quad (2.29)$$

Two other very useful state functions called "Gibbs Free Energy" G and "Helmholtz Free Energy" A were defined by Joshua Willard Gibbs and Hermann von Helmholtz, respectively, in the early 1800s.

$$G = H - TS$$
$$A - U - TS \quad (2.30)$$

We can derive the following equations for differential changes in G and A:

$$dG = -SdT + VdP$$
$$dA = -SdT - PdV \quad (2.31)$$

Most environmental processes are functions of T and P and hence Gibbs free energy is a natural state function for those cases. These differential equations are the basis for deriving useful physicochemical thermodynamic relationships.

2.1.2.6 Enthalpy and Heat Capacity

For most chemical processes that are carried out at constant volume, the PV work is zero since dV is zero. However, there are several environmentally relevant processes for which the pressure is constant, but volume is not. In such cases, the work term is nonzero. It is suitable for such processes to define a new term called "enthalpy," which is denoted by the symbol H. It is also a state function that does not depend on the path taken by a system to arrive at that state. If only expansion work is considered against a constant external pressure P in Equation 2.2, then we have the relation $\Delta U = q - P\Delta V$. If the two states of the system are denoted by A and B, then we have

$$\left(U_B + PV_B\right) - \left(U_A + PV_A\right) = q \tag{2.32}$$

If we define $H = U + PV$, then we see that $\Delta H = q$ at constant P. The importance of H in dealing with systems involving material flow will become apparent later as we combine several of the thermodynamic laws. We have seen thus far that the total energy of the system at constant volume is given by $\Delta U = (q)_V$ and the total energy at constant pressure is given by $\Delta H = (q)_P$.

Closely related to the internal energy U and enthalpy H are the terms heat capacity at constant volume C_V and heat capacity at constant pressure C_P. The concept of temperature resulting from the first law states that the temperature differences provide the driving force for heat flow between bodies. The ratio of the amount of heat transferred to the body and the resulting temperature change is called the "heat capacity." For an infinitesimal change, the ratio is given by

$$C = \frac{dq}{dT} \tag{2.33}$$

$$C_V = \left(\frac{\partial U}{\partial T}\right)_V ; \quad C_P = \left(\frac{\partial H}{\partial T}\right)_P \tag{2.34}$$

Heat capacities are tabulated for several compounds in the literature usually as empirical equations. These are given as polynomial fits with temperature as the dependent variable and are valid only within the given range of temperature. A short compilation for some important compounds is given in Table 2.2. Note that for very small ranges of temperature, the heat capacity can be assumed to be a constant. For solids and liquids the heat capacities are similar in magnitude, but for gases, there is a relationship between the two, given by

$$C_P - C_V = R \tag{2.35}$$

2.1.2.7 Thermodynamic Standard States, Enthalpies of Reaction, Formation, and Combustion

The specification of "standard states" in thermodynamics results from the need to develop charts and tables for thermodynamic properties. The fundamental properties such as temperature, pressure, etc., are by themselves based on some

TABLE 2.2

Empirical Equations for Specific Molar Heat Capacities at Constant Pressure $C_p = a + bT + cT^2 + dT^3$ Applicable for $273 \leq T \leq 1500$ K (Units: cal K^{-1} mol^{-1})

Compound	a	b × 10²	c × 10⁵	d × 10⁹
H_2	6.95	−0.045	0.095	−0.208
O_2	6.08	0.36	−0.17	0.31
H_2O	7.70	0.046	0.25	−0.86
CH_4	4.75	1.2	0.30	−2.63
$CHCl_3$	7.61	3.46	−2.67	7.34

accepted standards. The thermodynamic properties derived from them also require the specification of some standard states. The definition of the standard state is to be noted whenever U or H or any other thermodynamic quantities are used in a calculation. For convenience, we choose some arbitrary state of a substance, at a specified temperature T, standard pressure P_0, standard molality m_0, or standard concentration C_0. Although the choice of P_0, m_0, or C_0 will depend on the system, the most common choices are the following (Kondepudi, 2008):

$$T = 298.15 \text{ K}; \quad P_0 = 1 \text{ atm } \left(= 1.013 \times 10^5 \text{ Pa}\right); \quad m_0 = 1 \text{ mol kg}^{-1}; \quad C_0 = 1 \text{ mol dm}^{-3}.$$

For any pure substance at a given temperature, the standard state is the most stable state (gas, liquid, or solid) at a pressure of 1 atm (= 1.013×10^5 Pa). In the gas phase, the standard state of any compound, either pure or as a component in a mixture, is the hypothetical state of ideal gas behavior at P = 1 atm. For a condensed phase (liquid or solid), the standard state, either as pure or as a component in a mixture, is that of the pure substance in the liquid or solid form at the standard pressure of P_0. For a solute in solution, the standard state is a hypothetical state of ideal solution at standard concentration C_0 at the standard pressure of P_0.

There are three important enthalpy terms that merit discussion. These are "standard enthalpy of reaction" (ΔH_r^o), "standard enthalpy of formation" (ΔH_f^o), and "standard enthalpy of combustion" (ΔH_c^o). The standard enthalpy of reaction is the enthalpy change for a system during a chemical reaction and is the sum of the standard enthalpy of formation of the products minus the sum of the standard enthalpy of formation of the reactants. The standard enthalpy of formation is the enthalpy change to produce 1 mol of the compound from its elements, all at standard conditions. The standards enthalpies of reaction may be combined in various ways. This is succinctly expressed in the "Hess's law of heat summation: The standard enthalpy of a reaction is the sum of the standard enthalpies of reactions into which an overall reaction may be divided, and holds true if the referenced temperatures of each individual reaction are the same." The standard enthalpy of combustion is that required to burn or oxidize 1 mol of the material to a final state that contains only H_2O (l) and CO_2.

Example 2.4   Enthalpy of an Acid-Base Reaction

Determine the enthalpy of reaction at 298 K for the neutralization of hydrochloric acid with sodium hydroxide.

$$HCl\ (aq) + NaOH\ (aq) \rightleftharpoons NaCl\ (aq) + H_2O\ (aq) \qquad (2.36)$$

First calculate the heat of formation of each species (ΔH_f^0) from the heat of formation of the constituent ions (ΔH_f^0). These values are obtained from Appendix B:

$$\Delta H_f^0\ (HCl,\ aq) = \Delta H_f^0\ \left(H^+,\ aq\right) + \Delta H_f^0\ \left(Cl^-,\ aq\right) = 0 - 167.2 = -167.2\ kJ\ mol^{-1}$$

$$\Delta H_f^0\ (NaOH,\ aq) = \Delta H_f^0\ \left(Na^+,\ aq\right) + \Delta H_f^0\ \left(OH^-,\ aq\right)$$
$$= -239.7 - 229.7 = -469.4\ kJ\ mol^{-1}$$

$$\Delta H_f^0\ (NaCl,\ aq) = \Delta H_f^0\ \left(Na^+,\ aq\right) + \Delta H_f^0\ \left(Cl^-,\ aq\right)$$
$$= -239.7 - 167.2 = -406.9\ kJ\ mol^{-1}$$

$$\Delta H_f^0\ (H_2O,\ aq) = -285.8\ kJ\ mol^{-1}$$

Hence, for the overall reaction,

$$\Delta H_r^0 = \Delta H_f^0\ (NaCl,\ aq) + \Delta H_f^0\ (H_2O,\ aq)$$
$$-\Delta H_f^0\ (HCl,\ aq) - \Delta H_f^0\ (NaOH,\ aq) = -56.1\ kJ\ mol^{-1}.$$

The enthalpy of reaction at temperatures other than standard temperature can be estimated from the following equation called "Kirchoff's law":

$$\Delta H_f\ (T) = \Delta H_f^0\ (T_0) + \int_{T_0}^{T} \Delta C_p dT \qquad (2.37)$$

2.1.3 CHEMICAL EQUILIBRIUM AND GIBBS FREE ENERGY

Gibbs free energy G is of significance for most environmental problems since G is sensitive to temperature T and pressure p. Many environmental processes occur at atmospheric pressure, which is generally a constant. Hence, it is important to develop the concept of equilibrium in terms of Gibbs free energy. J Willard Gibbs, considered by many to be the father of modern chemical thermodynamics, invented the function. It has units of kilojoules (kJ). From the fundamental definition of $G = H - TS = U + PV - TS$, and further if P and T are constant, we have

$$dG = \delta q - \delta w + PdV - TdS = \delta q - TdS \qquad (2.38)$$

Since we know that for spontaneous processes, $\delta q \leq TdS$ from second law, at constant T and P if only PV work of expansion is considered, we have the following criterion for any change in the independent variables of the system:

$$dG \leq 0 \qquad (2.39)$$

For a reversible process in a closed system at constant temperature and pressure, if only PV expansion work is allowed, there is no change in Gibbs function ($dG = 0$) at equilibrium. Spontaneous processes will occur in such a system at constant T and P when it is not at equilibrium with a consequent decrease in free energy ($dG < 0$).

This inequality for a spontaneous change is the criterion used to develop equilibrium models in environmental science and engineering. For a finite change in a system at constant temperature T, we can write

$$\Delta G = \Delta H - T\Delta S \qquad (2.40)$$

In order to assess whether a process in the natural environment is spontaneous or not, one estimates the difference in Gibbs free energies between the final and initial states of the system, that is, ΔG. If ΔG is negative, the process is spontaneous. In other words, the tendency of any system toward an equilibrium position at constant T and P is driven by its desire to minimize its free energy.

The free energy change for any system is, therefore, given by the inequality:

$$\Delta H - T\Delta S \leq 0 \qquad (2.41)$$

For an exothermic process, there is release of heat to the surroundings (enthalpy change, $\Delta H < 0$). For an endothermic process, since $\Delta H > 0$, $T\Delta S$ has to be greater than zero and positive and larger than ΔH. Thus, an endothermic process is accompanied by a very large increase in entropy of the system.

2.1.3.1 Free Energy Variation with Temperature and Pressure

Environmental processes that occur are generally functions of temperatures. Therefore, it is useful to know how the criterion for spontaneity and equilibrium for chemical processes vary with temperature. At constant pressure $(\partial G/\partial T)_p = -S$, and S is always positive; thus, it is clear that at constant pressure with increasing temperature, G should decrease. If S is large the decrease is faster. For a finite change in system from state A to state B, the differential $\Delta G = G_B - G_A$ and $\Delta S = S_B - S_A$ and hence we have

$$\left(\frac{\partial(\Delta G)}{\partial T}\right)_P = -\Delta S \tag{2.42}$$

To arrange this equation in a more useful manner, we need a mathematical manipulation, viz., first differentiating the quantity $(\Delta G/T)$ with respect to T using the quotient rule. This gives

$$\frac{\partial\left(\dfrac{\Delta G}{T}\right)}{\partial T} = -\frac{\Delta G}{T^2} + \frac{1}{T}\cdot\left(\frac{\partial \Delta G}{\partial T}\right) \tag{2.43}$$

Substitution and rearrangement gives us

$$\left(\frac{\partial(\Delta G/T)}{\partial T}\right)_P = -\frac{\Delta H}{T^2} \tag{2.44}$$

This is the "Gibbs-Helmholtz" equation and fundamentally describes the relationship between temperature and equilibrium in environmental processes.

Gibbs free energy also depends on pressure since at constant temperature we have $(\partial G/\partial P)_T = V$. V being a positive quantity, the variation in G is always positive with increasing pressure. Thus, the Gibbs free energy at any pressure P_2 can be obtained if the value at a pressure P_1 is known, using the equation

$$G(P_2) = G(P_1) + \int_{P_1}^{P_2} V dP \tag{2.45}$$

For most solids and liquids, the volume changes with pressure are negligible and hence G is only marginally dependent on pressure. However, for geochemical processes in the core of earth's environment, appreciable changes in pressures as compared to surface pressures are encountered, and hence pressure effects on Gibbs functions are important. Usually, because of the huge modifications in G, most materials undergo a phase change as they enter the earth's core. For gases, however, significant changes in volume can occur. For example, for an ideal gas, $V = nRT/P$, and hence integrating Equation 2.27 using this we see that the change in Gibbs function with pressure is given by

$$G(P_2) = G(P_1) + nRT \cdot \ln\left(\frac{P_2}{P_1}\right) \tag{2.46}$$

The standard Gibbs free energy of gases increases with increase in pressure.

Example 2.5   Free Energy and Temperature

Problem: The free energy change for a reaction is given by

$$\Delta G^0 = -RT \cdot \ln K_{eq} \tag{2.47}$$

where K_{eq} is the equilibrium constant for a reaction. If K_{eq} for the reaction H_2CO_3 (aq) → H^+ (aq) + HCO_3^- (aq) is 5×10^{-7} mol L^{-1}, at 298 K, what is the equilibrium constant at 310 K? ΔH^0 is 7.6 kJ mol^{-1}.
    Solution: Using Gibbs-Helmholtz relationship, we have

$$\frac{\partial \ln K_{eq}}{\partial T} = \frac{\Delta H^0}{RT^2} \tag{2.48}$$

and hence,

$$\ln\left(\frac{K_{eq}(T_2)}{K_{eq}(T_1)}\right) = -\frac{\Delta H^0}{R} \cdot \left(\frac{1}{T_2} - \frac{1}{T_1}\right) \tag{2.49}$$

Substituting, we get K_{eq} (at 310 K) = 5.63×10^{-7} mol L^{-1}.
    Other properties like vapor pressure, aqueous solubility, air/water partition constant, and soil/water partition constant show similar changes depending on the sign and magnitude of the corresponding standard enthalpy of the process. These are discussed later in this chapter.

2.1.4   CONCEPT OF MAXIMUM WORK

Thus, we have the following equation for the total free energy change for any process,

$$\Delta G = q - w_{tot} + P\Delta V - T\Delta S = -w_{tot} + P\Delta V \tag{2.50}$$

since $T\Delta S = q$. Thus, for any "permissible" process at constant T and P it follows that

$$-\Delta G = w_{tot} - P\Delta V = w_{useful} \tag{2.51}$$

Since $P\Delta V$ is the expansion work done or otherwise called the "wasted work," the difference between the total work and $P\Delta V$ is the "useful work" that can be done by the system. *At constant temperature and pressure, the useful work that can be done by a system on the surroundings is equivalent to the negative change in Gibbs free energy of a spontaneous process within the system.* If the system is at equilibrium, then $\Delta G = 0$, and hence no work is possible by the system.

Example 2.6   Calculation of Maximum Useful Work

Problem statement: Humans derive energy by processing organic polysaccharides. For example, glucose can be oxidized in the human body according to the reaction: $C_6H_{12}O_6$ (s) + $6O_2$ (g) → $6CO_2$ (g) + $6H_2O$ (l). If the standard heat of combustion of glucose at room temperature is -2808 kJ mol^{-1} and the standard entropy of the reaction is $+182$ J K^{-1} mol^{-1}, calculate the maximum amount of useful non-PV work available from 1 g of glucose. Assume that heat of combustion is independent of temperature.

Solution: Since $\Delta G^0 = \Delta H^0 - T\Delta S^0 = -2808 - (310)(0.182) = -2864$ kJ mol^{-1}. T is 310 K, the blood temperature of a normal human being. Hence from 1 g ($\equiv 0.0055$ mol) of glucose, the total free energy change for oxidation is -15.9 kJ. Since the negative of the free energy change gives the maximum non-PV work, the useful work available is also 15.9 kJ. Note that this is the basis for the term "number of calories" found on labels for foodstuffs.

In environmental bioengineering, microorganisms are used to process many types of organic wastes. Similar calculations can be made of the maximum non-PV useful work performed by these organisms, if they consume oxygen as the electron-donating species.

Example 2.7   Maximum Voltage from a Fuel Cell

A typical fuel cell works on the principle of converting chemical energy into electrical energy. A proton exchange membrane fuel cell (PEMFC) works on the principle of hydrogen fuel being supplied to an anode where it oxidizes and the proton then moves through a membrane to the cathode where it combines with oxygen to produce water. The overall reaction is therefore $H_2 + 2O_2 → 2H_2O$. Calculate the maximum voltage obtained from the PEMFC.

We know that for the respective reactions given here, the free energy values can be obtained from the tables:

$$\text{Anode:}\quad H_2 → 2H^+ + 2e^-\quad \Delta G_1 = 0 \text{ kJ mol}^{-1}$$

$$\text{Cathode:}\quad \frac{1}{2}O_2 + 2H^+ + 2e^- → H_2O\quad \Delta G_2 = -237.2 \text{ kJ mol}^{-1}$$

Hence, $\Delta G_{tot} = -237.2$ kJ mol^{-1}. Therefore, maximum work, $W_{max} = 237.2$ kJ mol^{-1}. Now, since electrical work is defined as nFE_0, where n is the number of total electrons transferred in the reaction (= 2), F is Faraday's constant (= 96,485 Coulomb mol^{-1}), the voltage $E_0 = 237.2/(2 \times 96,485) = 1.23$ V. Thus, the maximum voltage from the PEMFC under ideal conditions is 1.23 V.

2.1.5 GIBBS FREE ENERGY AND CHEMICAL POTENTIAL

The equations described thus far are applicable to homogeneous systems composed of a single component. In environmental engineering, one is concerned with different components in each phase. Since this section is devoted to single-phase thermodynamics, the focus here is only on single and multicomponents in a homogeneous single phase. The discussion of multiphase equilibrium is reserved for the next section. As we have already seen, the fundamental equation for Gibbs free energy of a system states that

$$dG = -SdT + VdP \qquad (2.52)$$

In other words, $G = G(T, P)$. If the phase is composed of several components this equation must be modified to reflect the fact that a change in the number of moles in the system changes the Gibbs free energy of the system. This means G is a function of the number of moles of each species present in the system. Hence, $G = G(T, P, n_1, ..., n_j)$. The total differential for G then takes the form

$$dG = \left(\frac{\partial G}{\partial P}\right)_{T,n_i} dP + \left(\frac{\partial G}{\partial T}\right)_{P,n_i} dT + \sum_i \left(\frac{\partial G}{\partial n_i}\right)_{P,T,n_{j,i}} dn_i \qquad (2.53)$$

In this equation, we recognize that

$$\left(\frac{\partial G}{\partial P}\right)_{T,n_i} = V$$

$$\left(\frac{\partial G}{\partial T}\right)_{P,n_i} = -S \qquad (2.54)$$

while the third partial differential is a new term given the name "chemical potential" by Gibbs in his formalism of thermodynamics. It is given the symbol μ_i, and is applicable to each species in the system and has units of $kJ\,mol^{-1}$.

$$\left(\frac{\partial G}{\partial n_i}\right)_{P,T,n_{j\neq i}} = \mu_i \qquad (2.55)$$

Chemical potential is the "partial molar Gibbs free energy" representing the change in the total Gibbs energy due to addition of a differential amount of species i to a finite amount of solution at constant temperature and pressure. For a single pure substance, the chemical potential is the same as its molar Gibbs energy.

Thus, the chemical potential of benzene in pure water is different from that in a mixture of water and alcohol. As its very name indicates, μ_i is an indicator of the potential for a molecule (e.g., movement from one phase to another, or a chemical reaction). Thus, it is analogous to a hydrostatic potential for liquid flow, electrostatic potential for charge flow, and gravitational potential for mechanical work. The chemical potential is thus a kind of "chemical pressure," and it is an intensive property of the system such as T and P. When the chemical potential of a molecule is the same in states a and b, then equilibrium is said to exist. This satisfies the criterion for equilibrium defined earlier, that is, $\Delta G = 0$. If the chemical potential is greater in state a than in state b, then a transfer or reaction of species i occurs spontaneously to move from a to b. This satisfies the criterion for a spontaneous process, that is, $\Delta G < 0$.

This formalism suggests that at constant T and P, the total free energy of a system is given by

$$G = \sum_i \mu_i n_i \qquad (2.56)$$

Example 2.8   Significance of Chemical Potential

Problem: Calculate the change in chemical potential for the vaporization of water at 1 atm and 25°C.

Solution: The reaction is H_2O (l) → H_2O (g). The free energy of formation per mole is −237 kJ mol^{-1} for liquid water and −229 kJ mol^{-1} for water vapor. For a pure substance the free energy per mole is the same as chemical potential. Hence, $\Delta\mu = -229 - (-237) = 8$ kJ mol^{-1}. The chemical potential of the vapor is higher, indicating that there is useful work available. It is clear that water vapor is more "potent" compared to liquid water.

2.1.5.1   Gibbs-Duhem Relationship for a Single Phase

In equilibrium calculations, an important problem is the estimation of the chemical potential of different components in a single phase. Upon differentiation of the expression for total Gibbs free energy given by Equation 2.56,

$$dG = \sum_i \mu_i dn_i + \sum_i n_i d\mu_i \qquad (2.57)$$

The formal definition for dG is

$$dG = -SdT + VdP + \sum_i \mu_i dn_i \qquad (2.58)$$

Therefore, we have

$$SdT - VdP + \sum_i n_i d\mu_i = 0 \qquad (2.59)$$

This is called the "Gibbs-Duhem relationship." At constant temperature and pressure, this equation reduces to

$$\sum_i n_i d\mu_i = 0 \qquad (2.60)$$

Equation 2.60 is useful in estimating the variation in chemical potential of one component with composition if the composition and variation in chemical potential of the second component is known. By dividing the equation throughout with the total number of moles $\sum n_i$, we can rewrite Equation 2.60 in terms of mole fractions $\sum x_i d\mu_i$.

2.1.5.2 Standard States for Chemical Potential

Let us examine the chemical potential of an ideal gas. At a constant temperature T

$$\left(d\mu_i\right)_T = \left(\frac{V}{n_i}\right) dP_i \qquad (2.61)$$

where P_i is the "pressure exerted by the component i (partial pressure)" in a constant volume V. Note that for an ideal gas $V/n_i = (RT/P_i)$, μ_i at any temperature, T can now be determined by integrating this equation if we choose as the lower limit a starting chemical potential (μ_i^0) at the temperature T and a reference pressure P_i^0, that is,

$$\mu_i = \mu_i^0 + RT \ \ln\left(\frac{P_i}{P_i^0}\right) \qquad (2.62)$$

Thus, the chemical potential of an ideal gas is related to a physically real measurable quantity, namely its pressure. Whereas both μ_i^0 and P_i^0 are arbitrary, they may not be chosen independent of one another, since the choice of one fixes the other automatically. The standard pressure chosen is $P_i^0 = 1$ atm. Thus, for an "ideal gas" the chemical potential is expressed as

$$\mu_i = \mu_i^0\left(T; \ P_i^0 = 1\right) + RT \cdot \ln P_i \qquad (2.63)$$

The chemical potential for an "ideal liquid" is analogous to this expression for the ideal gas. It is given by

$$\mu_i = \mu_i^0 \left(T; \ x_i^0 = 1 \right) + RT \cdot \ln x_i \tag{2.64}$$

where x_i is the "mole fraction of component" i in the solution. The reference state for the solution is pure liquid, for which the mole fraction is 1.

2.1.6 THERMODYNAMICS OF SURFACES AND COLLOIDAL SYSTEMS

The boundary between any two contiguous phases is called a "surface" or an "interface." The term "surface" is reserved for those that involve air as one of the contiguous phases. In environmental chemistry we encounter several interfaces: liquid/gas (e.g., air-sea, air bubbles in water, fog droplets in air, foam), liquid/liquid (e.g., oil-water), liquid/solid (e.g., sediment-water, soil-water, colloid-water), and solid/air (e.g., soil-air, aerosols in air). Often when one phase is dispersed in another at submicron dimensions (e.g., air bubbles in sea, fog droplets in air, colloids in groundwater, activated carbon in waste-water treatment), very large surface areas are involved. Surface tension accounts for the spherical shape of liquid droplets, the ability of water to rise in capillaries as in porous materials (e.g., soils), and for a variety of other reactions and processes at interfaces. It is, therefore, important to discuss some basics of surface and interfacial chemistry.

2.1.6.1 Surface Tension

Surfaces and interfaces are different from bulk phases. Consider, for example, water in contact with air (Figure 2.4a). The number density of water molecules gradually decreases as one moves from bulk water into the air (Figure 2.4b). Similar gradual changes in all other physical properties will be noticed as one moves from water to air across the interface. At a molecular level, there is no distinct dividing surface at

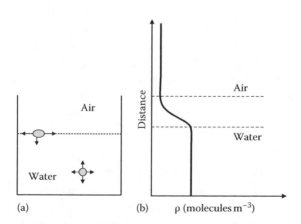

(a)   (b) ρ (molecules m^{-3})

FIGURE 2.4 (a) Isotropic forces on a bulk water molecule and anisotropic forces on a surface water molecule. (b) The number density of water molecules as a function of distance from the interface.

which the liquid phase ceases and the air phase begins. The so-called "interface" is therefore a "diffuse region" where the macroscopic properties change rather gradually across a certain thickness. The definition of this thickness is compounded when we realize that it generally depends upon the property considered. The thickness is at least a few molecular diameters (a few Angstroms). The ambiguity with respect to the location of an exact dividing surface makes it difficult to assign properties to an interface. However, Gibbs showed how this can be overcome.

The reason for the diffuse nature of the surface becomes clear when we consider the forces on a water molecule at the surface. Obviously, since there is a larger number density of molecules on the water side than on the air side, the molecule experiences different forces on either side of the interface. The pressure (force) experienced by a water molecule in the bulk is the time-averaged force exerted on it per unit area by the surrounding water molecules and is "isotropic" (i.e., same in all directions). For a molecule on the surface, however, the pressure is "anisotropic" (Figure 2.4a). It has two components, one normal to the surface, P_n, and one tangential to the surface, P_t. The net pressure forces in the lateral direction are substantially reduced in the interfacial region compared to the bulk region. Due to the fact that fewer liquid phase molecules than those in the bulk phase act upon the surface molecules, the surface molecules possess greater energy than the bulk molecules. It requires work to bring a bulk molecule to the surface since it means increasing the surface area. The net pressure forces in the surface will be negative, that is, the surface is said to experience a tension. The surface is said to have a contractile tendency, that is, it seeks to minimize the area, and is describable in thermodynamics in terms of a property called "surface or interfacial tension" and is designated by the symbol σ. It is the "force per unit length" on the surface ($N\,m^{-1}$ or more commonly $mN\,m^{-1}$) or equivalently the "free energy per unit surface area" ($J\,m^{-2}$).

The molecular picture of surface tension given here explains the variation in surface tension between different liquids (Table 2.3). It is clear that surface tension is a property that depends on the strength of the intermolecular forces between molecules. Metals such as Hg, Na, and Ag in their liquid form have strong intermolecular attractive forces resulting from metallic bonds, ionic bonds, and hydrogen bonds. They have, therefore, very high surface tension values. Molecules such as gases and liquid hydrocarbons have relatively weak van der Waals forces between them and have therefore low surface tensions. Water, on the other hand, is conspicuous in its strange behavior since it has an unusually high surface tension. The entire concept of "hydrophobicity" and its attendant consequences with respect to a wide variety of compounds in the environment is a topic of relevance to environmental engineering and will be discussed in Section 2.2.3.4.

2.1.6.2 Curved Interfaces and the Young-Laplace Equation

Curved interfaces are frequently encountered in environmental engineering. Examples are air bubbles in water, soap bubbles in water, fog droplets in air, aerosols, colloids, and particulates in air and water environments. Consider a curved interface, such as an air bubble in water, as shown in Figure 2.5. Surface tension is itself independent of surface curvature so long as the radius of curvature of the bubble is large in comparison to the thickness of the surface layer (a few Å). Young (1805) argued that in order to maintain the curved interface a finite pressure difference ought to

TABLE 2.3
Surface Tensions of Some Common Substances

Compound	Temperature (K)	σ (mN m^{-1})
Metals		
Na	403	198
Ag	1373	878.5
Hg	298	485.5
Inorganic salts		
NaCl	1346	115
NaNO$_3$	581	116
Gases		
H$_2$	20	2.01
O$_2$	77	16.5
CH$_4$	110	13.7
Liquids		
H$_2$O	298	72.13
CHCl$_3$	298	26.67
CCl$_4$	298	26.43
C$_6$H$_6$	293	28.88
CH$_3$OH	293	22.50
C$_8$H$_{17}$OH	293	27.50
C$_6$H$_{14}$	293	18.40

Source: Adamson, A.W., *Physical Chemistry of Surfaces*, 4th edn., John Wiley & Sons, Inc., New York, 1990.

Note: An excellent compilation of surface tensions of liquids is also available in Jasper (1972).

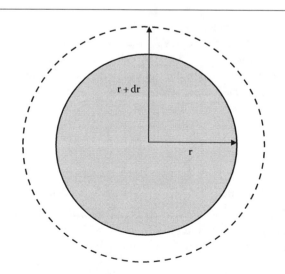

FIGURE 2.5 Cross section of an air bubble in water.

exist between the inside, P_i, and outside, P_o, of the bubble. The work necessary to move a volume dV of water from the bulk to the air bubble is given by $(P_i - P_o)dV$. This requires an extension of the surface area of the bubble by dA_σ, and the work done will be σdA_s. For a spherical bubble, $dV = 4\pi r^2 dr$ and $dA_\sigma = 8\pi r dr$. Therefore

$$\Delta P = \frac{2\sigma}{r} \qquad (2.65)$$

This is called the Young-Laplace equation. If we consider any curved interface that can be generally described by its two main radii or curvature, R_1 and R_2, then the Young-Laplace equation can be generalized as (Adamson, 1990; Adamson and Gast, 1997)

$$\Delta P = \sigma \cdot \left(\frac{1}{R_1} + \frac{1}{R_2} \right) \qquad (2.66)$$

This equation is a fundamental equation of "capillary phenomenon." For a plane surface, its two radii of curvature are infinite and hence ΔP is zero. Capillarity explains the rise of liquids in small capillaries, capillary forces in soil pore spaces, and vapor pressure above curved interfaces such as droplets. It also plays a role in determining the nucleation and growth rates of aerosols in the atmosphere.

2.1.6.3 Surface Thickness and Gibbs Dividing Surface

As stated earlier, a surface is not a strict boundary of zero thickness nor is it only a two-dimensional area. It has a finite thickness (a few Å). This poses a problem in assigning numerical values to surface properties. Fortunately, Gibbs, the architect of surface thermodynamics, came up with a simple proposition. It is appropriately termed the "Gibbs dividing surface."

Let us consider two phases of volume V_I and V_{II} separated by an "arbitrary" plane designated a. This is strictly a mathematical dividing plane. Gibbs suggested that one should handle all extensive properties (E, G, S, H, n, etc.) by ascribing to the bulk phases those values that would apply if the bulk phases continued uninterrupted up to the dividing plane. The actual values for the system as a whole and the total values for the two bulk phases will differ by the so-called "surface excess" or "surface deficiency" assigned to the surface region. For example, the "surface excess" in concentration for each component in the system is defined as the excess number of moles in the system over that of the sum of the number of moles in each phase. Hence,

$$n_i^\sigma = n_i - C_i^I V_I - C_i^{II} V_{II} \qquad (2.67)$$

The surface excess concentration (Γ_i, mol m^{-2}) is defined as the ratio of n_i^σ (moles) and the surface area A_σ (m^2), that is, $\Gamma_i = \dfrac{n_i^\sigma}{A_\sigma}$. The same equation holds for every other component in the system. If we now move the plane from a to b, then the volume of phase b decreases by $(b - a) A_s$ and phase a increases by an identical value. Since the total number of moles in the system is the same, the new surface excess value is

$$\Gamma_i^{new} = \Gamma_i^{old} + \left(C_i^{II} - C_i^I \right) \cdot (b - a) \qquad (2.68)$$

Thus, $\Gamma_i^{new} = \Gamma_i^{old}$ only if $C_i^{II} = C_i^{I}$. Thus, the location of the Gibbs dividing surface becomes important in the definition of surface excess. Gibbs proposed that to obtain the surface excess of all other components in a solution a "convenient" choice of the dividing plane is such that the "surface excess of the solvent is zero." For a pure component system, there is no surface excess. This is called the "Gibbs convention."

2.1.6.4 Surface Thermodynamics and Gibbs Equation and Gibbs Adsorption Equation

Consider a system where the total external pressure P and the system temperature T are kept constant. Let us assume that the system undergoes an increase in interfacial area while maintaining the total number of moles constant. The overall increase in free energy of the system is the work done in increasing the surface area. This free energy increase per unit area is the surface tension. Hence,

$$\sigma = \left(\frac{\partial G}{\partial A_\sigma} \right)_{T,P,n_i} \tag{2.69}$$

Analogous to dG for the bulk phase, we can define a surface free energy in the following form:

$$dG_\sigma = -S_\sigma dT + \sigma dA_\sigma + \sum_i \mu_i dn_i^\sigma \tag{2.70}$$

where we have replaced σ for P and dA_σ for dV. At constant T,

$$dG_\sigma = \sigma dA_\sigma + \sum_i \mu_i dn_i^\sigma \tag{2.71}$$

The total free energy at constant temperature for the surface is given by

$$G_\sigma = \sigma A_\sigma + \sum_i \mu_i dn_i^\sigma \tag{2.72}$$

Thus, the complete differential obtained for G_σ is given by

$$dG_\sigma = \sigma dA_\sigma + A_\sigma d\sigma + \sum \left(n_i^\sigma d\mu_i + \mu_i dn_i^\sigma \right) \tag{2.73}$$

Comparing the two expressions for dG_σ, we obtain the following equation:

$$A_\sigma d\sigma + \sum_i n_i^\sigma d\mu_i = 0 \tag{2.74}$$

This equation is fundamental to surface thermodynamics and is called the "Gibbs equation."

2.1.6.5 Gibbs Adsorption Equation

In environmental chemistry, we encounter systems composed of more than one species. If, for example, we consider a two-component system (solvent $\equiv 1$ and solute $\equiv 2$), the Gibbs equation can be rewritten using the Gibbs convention that surface concentration of solvent 1 with respect to solute 2, $\Gamma_{1(2)} = 0$,

$$\Gamma_{2(1)} = -\frac{d\sigma}{d\mu_2} \qquad (2.75)$$

Note that in most cases it is simply designated Γ. This is called the "Gibbs adsorption equation." It is the analogue of the "Gibbs-Duhem equation" for bulk phases. The Gibbs adsorption equation is used to calculate the surface excess concentration of a solute by determining the change in surface tension of the solvent with the chemical potential of the solute in the solvent. This equation is of particular significance in environmental engineering since it is the basis for the calculation of the amount of material adsorbed at air-water, soil (sediment)-water, and water-organic solvent interfaces.

Example 2.9 Surface Concentration from Surface Tension Data

The following data was obtained for the surface tension of a natural plant-based surfactant in water.

Aqueous Concentration (mol cm^{-3})	Surface Tension (ergs cm^{-2})
1.66×10^{-8}	62
3.33×10^{-8}	60
1.66×10^{-7}	50
3.33×10^{-7}	45
1.66×10^{-6}	39

The molecular weight of the surfactant is 300. Obtain the surface concentration at an aqueous concentration of 1.66×10^{-7} mol cm^{-3}.

Since $\Gamma_{2(1)} = -d\sigma/d\mu_2$ and for a dilute aqueous solution of solute concentration C_w, $d\mu_2 = RT \ln C_w$, we have surface concentration $\Gamma = \Gamma_{2(1)} = -(1/RT) (d\sigma/d \ln C_w)$. A plot of σ versus $\ln C_w$ gives a slope of $d\sigma/d \ln C_w = -6.2$ ergs cm^{-2} at $C_w = 1.66 \times 10^{-7}$ mol cm^{-3}. Hence the surface concentration is $\Gamma = -(-6.2)/(8.31 \times 10^7 \times 298) = 2.5 \times 10^{-10}$ mol cm^{-2}.

2.2 MULTICOMPONENT EQUILIBRIUM THERMODYNAMICS

Environmental systems are inherently complex and involve several phases, each containing many components. It is a characteristic of nature, and amply demonstrated through the science of thermodynamics that when two or more phases are in contact they tend to interact with each other via exchange of matter and energy. Mixtures of gases and liquids behave differently from their pure phases and deviate from ideality.

The phases interact with one another till a "state of chemical equilibrium" is reached. Two new concepts have to be introduced, viz., "fugacity" and "activity," to describe multicomponent heterogeneous systems that display varying degrees of nonideality.

There are three steps to understand complex heterogeneous multicomponent systems: (1) translation of the real problem into an abstract world of mathematics, (2) solution to the mathematical problem, and (3) projection of the solution to the real world of significant and measurable parameters (Prausnitz et al., 1999). The "chemical potential" discussed earlier in Chapter 2 is the appropriate mathematical abstraction to the physical problem. The definition of chemical equilibrium in terms of chemical potential is the framework for the solution of the physical problem in the abstract world of mathematics.

As noted earlier, the property called "chemical potential" introduced by Gibbs is a highly useful mathematical abstraction to physical reality. The chemical potential is measured indirectly. Hence, a new term was defined called "fugacity" (Lewis and Randall, 1961).

2.2.1 IDEAL AND NONIDEAL FLUIDS

The distinction between "ideal" and "real (nonideal)" gases is straightforward. If molecules in a gas do not interact with one another or if all interactions are completely elastic, the gas is considered "ideal." The molecules in an ideal gas are "point particles" and have no excluded volumes. Secondly, molecules of an ideal gas collide in an "elastic" fashion, that is, there is neither repulsive nor attractive energy between molecules. The pressure-volume-temperature (P-V-T) relationship for an ideal gas is given by the well-known "ideal gas law":

$$PV = nRT \qquad (2.76)$$

We know that most gases are not ideal since a gas cannot be cooled to zero volume, and even at moderate densities the molecules interact with one another. Molecules in real gases have definite excluded volumes. The P-V-T relationship for a real gas is the so-called "virial equation of state":

$$\frac{PV}{RT} = 1 + \frac{B(T)}{V} + \frac{C(T)}{V^2} + \cdots \qquad (2.77)$$

The terms $B(T)$ and $C(T)$ are called second and third virial coefficients.

The definition of an "ideal" solution is different from that of an ideal gas. For solutions one cannot neglect intermolecular forces. Consider a solvent ($\equiv A$) containing a solute ($\equiv B$). The possible solute-solvent interactions are A-A, A-B, and B-B. The solution is considered ideal if the three forces are identical and if there is no volume change or enthalpy change during mixing. There also exist situations in which there are so few B molecules among a large number of A molecules that B-B interactions are negligible. These are considered "ideally dilute solutions." For these solutions we can apply the limiting conditions $x_A \rightarrow 1$, $x_B \rightarrow 0$. The limiting condition is a thermodynamic definition of an ideal dilute solution. A practical definition of dilute solution in environmental engineering is somewhat less stringent than the thermodynamic definition.

2.2.1.1 Concentration Units in Environmental Engineering

Environmental concentrations (air, soil, water) are expressed in a variety of units (Appendix D). Traditional units such as parts per million (ppm), parts per billion (ppb), and parts per trillion (ppt) are still in use, although these units are inexact and of dubious applicability. Chemical thermodynamicists prefer to work in mole fraction, molality, and molarity units.

"Mole fraction" is defined as the ratio of the number of moles of a solute to the total number of moles of all species in the mixture, that is,

$$x_i = \frac{n_i}{\sum_i n_i} \tag{2.78}$$

"Molarity" is the moles of solute per liter (dm^3) of the mixture.
"Molality" is the moles of solute per kilogram of the pure phase.

The different concentration units are shown in Table 2.4. Other applicable and relevant units are derived from these basic units for all natural phases.

2.2.1.2 Dilute Solution Definition

Let us consider a dilute mixture of A in a pure phase B. If we consider a fixed volume V (m^3) of ideal dilute mixture, then addition of A and removal of B do not affect the total volume, and hence Equation 2.78 can be rewritten in terms of molarity as

$$x_A = \frac{C_A}{C_A + C_B} \tag{2.79}$$

TABLE 2.4
Concentration Units in Environmental Chemistry

Phase	Conventional Units	Preferred SI Units
Water or organic solvents	$mg\,L^{-1}$ water	$mol\,m^{-3}$ water
		$mol\,dm^{-3}$ water
Air	$\mu g\,m^{-3}$ air	$mol\,m^{-3}$ air
Soil or sediment	$mg\,kg^{-1}$ solid	$mol\,kg^{-1}$ solid

Note: In the aqueous phase, $mg\,L^{-1}$ is equivalent to ppm; in the air phase another equivalent unit to $\mu g\,m^{-3}$ is parts per million by volume (ppmv). See also Appendix D.

For dilute mixtures, $C_A \ll C_B$, and hence, $x_A \sim C_A/C_B$. For aqueous solutions, C_B is the molar density of water, which we shall designate ρ_w (= 55.5 $mol\,dm^{-3}$). Thus, mole fraction is proportional to molarity in a dilute aqueous solution. Relationships similar to this can be derived for solutions in gases and solids.

Example 2.10 Calculation of Concentrations in an Environmental Matrix

Problem statement: Calculate the mol fraction, molarity, and molality of the following solutes in water: (1) ethanol, 2 g in 100 mL water; (2) chloroform, 0.7 g in 100 mL water; (3) benzene, 0.1 g in 100 mL water; and (4) hexachlorobenzene, 5×10^{-7} g in 100 mL water.

Solution: First obtain the density and molecular weight of the compounds from standard CRC tables. Then calculate the volume of each compound and therefore the total volume of the solution. The final results are tabulated here:

Compound	ρ $(g\,cm^{-3})$	Mol. Wt.	Molarity $(mol\,dm^{-3})$	Mole Fraction	Molality $(mol\,kg^{-1})$
Ethanol	0.79	46	0.420	7.5×10^{-3}	0.430
Chloroform	1.48	119	0.058	1.0×10^{-3}	0.058
Benzene	0.87	78	0.013	2.3×10^{-4}	0.013
Hexachloro-benzene	1.57	285	1.7×10^{-8}	3.1×10^{10}	1.7×10^{-8}

Notice that only for ethanol is the molality different from molarity. At very low mol fractions, molarity and molality are identical. Since the molarities are all less than 2.773 $mol\,dm^{-3}$, the solutions can be considered ideal.

2.2.2 FUGACITY

The concept of fugacity is an invaluable tool in constructing equilibrium models for the fate and transport of chemicals in the environment (Mackay, 1991). "Fugacity" is derived from the Latin word *fugere*, which literally means "to flee." Thus, fugacity measures the "escaping or fleeing" tendency of a molecule from a phase. If the fugacity of a compound is the same in two phases, then the molecule is said to be in equilibrium with both phases. Chemical potential, although a useful quantity, is somewhat awkward to use to quantify equilibrium since it can only be estimated indirectly. For example, earlier we noted that μ_i varies nonlinearly with P_i for gases and x_i for solutions. For real, nonideal gases and solutions, the expressions for chemical potentials have to be modified and the concept of fugacity as corrected pressure was introduced. As will be shown later, the fugacity is directly related to equilibrium and can be obtained experimentally.

2.2.2.1 Fugacity of Gases

Fugacity is defined as an idealized pressure, f_i^g, for a real gas such that the expression for chemical potential can be written as

$$\mu_i^g = \mu_i^{g0} + RT \ \ln\left(\frac{f_i^g}{f_i^{g0}}\right) \tag{2.80}$$

The standard state fugacity is $f_i^{g0} = 1$ atm. For an ideal gas mixture, fugacity is the same as partial pressure, and $f_i^g = P_i$.

"Fugacity coefficient" $\chi_i = f_i^g / P_i$ indicates the degree of nonideality of the gas mixture. Note that by Dalton's law of partial pressure $P_i = y_i P_T$, where y_i denotes the mole fraction of i in the gas phase and P_T is the total pressure (1 atm for most environmental calculations). Fugacity has dimensions of pressure and is linearly related to pressure. The ratio f_i^g / f_i^{g0} is termed "activity," a_i.

2.2.2.2 Fugacity of Condensed Phases (Liquids and Solids)

The concept of fugacity can be extended to condensed phases such as liquids and solids. Both liquids and solids also exert vapor pressure, and hence their escaping tendency (fugacity) can be evaluated similar to that of a gas. Pure condensed phases exert partial pressure in the gas phase in equilibrium with it, which is termed the saturated vapor pressure. If the fugacity of the saturated vapor at temperature T and its saturated vapor pressure P_i^* is denoted by f_i^* and the fugacity of the condensed phase (liquid l or solid s) is denoted by f_i^c, then the following equation can be derived (Prausnitz et al., 1999):

$$f_i^c = P_i^* \chi_i \ \exp\left(\int_{P_i^*}^{P} \frac{V_i^c}{RT} dP\right) \tag{2.81}$$

where V_i^c is the partial molar volume of i in the condensed phase c, and $\chi_i = f_i^* / P_i^*$. The fugacity coefficient corrects for the departure from ideal gas behavior. The second correction, which is the exponential factor, is called the "Poynting correction" and is indicative of the fact that the pressure P of the condensed phase (liquid or solid) is different from the condensed phase saturation pressure P_i^s. The Poynting correction is nearly 1 under the low-pressure conditions of a few atmospheres encountered in most environmental engineering calculations. Thus, we have for both pure solids and liquids

$$f_i^c \approx P_i^* \tag{2.82}$$

2.2.2.3 Activities of Solutes and Activity Coefficients

The discussion of nonideal solutions (liquids and solids) is identical to that of real gases. For real gases, partial pressure should be replaced with the corrected pressure, namely fugacity in the expression for chemical potential. For nonideal solutions, a similar substitution was suggested for mol fraction x_i. This nondimensional quantity was called "activity" a_i. Thus, we have

$$\mu_i^l = \mu_i^{l0} + RT \ln\left(\frac{f_i^l}{f_i^{l0}}\right) \tag{2.83}$$

The term within the brackets is the "activity," a_i, which is defined as the ratio of the fugacity to the fugacity at some chosen standard state. For an "ideal" solution, $a_i = x_i$. Hence $f_i^l = x_i P_i^*$. Note that the ideal state functions are always defined at a standard pressure P and standard temperature T. For an "ideal solution,"

$$\mu_i^l(P, T, x_i) = \mu_i^{l0}(P, T) + RT \ln(x_i) \tag{2.84}$$

This also means that all chemical potentials are defined at a given pressure and temperature.

An "activity coefficient" γ_i (also dimensionless) can be defined to show the departure from ideality for solutions:

$$\gamma_i = \frac{a_i}{x_i} \tag{2.85}$$

The activity coefficient is an important term that we will use throughout this book and should be understood carefully. A value of 1 denotes ideal behavior and anything different represents nonideal behavior. Both positive and negative deviations from ideality are possible. A few words are now in order regarding the appropriate standard state chemical potentials. There are two standard conventions in chemical thermodynamics.

In the first convention (I), the standard state of each compound i is taken to be the pure liquid form at the temperature T and pressure P of the solution denoted as μ_i^*.

$$\mu_i^{l0} = \mu_i^* \tag{2.86}$$

In this convention,

$$\lim_{x_i \to 1} \gamma_i \to 1 \tag{2.87}$$

In the second convention (II), the standard state is that of the pure solvent (designated A) at temperature T and pressure P of the solution.

$$\mu_A^{l0} = \mu_A^* \tag{2.88}$$

We then have the following limit:

$$\lim_{x_A \to 1} \gamma_A \to 1; \quad \lim_{x_i \to 0} \gamma_i \to 1 \tag{2.89}$$

Table 2.5 summarizes the definitions of fugacity and activity coefficients for mixtures of real gases and solutions (liquids or solids).

TABLE 2.5

Standard States, Fugacity, and Activity Coefficients for Real Gas Mixtures and Solutions

	Fugacity Coefficient or Activity Coefficient	Standard States
Real gas mixture	$\chi_i = \dfrac{f_i^g}{P_i} = \dfrac{f_i^g}{y_i P_T}$	$f_i^g \to P_i$ as $P_T \to 0$ $\chi_i \to 1$ as $P_T \to 0$
Real solutions	$a_i = \gamma_i x_i = \dfrac{f_i^l}{f_i^{l0}}$	$\gamma_i \to 1$ as $x_i \to 0$ (convention I) $\gamma_i \to 1$ as $x_i \to 1$ (convention II) $f_i^{l0} = P_i^*$
Solid mixture	$a_i = \gamma_i x_i = \dfrac{f_i^s}{f_i^{s0}}$	$f_i^{s0} = P_i^*$

Notes:

1. For gases, $\mu_i^g = \mu_i^{g0} + RT \ln\left(\dfrac{f_i^g}{f_i^{g0}}\right)$ where $f_i^{g0} = 1$ atm, and for solutions, $\mu_i^l = \mu_i^{l0} + RT \ln\left(\dfrac{f_i^l}{f_i^{l0}}\right)$

2. P_i^* denotes saturated vapor pressure. P_T is the total pressure.

3. For a gaseous mixture $f_i^g = y_i f_i^{g,pure}$ and is called the "Lewis-Randall rule," that is, fugacity of i in a mixture is the product of its mole fraction in the gas mixture and the fugacity of pure gaseous component at the same temperature and pressure.

Example 2.11   Calculation of Chemical Potentials Using Different Conventions

Problem: Given the activity coefficient of chloroform measured using the two conventions described here in a chloroform/benzene mixture at 50°C, calculate the corresponding chemical potentials.

	Activity Coefficient, γ_i	
Mole Fraction of Chloroform, x_i	Convention I	Convention II
0.40	1.25	0.70

Solution: The standard state chemical potential is not given. Hence, we can only determine the difference in chemical potentials, $\Delta\mu_i = RT \ln(\gamma_i x_i)$

Convention I: $\Delta\mu_i = (8.314 \text{ J (mol K)}^{-1})(323 \text{ K}) \ln(1.25 \times 0.4) = 240 \text{ J mol}^{-1}$
Convention II: $\Delta\mu_i = (8.314 \text{ J (mol K)}^{-1})(323 \text{K}) \ln(0.7 \times 0.4) = -383 \text{ J mol}^{-1}$

Example 2.12   Calculation of the Activity of Water in Seawater

Problem statement: If the vapor pressure of a sample of seawater is 19.02 kPa at 291 K, calculate the activity of water in the solution.

Solution: The vapor pressure of pure water at 291 K is 19.38 kPa (*CRC Handbook of Chemistry and Physics*). Hence, $a_{water} = 19.02/19.38 = 0.9814$.

2.2.3 IDEAL, DILUTE SOLUTIONS

2.2.3.1 Vapor-Liquid Equilibrium: Henry's and Raoult's Laws

There are two important relationships that pertain to the equilibrium between an ideal solution and its vapor. These are called Henry's law and Raoult's law. We have already alluded to these laws, although not explicitly, in the selection of standard states for activity coefficients. These two laws have significant applications in environmental engineering, which are discussed in Chapter 3.

2.2.3.1.1 Henry's Law

On a purely experimental basis, the English chemist William Henry noted that the partial pressure of the gas above the liquid is proportional to the amount of the gas dissolved in the liquid. This proportionality constant is called Henry's constant and is denoted K_H. It has units of pressure (atmospheres, Torr, or Pascal). Mathematically,

$$P_i = K_H \cdot x_i \tag{2.90}$$

Henry's constant is expressed in a variety of other units since it generally represents the ratio between concentrations of gas and liquid phases of the compound. If the liquid and gaseous concentrations are expressed on a molar basis ($mol\,dm^{-3}$), then dimensionless Henry's constant (K_{AW}) can be obtained. If both the liquid and gas phase concentrations are expressed as mole fractions, another dimensionless value can be obtained for Henry's constant (K_X). If the gas phase concentration is denoted in pressure units (Pa) and the liquid phase concentration is in molarity ($mol\,dm^{-3}$), a different unit for Henry's constant (K'_{AW}, $Pa\,dm^3\,mol^{-1}$) can be obtained. One should be very careful in noting the correct units for Henry's constants obtained from the literature. Table 2.6 summarizes these definitions and their interrelationships.

TABLE 2.6
Henry's Constant Definitions for Vapor-Liquid Equilibrium and Their Interrelationships

Definition	Units	Relationship to K_H
$K_H = P_i/x_i$	Pa	—
$K_{AW} = C_i^a/C_i^w$	Dimensionless	$K_{AW} = (V_w/RT)\cdot K_H$
$K'_{AW} = y_i/x_i$	Dimensionless	$K'_{AW} = (V_a/RT)\cdot K_H$
$K''_{AW} = P_i/C_i^w$	$Pa\,m^3\,mol^{-1}$	$K''_{AW} = V_w \cdot K_H$

Notes: V_w is the partial molar volume of water ($= 0.018\ L\,mol^{-1}$ at 298 K) and V_a is the partial molar volume of air ($= 22.4\ L\,mol^{-1}$ at 298 K). C_i^a and C_i^w are the molar concentrations of solute in the air and water, respectively. y_i is the mole fraction of solute i in the air.

The constant K_H not only depends on the nature of the species i but also on temperature and pressure. The criterion for equilibrium is that the gas phase fugacity of i and the liquid phase fugacity of i should be same, that is,

$$f_i^g = f_i^l \tag{2.91}$$

Therefore,

$$P_i = \gamma_i x_i f_i^{l0} \tag{2.92}$$

Hence, we have

$$K_H = \frac{P_i}{x_i} = \gamma_i f_i^{l0} \tag{2.93}$$

In this equation, f_i^{l0} is the standard state liquid fugacity. Thus, if one knows the activity coefficient of i in the liquid phase and the standard state fugacity, then Henry's law constant can be obtained. If the liquid phase is ideal and the more general fugacity is used instead of partial pressure, then we can write a general expression for K_H as follows:

$$K_H = \lim_{x_i \to 0} \frac{f_i^g}{x_i} \tag{2.94}$$

The most general definition of Henry's law is

$$K_H = \frac{y_i \chi_i P_T}{\gamma_i x_i} \tag{2.95}$$

This definition recognizes nonideality in both liquid and gas phases.

2.2.3.1.2 Raoult's Law

Raoult's law is an important relationship that describes the behavior of ideal solutions. Its applicability is predicated upon the similar characters of a solute and a solvent, and hence a mixture of the two behaves ideally over the entire range of mole fractions. Let us consider a solvent i in a mixture in equilibrium with its vapor:

$$f_i^{vapor} = f_i^{liquid\ mixture} \tag{2.96}$$

$$P_i = \gamma_i x_i P_i^* \tag{2.97}$$

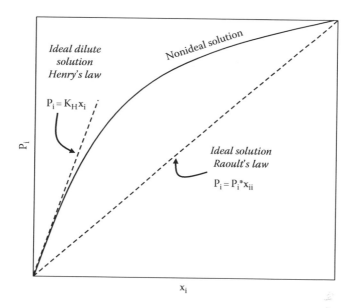

FIGURE 2.6 Illustration of Henry's law and Raoult's law. For a component that is pure (the solvent) and behaves ideally it follows Raoult's law, where the partial pressure is proportional to the mole fraction. When it is the minor component (the solute) in a dilute solution, its partial pressure is again proportional to its mole fraction but has a different proportionality constant, which is Henry's law.

Noting that for an ideal mixture $\gamma_i = 1$, we have

$$x_i = \frac{P_i}{P_i^*} \tag{2.98}$$

This is called Raoult's law. It states that the partial pressure of a compound P_i is given by the liquid phase mole fraction multiplied into the pure compound vapor pressure P_i^*.

Figure 2.6 is a description of the applicability of Henry's and Raoult's laws for mixtures. As shown in the figure, when the mole fraction of all solutes is very small, the solvent obeys Raoult's law while the solute obeys Henry's law. Solute obeying both the laws is ideal in nature. Over the range when the solvent obeys Raoult's law, the solute will obey Henry's law.

Example 2.13   Different units for Henry's Law Constant

Henry's law constant (molar concentration ratio, K_{AW}) for benzene is 0.225. Calculate the value in other units (K_H, K_{AW}'', and K_{AW}').

Use Table 2.6. Note that $V_w = 0.018\,L\,mol^{-1}$ and $V_a = 22.4\,L\,mol^{-1}$. Hence $K_H = (0.225)$ $(0.08205*98/0.018) = 306$ atm $(= 3.1 \times 10^7$ Pa), $K_x = 22.4*306/(0.082205*298) = 280$, $K_{AW}'' = 0.018 * 306 = 5.5\,L$ atm mol^{-1} $(= 5.6 \times 10^5$ Pa dm^3 mol^{-1}).

Example 2.14   Activity Coefficient and Raoult's Law

Consider a mixture of chloroform and propanone for which the partial pressure of chloroform at a liquid mole fraction of 0.2 is 35 Torr. If the pure compound vapor pressure of chloroform is 293 Torr, calculate the activity coefficient of chloroform in the mixture.

$$a_i = P_i/P_i^* = 35/293 = 0.12.$$

Hence,

$$\gamma_i = a_i/x_i = 0.12/0.20 = 0.597.$$

The discussion of nonideal solutions (liquids and solids) is identical to that of real gases. For real gases, partial pressure should be replaced with the corrected pressure, namely, fugacity in the expression for chemical potential. For nonideal solutions, a similar substitution was suggested for mol fraction x_i. This nondimensional quantity was called "activity" a_i. Thus, we have

$$\mu_i^l = \mu_i^{l0} + RT \ \ln\left(\frac{f_i^l}{f_i^{l0}}\right) \tag{2.99}$$

The term within the brackets is the "activity" a_i, which is defined as the ratio of the fugacity to the fugacity at some chosen standard state. For an "ideal" solution, $a_i = x_i$. Hence, $f_i^l = x_i P_i^*$. Note that the ideal state functions are always defined at a standard pressure P and standard temperature T. For an "ideal solution,"

$$\mu_i^l\left(P,T,x_i\right) = \mu_i^{l0}\left(P,T\right) + RT \ln\left(x_i\right) \tag{2.100}$$

This also means that all chemical potentials are defined at a given pressure and temperature.

An "activity coefficient" γ_i (also dimensionless) can be defined to show the departure from ideality for solutions:

$$\gamma_i = \frac{a_i}{x_i} \tag{2.101}$$

The activity coefficient is an important term that we will use throughout this book and should be understood carefully. A value of 1 denotes ideal behavior and anything different represents nonideal behavior. Both positive and negative deviations from ideality are possible. A few words are now in order regarding the appropriate standard state chemical potentials. There are two standard conventions in chemical thermodynamics.

In the first convention (I), the standard state of each compound i is taken to be the pure liquid form at the temperature T and pressure P of the solution denoted as μ_i^*.

$$\mu_i^{l0} = \mu_i^* \tag{2.102}$$

In this convention,

$$\lim_{x_i \to 1} \gamma_i \to 1 \qquad (2.103)$$

In the second convention (II), the standard state is that of the pure solvent (designated A) at temperature T and pressure P of the solution.

$$\mu_A^{10} = \mu_A^* \qquad (2.104)$$

We then have the following limit:

$$\lim_{x_A \to 1} \gamma_A \to 1; \quad \lim_{x_i \to 0} \gamma_i \to 1 \qquad (2.105)$$

Table 2.5 summarizes the definitions of fugacity and activity coefficients for mixtures of real gases and solutions (liquids or solids).

Example 2.15   Calculation of Chemical Potentials Using Different Conventions

Problem: Given the activity coefficient of chloroform measured using the two conventions described here in a chloroform/benzene mixture at 50°C, calculate the corresponding chemical potentials.

Mole Fraction of Chloroform, x_i	Activity Coefficient, γ_i	
	Convention I	Convention II
0.40	1.25	0.70

Solution: The standard state chemical potential is not given. Hence, we can only determine the difference in chemical potentials, $\Delta\mu_i = RT \ln(\gamma_i x_i)$

Convention I: $\Delta\mu_i = (8.314 \text{ J (mol K)}^{-1}) (323 \text{ K}) \ln(1.25 \times 0.4) = 240 \text{ J mol}^{-1}$.
Convention II: $\Delta\mu_i = (8.314 \text{ J (mol K)}^{-1}) (323 \text{ K}) \ln(0.7 \times 0.4) = -383 \text{ J mol}^{-1}$

Example 2.16   Calculation of the Activity of Water in Seawater

Problem statement: If the vapor pressure of a sample of seawater is 19.02 kPa at 291 K, calculate the activity of water in the solution.
Solution: The vapor pressure of pure water at 291 K is 19.38 kPa (*CRC Handbook of Chemistry and Physics*). Hence, $a_{water} = 19.02/19.38 = 0.9814$.

2.2.3.2   Vapor Pressure of Organic Compounds: Clausius-Clapeyron Equation

Vapor pressure P_i^* is the equilibrium pressure exerted by the vapor of a compound in contact with its pure condensed phase (viz., a liquid or solid). If a pure liquid is in equilibrium with its vapor, one intuitively pictures a static system. From a macroscopic point of view, this is indeed correct. But from a molecular thermody-namic point of view, the situation is one where there is a continuous interchange of molecules at the surface, which is in a state of "dynamic equilibrium." However, there is no net evaporation at the surface. Temperature will greatly influence the dynamic equilibrium and hence P_i^* is sensitive to temperature. Moreover, it should be obvious that since the intermolecular forces are vastly different for different compounds, the range of P_i^* should be large. For typical compounds of environmental significance, the range is between 10^{-12} bar to 1 bar at room temperature (see Appendix A).

In order to formulate the thermodynamic relationships involving P_i^* for solids and liquids, we shall first study how a pure condensed phase (e.g., water) behaves as the pressure and temperature are varied. This variation is usually represented on a P-V-T plot called a "phase diagram." Figure 2.7 is the phase diagram for water. Each line in the diagram is a representation of the equilibrium between the adjacent phases. For example, line AC is the equilibrium curve between the vapor and liquid phases. Point A is called the "triple point," which is the coexistence point of all three phases (ice, liquid water, and water vapor) in equilibrium. By definition, the triple point of water is at 273.16 K. The pressure at this point for water is 4.585 Torr. The boiling point of the liquid T_b at a given pressure is the temperature at which $P = P_1^*$ while the normal boiling point is that at which $P_1^* = 1$ atm. BA is the solid-vapor equilibrium line. Similar to the normal boiling point for liquids, the normal melting point is the temperature at which $P_s^* = 1$ atm. Beyond point C, liquid and vapor phases cannot coexist in equilibrium; this is called the "critical point." The critical temperature, T_c; critical pressure, P_c; and critical volume, V_c are unique to a compound. For water, T_c is 647 K and P_c is 218 atm. The phase above the critical point of a compound is

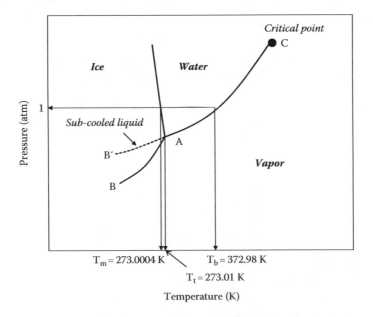

FIGURE 2.7   Schematic of the phase diagram for water.

called "supercritical state." Extending the line BA beyond A to BA′, one obtains the hypothetical "subcooled liquid state," which is important in estimating the solubility of a solid in a liquid. The ratio of the subcooled liquid vapor pressure to the solid vapor pressure is the fugacity of the solid.

The equilibrium at any point along the equilibrium P-T line in Figure 2.7 can be described in terms of chemical potentials. If I and II represent two phases in equilibrium, then we have the following criteria:

$$-S^I dT + V^I dP = -S^{II} dT + V^{II} dP \tag{2.106}$$

Hence,

$$\frac{dP}{dT} = \frac{\Delta S}{\Delta V} = \frac{\Delta H}{T\Delta V} \tag{2.107}$$

where ΔS and ΔV are the entropy and volume changes for phase transition I and II. We also used the definition of entropy change, $\Delta S = \Delta H/T$.

For molar changes, we have the following equation:

$$\frac{dP}{dT} = \frac{\Delta H_m}{T\Delta V_m} \tag{2.108}$$

This is the "Clausius-Clapeyron equation." For both vapor-liquid and vapor-solid equilibrium, it gives the change in vapor pressure with temperature. Since ΔH_m and ΔV_m are positive for liquid $\rightarrow$ vapor and solid $\rightarrow$ vapor transitions, dP/dT is always positive, that is, both liquid and solid vapor pressure increase with temperature. Both for liquid $\rightarrow$ vapor and solid $\rightarrow$ vapor transitions, molar volume of the gas is larger than that of the liquid or solid. Hence, $\Delta V_m \approx V_m^g = RT/P$. Thus, Equation 2.110 becomes

$$\frac{d\ln P}{dT} = \frac{\Delta H_m}{RT^2} \tag{2.109}$$

The quantity ΔH_m is not independent of T. However, over small ranges of temperature, it can be assumed to be constant. Then we can integrate this equation to get

$$\ln\left(\frac{P_2}{P_1}\right) = -\frac{\Delta H_m}{R}\left(\frac{1}{T_2} - \frac{1}{T_1}\right) \tag{2.110}$$

If $P_1 = 1$ atm (as for most environmental engineering problems), $T_1 = T_b$, the normal boiling point of the liquid, and hence,

$$\ln P^* = -\frac{\Delta H_m}{RT} + \frac{\Delta H_m}{RT_b} \tag{2.111}$$

Within our approximation if we plot ln P versus 1/T, we should get a straight line with a slope $-\Delta H_m/R$ and a constant intercept of $\Delta H_m/RT_b$, that is, vapor pressure of liquids and solids varies as $\ln P_i^* \approx A - B/T$. If the temperature range is significant such that $\Delta H \neq$ constant, then the acceptable form of fitting the $\ln P_i^*$ versus T data is given by the "Antoine equation."

$$\log P_i^* = A - \frac{B}{t+C} \qquad (2.112)$$

where t is usually expressed in Celsius. A, B, and C are constants specific for a compound. Table 2.7 lists the values for some selected compounds.

For the solid-vapor equilibrium involving the direct transition from solid to vapor (process called "sublimation"), the heat of sublimation is composed of two parts, viz., the heat of melting (solid to subcooled liquid) and the heat of vaporization of the subcooled liquid (In Figure 2.7, this is the curve B'-A-C). If the temperature over which the vapor pressure of a solid is being monitored includes the melting point (T_m) of the solid, then the $\ln P_i^*$ versus $1/T$ curve will show a change in slope at T_m. We consider the pure liquid as the reference phase for chemical potential and fugacity when calculating free energy changes for phase transfer processes. For compounds that are solids at room temperature, the appropriate choice is the hypothetical standard state of the subcooled liquid.

The entropy of vaporization ΔS_v is nearly constant ($\approx 88\ \mathrm{J\,mol^{-1}\,K^{-1}}$) for a variety of compounds. This is called the "Trouton's rule." The constancy of the entropy of vaporization is due to the fact that the standard boiling points of liquids are roughly equal fraction of their critical temperatures. Most liquids behave alike not only at

TABLE 2.7
Antoine Constants for Some Environmentally Significant Compounds

Compound	Range of T (°C)	A	B	C
Organic compounds				
Chloroform	−35–61	6.493	929.4	196.0
Benzene	8–103	6.905	1211.0	220.8
Biphenyl	69–271	7.245	1998.7	202.7
Naphthalene	86–250	7.010	1733.7	201.8
Tetrachloroethylene	37–120	6.976	1386.9	217.5
Pyrene	200–395	5.618	1122.0	15.2
p-Dichlorobenzene	95–174	7.020	1590.9	210.2
Pentafluorobenzene	49–94	7.036	1254.0	216.0
Inorganic compounds				
Nitrogen		7.345	322.2	269.9
Carbon dioxide		9.810	1347.7	273.0
Hydrogen peroxide		7.969	1886.7	220.6
Ammonia		9.963	1617.9	272.5
Sulfur dioxide		7.282	999.9	237.2

Note: Calculated P in mmHg.

their critical temperatures but also at equal fractions of their critical temperatures. Hence, different liquids should have about the same entropy of vaporization at their normal boiling points. Therefore we have a convenient way of estimating the molar heat of vaporization of a liquid from $\Delta S_v = \Delta H_v/T_b = 88 \ \mathrm{J\,mol^{-1}\,K^{-1}}$. Since the molar enthalpy of vaporization is an approximate measure of the intermolecular forces in the liquid, Kistiakowsky (1923) obtained the following equation:

$$\Delta S_v = 36.6 + 8.3 \ln T_b \tag{2.113}$$

Notice from Table 2.8 that the Trouton's rule fails for highly polar liquids and for liquids that have $T_b < 150 \ \mathrm{K}$.

For a considerable number of compounds of environmental significance, particularly high-molecular-weight compounds, reliable vapor pressure measurements are lacking. As a consequence, we have to resort to correlations with molecular and structural parameters.

Using the Clausius-Clapeyron equation as the starting point, the constancy of heat capacity, C_p, and the Trouton's rule, the following equation can be derived for the vapor pressure of liquids, P_i^* (Schwarzenbach et al., 1993):

$$\ln P_i^* = 19 \cdot \left(1 - \frac{T_b}{T}\right) + 8.5 \cdot \ln\left(\frac{T_b}{T}\right) \tag{2.114}$$

where P_i^* is in atmospheres.

For solids, the vapor pressure P_i^* is related to the subcooled liquid vapor pressure $P_{s(l)}^*$ as follows (Mackay, 1991; Prausnitz et al., 1999):

$$\ln\left(\frac{P_i^*}{P_{s(l)}^*}\right) = -6.8 \cdot \left(\frac{T_m}{T} - 1\right) \tag{2.115}$$

TABLE 2.8
Thermodynamic Transfer Functions for Methane from Different Solvents (A) to Solvent B (Water, W) at 298 K

Solvent A	ΔG_i^0 (kJ mol^{-1})	ΔH_i^0 (kJ mol^{-1})	ΔS_i^0 (kJ mol^{-1})
Cyclohexane	7.61	−9.95	−17.56
1,4-Dioxane	6.05	−11.89	−17.94
Methanol	6.68	−7.97	−14.65
Ethanol	6.72	−8.18	−14.90
1-Propanol	6.69	−8.89	−15.56
1-Butanol	6.57	−7.14	−13.72
1-Pentanol	6.48	−8.35	−14.85

Sources: Franks, F., *Water*, Royal Society of Chemistry Paperbacks, Royal Society of Chemistry, London, England, 1983; Ben Naim, A., *Hydrophobic Interactions*, Plenum Press, New York, 1980.

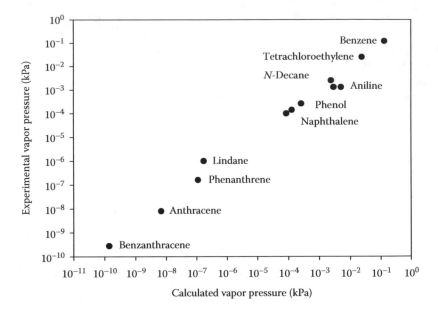

FIGURE 2.8 Parity plot of the experimental and predicted vapor pressure of selected liquid and solid organic molecules. (Data from Schwarzenbach, R.P. et al., *Environmental Organic Chemistry*, John Wiley & Sons, Inc., New York, 1993.)

Figure 2.8 is a comparison of experimental vapor pressures plotted against the predicted values using Equations 2.114 and 2.115 for some of the compounds.

Example 2.17   Vapor Pressure Estimations

Estimate the vapor pressure of benzene (a liquid) and naphthalene (a solid) at room temperature (25°C).

For benzene, which is a liquid (T_b = 353 K), $\ln P_i^* = 19(1 - 353/298) + 8.5\ln(353/298) = -3.50 + 1.44 = -2.06$. Hence, the vapor pressure is 0.13 atm.

For naphthalene, which is a solid (T_m = 352 K, T_b = 491 K), $\ln\left(P_i^* / P_{s(l)}^*\right) = -6.8(353/298 - 1) = -1.255$. For T < T_b, $\ln P_{s(l)}^* = 19(1 - 491/298) + 8.5\ln(491/298) = -8.06 \cdot P_{s(l)}^* = 3.2 \times 10^{-4}$ atm. Hence, $P_i^* = 9.2 \times 10^{-5}$ atm.

2.2.3.3   Vapor Pressure over Curved Surfaces

Consider the formation of fog droplets in the atmosphere, condensation (nucleation) of small clusters leading to formation of aerosols, cloud droplets, and raindrops. All of these involve highly curved interfaces. The development of equilibrium thermodynamic quantities (free energy and chemical potential) for these systems involves modifications to the vapor pressure relationships for solutes distributed between liquid and vapor phases. The vapor pressure over a curved surface depends on its radius of curvature. Let us consider the curved surface of a liquid in contact with its vapor. From earlier discussions on surfaces (Section 2.1.6), we have at constant T the following expression for the molar free energy change:

$$\Delta G = V_m \Delta P = V_m \cdot \sigma \left(\frac{1}{r_1} + \frac{1}{r_2} \right) \tag{2.116}$$

where
   V_m is the molar volume of the liquid
   r_1 and r_2 are the principal radii of curvature of the surface

Over a plane surface, we can use the Young-Laplace equation for ΔP. Further, the chemical potential between the curved surface (with vapor pressure P_i^{*c}) and a plane surface (with saturation vapor pressure P_i^*) is given by

$$\Delta G = \mu_i^c - \mu_i = RT \cdot \ln \left(\frac{P_i^{*c}}{P_i^*} \right) \tag{2.117}$$

Equating the two free energy differences, we have the following equation:

$$\frac{P_i^{*c}}{P_i^*} = \exp \left[\frac{\sigma V_m}{RT} \left(\frac{1}{r_1} + \frac{1}{r_2} \right) \right] \tag{2.118}$$

In environmental engineering, we are particularly interested in spherical surfaces (e.g., fog, rain, cloud, mist) for which $r_1 = r_2 = r$. Hence, we have

$$\frac{P_i^{*c}}{P_i^*} = \exp \left[\frac{2}{r} \cdot \frac{\sigma V_m}{RT} \right] \tag{2.119}$$

This is the "Kelvin equation," which gives the vapor pressure over a curved surface, P_i^{*c}, relative to that over a plane surface, P_i^*, given the surface tension of the liquid, radius of the drop, and temperature. For a solid crystal in equilibrium with a liquid, the Kelvin equation will also apply if the vapor pressures are replaced with the activity of the solute in the solvent.

Consider water, the most ubiquitous of phases encountered in environmental engineering. The surface tension of water at 298 K is 72 mN m^{-1}, and its molar volume is 18×10^{-6} m^3 mol^{-1}. Figure 2.9 shows the value of P_i^{*c} / P_i^* for various size water drops. When $r \geq 1000$ nm, the normal vapor pressure is not affected, whereas for $r \leq 100$ nm, there is an appreciable increase in vapor pressure. For liquids of large molar volume and surface tension, the effect becomes even more significant. An example is mercury, which vaporizes rapidly when comminuted. The conclusion is that in atmospheric chemistry and water chemistry, for very small sizes, the effect of radius on vapor pressure should not be neglected.

The Kelvin effect was experimentally verified for several liquids down to dimensions as small as 30 Å (Israelchvili, 1992). It provides the basic mechanism for the super saturation of vapor that leads to nucleation. The nucleation and formation of

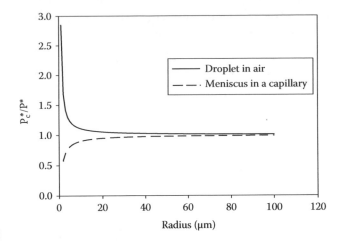

FIGURE 2.9   Application of the Kelvin equation for water droplets in air and water confined in a capillary.

clusters from the vapor phase starts with small nuclei that grow to macroscopic size in stages. The presence of dust or other foreign particles augments the early stages of nucleation. Without dust, the enhanced vapor pressure over curved surfaces provides an energy barrier and hence the early stage of nucleation will require activation energy. These and other implications of the Kelvin equation in environmental engineering will become clear when we discuss the theory of nucleation of atmospheric particles later in this chapter.

Now consider the reverse situation of the vapor pressure of a liquid confined in small capillaries or pore spaces such as soils and sediments (Figure 2.10). The situation is opposite to that of the liquid drops mentioned earlier. The curvature of the surface is of opposite sign, and the vapor pressure is "reduced" relative to that at a flat surface. Therefore, we have

$$\frac{P_i^{*c}}{P_i^*} = \exp\left[-\frac{2}{r} \cdot \frac{\sigma V_m}{RT} \right] \tag{2.120}$$

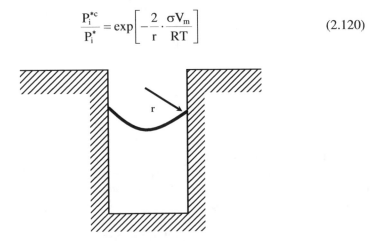

FIGURE 2.10   Meniscus for water confined in a capillary.

Figure 2.10 also shows this relationship for different pore diameters. Liquids that wet the solid will therefore condense into pores at pressures below the equilibrium vapor pressure corresponding to a plane surface. This is termed "capillary condensation" and is an important process in soil matrices. It is important in understanding the infiltration of nonaqueous phase liquids into subsurface soil. The phenomenon is also important in understanding the nucleation of bubbles in a liquid. To support a vapor bubble of radius r in water, the pressure must exceed that of the hydrostatic pressure by $2\sigma/r$. For a 100 nm radius bubble in water at room temperature, this gives a pressure of 14.6 atm. To nucleate a bubble of zero radius (i.e., to start boiling), therefore, we need infinite pressure. This is one of the reasons for the significant superheating required for boiling liquids.

2.2.3.4 Liquid-Liquid Equilibrium

2.2.3.4.1 Octanol-Water Partition Coefficient

If two liquid phases are in contact, and a solute is present in both, then at equilibrium we have a distribution of solute between the two phases that are consistent with equal chemical potentials or fugacity values. In environmental engineering, liquid-liquid equilibrium is common, for example, the distribution of organic chemicals in the water environment, such as occurring during oil spills at sea and inland waterways, floating oils in wastewater treatment plants, and subsurface organic gasoline spills in contact with groundwater. A solvent extraction is another well-known operation in environmental engineering separation processes where liquid-liquid equilibrium becomes important. A specific liquid-liquid system (viz., octanol-water) has special relevance to environmental engineering.

Consider a solute i that is distributed between two solvents (octanol $\equiv$o and water $\equiv$w). At equilibrium the solute i should have equal fugacity in both octanol and water phases. Thus,

$$f_i^o = f_i^w \tag{2.121}$$

Further, we have

$$x_i^o \gamma_i^o f_i^{lo} = x_i^w \gamma_i^w f_i^{lo} \tag{2.122}$$

and therefore

$$K_{ow}^* = \frac{x_i^o}{x_i^w} = \frac{\gamma_i^w}{\gamma_i^o} \tag{2.123}$$

is the "partition constant" for a solute between two phases defined as the ratio of mole fractions of compound i between the octanol and water phases.

A large body of literature exists on the partitioning of a variety of environmentally significant compounds between the organic solvent, 1-octanol and water (Leo et al., 1971). The octanol-water partition constant defined in terms of the ratio of the molar concentrations of solute i in both phases is designated K_{ow}.

$$K_{ow} = \frac{C_i^o}{C_i^w} = \frac{x_i^o V_w}{x_i^w V_o} = \left(\frac{\gamma_i^w}{\gamma_i^o}\right) \cdot \left(\frac{V_w}{V_o}\right) \tag{2.124}$$

where V_w and V_o are the partial molar volumes of water and 1-octanol, respectively. The availability of such a large database on K_{ow} is not entirely fortuitous. It has long been a practice in pharmaceutical sciences to seek correlations of the various properties of a drug with its K_{ow}; the reason for which is that 1-octanol appears to mimic the lipid content of biota very well. A similar reasoning led to the acceptance of K_{ow} as a descriptor of chemical behavior in the environment. 1-octanol has the same ratio of carbon to oxygen as the lipids and represents satisfactorily the organic matter content in soils, sediments, and atmospheric aerosols. 1-octanol is also readily available in pure form and is only sparingly soluble in water. Appendix A lists the log K_{ow} values for a variety of compounds.

It is important to note that generally large K_{ow} values are associated with compounds that have low affinity with the aqueous phase. This becomes clear when one notes that most organic solutes behave ideally in octanol ("like" dissolves "like"), and hence γ_i^o varies only little (from about 1 to 10) whereas γ_i^w varies over several orders of magnitude (0.1 to 10^7) (see Table 2.9). Since both V_o and V_w are constants, the variation in K_{ow} is entirely due to variations in γ_i^w. In other words, *K_{ow} is a measure of the relative nonideality of the solute in water as compared to that in octanol. Hence, K_{ow} is taken to be a measure of the hydrophobicity or the incompatibility of the solute with water.*

K_{ow} values in the literature are reported at "room temperature." This means the temperature is 298 K, with occasional variability of about $5°$. It is important to note that the temperature dependence of K_{ow} is nearly negligible for these temperature variations.

There are many cases when reliable experimental K_{ow} values are not available. Under these circumstances it is possible to estimate K_{ow} from basic structural parameters of the molecule (Lyman et al., 1990). Langmuir (1925) first suggested that the interaction of a

---

TABLE 2.9
Values of γ_i^w and γ_i^o for Typical Organic Compounds at 298 K

Compound	γ_i^o	γ_i^w
Benzene	2.83	2.4×10^3
Toluene	3.18	1.2×10^4
Naphthalene	4.15	1.4×10^5
Biphenyl	5.30	4.2×10^5
p-Dichlorobenzene	3.54	6.1×10^4
Pyrene	8.66	9.6×10^6
Chloroform	1.40	8.6×10^2
Carbontetrachloride	3.83	1.0×10^4

Sources: Mackay, D., Volatilization of organic pollutants from water, EPA Report No.: 600/3-82-019, NTIS No.: PB 82-230939, National Technical Information Service, Springfield, VA, 1982; Chiou, C.T., Partition coefficient and water solubility in environmental chemistry, in: Saxena, J. and Fisher, F. (eds.), *Hazard Assessment of Chemicals*, Vol. 1, Academic Press, New York, 1981; Yalkowsky, S.H. and Banerjee, S., *Aqueous Solubilities—Methods of Estimation for Organic Compounds*, Marcel Dekker Inc., New York, 1992.

molecule with a solvent could be obtained by summing the interactions of each fragment of the molecule with the solvent. The same principle was extended to octanol-water partition ratios by Hansch and Leo (1979). This method involves assigning values of interaction parameters to the various fragments that make up a molecule. For example, an alkane molecule (ethane, CH_3-CH_3) is composed of two $-CH_3$ groups, each contributing equally toward the K_{ow} of the molecule. There are two parts to this type of calculation: a "fragment constant (designated b)" and "a structural factor (designated B)." It is presumed that from chemical to chemical these constants are the same for a specific subunit and that they are additive. Hence we can write

$$\log K_{ow} = \sum_j b_j + \sum_k B_k \qquad (2.125)$$

The fragment constants are fundamental to the subunit j while the structural factors relate to specific intermolecular forces between the subunits. The value of b_j for a specific subunit will be different based on which other atom or subunit it is attached to. For example, a $-Cl$ atom attached to an alkane C has a different b_j from the one attached to an aromatic C. Furthermore, a $-Cl$ attached to an unsaturated $-C\equiv$ unit will be different from the one attached to a $-C=$ unit. In effect, the π-electron cloud of the $-C\equiv$ unit reacts differently from the one in a $-C=$ unit. Substituents attached to an aromatic C unit typically make less of a contribution to $\log K_{ow}$ than those attached to an alkane C atom. One should also expect widely different contributions if a nonpolar group such as $-CH_3$ replaces the $-Cl$ atom in the molecule. The intermolecular forces among fragments and subunits are characterized by the B factor. The more complex the stereochemistry of the molecule, the less contribution it makes toward $\log K_{ow}$. The interactions between polar moieties in the molecule give rise to electronic factors. Increasing polarity invariably decreases the contribution toward $\log K_{ow}$.

For almost any new compound manufactured today one can obtain $\log K_{ow}$ via computation. Computer software exists to estimate $\log K_{ow}$ using the method described without even knowing the structure of the molecule. The reliability of the method is however questionable for structurally complex molecules. Nevertheless, it is an acceptable method of estimation for many environmentally significant compounds for which the structures are well established. Appendix D lists the b and B factors for some typical fragments. A detailed listing of the parameters is available in Lyman et al. (1990).

Since experimental values are always the most useful ones, another approach to obtaining $\log K_{ow}$ exists that utilizes the $\log K_{ow}$ of known compounds. This is done by adding or subtracting appropriate b and B values to or from the parent compound.

$$\log K_{ow}(\text{new}) = \log K_{ow}(\text{parent}) \pm \sum_j b_j \pm \sum_k B_k \qquad (2.126)$$

There are a set of specific rules that must be followed to estimate the octanol-water partition coefficients for a given molecule based on the specific values of b and B listed in the literature. The books by Hansch and Leo (1979) or Lyman et al. (1990) give more details on complex structures and a greater depth of discussion.

Example 2.18   Estimation of Log K_{ow} Using Fragment Constant Method

Let us determine the log K_{ow} of the following compounds: Hexane, chloroben-zene, cyclopentane, and chlorobiphenyl using data from Appendix C. The first information we need in each case is the correct molecular structure.

Hexane: $H_3C-(CH_2)_4-CH_3$. We have 6 C atoms contributing to $6b_c$, 14 H atoms contributing $14b_H$. Therefore, $\Sigma b_j = 6(0.20) + 14(0.23) = 4.42$.

If there are n bonds, they contribute to $(n-1)B_b$ factors. In this case, there are 5 C–C bonds and hence, $\Sigma B_k = (5-1)(-0.12) = -0.48$. Hence, log $K_{ow} = 4.42 - 0.48 = 3.94$. The experimental value is 4.11. Hence, we have a 4% error in our estimation.

Chlorobenzene: C_6H_5Cl. Since the log K_{ow} of benzene (C_6H_6) is reported to be 2.13, we only need to subtract the fragment constant for 1H bonded to an aromatic ring (b_H^φ) and add that for 1 Cl atom bonded to an aromatic ring (b_H^φ). Thus, log $K_{ow} = 2.13 - 0.23 + 0.94 = 2.84$. The experimental value is 2.98. The error is 5%.

Cyclopentane: C_5H_{10}. Fragment factors: $5b_c + 10\ b_H = 3.30$. Structural factors: $(5-1)B_b = -0.36$. Hence, log $K_{ow} = 2.94$. The experimental value is 3.00.

Chlorobiphenyl: $H_5C_6-C_6H_4Cl$. Since for biphenyl the value is 4.09, for chloro-biphenyl it is $4.09 - 0.23 + 0.94 = 4.8$.

2.2.3.4.2 Linear Free Energy Relationships

It is important to note that other solvent-water partition constants (K_{sw}) can be related to K_{ow} (Collander, 1951). The partition constant between octanol and water is related to other partition constants as

$$\log\ K_{sw} = a\ \log K_{ow} + b \qquad (2.127)$$

where a and b are constants. This is called a "linear free energy relationship" (LFER). The slope a is a measure of the relative variability of the activity coefficient of the solute between the two solvents. The intercept b is a constant and can be regarded as the value of log K_{sw} for a hypothetical compound whose log K_{ow} is zero. A number of solvent-water partition constants have been related to K_{ow}. If the solvent is similar in nature to octanol, a high degree of correlation can be expected. It should be remembered that using K_{ow} as the reference emphasizes that it is the high activity coefficient of the compound in water that accounts for the unique partitioning behavior. The changes in log K_{ow} and log K_{sw} track one another only for homologous series of compounds. If the compounds vary in their characteristics (polarity, stereochemistry), then the linear relationship will be less than satisfactory. The situation then calls for separate LFERs for classes of compounds that resemble one another. In other words, a single correlation for all types of compounds in untenable. The correlation varies if the organic compound that partitions differs in its hydrogen donor or acceptor ability and also if they contain heteroatoms such as N, S, etc.

Example 2.19   Estimation of Log K_{ow} from Log K_{sw}

For *n*-heptane as a solvent, the following correlation is applied for organic compounds that are hydrogen donors: log $K_{ow} = 0.541$ log $K_{sw} + 1.203$ with $r^2 = 0.954$. For *o*-Chloroaniline, the log K_{sw} is 1.12. Hence, the predicted value is log $K_{ow} = (0.541)(1.12) + 1.203 = 1.81$. The observed value is 1.90.

2.2.4 NONIDEAL SOLUTIONS

This section is devoted to nonideal behavior that is common to environmental systems. Many solutes even when present in dilute concentrations behave nonideally because of unfavorable molecular interactions between the components of the medium (solid, liquid, or gas) in which they are mixed and the solute.

2.2.4.1 Activity Coefficients for Nonideal Systems

Mixtures of real fluids (gases or liquids) do not form ideal solutions, although similar fluids approach ideal behavior. All nonelectrolytes at their "infinite dilution limit" (lim) in solution follow "ideal" behavior. As mentioned in the previous sections, the $\lim_{x_i \to 0}$ activity coefficient γ_i represents the nonideal behavior in aqueous systems. The activity coefficient can be obtained experimentally, and where it cannot be determined directly, chemical engineers have devised theoretical models to compute the activity coefficients from correlations with other solute parameters (e.g., surface area, volume, octanol-water partition constants) or group-interaction parameters (e.g., UNIFAC, NRTL).

2.2.4.1.1 Excess Functions and Activity Coefficients

Apart from the activity coefficient that is a measure of nonideality, another quantity that is also indicative of the same is the "excess partial molar Gibbs energy" denoted by g^E and defined as the difference between the actual and ideal free energy, $G_{actual} - G_{ideal}$. This is an experimentally accessible quantity. Other thermodynamic functions such as excess molar enthalpy and entropy can also be defined in a similar fashion. By definition, all excess functions are zero for $x_i = 0$ and $x_i = 1$. Some authors use excess Gibbs function as a more appropriate measure of deviation from ideality than activity coefficient. The relationship between the excess function and activity coefficient is easily derived. For component i in solution, the excess molar Gibbs free energy is defined as

$$g_i^E = RT \ln\left(\frac{f_i(\text{actual})}{f_i(\text{ideal})}\right) = RT \ln\gamma_i \tag{2.128}$$

where $f_i(\text{actual}) = x_i\gamma_i f_i^{lo}$ and $f_i(\text{ideal}) = x_i f_i^{lo}$.

The excess molar Gibbs function for solution is then given by summing over all components:

$$g^E = \sum_i x_i g_i^E = RT \sum_i x_i \ln \gamma_i \tag{2.129}$$

Example 2.20   Excess Gibbs Energy of Solution

An aqueous solution of benzene has mole fraction benzene of 1×10^{-5} and a benzene activity coefficient of 2400. What is the excess Gibbs free energy of benzene in solution?

$$g_i^E = RT \ln\gamma_i = 8.314 \times 10^{-3} \left(kJ \ (K \ mol)^{-1}\right) 298 (K) 7.78 = 19.3 \ kJ \ mol^{-1}.$$

2.2.4.2 Activity Coefficient and Solubility

Solubility in liquids (especially water) is of special relevance in environmental engineering since water is the most ubiquitous of all solvents on earth. Hence, considerable effort has gone into elucidating the solubility relationships of most gases, liquids, and solids in water. Water is called a "universal solvent," and it richly deserves that name. Some compounds are easily soluble in water to very high concentrations; some are completely miscible with water at all proportions, while some others have very limited solubility in it. These extremes of solubility are of special interest to an environmental scientist or engineer. As we shall see in the next section, the structure of water still evokes considerable debate.

The solubility of gases in liquids is obtained from Henry's law, which was defined in Section 2.2.3.1.

Let us now consider a sparingly soluble liquid (say a hydrocarbon H) in contact with water. The solubility of the liquid hydrocarbon i in water may be considered to be an equilibrium achieved between pure hydrocarbon phase (l) and the aqueous phase (W). Applying the criterion of equal fugacity at equilibrium,

$$\begin{aligned} f_i^w &= f_i^l \\ \gamma_i^w x_i^w f_i^{l0} &= \gamma_i^l x_i^l f_i^{l0} \end{aligned} \tag{2.130}$$

where f_i^{l0} is the pure component reference fugacity of i at system temperature. For pure hydrocarbon, l, we have $x_i^l = 1$ (pure phase mole fraction is 1) and $\gamma_i^l = 1$ (solute in its own pure form is ideal). Denoting $x_i^w = x_i^*$, the saturation solubility of the hydrocarbon i in water, we can write

$$x_i^* = \frac{1}{\gamma_i^*} \tag{2.131}$$

This equation states that the activity coefficient of a saturated solution of a "sparingly soluble" compound in water is the reciprocal of its saturation mole fraction solubility in water. Therefore, large deviation from ideality translates to very low mole fraction solubility in water.

For the compound i, which is only sparingly soluble in water, we also have the following Henry's law expression:

$$K_H = \gamma_i^* P_i^* = \frac{1}{x_i^*} \cdot P_i^* \tag{2.132}$$

For solutes that are solids at the temperature of measurement, the basic condition of equal fugacity still holds at equilibrium. If solid (s) is a pure species in contact with water (w),

$$f_i^{s0} = \gamma_i^w x_i^w f_i^{l0} \tag{2.133}$$

where
f_i^{s0} is the fugacity of pure solid i
f_i^{l0} is the fugacity of the pure liquid i

This is the hypothetical subcooled liquid state for compounds that are solids at room temperature (see discussion on phase diagrams). Thus, the saturation solubility of i in water is given by

$$x_i^* = \left(\frac{1}{\gamma_i^*}\right) \cdot \left(\frac{f_i^{s0}}{f_i^{10}}\right) = \left(\frac{1}{\gamma_i^*}\right) \cdot \exp\left[-\frac{\Delta H_m}{R}\left\{\frac{1}{T} - \frac{1}{T_m}\right\}\right] \qquad (2.134)$$

The ratio of fugacity coefficients was obtained from Prausnitz et al. (1999). In this equation, ΔH_m is the molar enthalpy of fusion or melting (J mol^{-1}) and T_m is the melting point of the solid (K). Since, according to Trouton's rule, $\Delta S_m = \Delta H_m/T_m$, and for most organic compounds, $\frac{\Delta S_m}{R} \sim 13.6$ entropy units, we have

$$x_i^* = \left(\frac{1}{\gamma_i^*}\right) \cdot \exp\left[6.8\left(1 - \frac{T}{T_m}\right)\right] \qquad (2.135)$$

For sparingly soluble organic compounds in water, it is possible to equate γ_i^* with γ_i^∞, the so-called "infinite dilution activity coefficient." The condition of infinite dilution is the limit as $x_i \to 0$. For a large number of sparingly soluble organics in water, this condition is satisfied even at their saturation solubility. Hence, the two activity coefficients are indistinguishable. The mole fraction at saturation for most compounds of environmental significance lies between 10^{-6} and 10^{-3}. The mole fraction of 10^{-6} is usually taken as the "practical limit of infinite dilution." For those compounds for which the solubility values are very large, γ_i^* is significantly different from γ_i^∞.

Using the Gibbs-Helmholtz relation obtained earlier and replacing the Gibbs free energy change in that expression by the "excess" molar Gibbs free energy of dissolution of liquid i, we can derive the following fundamental relationship for activity coefficient:

$$\frac{d \ln \gamma_i^*}{dT} = -\frac{h_i^E}{RT^2} \qquad (2.136)$$

h_i^E is called the excess molar enthalpy of solution for component i. If the excess molar enthalpy of dissolution is constant over a small range of temperature we should get a linear relationship

$$\ln \gamma_i^* = \frac{h_i^E}{RT} + c \qquad (2.137)$$

where c is a constant. For liquid solutes, since no phase change is involved, the excess enthalpy is identical to the enthalpy change for solution. Experimental evidence for the effect of temperature on the solubility of several organic liquids in water shows that the decrease in activity coefficient (increase in solubility) over

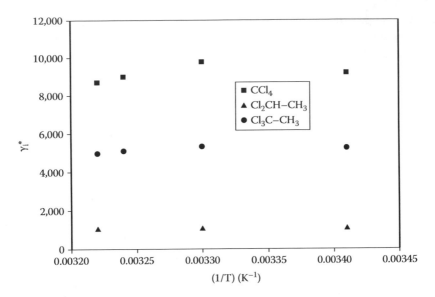

FIGURE 2.11 Variation of activity coefficients at infinite dilution in water for typical environmentally significant compounds versus temperature. (Data from Tse, G. et al., *Environ. Sci. Technol.*, 26, 2017, 1992.)

a 20° rise in temperature is in the range of 1–1.2 (Tse et al., 1992). This is shown in Figure 2.11. For some of the compounds the value of activity coefficient decreases slightly with increase in temperature, particularly if the excess molar enthalpy is near zero or changes sign within the narrow range of temperature. In conclusion, we may state that within the narrow ranges of temperatures encountered in the environment, the activity coefficients of liquid solutes in water do not change appreciably with temperature.

For compounds that are solids or gases at the temperature of dissolution in water, the total enthalpy of solution will include an additional term resulting from a phase change (subcooled liquids for solid solutes and superheated fluids for gases) that will have to be added to the excess enthalpy of solution. For most solids, the additional term for phase change (enthalpy of melting) will dominate and hence the effect of temperature on activity coefficient will become significant. The mol fraction solubility of naphthalene at 281 K was 2.52×10^{-6} whereas at 305 K it increased to 5.09×10^{-6}. Thus, in environmental engineering the effects of temperature on activity coefficients for solids in water cannot be ignored.

2.2.4.3 Ionic Strength and Activity Coefficients

The activities of dissociating (ionic) and nondissociating (neutral) species in aqueous solutions are influenced by the "ionic strength" of the solution. Ionic strength is the combined effect of all ionic species in water.

The ionic strength (denoted I) is given by

$$I = \frac{1}{2} \cdot \sum_i m_i z_i^2 \qquad (2.138)$$

where

  m_i is the molality of species i (mol kg^{-1})
  z_i is the charge of species i in solution

Ionic strength has units of mol kg^{-1}. For example, if we consider a 1:1 electrolyte ($z_+ = 1$; $z_- = -1$), then $I = 0.5 (m_+ + m_-) = m$. Note that I is always a positive quantity and is additive for each species in solution.

The activity of dissociating (ionic) species is related to I through the Debye-Huckel equation, which is described in Section 2.2.4.4.3. It is based on the fact that long-range and Coulombic forces between ions are primarily responsible for departures from ideality in solutions. It suffices to summarize the final equation in this context. The Debye-Huckel theory has been modified and extended to higher ionic strength values. These are summarized in Table 2.10 (Pankow, 1992; Stumm and Morgan, 1996).

For nondissociating (neutral organic) species, the effect of ionic strength I on activity coefficient is given by the McDevit-Long theory, which predicts an equation whose general form is $\log \gamma_i = k_{M-L} I$, where k_{M-L} depends on the type of ionic species in solution and also on the temperature and pressure. k_{M-L} is specific to the species i of concern. Because of the dependence of k_{M-L} on both the nature of the species i and the ions, a universal equation for the effect of I on the activity coefficients of neutral species is not likely. Positive k_{M-L} values indicate that salts tend to make the solvent less favorable for the solute. This is called the "salting out" process and is the basis of the widely used concept of purifying organic compounds by crystallization from their mother liquor. Since ions tend to bind water molecules in their hydration layer they make less water molecules available for solubilizing organics and hence the organics tend to fall out of water. The opposite effect of "salting in" is caused when k_{M-L} is negative.

TABLE 2.10
Relationships between Ionic Strength (I) and Mean Ionic Activity Coefficient ($\gamma_{\pm}$)

Name	Equation	Range of I
Debye-Huckel extended	$\log \gamma_{\pm} = -Az_i^2 \sqrt{I}$	$I < 10^{-2.3}$
Debye-Huckel	$\log \gamma_{\pm} = -Az_i^2 \left(\dfrac{\sqrt{I}}{1 + B \cdot \sqrt{I}} \right)$	$I < 10^{-1}$
Guntelberg	$\log \gamma_{\pm} = -Az_i^2 \left(\dfrac{\sqrt{I}}{1 + \sqrt{I}} \right)$	$I < 10^{-1}$
Davies	$\log \gamma_{\pm} = -Az_i^2 \left[\left(\dfrac{\sqrt{I}}{1 + \sqrt{I}} \right) - 0.2 \cdot I \right]$	$I < 0.5$

Source: Pankow, J.F., *Aquatic Chemistry Concepts*, Lewis Publishers, Chelsea, MI, 1992.

Notes: Parameters A and B depend on temperature and dielectric constant of the liquid. For water at 298 K, A = 0.51 and B = 0.33. The parameter a is an ion-size parameter and is listed by Pankow (1992). Note that $z_i^2 = |z_+| \cdot |\zeta_-|$.

TABLE 2.11

ϕ Values for Some Organic Compounds in Seawater

Organic Compound	Molar Volume ($cm^3 mol^{-1}$)	ϕ
Naphthalene	125	0.00242
Biphenyl	149	0.00276
Phenanthrene	182	0.00213
Dodecane	228	0.000962
Tetradecane	259	0.000964
Hexadecane	292	0.00233
Octadecane	327	0.00290
Eicosane	358	0.00190
Hexacane	456	0.00488

Source: Aquan-Yeun, M. et al., *J. Chem Eng. Data*, 24, 30, 1979.

Aquan-Yeun et al. (1979) suggested that the effect of ionic strength on the activity coefficient γ_i of the neutral organic species i can be correlated using the following form of the Setschenow equation:

$$\log\left(\frac{\gamma_i}{\gamma_0}\right) = \Phi V_i C_s \tag{2.139}$$

where γ_0 is the activity coefficient of the neutral solute species i in pure water. Φ is a parameter that depends on the partial molar volume of the salt in solution (V_o) and the molar volume of the liquid salt (V_s). In this equation, V_i is the partial molar volume of the organic solute species in solution ($cm^3 mol^{-1}$) and C_s is the molar concentration ($mol\, dm^{-3}$) of the salt in solution. The values of ϕ for several hydrocarbon compounds found in seawater are listed in Table 2.11. The values are found to range from a low of 0.000962 for dodecane to as high as 0.00488 for hexadecane. A mean value of 0.0025 is used in most estimation of activity coefficients. Most environmental waters have salt concentrations varying from 0 to 0.5 $mol\, dm^{-3}$ where the latter is the mean seawater salt concentration. In seawater, the increase in activity coefficient over that of pure water is seen to be from 1.2 to 2.3 for compounds of molar volumes ranging from 7.5×10^{-5} to 2.0×10^{-4} $m^3 mol^{-1}$. The ionic strength effects are therefore not entirely negligible for saline waters.

Example 2.21   Mean Ionic Activity Coefficient Calculation

Problem statement: Determine $\gamma_\pm$ for a 0.002 molal solution of NaCl in water at 298 K.

    Solution: For a 1:1 electrolyte such as NaCl, $m_+ = m_- = m$ and hence $I = 0.002$ $mol\, kg^{-1}$. Since $I < 0.002$, we can use the Debye-Huckel limiting law:

$$\log \gamma_\pm = -(0.509)(0.002)^{1/2} = -0.023$$

$$\gamma_\pm = 0.949$$

2.2.4.4 Correlations with Hydrophobicity

As we stated earlier, organic compounds of low aqueous solubility in water have generally large activity coefficients. In order to explain large activity coefficients, we need to know the structure of water both in the liquid and solid states. The following discussion focuses on the special structural features of water that are of relevance in understanding the hydration of neutral and polar compounds.

2.2.4.4.1 Special Structural Features of Water

Water is an inorganic compound. It is the only inorganic compound that exists on earth in all three physical forms at the same time, viz., gas, liquid, and solid. It is a remarkable solvent and there is practically no compound on earth that is "insoluble" in water. Large molecules such as proteins are hydrated with water molecules. Small molecules such as helium and argon (rare gases) have low solubilities in water. Ionic compounds (e.g., NaCl) are highly soluble in water. Particulate and suspended materials in water exist in the atmosphere as aerosols, fog, and mist. In spite of its ubiquitous nature in our environment, the structure of water is far from being completely resolved.

Water displays anomalous properties. It expands in volume on freezing unlike the other compounds. It exists over a wide range of temperatures as a liquid, which indicates that long-range intermolecular forces are dominant. That the molar enthalpy of fusion (5.98 kJ mol^{-1}) is only 15% of the molar enthalpy of vaporization (40.5 kJ mol^{-1}) is indicative of the fact that water retains much of its ordered ice-like structure even as a liquid, which disappears only when it is boiled. The melting point, boiling point, and the enthalpy of vaporization are unexpectedly high for a compound of such a low molecular weight. Not that these high values are unusual, but they are mostly exhibited by metallic and ionic crystals. Although ice is less dense than liquid water, the isothermal compressibility of liquid water is remarkably low, indicating that the repulsive forces in liquid water are of low magnitudes. The surface tension of water (72 mN m^{-1}), an indication of the forces at the surface of the liquid, is also unusually high. The most anomalous property is its heat capacity. For a compound of such low molecular weight, it is abnormally high. It is also interesting that the heat capacity reduces to about half when water is frozen or when it is boiled. It is this particular property of liquid water that helps maintain the ocean as a vast storehouse of energy. Although the pressure-volume-temperature (P-V-T) relationship for water is abnormal, giving rise to the anomalies mentioned, its transport properties such as diffusion constant and viscosity are similar to those of most molecular liquids.

We, therefore, conclude that peculiar intermolecular forces are at play to give water its anomalous properties. To gain an understanding, we need to focus on the structure of water and ice lattice. Figure 2.12 shows the structure of an ice lattice wherein the central water molecule is tetrahedrally linked to four other water molecules. The O–H intramolecular bond distance is 0.10 nm, whereas the intermolecular O–H bond distance is only 0.176 nm. At first glance, it may seem that the O–H bond is also covalent in nature. It is indeed larger than the true covalent O–H bond distance of 0.1 nm, but it is smaller than the combined van der Waals radii of the two molecules (0.26 nm). Bonds that have these intermediate characters are called

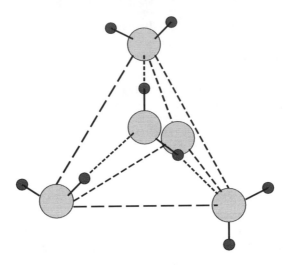

FIGURE 2.12　Three-dimensional ice lattice structure.

"hydrogen bonds." For many years, it was thought that H-bond had a predominant covalent nature. However, more recently it has been convincingly shown that it is an electrostatic interaction. The H atom in water is covalently linked to the parent O atom but enters into an electrostatic linkage with the neighboring O atom. Thus, H acts as a mediator in bonding between two electronegative O atoms and is represented as O–H–O. Although not covalent, H-bonds show some characteristics of weak covalent bonds; for example, they have bond energies of 10–40 kJ mol^{-1} and are directional in nature. Actual covalent bonds have energies of the order of 500 kJ mol^{-1}, and weak van der Waals bonds are only of the order of 1 kJ mol^{-1}. The directional character of H-bond in water allows it to form a three-dimensional structure wherein the tetrahedral structure of the ice lattice is propagated in all three dimensions.

Although the actual structure of liquid water is debatable, it is generally accepted that the tetrahedral geometry of ice lattice is maintained in liquid water to a large extent shown by IR and Raman spectroscopic investigations. An accepted model for water is called the ST2 model (Stillinger and Rahman, 1974). In this model, each H atom in the tetrahedron covalently bonded to an O carries a net charge of +0.24e (e is the electron unit), while the two H atoms on the opposite side of the O atom that participate in H-bonds carry a net compensating charge of –0.24e each. The H–O–H bond angle is 109°. Molecular computer simulations of the ST2 model confirm that the tetrahedral coordination of the O atom with other H atoms is the cause for the many unusual properties of water. In liquid water, although the ice-like structure is retained to a large extent, it is disordered and somewhat open and labile. The interesting fact is that the number of nearest neighbors in the lattice in liquid water increases to five on the average, whereas the average number of H-bonds per molecule decreases to 3.5. The mean lifetime of a H-bond in liquid water is estimated to be 10^{-11} s. Generally, it can be said that only a tetrahedral structure in liquid water can give rise to this open three-dimensional structure, and it is this property more than even the H-bonds themselves that imparts strange properties to a low-molecular-weight compound such as water.

2.2.4.4.2   Hydrophobic Hydration of Nonpolar Solutes

The introduction of a solute changes the intermolecular forces in water. The solute-water interactions are called "hydration forces." Most of our current understanding of hydration phenomena is based on indirect experimental evidence. An understanding of hydration (or more generally solvation) starts with a thermodynamic cycle, which incorporates all of the so-called standard thermodynamic transfer functions. It is useful and instructive to compare the thermodynamic functions for transfer of a solute molecule (say i) at a specified standard state from solvent A to solvent B. Solvent B is water while solvent A can be any other liquid or even air. Both solvents A and B can be either pure or mixtures. Let us consider a solute (methane) that is transferred from different organic solvents (A) to solvent water (B = W). The values of molar free energy, enthalpy, and entropy have been obtained and tabulated (Franks, 1983). These values are shown in Table 2.8. The free energy change is positive in all the cases. It is less positive for a more nonpolar solvent (e.g., methanol versus cyclohexane). However, the enthalpy change in all the cases is negative, indicating exothermic transfer of the molecule from solvent to water, that is, the enthalpic contribution to the transfer of methane to water is highly favorable. The entropic contribution is, however, positive in all these cases. Thus, we arrive at the remarkable conclusion that the dissolution of a nonpolar solute such as methane in water is "entropically unfavorable." Methane is not the only solute for which this behavior has been observed. Most nonpolar compounds or slightly polar compounds of environmental significance show this feature. A majority of polar compounds that have only one polar group such as alcohols, amines, ketones, and ethers also show this behavior.

How can we explain the large entropic contribution to solution of nonpolar compounds in water? The first convincing explanation was given by Frank and Evans (1945), and further expanded by Franks (1983) and Israelchvili (1992). The essential tenets of the theory are summarized here. Most nonpolar compounds are incapable of forming H-bonds with water. This means that in the presence of such a solute, water molecules lose some of their H-bonds among themselves. The charges on the tetrahedral water structure will then have to be pointed away from the foreign molecule so that at least some of the H-bonds can be reestablished. If the solute molecule is of very small size, this may be possible without loss of any H-bonds, since water has an open flexible structure. If the molecule is large, thanks to the ability of the tetrahedrally coordinated water to rearrange themselves, there is, in fact, more local ordering among the water molecules in the hydration layer (Figure 2.13). Thus, the introduction of a nonpolar or a polar molecule with a nonpolar residue will reduce the degrees of freedom for the water molecules surrounding it. Spatial, orientational, and dynamic degrees of freedom of water molecules are all reduced. As stated earlier, the average number of H-bonds per molecule in water is about 3.5. In the presence of a nonpolar solute, this increases to four. If there is local ordering around the solute, we should expect from thermodynamics that the local entropy of the solvent is reduced; this is clearly "unfavorable." Such an interaction is called a "hydrophobic hydration." It should be remembered that in transferring to water, although the solute

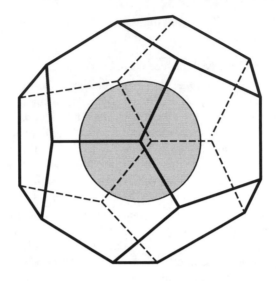

FIGURE 2.13   Solute molecule within the tetrahedral cage of water molecules.

TABLE 2.12
Thermodynamic Functions for Transfer of Nonpolar Molecules
from the Vapor Phase (A) to Water (W) at 298 K

Solute	ΔG_i^0 (kJ mol^{-1})	ΔH_i^0 (kJ mol^{-1})	ΔS_i^0 (kJ mol^{-1})
Methane	26.15	−12.76	−38.91
Ethane	25.22	−16.65	−41.89
Propane	26.02	−23.85	−49.87
n-Butane	26.52	−25.10	−51.62

Source: Nemethy, G. and Scheraga, H.A., *J. Chem. Phys.*, 36, 3401, 1962.

---

entropy increases, the solvent entropy decrease is so much larger that it more than offsets the former, and hence, the overall entropy of the process is negative. None of the present theories of solute-solvent interactions can handle this phenomenon satisfactorily. In Table 2.12, the entropy contribution is about 60%–70% of the overall free energy of transfer.

Table 2.12 gives the thermodynamic functions of solution of several gases from the vapor phase into water. Once again we see that the entropic contribution toward the overall free energy of dissolution is large. The incompatibility of nonpolar and slightly polar compounds with water is called the "hydrophobic effect"; this property is characterized by the activity coefficient of the solute in water or the octanol-water partition constant of the solute.

The entropic contribution toward the free energy of solution increases as the solute size increases. It has become evident from the data that although for smaller molecules the entropic contribution predominates, for larger molecules the enthalpic contribution is equally important in making the excess free energy of solution positive.

Example 2.22   Excess Thermodynamic Functions of Solution of Large Hydrophobic Molecules in Water

1. The following data were obtained by Biggar and Riggs (1974) for the aqueous solubility of a chlorinated insecticide, viz., heptachlor. It has a molecular weight of 373 and a melting point of 368 K.

t (°C)	Solubility (μg L^{-1})
15	100
25	180
35	315
45	490

The enthalpy of melting of heptachlor is 16.1 kJ mol^{-1}. Calculate the excess functions for solution of heptachlor in water at 298 K.

Since heptachlor is a solid at 298 K we need to properly account for the enthalpy of melting of the solid. The equation for the given solid solubility should be used. The data required are x_i^* and 1/T.

1/T (K^{-1})	x_i^*
0.00347	4.82×10^{-9}
0.00335	8.68×10^{-9}
0.00324	1.52×10^{-8}
0.00314	2.36×10^{-8}

A plot of ln $(1/x_i^*)$ versus (1/T) is then obtained (Figure 2.14). The slope of the plot is 4845 K, the intercept is 2.328, and a correlation coefficient of 0.999. From the slope we obtain the total enthalpy change ΔH_i^s as $4845 \times 8.314 \times 10^{-3} = 40.3$ kJ mol^{-1}. The excess enthalpy of solution is then given by $h_i^E = 40.3 - 16.1 = 24.2$ kJ mol^{-1}. The excess free energy at 298 K is given by $g_i^E = RT \ln \gamma_i^* = 42$ kJ mol^{-1}. The contribution from excess entropy is $TS_i^E = -17.8$ kJ mol^{-1}. The entropy of solution is therefore $S_i^E = -59.8$ J $\left(K \, mol\right)^{-1}$.

2. The solubility data for benzene is given as follows (May et al., 1983).

T (K)	x_i^*
290.05	4.062×10^{-4}
291.75	4.073×10^{-4}
298.15	4.129×10^{-4}
298.95	4.193×10^{-4}

Benzene has a melting point of 5.5°C and a boiling point of 80.1°C. Its molecular weight is 78.1. Determine the excess functions of solution of benzene in water.

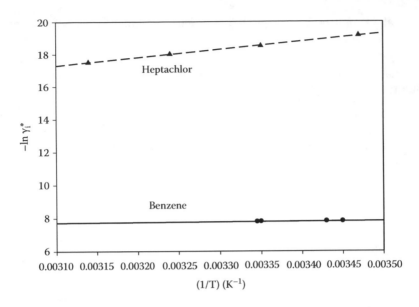

FIGURE 2.14   A plot of ln $(1/x_i^*)$ versus 1/T for heptachlor and benzene.

Since benzene is a liquid at the temperatures given we can ignore the enthalpy contribution from phase changes. Hence the equation for solubility of liquids given earlier can be used. A plot of ln $(1/x_i^*)$ versus $(1/T)$ can be made (Figure 2.14). The slope is 261.8 K, the intercept is 6.907, and the correlation coefficient obtained is 0.861. The excess enthalpy of solution is then given by $h_i^E = (261.8)(8.314 \times 10^{-3}) = 2.2$ kJ mol^{-1}. The free energy of solution at 298.15 K is $g_i^E = 18.9$ kJ mol^{-1}. Hence, the excess entropy of solution of benzene is $S_i^E = -56$ J $(K\ mol)^{-1}$.

It should be noted that in addition to the unfavorable entropy contribution, the enthalpy contribution also contributes toward the unfavorable excess Gibbs free energy of solution for large molecules in water.

2.2.4.4.3   Hydrophobic Interactions between Solutes

It is natural to ask what the consequences of bringing two hydrophobic solutes near one another in water are. This process is called "hydrophobic interaction." Since the introduction of a nonpolar moiety in water is an unfavorable process, it is likely that a hydrophobic compound in water will seek out another of its own kind to interact with. The solute-solute interaction for hydrophobic compounds in water is an attempt to partially offset the entropically unfavorable hydration process. Accordingly, the free energy, enthalpy, and entropy of hydrophobic interactions should all be of opposite sign to that of hydrophobic hydration. It is also notable that hydrophobic interactions between two solutes in water are larger than in free space (vacuum). For example, Israelchvili (1992) calculated that for two methane molecules in free space the interaction energy is -2.5×10^{-21} J but augments to -14×10^{-21} J in water. A hydrophobic interaction is not an affirmative bond between solute species in water; albeit earlier thoughts on this led to the formulation of terms like hydro-phobic bonds that are misleading. Interactions between two solutes in water occur between overlapping hydration layers and are long-range compared to other types of bonds. Thermodynamic databases for hydrophobic interactions are not large because of property measurement difficulties at the inherent low solubilities of hydrophobic compounds. Tucker and Christian (1979) determined that the interac-tion energy between benzene molecules in water to form a dimer is -8.4 kJ mol^{-1}, whereas Ben Naim (1980) calculated a value of -8.5 kJ mol^{-1} between two meth-ane molecules in water. A satisfactory conjecture of hydrophobic interaction for sparingly soluble organics in water is still lacking, but theorists are making prog-ress in this field. The same hydrophobic interactions play a major role in the formation of associated structures from surfactants in water, biological membrane structures, and conformations of proteins in biological fluids. For these cases, thermodynamic parameters are well established and hydrophobic interactions are pretty well understood.

2.2.4.4.4 Hydrophilic Interactions for Solutes in Water

Noninteracting solutes in water experience repulsive forces between one another. Polar compounds, some of which are ionic, exhibit this behavior. Many strongly hydrated ions and some zwitterions (those that have both anionic and cationic char-acteristics) are hydrophilic. Some nonpolar compounds are also hydrophilic if they contain electronegative atoms capable of interacting with the H-bonds in water. In contrast to the structuring imposed by a hydrophobic solute on neighboring water molecules, a hydrophilic compound tends to disorder the water molecules around it. For instance, urea dissolved in water tends to make the water environment so differ-ent that it can unfold a hydrophobic protein molecule in water. In summary, for polar molecules in water, primarily solvation effects determine the free energy contribu-tion to interaction with water.

Aqueous electrolytes behave nonideally even at very low concentrations. Fortunately, the limiting behavior (infinitely dilute solution) for electrolytes is well understood and is called the "Debye-Huckel theory." The final result from this theory was presented in Section 2.2.4.3 where the effects of ionic strength on the mean activity coefficient of ions and on the activity coefficient of neutral solutes in water were discussed.

Using the Debye-Huckel theory, the mean activity coefficient of ions ($\gamma_\pm$) is given by (Bockris and Reddy, 1970):

$$\ln \gamma_\pm = -\frac{N_A \left(z_+ z_-\right) e^2}{2\varepsilon RT} \cdot \kappa \tag{2.140}$$

where $1/\kappa$ is called the Debye length, and κ given by

$$\kappa = \left(\frac{4\pi}{\varepsilon kT} \cdot \sum_i n_i^0 z_i e^2 \right)^{1/2} \tag{2.141}$$

Rewriting the number of moles of species i, n_i^0, in terms of molality, we have

$$\sum_i n_i^0 z_i e^2 = \frac{N_A e^2}{1000} \cdot \sum_i m_i z_i^2 \qquad (2.142)$$

Since ionic strength $I = (1/2)\sum m_i z_i^2$, we have $\kappa = BI^{1/2}$ where $B = (8\pi N_A e^2/1000\varepsilon kT)^{1/2}$. Therefore we have the familiar equation given in Section 2.2.4.3 as follows:

$$\log \gamma_\pm = -A(z_+ z_-) \cdot I^{1/2} \qquad (2.143)$$

where $A = [(N_A e^2)/(2.303\ 2\varepsilon RT)]\ B$. Note that A and B are functions of temperature and, most importantly, the dielectric constant of the solvent. For water, at 298 K, $A = 0.511$ $(L\,mol^{-1})^{1/2}$ and $B = 0.3291$ $Å^{-1}(L\,mol^{-1})^{1/2}$. This equation is called the "Debye-Huckel limiting law" and confirmed for many dilute solutions of electrolytes. Many applications in environmental engineering involve dilute solutions and the equation is adequate. The theory is untenable at high NaCl concentrations (i.e., high I values or small values of κ^{-1}). The Debye-Huckel law needs correction at high ionic strengths. Without enumerating the detailed discussion of these attempts, it is sufficient to summarize the final results (see Section 2.2.4.3, Table 2.10).

2.2.4.4.5 Molecular Theories of Solubility—An Overview

A molecular model for the dissolution of nonpolar solutes in water contemplates the energy associated with the various stages of bringing the solute molecule from another phase (gas or liquid) into water. The Gibbs free energy for the process has two components (Figure 2.15). It is the sum of the cavity term G_c and the solute-solvent interaction G_t term. The first component is the energy associated with the formation of a cavity in water capable of accommodating the solute. Once the solute is in the cavity, it establishes the requisite solute-solvent interaction. The water then rearranges around the solute to maximize its favorable disposition about the solute. Uhlig (1937) and Eley (1939) suggested this concept of solubility. The work required to stretch a surface is the work done against the surface tension of the solvent. Therefore, the work required to form a cavity in water should be the product of the cavity area and surface tension of water. Thus, G_c is directly proportional to surface tension. Saylor and Battino (1958) verified this for a nonpolar solute (Argon). Choi et al. (1970) showed that it is not the bulk surface tension of the solvent that should be considered but the microscopic surface tension of the highly curved cavity. The surface tension of a microscopic cavity was approximately one third of the value for a plane surface.

Hermann (1972) obtained the free energy associated with the formation of a cavity and the introduction of various hydrocarbons into water. Table 2.13 summarizes his calculations. It gives the values of G_c and G_t for a variety of hydrocarbons in water. The accord between the experimental values and predicted values is satisfactory, considering the fact that an approximate structure for

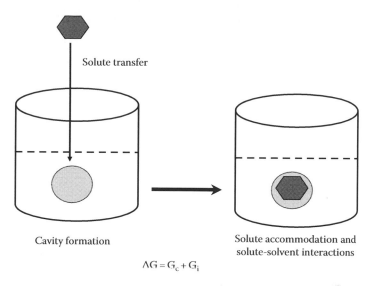

Solute transfer

Cavity formation

Solute accommodation and
solute-solvent interactions

$$\Delta G = G_c + G_i$$

FIGURE 2.15 Molecular description of the solubility of a nonpolar solute in water. The first stage is the formation of a cavity to accommodate the solute, and the second stage involves establishing molecular interactions between the solute and water molecules.

TABLE 2.13
Free Energy Values for the Dissolution of Hydrocarbons in Water

Compound	Cavity Surface Area (Å^2)	G_c (kJ mol^{-1})	G_t (kJ mol^{-1})	ΔG_{calc} (kJ mol^{-1})	ΔG_{expt} (kJ mol^{-1})
Methane	122.7	22.8	−16.1	6.7	8.1
Ethane	153.1	31.7	−24.3	7.4	7.5
Propane	180.0	39.8	−31.7	8.2	8.3
n-Butane	207.0	48.4	−39.0	9.4	9.0
n-Pentane	234.0	57.0	−46.2	10.8	9.8
2,2,3-Trimethylpentane	288.5	75.1	−62.7	12.4	11.9
Cyclopentane	207.1	48.4	−41.3	7.0	5.0
Cyclohexane	224.9	54.1	−47.2	6.9	5.2

Source: Hermann, R.B., *J. Phys. Chem.*, 79, 163, 1972.

water was assumed. The most important aspect is that the values of cavitation free energies are all positive, whereas those for the interactions are all negative. The major contribution to the unfavorable free energy of dissolution of hydrophobic molecules results from the work that has to be done against perturbing the structure of water. Practically all molecules have interaction energies that are favorable to dissolution. *Thus, the term hydrophobic molecule is a misnomer. It is not that the molecules have any phobia to water, but it is the water that rejects the solute molecule.*

Example 2.23   Henry's Constant from Free Energy of Solution

$$K_{AW} = \exp\left(\frac{\Delta G}{RT}\right) \tag{2.144}$$

Knowledge of the free energy of transfer of a mole of solute i from the gas phase to water from theory allows a direct estimation of Henry's constant K_{AW} from the equation.

For methane, the free energy change at 298 K is 6.7 kJ mol^{-1}. Hence, Henry's constant is estimated to be 14.9. The experimental value is 28.6.

2.2.4.5   Structure-Activity Relationships and Activity Coefficients in Water

This section deals with methods for estimation of activity coefficients from solute structure. In cases where reliable experimental values are not available there will arise the need to procure estimates of activity coefficients from correlations. Since activity coefficients are inversely related to mole fraction solubility (or molar solubility), it is enough to know one parameter to obtain the other. Occasionally, in this section, we will use these interchangeably.

The activity coefficients of solutes in water can be related to several solute parameters such as molecular surface area, molar volume, octanol-water partition constants, and normal boiling points. This technique of estimation is called the structure activity relationship (SAR). These correlations are given in Table 2.14.

Example 2.24   Estimating Activity Coefficient from Molecular Area

Assess the applicability of the equation $\gamma_i^* = \exp(DG_m/RT)$ for the solubility of benzene in water. Explain any discrepancy.

Let us determine the molecular surface area of benzene by assuming a spherical shape for the molecule. This is clearly an approximation. The molecular radius can be determined from the molar volume as $r = (3M/4\pi\rho N)^{1/3} = 3.9$ Å. Hence, surface area of a molecule is given by $4\pi r^2 = 149$ Å^2. Since $\sigma_{i/w}$ for benzene-water system is 35 mN m^{-1}, we obtain $\Delta G_m = 23$ kJ mol^{-1}. Hence, $\gamma_i^* = 3.35 \times 10^5$.

Now let us use the actual surface area of benzene obtained via a computer calculation. The value is 109.5 Å^2. We then obtain $\Delta G_m = 23$ kJ mol^{-1}. Hence $\gamma_i^* = 1.11 \times 10^4$.

The actual experimental value at 298 K is 2.43 × 10^3.

Note that if the free energy were 19.3 kJ mol^{-1} instead of the calculated value of 23 kJ mol^{-1}, we would have predicted the actual experimental value of the activity coefficient. This calculation shows how sensitive the estimates are to the free energy of solution resulting from its exponential dependence. The free energy depends on both surface energy and molecular area, and hence both of these contribute to uncertainty in the free energy calculation. The dilemma arises on what is the exact value of surface energy that should be used, since at small radius of curvature the surface energy is dependent on curvature. The concept of surface tension does not hold for a single molecule; however, the concept of interfacial energy remains valid even for an isolated molecule (Israelchvili, 1992). Sinanoglu (1981) showed that clusters of gaseous carbon tetrachloride composed of even as small as seven molecules show surface energies analogous to those of a planar macroscopic surface.

TABLE 2.14
Correlations for Aqueous Solubility Using SARs

Solubility and Molecular Surface Area, A_m (Å^2): $\text{Log } C_i^* = a_1 + a_2 A_m + a_3 T_m$

Compound Class	Type	a_1	a_2	a_3	r^2
Aliphatic alcohols	(l)	3.80	−0.0317	—	0.986
Aliphatic hydrocarbons	(l)	0.73	−0.0323	—	0.980
Alkylbenzenes	(l)	2.77	−0.0184	—	0.988
Polycyclic aromatics	(s)	1.42	−0.0282	−0.0095	0.988
Halogenated benzenes	(s)	3.29	−0.0422	−0.0103	0.997
Polychlorinated biphenyls	(s)	1.21	−0.0354	−0.0099	0.996

Solubility and Molecular Volume

Compound Class	Type	a	b	c	r^2
$\log C_i^* = a + b V_m$, where volume is in 0.01 Å^3					
Variety of compounds	Liquids	1.22	−2.91	—	0.506
$\log C_i^* = a + b V_m + c V_H$, where V_H accounts for the polarity					
Variety of compounds	Liquids	0.72	−3.73	4.10	0.960
$\log C_i^* = a + b V_m + c T_m$, where T_m is the melting point in °C					
PAHs	Solids and liquids	3.00	−0.024	−0.010	0.975

Solubility and Octanol-Water Partition Constants: $-\text{Log } C_i^* = a \text{ Log } K_{ow} + b T_m + c$, Where T_m is the Melting Point in °C

Compound Class	a	b	c	r^2
Alcohols	1.113	—	−0.926	0.935
Ketones	1.229	—	−0.720	0.960
Esters	1.013	—	−0.520	0.980
Ethers	1.182	—	−0.935	0.880
Alkyl halides	1.221	—	−0.832	0.861
Alkynes	1.294	—	−1.043	0.908
Alkenes	1.294	—	−0.248	0.970
Benzene and derivatives	0.996	—	−0.339	0.951
Alkanes	1.237	—	+0.248	0.908
Halobenzenes	0.987	0.0095	0.718	0.990
Polynuclear aromatics	0.880	0.01	0.012	0.979
Polychlorinate biphenyls	1.000	0.01	0.020	—
Alkyl benzoates	1.14	0.005	−0.51	0.991
Mixed drugs	1.13	0.012	−1.02	0.955
Priority pollutants	1.12	0.017	−0.455	0.960
Acids, bases, and neutrals	0.99	0.01	−0.47	0.912
Steroids	0.88	0.01	0.17	0.850

Note: C_i^* is in mol L^{-1}.

If we use the correlation given in Table 2.14, the solubility in water for benzene is $C_i^* = 0.0467$ mol dm^{-3}, which gives a $\gamma_i^* = 1.2 \times 10^3$.

Example 2.25   Aqueous Solubility from Molecular Parameters

Calculate the aqueous solubility of naphthalene from molecular area.
For naphthalene, A_m is 156.76 Å^2. Hence, $\log C_i^* = -0.0282(156.76) + 1.42 - 0.0095(80.3) = -3.76$. $C_i^* = 0.00017$ mol L^{-1}.

Example 2.26   Calculations of Aqueous Solubility from K_{ow}

1. *Naphthalene*: Melting point = 80.3°C (a solid at room temperature), $\log K_{ow}$ = 3.25. Hence, from Table 2.14 we have $-\log C_i^* = (0.880)(3.25) + (0.01)(80.3) + 0.012 = 3.675$. Hence, $C_i^* = 2.11 \times 10^{-4}$ mol dm^{-3} = 27 mg dm^{-3}. The experimental value is 30 mg dm^{-3}. Hence, the error in the estimation is −10%.
2. *1,2,4-Trichlorobenzene*: Melting point = 17°C (a liquid at room temperature), $\log K_{ow}$ = 4.1. Hence, from Table 2.14 we have $-\log C_i^* = (0.987)(4.1) + 0.718 = 4.76$. Hence, $C_i^{*t} = 1.72 \times 10^{-5}$ mol dm^{-3} = 3.1 mg dm^{-3}. The experimental value is 31 mg dm^{-3}. The error in the estimation is −90%. Such large errors are not unusual and hence care should be taken in accepting the estimated solubilities at their face values.

2.2.4.6   Theoretical and Semiempirical Approaches to Aqueous Solubility Prediction

2.2.4.6.1   First-Generation Group Contribution Methods

Langmuir (1925) suggested that each group in a molecule interacts separately with the solvent molecules around it, and hence the complete interaction can be determined by summing the group contributions. The basic idea behind the group contribution approach is that whereas there are thousands of compounds, there are only a small number of functional groups that form these compounds through a variety of permutations and combinations. An excellent example of such a principle is the variation in solubility among a homologous series of aliphatic hydrocarbons. McAuliffe (1966) reported careful studies of the solubility of homologous series of *n*-alkanes, *n*-alkenes, and *n*-dienes. For *n*-alkanes, each CH_2 group added to the molecule increases the logarithm of the aqueous activity coefficient of the solute by a constant factor of 0.44 units. The increase in activity coefficients for aromatic hydrocarbons, aliphatic acids, alcohols, and substituted phenols also showed an analogous trend. Thus, we can separately assess the solubility (activity coefficient) of different substituent groups on a parent compound. This would then enable one to predict the solubility of other members of the homologous series. This rationalization led Tsonopoulos and Prausnitz (1971) to develop a set of group contribution parameters for obtaining the infinite dilution activity coefficients of aromatic and aliphatic substituent groups on benzene.

They noted that log γ_i^∞ of a series of alkylbenzenes can be expressed in the form of a linear equation, viz.,

$$\log \gamma_i^\infty = a + b \cdot (n - 6) \qquad (2.145)$$

where n is the total number of carbon atoms in the molecule. (n − 6) is therefore the number of carbon atoms in addition to the benzene ring. The constant (designated a) represents the contribution from the parent group (in this case the aromatic ring). Additional atoms or groups attached to the aromatic ring or the aliphatic side chains then make specific contributions to the activity coefficient. The same general form of the equation (with different values of a and b) also represents the activity coefficients of polyaromatic hydrocarbons in water.

Example 2.27   Solubility from Group Contributions

Tsonopoulos and Prausnitz (1971) reported the following group contributions for aromatic and aliphatic substitutions on the benzene ring.

Group or Atom	Substituent Position on Ring	Aliphatic Side Chain
–Cl	0.70	
–Br	0.92	
–OH	−1.70	−1.90
–C=C–		−0.30

The constants a and b for benzene derivatives and polyaromatic hydrocarbons are given here:

Compound Class	a	b
Alkylbenzenes	3.39	0.58
Polyaromatics	3.39	0.36

Let us estimate the infinite dilution activity coefficients of some compounds using this information. In each case, we start with the parent group, in this case the aromatic ring for which the contribution is the value of a from the linear correlation. To this we then add the contributions of the functional groups.

1. *Ethylbenzene*: For this case, we use the linear correlation directly with n = 8, giving log $\gamma_i^\infty = 3.39 + 2 \times 0.58 = 4.55$.
2. C_6H_5–CH_2–CH_2–OH: Since this is a derivative of ethylbenzene, we add to it the contribution from an OH group, giving log $\gamma_i^\infty = 4.55 - 1.90 = 2.65$
3. *Hexachlorobenzene*: In this case, we start with the basic group, viz., the aromatic ring for which the contribution is 3.39, and then add to it the contributions from the six Cl atoms. Hence, log $\gamma_i^\infty = 3.39 + 6 \times 0.70 = 7.59$
4. *Naphthalene*: The basic building block in this case is again the benzene ring, but we use the correlation for polyaromatics with n = 4. Hence, log $\gamma_i^\infty = 3.39 + 4 \times 0.36 = 4.83$.

This calculation is an illustration of what is available in the existing chemical engineering literature for estimating activity coefficients through group contributions. We choose the method of Tsonopoulos and Prausnitz (1971) for illustrative purposes. It should be borne in mind that other investigators also reported the same general method of obtaining group contributions. Pierotti et al. (1959) correlated the activity coefficient group contributions based on existing liquid-liquid and vapor-liquid equilibrium data. Wakita et al. (1986) reported a set of group contribution parameters for a variety of aliphatic and aromatic contributions.

We should note that the group contribution approach has inherent drawbacks. It is limited by the lack of availability of the requisite fragment values. In practice, there are several groups for which group contributions are unavailable. The method also has limitations when it comes to distinguishing between geometric isomers of a particular compound. Any group contribution approach is essentially approximate since the contribution of any group in a given molecule is not always exactly the same in another molecule. Moreover, the contribution made by one group in a molecule is constant and nonvarying only if the rest of the groups in the molecule do not exert any influence on it. This is a major drawback of any group contribution scheme.

2.2.4.6.2   Excess Gibbs Free Energy Models

There are several equations to estimate activity coefficients of liquid mixtures based on excess Gibbs free energy of solution. These are summarized in Table 2.15. For details of their derivation, see Sandler (1999).

2.2.4.6.3   Second-Generation Group Contribution Methods: The UNIFAC Method

The most useful and reliable method of activity coefficient estimation resulted from the need for chemical engineers to procure activity coefficients of liquid mixtures. In the chemical process industry (CPI), the separation of components from complex mixtures is a major undertaking. Chemical engineers recognized early on that a large database for activity coefficients of a vast number of binary and ternary liquid-liquid and vapor-liquid systems exist. Also, there exists a sound theory of liquid mixtures based on the Guggenhcim quasi-chemical approximation originating in statistical thermodynamics. This is called UNIversal QUAsi Chemical (UNIQUAC) equation. This theory incorporates the solvent cavity formation and solute-solvent interactions and is useful for establishing group contribution correlations where the independent variables are not the concentrations of the molecules themselves but those of the functional groups. The basic tenet of this conjecture was combined with the large database on activity coefficients to obtain group contribution parameters for a variety of molecular groups. This approach was called UNIversal Functional group Activity Coefficient (UNIFAC) method. Though developed by chemical engineers, it has lately found extensive applications in environmental engineering.

The UNIFAC method is especially suitable for activity coefficients of complex mixtures. It was developed for nonelectrolyte systems and should be used only for such systems. Most of the data for group contributions were procured at high mole fractions, and therefore extrapolation to very small mole fractions (infinitely dilute) for environmental engineering calculations should be made with circumspection.

TABLE 2.15
Correlation Equations for Activity Coefficients in a Binary Liquid Mixture Based on Excess Gibbs Free Energy

Description	Equation
One constant Margules[a]	$\gamma_1 = \exp\left(\dfrac{Ax_2^2}{RT}\right); \quad \gamma_2 = \exp\left(\dfrac{Ax_1^2}{RT}\right)$
Two constant Margules[b]	$RT \ln \gamma_1 = \alpha_1 x_2^2 + \beta_1 x_2^3; \quad RT \ln \gamma_2 = \alpha_2 x_1^2 + \beta_2 x_1^3$
van Laar[c]	$\ln \gamma_1 = \dfrac{\alpha}{\left(1 + \dfrac{\alpha}{\beta}\cdot\dfrac{x_1}{x_2}\right)^2}; \quad \ln \gamma_2 = \dfrac{\beta}{\left(1 + \dfrac{\beta}{\alpha}\cdot\dfrac{x_2}{x_1}\right)^2}$
Flory-Huggins[d]	$\ln \gamma_1 = \ln\left(\dfrac{\phi_1}{x_1}\right) + \left(1 - \dfrac{1}{m}\right)\phi_2 + \chi\phi_2^2$
	$\ln \gamma_2 = \ln\left(\dfrac{\phi_2}{x_2}\right) + (m-1)\phi_1 + \chi\phi_1^2$
Wilson[e]	$\ln \gamma_1 = -\ln\left(x_1 + \Lambda_{12}x_2\right) + x_2\left(\dfrac{\Lambda_{12}}{x_1 + \Lambda_{12}x_2} - \dfrac{\Lambda_{21}}{\Lambda_{21}x_1 + x_2}\right)$
NRTL[f]	$\ln \gamma_1 = x_2^2\left[\tau_{21}\left(\dfrac{G_{21}}{x_1 + x_2 G_{21}}\right)^2 + \dfrac{\tau_{12}G_{12}}{\left(x_2 + x_1 G_{12}\right)^2}\right]$

[a] Only applicable for liquid mixtures when the constituents (1 and 2) are both of equal size, shape, and chemical properties.

[b] $\alpha_i = A + (3-1)^{i+1}B$ and $\beta_i = 4(-1)^i B$; $i = 1$ or 2.

[c] $\alpha = 2\,q_1 a_{12}$, $\beta = 2\,q_2 a_{12}$. a_{12} is the van Laar interaction parameter, q_1 and q_2 are liquid molar volumes and are listed in Perry's *Chemical Engineer's Handbook*. Note also that as $x_1 \to 0$, $\ln\gamma_1 = \gamma_1^\infty$, and as $x_2 \to 0$, $\ln\gamma_2 = \gamma_2^\infty$.

[d] $\varphi_1 = x_1/(x_1 + mx_2)$ and $\varphi_2 = mx_2/(x_1 + mx_2)$ are volume fractions. $m = v_2/v_1$. χ is the Flory parameter.

[e] Λ_{12} and Λ_{21} are given in *J. Am. Chem. Soc.*, 86, 127, 1964.

[f] G_{12}, G_{21}, τ_{12}, and τ_{21} are given in *AIChE J.*, 14, 135, 1968.

Arbuckle (1983) noted that when the mole fractions were very small, UNIFAC underpredicted the solubilities. Banerjee and Howard (1988) also noted this and have suggested useful empirical corrections. Despite these observations, it is generally recognized that for most complex mixtures UNIFAC can provide conservative estimates of activity coefficients. Its remarkable versatility is evident when mixtures are considered. As more and more data on functional group parameters in UNIFAC become available, this may eventually replace most other methods of estimation of activity coefficients.

Since UNIFAC activity coefficients of compounds in octanol can be analogously determined, it is possible to obtain directly UNIFAC predictions of octanol-water partition constants. Adequate corrections should be made for the mutual solubilities of octanol and water if UNIFAC predictions should be attempted in this manner.

It is unnecessary to perform detailed calculations nowadays, since sophisticated computer programs are available to calculate UNIFAC activity coefficients for complex mixtures.

2.2.4.7 Solubility of Inorganic Compounds in Water

One area that we have not considered is the solubility of metal oxides, hydroxides, and other inorganic salts. Because of the preponderance of the organic over inorganic compounds, there is a general bias to the study of environmentally significant organic compounds. Therefore, less of an emphasis has been placed on the study of inorganic reactions in environmental chemistry. Perhaps the most significant aspect of environmental inorganic chemistry is the study of the aqueous solubility of inorganic compounds. As observed earlier, the water molecule possesses partial charges at its O and H atoms, which make a permanent dipole moment of 1.84×10^{-8} esu. The partial polarity of water molecules allows it to exert attractive or repulsive forces to other charged particles and ions in the vicinity. The dipoles between water molecules are attracted to one another by the H-bonds. In the presence of an ion, the water dipoles orient themselves such that a slight displacement of the oppositely charged parts of each molecule occurs, thereby lowering the potential energy of water. The H-bonds between water molecules are slightly depleted, facilitating the entry of the ion into the water structure. This discussion points to the importance of understanding the charge distribution around the central ion in water.

There are important items to be understood if the chemistry of metals, metal oxides, and hydroxides in the water environment has to be studied. The development of these ideas requires some cognition of the kinetics of reactions and reaction equilibria. Elaborate discussion of ionization at mineral-water interfaces, complexation, effects of pH and other ions, and other reactions are relegated to Chapter 4.

2.2.5 ADSORPTION ON SURFACES AND INTERFACES

In Section 2.1.6, we explored the thermodynamics of surfaces and interfaces using the concepts of surface excess properties defined by Gibbs. The Gibbs equation is the basic equation of surface thermodynamics that forms the basis for the relationships between surfaces or interfaces and the bulk phases in equilibrium. Examples of phase boundaries of relevance in environmental chemistry include the following: air/water, soil (sediment)/water, soil(sediment)/air, colloids/water, and atmospheric particulate (aerosols, fog)/air. Typically, an "adsorption isotherm" defines the relationship between the bulk phase and surface phase concentrations.

2.2.5.1 Gibbs Equation for Nonionic and Ionic Systems

Let us apply the Gibbs equation to two different systems, viz., compounds that are neutral (nonionic) and those that are ionic (dissociating). Consider a binary system (solute $\equiv$ i, water $\equiv$ w) for which the Gibbs adsorption equation was derived in Section 2.1.6.5.

$$\Gamma_i = -\frac{d\sigma_{wa}}{d\mu_i^w} \qquad (2.146)$$

where the surface excess is defined relative to a zero surface excess of solvent (water). σ_{wa} represents the surface tension of water. For a solid-water interface, the surface tension is replaced by the interfacial tension of the solid-water boundary. For a solid-air interface, the analogous term is the interfacial tension (energy) of the solid-air boundary.

If the solute in water is nondissociating (neutral and nonionic), then the equation for chemical potential is

$$\mu_i^w = \mu_i^{w0} + RT \ln a_i \qquad (2.147)$$

where a_i is the activity of solute i in bulk water. Using this expression, we obtain the following:

$$\Gamma_i = -\frac{1}{RT} \cdot \frac{d\sigma_{wa}}{d \ln a_i} \qquad (2.148)$$

If the solute dissociates in solution to give two or more species in the aqueous phase, then this equation has to be modified:

$$\Gamma_i = -\frac{1}{\nu RT} \cdot \frac{d\sigma_{wa}}{d \ln a_\pm} \qquad (2.149)$$

where ν is the number of dissociated ionic species in water resulting from the solute. Notice the distinction between the expressions for ionic and nonionic systems. The surface excess in both cases comes from the solution surface tension (interfacial energy for solids) as a function of the activity of solute i in solution. If $(d\sigma/d \ln a) > 0$, Γ_i is negative, and we have a "net depletion" at the surface. If, however, $(d\sigma/d \ln a) < 0$, then Γ_i is positive, and we have a "net surface excess" (positive adsorption) on the surface. Although Gibbs equation applies theoretically to solid-water interfaces as well, the direct determination of the interfacial tension at the solid-water boundary is impractical. However, adsorption of both molecules and ions at this boundary leads to a decrease in surface energy of the solid. In principle, however, there is an analogous term called adhesion tension, $A_{sw} = \sigma_{sa} - \sigma_{sw}$ (s represents solid), which can be used to replace the interfacial tension term. The solid-air interfacial tension is assumed to be a constant, and hence $dA_{sw} = -d\sigma_{\sigma sw}$ appears in the Gibbs adsorption equation.

2.2.5.2 Equilibrium Adsorption Isotherms at Interfaces

The same general principle of equality of fugacity (or chemical potential) as we discussed earlier for bulk phases determines the equilibrium between the surface or interface and the bulk phase. Figure 2.16 shows some common types of adsorption equilibria in environmental engineering and their examples.

At equilibrium, we can write $f_i^\Gamma = f_i^l$ for interface/liquid and $f_i^\Gamma = f_i^g$ for interface/gas phase equilibrium. The superscript Γ represents the interface. In either case, this gives a relation between the surface concentration Γ_i and the bulk phase concentration

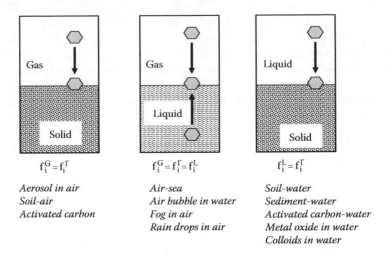

$$f_i^G = f_i^\Gamma \qquad\qquad f_i^G = f_i^\Gamma = f_i^L \qquad\qquad f_i^L = f_i^\Gamma$$

Aerosol in air   *Air-sea*   *Soil-water*
Soil-air   *Air bubble in water*   *Sediment-water*
Activated carbon   *Fog in air*   *Activated carbon-water*
      Rain drops in air   *Metal oxide in water*
      Colloids in water

FIGURE 2.16 Examples of adsorption equilibria encountered in natural and engineered environmental systems.

C_i (or activity a_i). This relationship is called an "adsorption isotherm" and is a convenient way to display experimental adsorption data. To utilize the equality of chemical potentials, we need to obtain an expression for f_i^Γ in terms of Γ_i. Let us consider the surface as a two-dimensional liquid phase with both solvent molecules and solute molecules in it. The solute i, which has an excess concentration on the surface, is assumed to be dilute. The appropriate expression for fugacity is then

$$f_i^\Gamma = \theta_i \gamma_i^\Gamma f_i^{\Gamma 0} \qquad\qquad (2.150)$$

where θ_i is the fractional surface coverage of solute i, that is, $\theta_i = \Gamma_i / \Gamma_i^{max}$, with Γ_i^{max} as the maximum surface excess for solute i. Note that the standard fugacity is defined as the chemical potential of the adsorbate when $\theta_i = 1$. It has to be so defined since the liquid near the surface is different from that of the bulk liquid. Note that if the surface phase was nonideal, then we need to replace the Γ_i in θ_i for the surface activities to get a_Γ / a_Γ^0. Thus, the analogy of the definition of fugacity of surface states of molecules with those in the bulk is evident. For the adsorbate in equilibrium with the liquid phase, we obtain

$$\theta_i \gamma_i^\Gamma f_i^{\Gamma 0} = x_i \gamma_i^l f_i^{l 0} \qquad\qquad (2.151)$$

For ideal dilute solutions, the activity coefficients are unity. Hence, we have

$$\frac{\theta_i}{x_i} = \frac{f_i^{l 0}}{f_i^{\Gamma 0}} = K_\Gamma' \qquad\qquad (2.152)$$

Further, for dilute solutions, $x_i = C_i V_1$, and we have upon rearranging

$$\Gamma_i = K_\Gamma C_i \qquad\qquad (2.153)$$

This is called the "linear (Henry's) adsorption equation" with the linear adsorption constant (units of length) given by K_{Γ}.

In several environmental systems, where dilute solutions are considered, Equation 2.153 will represent adsorption from the liquid phase on both liquid and solid surfaces. If a solid-gas interface is considered where adsorption occurs from the gas phase, the concentration (Γ_i^{max}) in the equation may be replaced by the corresponding partial pressure. If we define a surface layer thickness δ (unit of length), and express both Γ_i and Γ_i^{max} as concentration units ($C_{\Gamma} = \Gamma_i/\delta$), we can express the linear adsorption constant as a dimensionless value, K_{ads}. However, since the definition of an interface thickness is difficult, this approach is less useful.

It is conventional, both in soil chemistry and in environmental engineering, to express adsorbed phase concentrations on the solid surface as amount adsorbed per mass of the solid. This means that the surface concentrations Γ_i and Γ_i^m, which are in moles per unit area of the solid, are converted to moles per gram of solid, which is the product $W_{ads} = \Gamma_i a_m$, where a_m is the surface area per unit mass of the solid. This definition then changes the units of the linear adsorption constant, designated K_{ads}, which is expressed in volume per unit mass of solid.

$$W_{ads} = K_{ads} C_i \qquad (2.154)$$

The linear adsorption constant is not applicable for many situations for a variety of reasons. At high concentrations of molecules on the surface, the assumption of no adsorbate-adsorbate interactions fails. Lateral interactions between adsorbed molecules necessitate that we assume a limited space-filling model for the adsorbed phase. This can be easily achieved by assuming that the expression for surface coverage be written as ($\theta_i/1 - \theta_i$) to account for the removal of an equivalent amount of solvent from the interface to accommodate the adsorbed solute i. In other words, the adsorption of a solute i on the surface can be considered as an exchange process involving the solute and solvent molecules. Equating fugacity as described, for the liquid-gas interface with the liquid chemical potential, we obtain

$$\frac{\theta_i}{1 - \theta_i} = x_i \cdot \left(\frac{f_i^{l0}}{f_i^{\Gamma 0}} \right) = x_i \cdot K_{\Gamma w}'' \qquad (2.155)$$

Noting further the dilute solution approximation for x_i, we have

$$\frac{\theta_i}{1 - \theta_i} = K_{Lang} \cdot C_i \qquad (2.156)$$

The adsorption isotherm given here is called the "Langmuir adsorption isotherm." It is particularly useful in a number of situations to represent the adsorption data. In the case of the solid-gas interface, the concentration term is replaced by the partial pressure of solute i in the gas phase. In fact, Langmuir first suggested this equation to represent gas phase adsorption data in catalysis.

It is important that one should know how to use experimental data to obtain appropriate adsorption isotherm parameters. In the case of a linear adsorption isotherm,

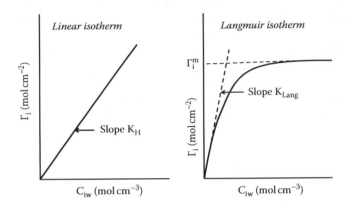

FIGURE 2.17   Schematic of linear and Langmuir adsorption isotherms.

it is obvious that a plot of Γ_i versus C_i should be linear with a slope of K_Γ. However, when a plot of the same is made for a Langmuir isotherm, a linear behavior is displayed at low C_i and approaches an asymptotic value at high C_i values. The linear region is characterized by $\Gamma_i = K_{Lang}C_i$, and we have $\Gamma_i = \Gamma_i^m$ (Figure 2.17). In practice, for the Langmuir isotherm, one plots $1/\Gamma_i$ versus $1/C_i$, the slope of which gives and an $(K_{Lang}\Gamma_i^m)^{-1}$ and an intercept of $(\Gamma_i^m)^{-1}$. If one takes the intercept over the slope value, one obtains directly K_{Lang}. The mere fact that a given set of data fits the Langmuir plot does not necessarily mean that the adsorption mechanism follows that of the Langmuir isotherm. On the contrary, other mechanisms such as surface complex formation or precipitation may also lead to similar plots.

An empirical relationship that represents any set of data on adsorption at low concentrations is called the "Freundlich adsorption isotherm." Apart from its universality in data representation, it was thought to have little theoretical value for a long time. More recently, it has been shown that it can be derived theoretically by considering the heterogeneous nature of adsorption sites (Sposito, 1984; Adamson, 1990). The Freundlich isotherm is expressed as follows:

$$\Gamma_i = K_{Freun}\left(C_i\right)^{1/n} \tag{2.157}$$

where K_{Freun} and $1/n$ are empirically adjusted parameters. K_{Freun} is an indicator of the adsorption capacity and $1/n$ is an indicator of the adsorption intensity. If n is 1, there is no distinction between the Freundlich and linear adsorption isotherms.

Example 2.28   Use of Adsorption Isotherms to Analyze Experimental Data

Three important examples are chosen to illustrate the use of adsorption equations to analyze experimental data in environmental engineering.

The first example is a "solid-liquid" system and is an important component in the design of an activated carbon reactor for wastewater treatment. The requisite first step is to obtain the isotherm data for a compound from aqueous solution onto granular activated carbon (GAC) in batch shaker flasks. In these experiments, a known amount of the pollutant is left in contact with a known weight of GAC under stirred conditions for an extended period of time, and the amount of pollutant left in the aqueous phase at equilibrium is determined using chromatography or other methods. Dobbs and Cohen (1980) produced extensive isotherm data at 298 K on a number of priority pollutants. Consider the case of an insecticide (chlordane) on activated carbon.

Amount Adsorbed (mg g^{-1} carbon)	Equilibrium Aqueous Phase Conc. (mg L^{-1})
87	0.132
79	0.06
64	0.026
53	0.0071
43	0.0032
31	0.0029
22	0.0021
18	0.0016
12	0.0006
11	0.0005

The molecular weight of chlordane is 409. Hence, this data can be divided by the molecular weight to obtain Γ_i in mol g^{-1} and C_i in mol L^{-1}. Since it is not clear as to which isotherm will best represent this data, we shall try both Langmuir and Freundlich isotherms.

Figure 2.18a is a Langmuir plot of $1/\Gamma_i$ versus $1/C_i$. The fit to the data at least in the mid region of the isotherm appears good, although at low $1/\Gamma_i$ the

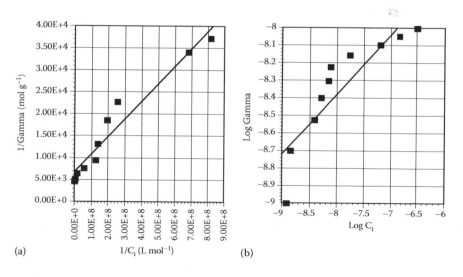

FIGURE 2.18   (a) Langmuir and (b) Freundlich isotherm plots for the adsorption of an insecticide (chlordane) on granular activated carbon (GAC).

percent deviation is considerable. The correlation coefficient (r^2) is 0.944. The slope is 4.03×10^{-5} with a standard error of 3.46×10^{-6}, and the intercept is 6655 with a standard error of 2992. Hence, $K_{Lang} = 1.65 \times 10^8$ L mol^{-1} and $\Gamma_i^m = 1.5 \times 10^{-4}$ mol g^{-1}.

Figure 2.18b is a Freundlich plot of log Γ_i versus log C_i. The correlation coefficient is (r^2) 0.897. The intercept is -1.0821 with a standard error of 0.112, and the slope is 0.381 with a standard error of 0.045. Hence, $K_{Freun} = 0.0827$ and $1/n = 0.381$.

Although r^2 appears to be better for the Langmuir plot, the error involved in the estimated slopes and intercepts is considerably larger. The adsorption isotherms obtained from these parameters are shown in Figure 2.19. Both the isotherms fit the data rather satisfactorily at low C_i values, whereas significant deviations are observed at high C_i values. The Langmuir isotherm severely underpredicts adsorption at high C_i values, whereas the Freundlich isotherm severely overpredicts the adsorption in the same region. Thus neither of the isotherms can be used over the entire region of concentrations in this particular case. It therefore does not sufficiently justify choosing the Langmuir model. Very small errors in the data can throw the prediction of the adsorption maximum off by as much as 50%, and hence the isotherm fit may be weakened considerably. If several measurements are available at each concentration value, then an F-statistic can be done on each point and the lack-of-fit sum-of-squares determined to obtain quantitative reinforcement of the observations. The necessity for multiple determinations of multipoint adsorption data needs no emphasis.

The second example is that of a "liquid-gas" system. In partitioning experiments at the air-water interface, it is important to obtain the ability of surface-active molecules to enrich at the interface. I will choose, as an example, the adsorption of n-butanol at the air-water interface. The measurement is carried out

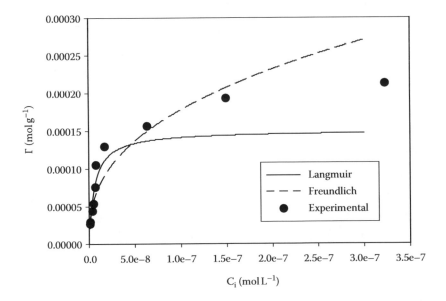

FIGURE 2.19   Adsorption isotherms for chlordane on GAC.

by determining the surface pressure $\Pi = \sigma_{aw}^0 - \sigma_{aw}$ and using the Gibbs equation to get Γ_i. The following data were obtained at 298 K (Kipling, 1965).

Equilibrium Aqueous Concn. C_i (mol L^{-1})	Adsorbed Concentration Γ_1 (mol cm^{-2})
0.0132	1.26×10^{-10}
0.0264	2.19
0.0536	3.57
0.1050	4.73
0.2110	5.33
0.4330	5.85
0.8540	6.15

A plot of $1/\Gamma_1$ versus $1/C_i$ was made. The slope of the plot was 8.5×10^7 and the intercept was 1.41×10^9 with an r^2 of 0.9972. Hence, $\Gamma_i^m = 7.1 \times 10^{-10}$ mol cm^{-2} and $K_{Lang} = 16.6$ L mol^{-1}. Figure 2.20 represents the experimental data and the Langmuir isotherm fit to the data. There is good agreement, although at large C_i values the adsorption is somewhat overpredicted. The maximum adsorption capacity Γ_i^m can be used to obtain the area occupied by a molecule on the surface, $a_m = 1/(N\Gamma_i^m) = 23 \times 10^{-16}$ cm^2. The total molecular surface area A_m is 114×10^{-16} cm^2. The latter area is closer to the cross-sectional area occupied by the $-(CH_2)_n-$ group on the surface when oriented perpendicular to the surface. Therefore, n-butanol is likely oriented normal to the surface with its OH group in the water and the long-chain alkyl group away from the water surface. Most other surfactants also occupy such an orientation at the air-water interface at low concentrations. One can explain this by invoking the concepts of hydrophobicity and hydrophobic interactions described in the earlier section.

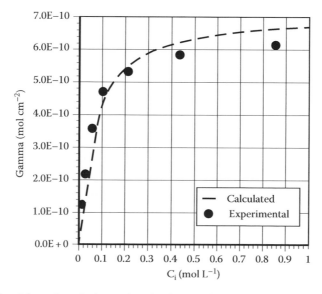

FIGURE 2.20   Adsorption of n-butanol at the air-water interface.

The third example is the adsorption of two neutral organic compounds, namely, 1-2 dichloroethane (DCE) and 1,1,1-trichloroethane(TCE), on soils. These compounds are significant soil and groundwater pollutants resulting from leaking storage tanks and improperly buried hazardous waste. The data is from Chiou et al. (1979) and is for a typical silty loam soil.

Equilibrium Aqueous Conc. ($\mu g\ L^{-1}$)		Amount Adsorbed ($\mu g\ g^{-1}$)	
TCE	DCE	TCE	DCE
1.0E05		100	
1.5E05		280	
3.0E05		450	
3.7E05		600	
4.0E05		740	
5.2E05		880	
	3.7E05		100
	9.8E05		300
	1.4E06		350
	2.3E06		700

It is generally true in environmental engineering literature that most data for sparingly soluble neutral organics is plotted as linear isotherms. The linear isotherm concept is said to hold for these compounds even up to their saturation solubilities in water. For these compounds, the saturation solubilities are 8.4E06 for DCE and 1.3E06 for TCE, respectively. A plot of the amount adsorbed versus the equilibrium solution concentration can be made for both solutes (Figure 2.21). The linearity is maintained in both cases with correlation coefficients of 0.9758 and 0.9682 for DCE and TCE, respectively. The linear adsorption constants are 0.291 and 1.685 L kg^{-1} for DCE and TCE, respectively. Since TCE is less soluble

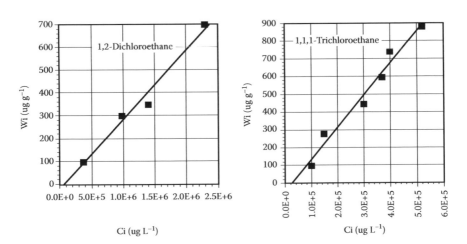

FIGURE 2.21 Linear adsorption isotherms for two organic compounds on a typical soil. (Data from Chiou, C.T. et al., *Science*, 206, 831, 1979.)

and more hydrophobic (cf. discussion of hydrophobicity in the earlier section), it is no surprise that it has a large adsorption constant. For a majority of hydrophobic organic compounds, the linear adsorption isotherms on both soils and sediments afford a good representation of adsorption data.

It is well documented in both the surface chemistry and environmental chemistry literature that various forms of adsorption isotherms are possible. Most of the isotherm shapes can be explained starting from what is known as the "Brauner-Emmett-Teller (BET) isotherm." The Langmuir isotherm limits the adsorption capacity to a single molecular layer on the surface. The BET approach relaxes the assumption in the Langmuir model and suggests that adsorption need not be restricted to a single mono-layer and that any given layer need not be complete before the subsequent layers are formed. The first layer adsorption on the surface occurs with an energy of adsorption equivalent to the heat of adsorption of a monolayer, just as in the case of the Langmuir isotherm. The adsorption of subsequent layers on the monolayer occurs via vapor con-densation. Figure 2.22 is a schematic of the multilayer formation. Multilayer can be formed in the case of adsorption of solutes from the gas phase and of solutes from the liquid phase onto both solid and liquid surfaces. The theory underlying the BET equa-tion is well founded and can be derived either from a purely kinetic point of view or from a statistical thermodynamic point of view (Adamson, 1990). Without elaborating on the derivation, we shall accept the final form of the BET equation. If it is assumed that (1) the number of layers on the surface is n64, and (2) the energy of adsorption for any molecule in all layers n = 2, 3 ,4, ..., is the same but is different for a molecule in the layer n = 1, one obtains

$$\theta_i = \frac{\Gamma_i}{\Gamma_i^m} = \frac{K_{BET}\psi_i}{\left(1-\psi_i\right)\left\{1+\left(K_{BET}-1\right)\cdot\psi_i\right\}} \qquad (2.158)$$

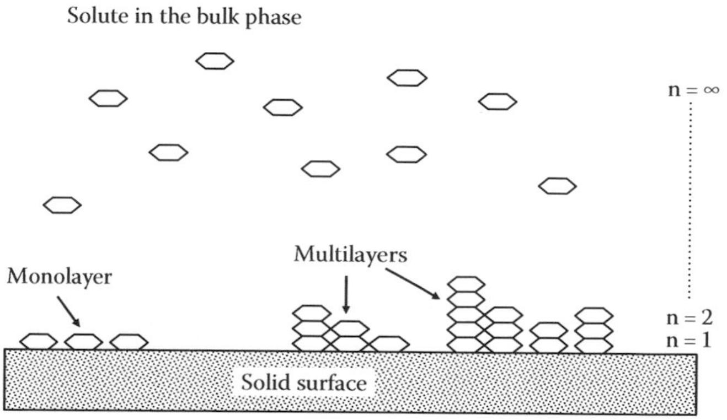

FIGURE 2.22 Schematic of the adsorption of a compound from either the gas phase or liquid phase onto a solid. Simultaneous formation of mono- and multilayers of solute on the surface is shown. Note that the same phenomenon can also occur on a liquid surface instead of the solid surface. The BET isotherm applies in either case.

where Γ_i^m is the monolayer capacity, $\psi_i = P_i / P_i^*$ for adsorption from the gas phase, and $\psi_i = C_i / C_i^*$ for adsorption from the liquid phase. P_i^* is the saturation vapor pressure of solute i in the gas phase, whereas C_i^* is the saturated concentration of solute i in the liquid phase. Notice that θ_i can now be greater than 1 indicating multilayer adsorption. K_{BET} is the BET adsorption constant. Figure 2.23 illustrates the isotherm shapes for different values of K_{BET}. For values of 10 and 100, a clear transition from a region of monolayer saturation coverage to multilayer is evident. These isotherms are characterized as Type II. The Langmuir isotherm is characterized as a Type I isotherm. When the value of K_{BET} is very small (0.1 and 1 in Figure 2.23), there is no such clear transition from monolayer to multilayer coverage. These isotherms are called Type III isotherms. There are numerous examples in environmental engineering where Type II isotherms have been observed; examples include the adsorption of volatile organic compounds (VOCs) on soils (Chiou and Shoup, 1985; Valsaraj and Thibodeaux, 1988) and aerosol particulates (Thibodeaux et al., 1991). Type III isotherms are also common in environmental engineering, for example, the adsorption of hydrocarbon vapors on water surfaces (Hartkopf and Karger, 1973; Valsaraj, 1988) and in soil-water systems (Pennel et al., 1992).

The BET isotherm is often used to obtain the surface areas of soils, sediments, and other solid surfaces by monitoring the adsorption of nitrogen and other inert gases. The method has also enjoyed use in the environmental engineering area. The BET equation can be recast into the following form

$$\frac{1}{\Gamma_i} \cdot \frac{\psi_i}{1-\psi_i} = \frac{1}{K_{BET}\Gamma_i^m} + \frac{(K_{BET}-1)}{K_{BET}} \cdot \psi_i \qquad (2.159)$$

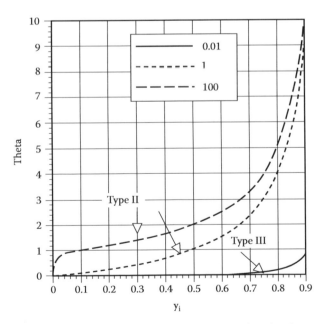

FIGURE 2.23 Different shapes of the BET isotherm. Note that the shapes vary with the isotherm constant, K_B.

so that both K_{BET} and Γ_i^m can be obtained from the slope and intercept of a linear regression of the left-hand side versus y_i. The linear region of the plot typically lies between ψ_i of 0.05 and 0.3 and extrapolation below or above this limit should be approached with caution. If the specific area of the adsorbate, σ_m, is known then the total surface area of the adsorbent can be obtained from the equation

$$\sigma_m N \Gamma_i^m = S_a \tag{2.160}$$

Pure nitrogen, argon, or butane with specific surface areas of 16.2, 13.8, and 18.1 × 10^{-16} cm^2 molecule^{-1} are used for this purpose.

Example 2.29   Use of the BET Equation

Poe et al. (1988) reported that the vapor adsorption of an organic compound (ethyl ether) on a typical dry soil (Weller soil from Arkansas) showed distinct multilayer formation. The following data were obtained:

Relative Partial Pressure, ψ_i	Ether Adsorbed (mg g^{-1} soil)
0.10	6.74
0.28	13.27
0.44	15.32
0.56	18.04
0.63	23.49
0.80	32.28
0.90	46.09

Use these data to obtain (1) the BET isotherm constants and (2) the surface area of the soil (in m^2 g^{-1}).

Since the data is given directly in terms of ψ_i and Γ_i, a plot of $(1/\Gamma_i)$ $(\psi_i/1 - \psi_i)$ versus ψ_i can be made as shown in Figure 2.24a. Only the data between ψ_i of 0.01 and 0.63 were considered for this plot. A good linear fit with an r^2 of 0.9776 was obtained. The slope S was 0.1142 and the intercept I was 0.00211. Hence, $\Gamma_i^m = 1/(S+I) = 8.59$ mg g^{-1}, $K_{BET} = (S/I) + 1 = 55.1$. Using these parameters, the BET isotherm was constructed and compared to the experimental plot of Γ_i versus ψ_i. An excellent fit to the data was observed as in Figure 2.24b. The transition from a monolayer to a multilayer adsorption is clearly evident in this example.

From the value of Γ_i^m one can calculate the total surface area of the soil if the molecular area occupied by ether on the surface is known. In order to do this, we shall first express Γ_i^m as $8.59 \times 10^{-3}/74 = 1.16 \times 10^{-4}$ mol g^{-1}. The surface area occupied by a molecule of ether in the monolayer is given by $A_m = \varphi(M/\rho N)^{2/3}$, where φ is the packing factor (1.091 if there are 12 nearest neighbors in the monolayer), M is the molecular weight of ether (= 74), ρ is the liquid density of ether (= 0.71 g cm^{-3}), and N is Avogadro's number (= 6.023×10^{23}). In this example, substituting these values we obtain $A_m = 28 \times 10^{-16}$ cm^2. Therefore, the total surface area of the soil $S_a = \Gamma_i^m N A_m = 20$ m^2 g^{-1}. For large polyatomic molecules, φ will be considerably different from 1.091. Moreover, for ionic compounds the

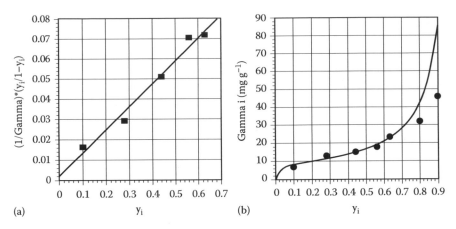

FIGURE 2.24   (a) Linear BET plot to obtain isotherm constants; (b) BET isotherm and the experimental points.

lateral interactions between the adsorbates will affect the arrangements in the monolayer. It is also known that ionic compounds will affect the interlayer spacing in the soil and hence will overestimate the surface areas. For these reasons, ideally, the BET surface areas are measured via the adsorption of neutral molecules (e.g., nitrogen and argon).

2.2.5.3   Adsorption at Charged Interfaces

There are numerous examples in environmental engineering where charged (electrified) interfaces are important. Most reactions in the soil-water environment are mediated by the surface charges on oxides and hydroxides in soil. Colloids used for settling particulates in wastewater treatment work on the principle that modifying the charge distribution on particulates facilitates flocculation. Surface charges on air bubbles and colloids help in removing metal ions by foam flotation. The aggregation and settling of atmospheric particulates and the design of air pollution control devices also involve charged interfaces. Most geochemical processes involve the adsorption and/or complexation of ions and organic compounds at the solid/water interface. It is, therefore, obvious that a quantitative understanding of adsorption at charged surfaces is imperative in the context of this chapter. Two notable books deal extensively with these aspects (Morel and Hering, 1993; Stumm, 1993). In this section, we shall explore how the distribution of charges in the vicinity of an electrified interface will be computed. The applications of this to the problems cited are delegated to Chapter 4. In the aqueous environment, the problem of applying these simple principles of adsorption leads to a more complex theory that incorporates chemical binding to adsorption sites besides the electrical interactions. This is called the "surface complexation model." We shall discuss this model and other reaction kinetics applications in Chapter 4.

For the most part, the approach is similar to the one used to obtain the charge distribution around an ion (Section 2.2.4.4.4 on Debye-Huckel theory). The Poisson

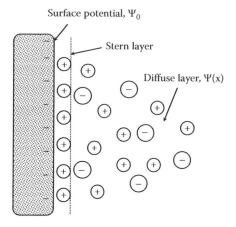

FIGURE 2.25   Distribution of charges in a solution near a negatively charged surface.

equation from electrostatics is combined with the Boltzmann equation from thermo-dynamics to obtain the charge distribution function. This is known as the "Guoy-Chapman theory." The essential difference is that here we are concerned with the distribution of charged particles near a surface instead of near an ion. For simplicity we choose an infinite planar surface, as shown in Figure 2.25. The electrical potential at the surface, Ψ_0, is known. The planar approximation for the surface holds in most cases when the size of the particles near the charged surface is much larger than the distance over which the particle-surface interactions occur in the solution. As shown in Figure 2.25, we have a double layer of charges—a localized charge density near the surface compensated for by the charge density in the solution. This is called the "diffuse-double layer model." The final equation for charge distribution is (Bockris and Reddy, 1970)

$$e^{u/2} = \frac{e^{u_0/2}+1+\left(e^{u_0/2}-1\right)e^{-\kappa x}}{e^{u_0/2}+1-\left(e^{u_0 2}-1\right)e^{-\kappa x}} \tag{2.161}$$

where $u = ze\Psi/kT$ and κ is the inverse of the Debye length defined earlier. For very small values of Ψ_0 (-25 mV or less for a 1:1 electrolyte), that is, $u_0 \ll 1$, we can derive the simple relationship

$$\Psi = \Psi_0 e^{-\kappa x} \tag{2.162}$$

The significance of κ is easily seen in this equation. It is the value of κ at which $\Psi = \Psi_0/e$, and hence, $1/\kappa$ is taken as the "effective double layer thickness."

If $u_0 \gg 1$, that is, $\Psi_0 \gg 25$ mV (for a 1:1 electrolyte), the result is

$$\Psi = \frac{4kT}{ze} \cdot e^{-\kappa x} \tag{2.163}$$

which tends to a value of $\Psi_0 = 4kT/ze$ as $x \to 1/\kappa$.

The expression for Ψ obtained here can be related to the surface charge density σ through the equation

$$\sigma = \frac{\varepsilon}{4\pi} \int_{0}^{\infty} \frac{d^2\Psi}{dx^2} \cdot dx = \sqrt{\frac{2n_0\varepsilon kT}{\pi}} \cdot \sinh\left(\frac{u_0}{2}\right) \qquad (2.164)$$

For small values of u_0 we get $\sigma = (\varepsilon\kappa/4\pi)\Psi_0$, which is similar to that for a charged parallel plate condenser (remember undergraduate electrostatics!).

If the definition of Debye length is cast in terms of the following equation: $\kappa^2 = (4\pi e^2/\varepsilon kT)\Sigma n_i z_i^2$, then Equation 2.164 becomes generally applicable to any type of electrolyte. Note that the ionic strength I is related to $\Sigma n_i z_i^2$, since $n_i = (N_A C_i/1000)$ with C_i in $mol\,L^{-1}$. It is useful to understand how the value of Ψ changes with κ, which in turn is effected by the ionic strength I. Since I is proportional to κ, this suggests that at high ionic strength the compressed double layer near the surface reduces the distance over which long-range interactions due to the surface potential is felt in solution. This is the basis for flocculation as a wastewater treatment process to remove particles and colloids. It is also the basis for designing some air sampling devices that work on the principle of electrostatic precipitation. A number of other applications exist, which we shall explore in Chapter 4.

The Guoy-Chapman theory is elegant and easy to understand, but it has a major drawback. Since it considers particles as point charges, it fails when $x \rightarrow r$, the radius of a charged particle. Stern modified the inherent assumption of zero volume of particles assuming that near the surface there is a region excluded for other particles. In other words, there are a number of ions "stuck" on the surface that have to be brought into solution before other ones from solution replace them. Thus, the drop-off in potential near the surface is very gradual and almost flat till x reaches r, beyond which the Guoy-Chapman theory applies. The region is called the "Stern layer." Further modifications to this approach have been made but are beyond the scope of this book. Interested students are encouraged to consult Bockris and Reddy (1970) for further details.

Example 2.30   Calculation of Double Layer Thickness

Calculate the double layer thickness around a colloidal silica particle in a 0.001 M NaCl aqueous solution at 298 K.

$\kappa^2 = (4\pi e^2/\varepsilon kT)\Sigma n_i z_i^2$. We know that $\Sigma n_i z_i^2 = 2n$, $e = 4.802 \times 10^{-10}$ C, $\varepsilon = 78.5$, $k = 1.38 \times 10^{-16}$ ergs molecule^{-1}. K, and T = 298 K. Noting the relationship between C_i and n_i given earlier, we can write $\kappa^2 = 1.08 \times 10^{15}$ C_i; hence, $\kappa = 3.28 \times 10^7$ $C^{1/2} = 1.038 \times 10^6$ cm^{-1}. Hence, double layer thickness $1/\kappa = 9.63 \times 10^{-7}$ cm (= 96.3 Å).

2.3   CONCEPTS FROM CHEMICAL REACTION KINETICS

Environmental systems are dynamic in nature. Changes with time are important in understanding the fate and transport of chemicals and in process design for waste treatment. The time scales of change in natural systems (weathering of rocks, atmospheric reactions, biological- or thermal-induced reactions) range from a few femto seconds (10^{-15} s) to as large as billions of years. The subject of thermodynamics does not deal with the questions regarding time-varying properties in environmental systems. Chemical kinetics plays the key role in determining the time-dependent behavior of environmental systems.

Chemical kinetics is the study of changes in chemical properties with time resulting from reaction in a system. For example, the following questions are answered using the principles from chemical kinetics:

1. At what rate does a pollutant disappear from or transform in an environmental compartment, or a waste treatment system?
2. What is the concentration of the pollutant in a given compartment at a given time?
3. At what rate will a chemical move between different environmental compartments, and how fast will it be transferred between phases in a waste treatment system?

In order to answer questions (1) and (2), rates and mechanisms of reactions are needed. However, question (3) requires knowledge of kinetic data as well as momentum and mass transfer information between different phases.

In this chapter, we will discuss concepts from chemical kinetics that form the basis for the discussion of rates and mechanisms in environmental engineering. Chemical kinetics covers a broad range of topics. At its simplest level, it involves empirical studies of the effects of variables (temperature, concentration, and pressure) on various reactions in the environment. At a slightly advanced level, it involves elucidation of reaction mechanisms. At its most advanced level, it involves the use of powerful tools from statistical and quantum mechanics to understand the molecular rearrangements accompanying a chemical reaction. We shall not deal with this last issue since it falls beyond the scope of this book. Generally, the applications of chemical kinetics in environmental engineering are limited to the following: (1) experimentally establishing the relationship between concentration, temperature, and pressure in chemical reactions, (2) using the empirical laws to arrive at the reaction mechanism, and (3) using the rate data in models for predicting the fate and transport of pollutants in the natural environment and in process models for designing waste treatment systems.

As mentioned in Chapter 1, processes in natural systems are generally driven by nonequilibrium conditions. True "chemical equilibrium" in natural environmental systems is rare, considering the complex and transient nature of energy and mass transport in natural systems. Although only local phenomena can be affected sometimes, they are coupled with global phenomena, and hence, any minor disturbance easily propagates and alters the rate of approach to equilibrium (Pankow and Morgan, 1981). If the rate of input of a compound equals its rate of dissipation in a system, it is said to be at "steady state." For most natural systems, this occurs for long periods of time interrupted by periodic offsets in system inflows and outflows. Such a behavior is characterized as "quasi-steady state." If the concentration of a compound changes continuously (either decreases or increases) with time because of reactions and/or continuous changes in inflows and outflows, the system is said to show "unsteady state" behavior. The time rate of change of concentrations of metals and organic compounds in natural systems can be ascertained by applying the suitable equations for one or the other of the mentioned states in environmental models.

2.3.1 PROGRESS TOWARD EQUILIBRIUM IN A CHEMICAL REACTION

A chemical reaction reaches equilibrium if there is no perceptible change with time for reactant and product concentrations. The process can then be characterized by a unique parameter called the "equilibrium constant" (K_{eq}) for the reaction. For a general reaction represented by the following stoichiometric equation

$$aA + bB \rightleftharpoons xX + yY \tag{2.165}$$

the equilibrium constant is defined by

$$K_{eq} = \frac{a_X^x a_Y^y}{a_A^a a_B^b}$$

where a denotes activity. Generally, a double arrow indicates a reversible reaction at equilibrium, whereas a single arrow indicates an irreversible reaction proceeding in the indicated direction. The general stoichiometric relation that describes a chemical reaction such as given in Equation 2.160 is

$$\sum_i v_i M_i = 0 \tag{2.166}$$

where
  v_i is the stoichiometric coefficient of the ith species
  M_i is the molecular mass of i

Note the convention that v_i is positive for products and negative for reactants. At constant T and P, the free energy change due to a reaction involving changes dn_i in each species is

$$dG = \sum_i \mu_i dn_i \tag{2.167}$$

We can define a term called the "extent of the reaction" ξ, as defined by

$$n_i = n_i^0 + v_i \xi \tag{2.168}$$

where n_i^0 is the number of moles of i when $\xi = 0$ (i.e., the initial condition). Note that when $\xi = 1$, all reactants have converted to products. From Equation 2.165, we have

$$dn_i = v_i d\xi \tag{2.169}$$

Therefore, the change in free energy for the reaction is given by

$$dG = \sum_i \mu_i v_i d\xi \qquad (2.170)$$

As noted earlier, the quantity $\sum_i v_i \mu_i$ is called the "free energy change of the reaction," ΔG:

$$\Delta G = \sum_i v_i \mu_i \qquad (2.171)$$

Thus, we have the relation

$$\Delta G = \frac{dG}{d\xi} \qquad (2.172)$$

The free energy change of a reaction is the rate of change of Gibbs free energy with the extent of the reaction. Figure 2.26 shows a typical change in free energy with the extent of reaction. Note that at equilibrium, $dG/d\xi = 0$, that is, $\Delta G = 0$, as required by the laws of thermodynamics. The entropy production is

$$dS_{int} = -\frac{\Delta G}{T} \cdot d\xi \qquad (2.173)$$

Since $\mu_i = \mu_i^0 + RT \ln a_i$,

$$\Delta G = \sum_i v_i \mu_i^0 + RT \sum_i v_i \cdot \ln a_i \qquad (2.174)$$

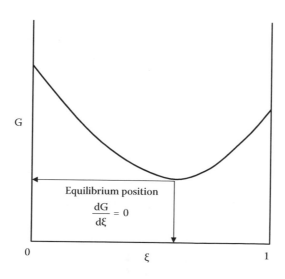

FIGURE 2.26 Gibbs function variation with the extent of reaction.

This can also be written as

$$\Delta G = \Delta G^0 + RT \ln \prod_i (a_i)^{v_i} \qquad (2.175)$$

For the general reaction (Equation 2.1), we have

$$\prod_i (a_i)^{v_i} = \frac{a_X^x a_Y^y}{a_A^a a_B^b} = K \qquad (2.176)$$

Hence,

$$\Delta G = \Delta G^0 + RT \ln K \qquad (2.177)$$

Notice that only at equilibrium (i.e., $\Delta G = 0$) does K equal K_{eq}, at which point we have

$$\Delta G^0 = -RT \ln K_{eq} \qquad (2.178)$$

K is called the "reaction quotient" and K_{eq} is the "equilibrium constant". Therefore, we can write

$$\Delta G = \frac{dG}{d\xi} = RT \ln \left(\frac{K}{K_{eq}} \right) \qquad (2.179)$$

The approach to equilibrium for a chemical reaction is measured by this equation. The activity of a compound is given by $a = \gamma[i]$, where [i] represents the concentration and γ is the activity coefficient. Note that [i] represents the molar concentration C_i (mol L^{-1}) for reactions in solution or the partial pressures P_i (atm or kPa) for reactions in the gas phase. Hence,

$$K_{eq} = \frac{[X]^x [Y]^y}{[A]^a [B]^b} \cdot \frac{\gamma_X^x \gamma_Y^y}{\gamma_A^a \gamma_B^b} \qquad (2.180)$$

From the relation for ΔG or $dG/d\xi$, one can conclude that if $\Delta G = \dfrac{dG}{d\xi} < 0$, $K < K_{eq}$, the reaction will be spontaneous and will proceed from left to right in Equation 2.159, whereas if $\Delta G = \dfrac{dG}{d\xi} > 0$,

2.3.2 REACTION RATE, ORDER, AND RATE CONSTANT

Since $n_i = n_i^0 + v_i \xi$, we have

$$d\xi = \frac{1}{v_i} \cdot dn_i \qquad (2.181)$$

As long as this equation represents a single reaction, it is immaterial as to the referenced species i. However, if the reaction occurs in a series of steps, then the rate at which one species is consumed will be different from the rate of production of another, and hence the rate has to be specified in concert with the species it refers to. For example, consider the reaction $H_2 + (1/2)O_2 \rightarrow H_2O$, for which the extent of reaction is

$$d\xi = \frac{dn_{H_2O}}{(1)} = \frac{dn_{O_2}}{\left(-\dfrac{1}{2}\right)} = \frac{dn_{H_2}}{(-1)} \tag{2.182}$$

Note that the appropriate use of sign for ν_i assures that ξ is always positive. Thus, the rate can be expressed either as the decrease in moles of the reactant with time or the increase in moles of product with time. The "rate of the reaction" is related to the "extent of the reaction" by

$$r = \frac{1}{V} \cdot \frac{d\xi}{dt} = \frac{1}{V} \cdot \frac{1}{\nu_i} \frac{dn_i}{dt} = \frac{1}{\nu_i} \cdot \frac{d[i]}{dt} \tag{2.183}$$

where $[i] = n_i/V$ is the concentration. It is important to bear in mind that since these definitions encompass all macroscopic changes in concentration of a given species with time, *the rate of a reaction at equilibrium should be zero*. Thus, the entire realm of chemical kinetics is geared toward understanding how fast a system approaches equilibrium.

In the study of chemical reaction kinetics, the first step is procuring the functional relationship between the rate of change in concentration of one of the species and the concentration of other species involved. Such a relationship is called a "rate equation." It is obtained through a series of experiments designed to study the effects of each species concentration on the reaction rate. For a general stoichiometric equation of the type $aA + bB \rightarrow cC + dD$, the empirically derived rate expression is written as

$$r = \frac{1}{(-a)} \cdot \frac{d[A]}{dt} = k \cdot [A]^\alpha [B]^\beta [C]^\gamma [D]^\delta \tag{2.184}$$

where the rate r is expressed as the disappearance of A with time. The rate is always a positive quantity and has units of concentration per unit time (e.g., $mol\,dm^{-3}\,s^{-1}$ or $mol\,L^{-1}\,s^{-1}$). For gas phase reactions, partial pressure replaces concentration and the rate is in pressure per unit time (e.g., $kPa\,s^{-1}$ or $atm\,s^{-1}$).

The "order" of this reaction is $n = \alpha + \beta + \gamma + \delta$, with the reaction being termed αth order in A, βth order in B, γth order in C, etc. Note that stoichiometry and order are not the same. The order of a reaction is more complicated to ascertain if the rate equation involves concentrations of A, B, etc., in the denominator as well. Such situations are encountered if a reaction proceeds in several steps where only some of the species take part in each step. The order should be distinguished from the aggregate number of molecules involved in the reaction; this is called the "molecularity" of the reaction. The constant k in the rate expression is called the "specific rate" or "rate constant." It is numerically the same as the rate if all reactants are present at unit concentrations. The unit of k will depend on the concentration units. Generally, it has dimensions of $(concentration)^{(1-\alpha-\beta-\gamma-\delta)}(time)^{-1}$.

2.3.3 KINETIC RATE LAWS

Any experiment designed to obtain a rate expression will require that one follows the change in concentration of a species with time. Starting from an initial (t = 0) value of $[i]_0$, the concentration will decay with time to its equilibrium value $[i]_\infty$. This is what is represented by the empirical rate expression given earlier. The experiments are designed to obtain the rate constant and the order of the reaction. This is achieved via the following methods.

2.3.3.1 Isolation Method

Consider a reaction involving two reactants A and B. Let the rate be r = k[A][B], where the overall order of the reaction is two. If, however, [B] ≫ [A], throughout the reaction, [B] remains constant in relation to [A]. Hence, k[B] ~ constant (k′) and the rate is r = k′[A]. This is called the "pseudo-first-order rate." If the rate were more complicated, for example,

$$r = \frac{k_1\left[A\right]^2}{k_2\left[B\right]+k_3} \tag{2.185}$$

we have r = k′[A]² where k′ is called the "pseudo-second-order rate." The dependence of r on each reactant can be isolated in turn to obtain the overall rate law.

2.3.3.2 Initial Rate Method

This is used in conjunction with the isolation method described earlier. The velocity or rate of an nth order reaction with A isolated may be generally expressed as r = k[A]n. Hence, we have log r = log k + n log [A]. The slope of the plot of log r versus log [A] will give n. This is most conveniently accomplished by measuring initial rate at different initial concentrations. In Figure 2.27a, the slope of [A] versus time as t → 0 gives the initial rate r_0. The log of initial rate is then plotted versus the log of initial concentration to obtain the slope n (Figure 2.27b).

2.3.3.3 Integrated Rate Laws

The most common method of obtaining the order and rate of a reaction is the method of integrated rate laws. The initial rates do not often portray the full rate law, especially if the reactions are complex and occur in several steps. Figure 2.28 represents the change in concentrations of reactant A and product B for a first-order reaction given by A → B. The derivative −d[A]/dt is the rate of disappearance of A with time (This is also equal to d[B]/dt at any time t). Since the reaction is first order in A, we have

$$r_A = \frac{1}{v_A}\cdot\frac{d\left[A\right]}{dt} = -\frac{d\left[A\right]}{dt} = k\left[A\right] \tag{2.186}$$

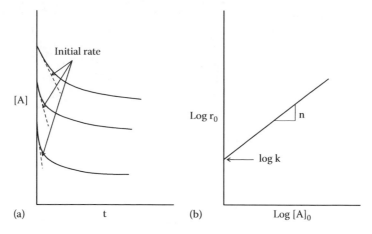

FIGURE 2.27   (a) Concentration versus time for various $[A]_0$ values. (b) Logarithm of initial rate versus logarithm of initial concentrations. The slope is the order of the reaction and the intercept gives the logarithm of the rate constant for the reaction.

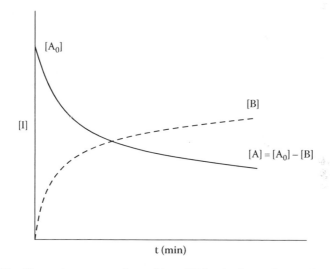

FIGURE 2.28   Change in concentrations of A and B for the first-order reaction A → B.

since $\nu_A = -1$. The rate is directly proportional to $[A]$. Since at $t = 0$, $[A] = [A]_0$, a constant, one can integrate this expression to get

$$\ln\left(\frac{[A]}{[A]_0}\right) = -kt \tag{2.187}$$

If the reaction is nth order in A, that is, nA → B, the integrated rate law is

$$\left(\frac{1}{n-1}\right)\cdot\left(\frac{1}{[A]^{n-1}} - \frac{1}{[A]_0^{n-1}}\right) = kt \tag{2.188}$$

Thus, for a first-order reaction, a plot of ln $[A]/[A]_0$ versus t will give k, the rate constant. Similar plots can be made for other values of n. By finding the most appropriate integrated rate expression to fit a given data, both n and k can be obtained.

If the reaction involves two or more components, the expression will be different. For example, if a reaction is second order (first order in A and first order in B), $A + B \rightarrow$ products, then $-d[A]/dt = k\ [A]\ [B]$. If x is the concentration of A that has reacted, then $[A] = [A]_0 - x$ and $[B] = [B]_0 - x$. Hence,

$$-\frac{d[A]}{dt} = \frac{dx}{dt} = k\left([A]_0 - x\right)\left([B]_0 - x\right) \tag{2.189}$$

The integrated rate law is

$$\left(\frac{1}{[A]_0 - [B]_0}\right)\ln\left(\frac{[B]_0\left([A]_0 - x\right)}{[A]_0\left([B]_0 - x\right)}\right) = kt \tag{2.190}$$

A plot of the term on the left-hand side versus t would give k as the slope.

Physical chemists have investigated a large variety of possible kinetic rate expressions over the years, and the integrated rate laws have been tabulated in the literature (Laidler, 1965; Moore and Pearson, 1981). Table 2.16 lists some of the rate laws most frequently encountered in environmental engineering.

An important parameter that is useful in analyzing rate data is the "half-life of a reactant," $t_{1/2}$. This is defined as the time required for the conversion of one half of the reactants to products. For a first-order reaction, this is (ln 2)/k and is independent of $[A]_0$. For a second-order reaction, the half-life is $1/(k[A]_0)$ and is inversely proportional to $[A]_0$. Similarly, for all higher-order reactions, appropriate half-lives can be determined.

Example 2.31   Reaction Rate Constant

The rate of loss of a volatile organic compound (chloroform) from water in an open beaker is said to be a first-order process. The concentration in water was measured at various times as given here:

t (min)	$[A]/[A]_0$
0	1
20	0.5
40	0.23
60	0.11
80	0.05
100	0.03

Find the rate constant for the loss of chloroform from water.

First obtain a plot of ln $[A]/[A]_0$ versus t as in Figure 2.29. The slope of the plot is 0.035 min^{-1}, with a correlation coefficient of 0.997. Hence, the rate constant is 0.035 min^{-1}.

TABLE 2.16

Integrated Rate Laws Encountered in Environmental Systems

Reaction Type	Order	Rate Law	$t_{1/2}$
$A \rightarrow B + \cdots$	0	$kt = x$	$[A]_0/2k$
	1	$kt = \ln([A]_0/[A]_0 - x)$	$(\ln 2)/k$
	≥ 2	$kt = \{1/n - 1\}\{1/([A]_0 - x)^{n-1} - 1/([A]_0)^{n-1}\}$	$(2^{n-1} - 1)/(n-1)k[A]_0^{n-1}$
$A + B \rightarrow C + D + \cdots$	2	$kt = \dfrac{1}{[B]_0 - [A]_0} \cdot \ln\left[\dfrac{[A]_0([B]_0 - x)}{[B]_0([A]_0 - x)}\right]$	$1/k[B]_0$
$A \leftrightarrow B$		$k_b t = \dfrac{[A]_{eq}}{[A]_0} \cdot \ln\left(\dfrac{[B]_{eq}}{[A] - [A]_{eq}}\right); \; k_f = k_b K_{eq}$	
$A + B \leftrightarrow C + D$		$k_f t = \dfrac{[B]_{eq}}{2[A]_0([A]_0 - [B]_{eq})}$ $\cdot \ln\left[\dfrac{[B]([A]_0 - 2[B]_{eq}) + [A]_0[B]_{eq}}{[A]_0([B]_{eq} - [B])}\right]$	
$A \xrightarrow{k_1} B \xrightarrow{k_2} X$		$[A] = [A]_0 e^{-k_1 t}$ $[B] = \dfrac{[A]_0 k_1}{k_2 - k_1}\left(e^{-k_1 t} - e^{-k_2 t}\right)$ $[C] = \dfrac{[A]_0}{k_2 - k_1}\left[k_2\left(1 - e^{-k_1 t}\right) - k_1\left(1 - e^{-k_2 t}\right)\right]$	
$A \xrightarrow{k_1} C$ $A \xrightarrow{k_2} D$		$[A] = [A]_0 e^{-(k_1 + k_2)t}$ $[C] = \dfrac{k_1[A]_0}{k_1 + k_2}\left[1 - e^{-(k_1 + k_2)t}\right]$ $[D] = \dfrac{k_2[A]_0}{k_1 + k_2}\left[1 - e^{-(k_1 + k_2)t}\right]$	

2.3.3.3.1   Reversible Reactions

The reaction rate laws delineated earlier discount the possibility of reverse reactions and will fail to give the general rate of a process near equilibrium conditions. In the natural environment, such reactions are common. When the product concentration becomes significant, the reverse reaction will also become significant near equilibrium. This is in keeping with the "principle of microscopic reversibility" enunciated by Tolman (1927), which states that *at equilibrium the rate of the forward reaction is the same as that of the backward reaction.* As an example, let us choose the reaction

$$A \underset{k_b}{\overset{k_f}{\rightleftharpoons}} B \qquad\qquad (2.191)$$

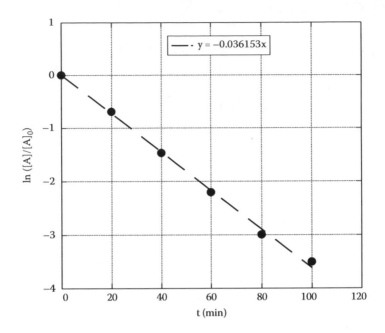

FIGURE 2.29 Experimental points and data fit to a first-order reaction rate law.

where the forward and backward reactions are both first order. The rate of the forward reaction is $r_f = k_f[A]$ and that of the backward reaction is $r_b = k_b[B]$. The net rate of change in [A] is due to the decrease in A by the forward reaction and the increase in the same by the reverse reaction. Thus,

$$r_A = -\frac{d[A]}{dt} = k_f[A] - k_b[B] \qquad (2.192)$$

If $[A]_0$ is the initial concentration of A, then by the mass conservation principle $[A]_0 = [A] + [B]$ at all times $(t > 0)$. Therefore we have

$$\frac{d[A]}{dt} = -(k_f + k_b)[A] + k_b[A]_0 \qquad (2.193)$$

This is a first-order ordinary differential equation, which can be easily solved to obtain

$$[A] = [A]_0 \cdot \left\{ \frac{k_b + k_f e^{-(k_f + k_b)t}}{k_b + k_f} \right\} \qquad (2.194)$$

As $t \to \infty$, $[A] \to [A]_{eq}$ and $[B] \to [B]_{eq}$.

$$[A]_{eq} = [A]_0 \cdot \frac{k_b}{k_f + k_b}$$

$$[B]_{eq} = [A] - [A]_{eq} = [A]_0 \cdot \frac{k_f}{k_f + k_b} \qquad (2.195)$$

The ratio $[B]_{eq}/[A]_{eq}$ is the equilibrium constant of the reaction, K_{eq}. It is important to note that

$$K_{eq} = \frac{[B]_{eq}}{[A]_{eq}} = \frac{k_f}{k_b} \tag{2.196}$$

The connection between thermodynamics and kinetics becomes apparent. In practice, for most environmental processes, if one of the rate constants is known, then the other can be inferred from the equilibrium constant. It should be noted that whereas the ratio k_f/k_b describes the final equilibrium position, the sum $(k_f + k_b)$ determines how fast equilibrium is established.

An example of a reversible reaction is the exchange of compounds between soil and water. Previously we showed that this equilibrium is characterized by a partition coefficient K_{sw}. Consider the transfer as a reversible reaction:

$$A_{water} \underset{k_b}{\overset{k_f}{\rightleftharpoons}} A_{soil} \tag{2.197}$$

The equilibrium constant is

$$K_{eq} = \frac{k_f}{k_b} = \frac{[A]_{soil}}{[A]_{water}} = \frac{W_A}{C_{Aw}} = K_{sw} \tag{2.198}$$

Similar analogies also apply to air/soil, aerosol/air, biota/water partition constants. Thus, an equilibrium partition constant is a ratio of the forward and backward rate constants for the processes. A large value of K_{sw} implies either a large value of k_f or a small value of k_b.

Example 2.32   Reversible Reaction

A reaction $A \rightleftharpoons B$ is said to occur with a forward rate constant, $k_f = 0.1$ h^{-1}. The concentration of A monitored with time is given:

t (h)	[A] (mM)
0	1
1	0.9
5	0.65
10	0.48
15	0.40
100	0.33
500	0.33

Find K_{eq} and k_b.

As $t \rightarrow 4$, $[A]_{eq} = [A]_0 (k_b/k_f + k_b)$. Since $[A]_{eq} = 0.33$, $k_b/k_f + k_b = 0.33/1 = 0.33$. Hence, $k_b = 0.05$ h^{-1} and $K_{eq} = 0.1/0.05 = 2$. $[B]_{eq} = [A]_0 (k_f/k_f + k_b) = 0.67$. Figure 2.30 plots the change in [A] and [B] with t.

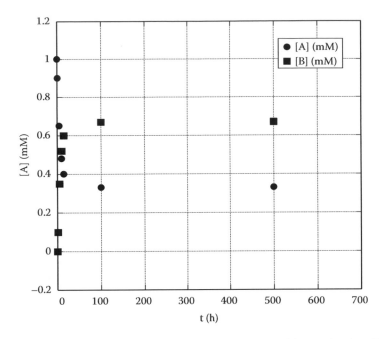

FIGURE 2.30   Change in concentrations of A and B for a reversible reaction A ↔ B.

Example 2.33   Equilibrium Constant from Standard Thermodynamic Data

Find the equilibrium constant at 298 K for the reaction: H_2S (gas) ⇔ H_sS (water) from the standard free energy of formation for the compounds

	G_f^0 (kJ mol^{-1})
H_2S (g)	−33.6
H_2S (aq)	−27.9

The overall free energy change for the reaction is the difference between the G_f^0 of the product and reactant. $\Delta G_f^0 = -27.9 + 33.6 = +5.7$ kJ mol^{-1}. Hence, $K_{eq} = \exp[-5.7/(8.314 \times 10^{-3})(298)] = 0.1$. Note that $K_{eq} = k_f / k_b = [H_2S]_w / P_{H_2S} = 1/K'_{AW}$, as defined previously in Chapter 2. Hence Henry's constant, $K'_{AW} = 9.98$ L atm mol^{-1}. From Appendix A, the value is 8.3 L atm mol^{-1}.

2.3.3.3.2   Series Reactions and Steady State Approximation

Many reactions in the environment occur either in series or in parallel. Reaction in series is of particular relevance to us since it introduces both the concepts of "steady state" and "rates determining steps" that are important in environmental chemical kinetics. Let us consider the conversion of A to C via an intermediate B:

$$A \xrightarrow{\ k_1\ } B \xrightarrow{\ k_2\ } C \tag{2.199}$$

The rates of production of A, B, and C are given by

$$\frac{d[A]}{dt} = -k_1[A]$$

$$\frac{d[B]}{dt} = k_1[A] - k_2[B] \qquad (2.200)$$

$$\frac{d[C]}{dt} = k_2[B]$$

Throughout the reaction, we have mass conservation such that $[A] + [B] + [C] = [A]_0$. The initial conditions are at $t = 0$, $[A] = [A]_0$, and $[B] = [C] = 0$. Solving the equations in succession and making use of this mass balance and initial conditions one obtains after some manipulation the values of $[A]$, $[B]$, and $[C]$, as given in Table 2.16. Figure 2.31 is a plot of $[A]$, $[B]$, and $[C]$ with time for representative values of $k_1 = 1$ min^{-1} and $k_2 = 0.2$ min^{-1}. We note that the value of $[A]$ decreases to zero in an exponential manner. The value of $[B]$ goes through a maximum and then falls off to zero. The value of $[C]$, however, shows a slow initial growth (which is termed the "induction period"), followed by an exponential increase to $[A]_0$.

In those cases where there are several intermediates involved in a reaction, such as occurs in most chemical reactions in the air and water environments, the derivation of the complete rate expression will not be quite as clear-cut as delineated here. The intermediate (e.g., B) is necessarily of low concentration and is assumed to be

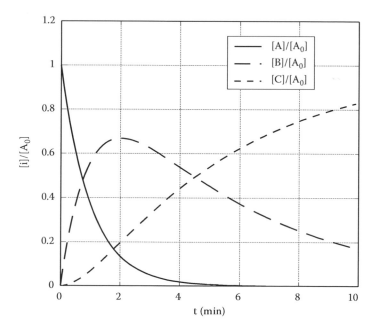

FIGURE 2.31 Change in concentrations of A, B, and C with time for a series reaction $A \rightarrow B \rightarrow C$.

constant during the reaction. This is called the "pseudo-steady state approximation" (PSSA). It allows us to set $d[B]/dt$ to zero. Thus, at pseudo-steady state we can obtain

$$-k_2\left[B^*\right]+k_1\left[A\right]=0$$

$$\left[B^*\right]=\frac{k_1}{k_2}\left[A\right]=\frac{k_1}{k_2}\left[A\right]_0 e^{-k_1 t} \tag{2.201}$$

Hence,

$$\left[C\right]=\left[A\right]_0\left(1-e^{-k_1 t}\right) \tag{2.202}$$

Comparing with the exact equations for [B] and [C] given in Table 2.16, we observe that the exact solutions approximate the steady state solution only if $k_2 \gg k_1$. In other words, when the reactivity of B is so large that it has little time to accumulate. The difference between the exact and approximate solutions can be used to estimate the departure from steady state.

Example 2.34   Series Reactions

The consumption of oxygen (oxygen deficit) in natural streams occurs due to biological oxidation of organic matter. This is called the biochemical oxygen demand, BOD. The oxygen deficit in water is alleviated by dissolution of oxygen from the atmosphere. These two processes can be characterized by a series reaction of the following form:

Organic matter $\xrightarrow[\text{oxidation}]{k_1}$ Oxygen deficit $\xrightarrow[\text{oxygenation}]{k_2}$ Oxygen restoration

$$A \xrightarrow{k_1} B \xrightarrow{k_2} C \tag{2.203}$$

In this case, the initial conditions are somewhat different from what was discussed earlier.

Here at $t = 0$, $[A] = [A]_0$, $[B] = [B]_0$, and $[C] = [C]_0$. The solution to the problem is given by

$$[A]=[A]_0 e^{-k_1 t}$$

$$[B]=\frac{k_1}{k_2-k_1}[A]_0\left(e^{-k_1 t}-e^{-k_2 t}\right)+[B]_0 e^{-k_2 t} \tag{2.204}$$

$$[C]=[C]_0\left\{1-\frac{\left(k_2 e^{-k_1 t}+k_1 e^{-k_2 t}\right)}{\left(k_2-k_1\right)}\right\}+[B]_0\left(1-e^{-k_2 t}\right)+[C]_0$$

Given a value of $k_1 = 0.1$ d^{-1}, $k_2 = 0.5$ d^{-1}, initial organic matter concentration $[A]_0 = 10$ mg L^{-1} and initial oxygen deficit $[B]_0 = 3$ mg L^{-1}, after 24 h (1 d), the oxygen deficit will be $[B] = (0.1) (10/0.4) (e^{-0.1} - e^{-0.5}) + (3) e^{-0.5} = 2.56$ mg L^{-1}.

A frequently encountered reaction type in environmental engineering is one in which a preequilibrium step precedes the product formation. The steady state concept is particularly useful in analyzing such a reaction.

$$A \underset{k_{bl}}{\overset{k_{fl}}{\rightleftharpoons}} B \xrightarrow{k_2} C \tag{2.205}$$

Most enzyme reactions follow this scheme. It is also of interest in many homogeneous and heterogeneous reactions (both in the soil and sediment environments), that is, those that occur at interfaces. For this reaction,

$$\frac{d[B]}{dt} = k_{fl}[A] - k_{bl}[B] - k_2[B] \tag{2.206}$$

Using the steady state approximation, we can set d[B]/dt =0. Hence,

$$[B^*] = \left(\frac{k_{fl}}{k_{bl} + k_2} \right) \cdot [A] \tag{2.207}$$

For species A, we obtain

$$\frac{d[A]}{dt} = -k_{fl}[A] + k_{bl}[B^*] = -\left(\frac{k_2 k_{fl}}{k_{b2} + k_2} \right) \cdot [A] \tag{2.208}$$

and for species C we have

$$\frac{d[C]}{dt} = k_2[B^*] = \left(\frac{k_2 k_{fl}}{k_{bl} + k_2} \right) \cdot [A] \tag{2.209}$$

These expressions can be readily integrated with the appropriate boundary conditions to obtain the concentrations of species A and C. If $k_{bl} \gg k_2$,

$$\frac{d[C]}{dt} = k_2 K_{eq}[A] \tag{2.210}$$

where $K_{eq} = k_{fl}/k_{bl}$ is the equilibrium constant for the first step in the reaction. This occurs if the intermediate B formed is converted to A more rapidly than to C. The rate of the reaction is then dominated by the value of k_2. Thus, B → C is said to be the "rate-determining step." Another situation is encountered if $k_2 \gg k_{bl}$ in which case, d[C]/dt = $k_{fl}[A]$. This happens if the intermediate is rapidly converted to C. Then k_{fl} determines the rate. The rate-determining step is then said to be the equilibrium reaction A ⇌ B.

Following are illustrations of how the concepts of integrated rate laws can be used to analyze particular environmental reaction schemes. An example from water chemistry and another one from air chemistry are chosen.

Example 2.35 Solution of Inorganic Gases in Water

A reaction of environmental relevance is the dissolution of an inorganic gas (e.g., CO_2) in water. The reaction proceeds in steps. The important step is the hydration of CO_2 followed by dissolution into HCO_3^- species in water. Stumm and Morgan (1996) and Butler (1982) have analyzed this reaction, and we shall adopt their approach here. In analyzing these reactions, we shall consider water to be in excess such that its concentration does not make any contribution toward the overall rate. The overall hydration reaction can be written as

$$CO_2(aq) + H_2O \underset{k_b}{\overset{k_f}{\rightleftharpoons}} HCO_3^-(aq) + H^+(aq) \qquad (2.211)$$

with $k_f = 0.03$ s^{-1} and $k_b = 7 \times 10^4$ mol L^{-1}s^{-1}. Since H_2O concentration is constant, we shall consider the functional reaction to be of the form $A \underset{k_b}{\overset{k_f}{\rightleftharpoons}} B + C$ where A represents CO_2, B represents HCO_3^-, and C represents H^+. Laidler (1965) provides the integrated rate law for this reaction:

$$\left(\frac{[B]_{eq}}{2[A]_0 - [B]_{eq}} \right) \cdot \ln\left\{ \frac{[A]_0[B]_{eq} + [B]([A]_0 - [B]_{eq})}{[A]_0([B]_{eq} - [B])} \right\} = k_f t \qquad (2.212)$$

where $[B]_{eq}$ is the concentration of B at equilibrium. If the pH of a sample of water is 5.7, then $[H]^+ = [C]_{eq} = 2 \times 10^{-6}$ mol L^{-1}. By stoichiometry, $[HCO_3^-]_{eq} = [B]_{eq} = [C]_{eq} = 2 \times 10^{-6}$ mol L^{-1}. If the closed system considered has an initial CO_2 of 5×10^{-5} mol L^{-1}, then $[A]_0 = [CO_2(aq)] = 5 \times 10^{-5}$ mol L^{-1}. Then we have

$$[B] = 2 \times 10^{-6} \left[\frac{(e^{1.5t} - 1)}{(e^{1.5t} - 0.96)} \right] \text{ and } [A] = [A]_0 - [B] = 5 \times 10^{-5} - [B].$$

If the equilibrium pH is 5, then $[B]_{eq} = 1 \times 10^{-5}$ mol L^{-1}, and hence

$$[B] = 1 \times 10^{-5} \left[\frac{(e^{0.27t} - 1)}{(e^{0.27t} - 0.8)} \right] \text{ and } [A] = 5 \times 10^{-5} - [B]. \text{ Figure 2.32 gives the con-}$$

centration of aqueous CO_2 with time as it is being converted to HCO_3^- in the system. The functional dependence on pH is shown. The equilibrium value of CO_2 (aq) is pH dependent and reaches 48 μM in ≈ 0.7 s at a pH of 5.7 and 40 μM in ≈ 14 s at a pH of 5. If the final equilibrium pH is to increase, more CO_2 (aq) has to be consumed, and hence the concentration falls to a lower equilibrium value.

In the case of SO_2 solution in water, we have

$$SO_2(aq) + H_2O \underset{k_b}{\overset{k_f}{\rightleftharpoons}} HSO_3^-(aq) + H^+(aq) \qquad (2.213)$$

where $k_f = 3.4 \times 10^{-6}$ s^{-1} and $k_b = 2 \times 10^8$ mol L^{-1}s^{-1}. Notice first of all that k_f in this case is much larger than for CO_2 and hence virtually instantaneous reaction can be expected. Let the initial SO_2 concentration be 5×10^{-5} mol L^{-1}. A similar analysis as this for CO_2 can be carried out to determine the approach to equilibrium for SO_2. Figure 2.33 displays the result at two pH values of 5.7 and 5.0. The striking differences in time to equilibrium from those for CO_2 dissolution reactions are evident. At a pH of 5.7 the equilibrium value is reached in ≈ 6 ns, whereas at a pH of 5.0 the characteristic time is ≈ 0.1 μs.

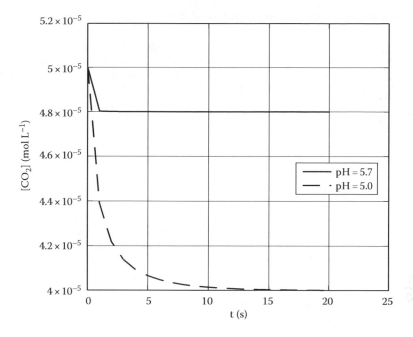

FIGURE 2.32 Kinetics of solution of CO_2 in water at different pH values.

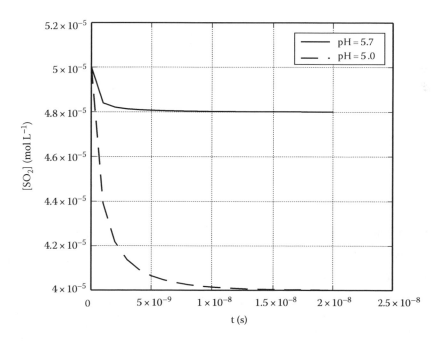

FIGURE 2.33 Kinetics of solution of SO_2 in water at different pH values.

Example 2.36 Atmospheric Chemical Reactions

A large variety of reactions between organic molecules in the atmosphere occur through mediation by N_2 or O_2 that are the dominant species in ambient air. If A and B represent two reactants and Z represents either N_2 or O_2, then the general reaction scheme consists of the following steps:

$$A + B \underset{k_{-1}}{\overset{k_1}{\rightleftharpoons}} A\text{--}B$$
$$A\text{--}B + Z \xrightarrow{k_2} AB + Z \tag{2.214}$$

This is an example of a series reaction with a preequilibrium step discussed earlier. The method of solution is similar. A--B represents an excited state of the AB species, which is the final product. Typically, these excited intermediates are produced by photo or thermal excitation. This short-lived intermediate transfers its energy to Z (N_2 or O_2) to form the stable AB complex. Thus, the overall reaction scheme is $A + B + Z \xrightarrow{k'} AB + Z$. Each step in this reaction is called an "elementary reaction." "Complex reactions" are composed of many such "elementary reactions."

The rate of disappearance of each species is given here:

$$-\frac{d[A]}{dt} = k_1[A][B] - k_{-1}[A\text{--}B]$$
$$\frac{d[A\text{--}B]}{dt} = k_1[A][B] - k_{-1}[A\text{--}B] - k_2[A\text{--}B][Z] \tag{2.215}$$
$$\frac{d[AB]}{dt} = k_2[A\text{--}B][Z]$$

In order to simplify the analysis we make use of the concept of steady state for [A--B]. Thus, $d[A\text{--}B]/dt = 0$, and hence, $[A\text{--}B]^* = k_1[A][B]/(k_{-1} + k_2[Z])$. Thus, we have the following differential equation for [A--B]:

$$\frac{d[A\text{--}B]}{dt} = \frac{k_1 k_2 [A][B]}{k_{-1} + k_2[Z]} \cdot [Z] = k'[A][B] \tag{2.216}$$

with $k' = \dfrac{k_1 k_2 [Z]}{k_{-1} + k_2[Z]}$ as a constant since [Z] is in excess and varies little. As $[Z] \to 0$, we have the "low pressure limit" $k_0'' = (k_1 k_2 / k_{-1})[Z] = k_0'[Z]$. In the "high-pressure limit," [Z] is very large, and $k_4' = k_1$ and is independent of [Z]. From k_0' and k_4', one can obtain $k' = k_0'' \left[1 + \dfrac{k_0''}{k_\infty'} \right]^{-1}$. Table 2.17 lists the rate constants for a typical atmospheric chemical reaction. Note that the values are a factor of two lower under stratospheric conditions. The general solution to this ordinary differential equation is the same as that for reaction $A + B \to$ products, as in Table 2.16. From the integrated rate law, one can obtain the concentration-time profile for species A in the atmosphere.

TABLE 2.17

Low and High Temperature Limiting k Values for the Atmospheric Reaction

$OH + SO_2 \xrightarrow{\ Z\ } HOSO_2$

T (K)	p (Torr)	[Z] (mole-cules cm^{-3})	k_0' (cm^6 mole-cule^{-2} s^{-1})	k_∞' (cm^3 mole-cule^{-1} s^{-1})	k' (cm^3 mole-cule^{-1} s^{-1})
300 (troposphere)	760	2.4×10^{19}	$(3.0 \pm 1.5) \times 10^{-31}$	$(2.0 \pm 1.5) \times 10^{-12}$	1.1×10^{-12}
−219 (stratosphere)	−39	1.7×10^{18}	8.7×10^{-31}	2.0×10^{-12}	5.2×10^{-13}

Source: Finlayson-Pitts and Pitts, 1986.

2.3.3.3.3 Parallel Reactions

In environmental reactions, there also exist cases where a molecule can simultaneously participate in several reactions. For example,

$$A \xrightarrow{k_1} C$$
$$A \xrightarrow{k_2} D \tag{2.217}$$

In this case, the rate of disappearance of A is given by

$$r = -\frac{d[A]}{dt} = k_1[A] + k_2[A]$$
$$\frac{d[C]}{dt} = k_1[A] \tag{2.218}$$
$$\frac{d[D]}{dt} = k_2[A]$$

Solutions for this case are also given in Table 2.16. First, both C and D increase exponentially with a rate constant $(k_1 + k_2)$, and A decreases exponentially with a rate constant $(k_1 + k_2)$. Second, the ratio of products $[C]/[D] = k_1/k_2$ at all times is called the "branching ratio."

2.3.4 ACTIVATION ENERGY

The rates of reactions encountered in nature are very sensitive to temperature. In the nineteenth century, the Swedish chemist Svänte Arrhenius proposed an empirical equation based on a large number of experimental observations. This is called the "Arrhenius equation":

$$k = A \cdot e^{\frac{E_a}{RT}} \tag{2.219}$$

where
   A is called the preexponential factor
   E_a is called the "activation energy"

The equation is also written in an alternative form by combining the two terms:

$$k \propto e^{-\frac{\Delta G^{\dagger}}{RT}} \qquad (2.220)$$

where $\Delta G^{\dagger}$ is called the "Gibbs activation energy." In this form, k bears a strong resemblance to the equilibrium constant K_{eq}, which depends on the Gibbs free energy. The integrated form of the Arrhenius equation is

$$\ln\left(\frac{k_2}{k_1}\right) = -\frac{E_a}{R}\left(\frac{1}{T_2} - \frac{1}{T_1}\right) \qquad (2.221)$$

Generally, a plot of ln k versus 1/T will give $-E_a/R$ as the slope and ln A as the intercept. The activation energy is interpreted as the minimum energy that the reactants must possess to convert to products. In the gas phase, according to the kinetic theory of collisions, reactions are said to occur if two molecules have enough energy when they collide. Albeit many collisions do occur per second, only a fraction leads to a chemical reaction. The excess energy during those collisions that lead to a reaction is equivalent to the activation energy and, hence, the term $\exp(-E_a/RT)$ is interpreted as the fraction of collisions with large enough energy to lead to reactions. The preexponential factor A is interpreted as the number of collisions that occur irrespective of the energy. The product $A \exp(-E_a/RT)$ is a measure of the "productive" collisions.

Example 2.37   Activation Energy for the Decomposition of an Organic Molecule in Water

Consider the organic molecule (dibromosuccinic acid) in water. Its rate of decomposition is a first-order process, for which the following rate constants were determined at varying temperatures:

T (K)	k (h⁻¹)
323	1.08×10^{-4}
343	7.34×10^{-4}
362	45.4×10^{-4}
374	138×10^{-4}

A plot of ln k versus 1/T can be made, as shown in Figure 2.34. The straight line has a slope $-E_a/R$ of $-11,480$ and an intercept of 26.34. The correlation coefficient is 0.9985. Therefore, $E_a = 95$ kJ mol⁻¹ and $A = 2.76 \times 10^{11}$ h⁻¹ for the given reaction.

2.3.4.1   Activated Complex Theory

The Arrhenius equation can be interpreted on a molecular basis. The theoretical foundation is based on the so-called "activated complex theory" (ACT).

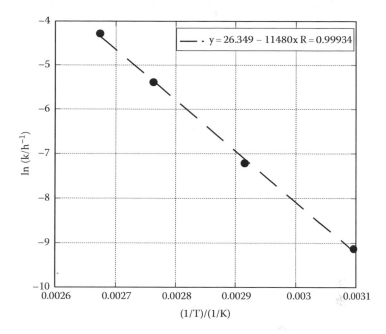

FIGURE 2.34 Temperature dependence of the rate constant for the decomposition of dibromosuccinic acid in aqueous solution. The plot of ln k versus 1/T is the linearized form of the Arrhenius equation. The slope gives the logarithm of the preexponential factor (ln A) and the slope is $-E_a/R$, where E_a is the activation energy. (Data from Laidler, K.J., *Chemical Kinetics*, McGraw-Hill, Inc., New York, 1965.)

We know that in order for the reactants to be converted to products, there should be a general decrease in the total energy of the system. For example, let us consider a bimolecular gas phase reaction where an H atom approaches an I_2 molecule to form HI. When H and I_2 are far apart, the total potential energy is that of the two species H and I_2. When H nears I_2, the I–I bond is stretched and an H–I bond begins to form. A stage will be reached when the H---I---I complex will be at its maximum potential energy and is termed the "activated complex." This is termed the "transition state." A slight stretching of the I---I bond at this stage will simultaneously lead to an infinitesimal compression of the H---I bond and the formation of the H–I molecule with the release of the I atom. The total potential energy of the HI and I species together will be less than that of the H and I_2 species that we started with. The progression of the reaction is represented by a particular position along the reaction path (i.e., intermolecular distance) and is termed the "reaction coordinate." A plot of potential energy versus reaction coordinate is called a "potential energy surface" and is shown in Figure 2.35. The transition state is characterized by such a state of closeness and distortion of reactant configurations that even a small perturbation will send them downhill toward the products. There is a distinct possibility that some of the molecules in the activated complex may revert to the reactants, but those that follow the path to the products will inevitably be in a different configuration from where they started. In actuality, the

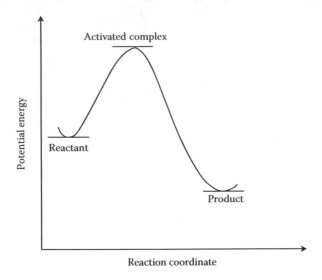

FIGURE 2.35 Formation of the activated complex in a reaction.

potential energy surface is multidimensional, depending on the number of intermolecular distances involved. Generally, simplifications are made to visualize the potential energy surface as a three-dimensional plot.

The activated complex theory presumes that an equilibrium between the reactants (A and B) and an activated complex $(AB)^\dagger$ is established. This complex further undergoes unimolecular decay into product P.

$$A + B \underset{}{\overset{K^\dagger}{\rightleftharpoons}} (AB)^\dagger \xrightarrow{k^\dagger} P \qquad (2.222)$$

The ACT tacitly assumes that even when the reactants and products are not at equilibrium, the activated complex is always in equilibrium with the reactants.

The rate of the reaction is the rate of decomposition of the activated complex. Hence,

$$r = k^\dagger P_{AB^\dagger} \qquad (2.223)$$

However, since equilibrium exists between A, B, and $(AB)^\dagger$

$$K^\dagger = \frac{P_{AB^\dagger}}{P_A P_B} \qquad (2.224)$$

with units of pressure^{-1} since the reaction occurs in the gas phase. Thus,

$$r = k^\dagger K^\dagger P_A P_B = k^* P_A P_B \qquad (2.225)$$

where k^* denotes the second-order reaction rate constant and is equal to $k^\dagger K^\dagger$. The rate constant for the decay of the activated complex is proportional to the frequency of vibration of the activated complex along the reaction coordinate. The rate constant $k^\dagger$ is therefore given by

$$k^\dagger = \kappa \cdot \nu \qquad (2.226)$$

where κ is called the transmission coefficient, which in most cases is ≈ 1. ν is the frequency of vibration and is equal to $k_B T/h$ where k_B is the Boltzmann constant and h is Planck's constant. ν has a value of 6×10^{12} s^{-1} at 300 K.

Concepts from statistical thermodynamics (which is beyond the scope of this textbook) can be used to obtain $K^\dagger$. A general expression is (Laidler, 1965)

$$K^\dagger = \frac{Q_{AB^\dagger}}{Q_A Q_B} \cdot e^{-\frac{E_0}{RT}} \qquad (2.227)$$

where Q is the "partition function," which is obtained directly from molecular properties (vibration, rotation, and translation energies). E_0 is the difference between the zero-point energy of the activated complex and the reactants. It is the energy to be attained by the reactants at 0 K to react, and hence is the activation energy at 0 K. Thus,

$$k^* = \kappa \cdot \frac{k_B T}{h} \cdot \frac{Q_{AB^\dagger}}{Q_A Q_B} \cdot e^{-\frac{E_0}{RT}} \qquad (2.228)$$

Note that $K^\dagger$ is the pressure units–based equilibrium constant. If molar concentrations are used instead of partial pressures, the appropriate conversion has to be applied. It should also be noted that the partition functions are proportional to T^n, and hence,

$$k^* \approx aT^n e^{-\frac{E_0}{RT}} \qquad (2.229)$$

Therefore, we have the following equation:

$$\frac{d \ln k^*}{dT} = \frac{E_0 + nRT}{RT^2} \qquad (2.230)$$

The experimental activation energy E_a was defined earlier:

$$\frac{d \ln k^*}{dT} = \frac{E_a}{RT^2} \qquad (2.231)$$

Hence, $E_a = E_0 + nRT$. This gives the relationship between the zero point activation energy and the experimental activation energy.

The statistical mechanical expressions of the ACT lead to significant difficulties since the structure of the activated complex is frequently unknown. This has led to a more general approach in which the activation process is considered on the basis of thermodynamic functions. Since $K^\dagger$ is the equilibrium constant, we can write

$$\Delta G^\dagger = -RT \ln K^\dagger \qquad (2.232)$$

as the "Gibbs free energy of activation." Hence, we have

$$k^* = \kappa \cdot \frac{k_B T}{h} \cdot e^{-\frac{\Delta G^{\dagger}}{RT}} \tag{2.233}$$

We can further obtain the components of $\Delta G^{\dagger}$, viz., the "enthalpy of activation" $\Delta H^{\dagger}$ and the "entropy of activation" $\Delta S^{\dagger}$. Hence,

$$k^* = \kappa \cdot \frac{k_B T}{h} \cdot e^{\frac{\Delta S^{\dagger}}{R}} \cdot e^{-\frac{\Delta H^{\dagger}}{RT}} \tag{2.234}$$

If k^* is expressed in L mol^{-1} s^{-1} (or dm^3 mol^{-1} s^{-1}), then the standard state for both $\Delta H^{\dagger}$ and $\Delta S^{\dagger}$ is 1 mol L^{-1} (or 1 mol dm^{-3}). The experimental activation energy is related to $\Delta H^{\dagger}$ as per the equation $E_a = \Delta H^{\dagger} - P\Delta V^{\dagger} + RT$. For unimolecular gas phase reactions, $\Delta V^{\dagger}$ is zero, and for reactions in solutions, $\Delta V^{\dagger}$ is negligible. Hence, we have

$$k^* = e \cdot \kappa \cdot \frac{k_B T}{h} \cdot e^{\frac{\Delta S^{\dagger}}{R}} \cdot e^{-\frac{E_a}{RT}} \tag{2.235}$$

In terms of the Arrhenius equation, $A = 2 \times 10^{13} \cdot e^{\frac{\Delta S^{\dagger}}{R}}$ L mol^{-1} s^{-1} at 298 K. For a bimolecular reaction in the gas phase, $P\Delta V^{\dagger} = \Delta n^{\dagger} RT = -RT$, $E_a = \Delta H^{\dagger} + 2RT$, and

$$k^* = e^2 \cdot \kappa \cdot \frac{k_B T}{h} \cdot e^{\frac{\Delta S^{\dagger}}{R}} \cdot e^{-\frac{E_a}{RT}} \tag{2.236}$$

Example 2.38   Activation Parameters for a Gas-Phase Reaction

The bimolecular decomposition of NO_2 by the following gas phase reaction is of significance in smog formation and in combustion chemistry of pollutants: $2NO_2 \rightarrow 2NO + O_2$. The reaction rate is $r = k^* P_{NO_2}^2$. An experiment was designed to measure the rate constant as a function of T:

T (K)	k* (L (mol s)$^{-1}$)
600	0.46
700	9.7
800	130
1000	3130

A plot of ln k* versus 1/T gives a slope of −13,310 and an intercept of 21.38 with a correlation coefficient of 0.9994. Therefore, E_a = 110 kJ mol^{-1} and A = 1.9 × 10^9 L (mol s)$^{-1}$. For a bimolecular gas phase reaction, $\Delta H^{\dagger} = E_a - 2RT$. At 600 K, $\Delta H^{\dagger}$ = 110 − 2(8.314 × 600)/1000 = 100 kJ mol^{-1}. Assume that the transmission coefficient κ = 1. Then, we have from the equation for rate constant,

$k^* = 0.46 = e^2 \left(1.2 \times 10^{13}\right) e^{\frac{\Delta S^\dagger}{R}} \left(2.65 \times 10^{-10}\right)$, from which $\Delta S^\dagger = -90$ J (mol K)$^{-1}$. Note that, in general, if $\Delta S^\dagger$ is negative, the formation of the activated complex is less probable and the reaction is slow. If $\Delta S^\dagger$ is positive, the activated complex is more probable and the reaction is faster. This is so since exp $(\Delta S^\dagger/R)$ is a factor that determines if a reaction occurs faster or slower than normal.

At 600 K, $\Delta G^\dagger = \Delta H^\dagger - T \Delta S^\dagger = 164$ kJ mol^{-1}. This is the positive free energy barrier that NO_2 must climb to react. The value of A calculated from $\Delta S^\dagger$ is A $=$ e(kT/h) exp($-\Delta S^\dagger/R$) $= 6.6 \times 10^8$ L (mol s)$^{-1}$. The value of $K^\dagger = \exp(-\Delta G^\dagger/RT) = 5.3 \times 10^{-15}$ L mol^{-1}. Note that the overall equilibrium standard free energy of the reaction will be $\Delta G^0 = 2(86.5) - 2(51.3) = 71$ kJ mol^{-1}. Hence, $\Delta G^\dagger \approx 2.3 \Delta G^0$.

2.3.4.2 Effect of Solvent on Reaction Rates

The activated complex theory can also be applied to reactions in solutions, but requires consideration of additional factors. Consider, for example, the reaction $A + B \rightleftharpoons C + D$ that occurs both in the gas phase and in solution. The ratio of equilibrium constants will be

$$\frac{K_{eq}^{sol}}{K_{eq}^{gas}} = \frac{K_{HA} K_{HB}}{K_{HC} K_{HD}} \cdot \left(\frac{RT}{V_0}\right)^{\Delta n} \frac{\gamma_A \gamma_B}{\gamma_C \gamma_D} \tag{2.237}$$

where

K_{Hi} is Henry's constant ($= P_i/x_i$)
V_0 is the volume per mole of solution
Δn is the change in moles of substance during the reaction
γ denotes the activity coefficient in solution

It can be argued that as per this equation, reactions in solution will be favored if the reactants are more volatile than products and vice versa if products are more volatile. The primary interaction in the solution phase is that with the solvent molecules for both reactants and products on account of the close packed liquid structure, whereas in the gas phase no such solvent-mediated effects are possible.

Consider the following activated complex formation in solution: $A + B \rightleftharpoons (AB)^\dagger \rightarrow P$. The equilibrium constant for activated complex formation in the solution phase must account for nonidealities due to the solvent phase. Hence,

$$K_{sol}^\dagger = \frac{\left[AB^\dagger\right]}{\left[A\right]\left[B\right]} \cdot \frac{\gamma_{AB^\dagger}}{\gamma_A \gamma_B} \tag{2.238}$$

and the rate in solution is

$$r_{sol} = \frac{k_B T}{h} \cdot \left[AB^\dagger\right] = \frac{k_B T}{h} K_{sol}^\dagger \left[A\right]\left[B\right] \frac{\gamma_A \gamma_B}{\gamma_{AB^\dagger}} \tag{2.239}$$

The rate constant for the reaction in solution is, therefore, given by

$$k_{sol} = \frac{k_B T}{h} K_{sol}^\dagger \frac{\gamma_A \gamma_B}{\gamma_{AB^\dagger}} \tag{2.240}$$

The activity coefficients are referred to the standard state of infinite dilution for solutes. The ratio of rate constants for solution and gas phase reactions is

$$\frac{k_{sol}}{k_{gas}} = \frac{K_{HA}K_{HB}}{K_{H,AB}} \cdot \left(\frac{V_0}{RT}\right) \cdot \frac{\gamma_A \gamma_B}{\gamma_{AB^\dagger}} \tag{2.241}$$

For a unimolecular reaction, both γ_A and $\gamma_{AB^\dagger}$ are similar if the reactant and activated complex are similar in nature, and hence $k_{sol} \approx k_{gas}$. Examples of these cases abound in the environmental engineering literature.

This discussion presupposes that the solvent merely modifies the interactions between the species. In these cases, since the solvent concentration is in excess of the reactants, it provides a medium for reaction. Hence, it will not appear in the rate expression. If the solvent molecule participates directly in the reaction, its concentration will appear in the rate equation. It can also play a role in catalyzing reactions. In solution, unlike the gas phase, the reaction must proceed in steps: (1) diffusion of reactants toward each other, (2) actual chemical reaction, and (3) diffusion of products away from one another. In most cases, steps (1) and (3) have activation energies of the order of 20 kJ, which is much smaller than the activation energy for step (2). Hence, diffusion is rarely the rate-limiting step in solution reactions. If the rate is dependent on either step (1) or (3), then the reaction will show an effect on the solvent viscosity.

We can rewrite the equation for k_{sol} as

$$k_{sol} = k_0 \frac{\gamma_A \gamma_B}{\gamma_{AB^\dagger}} \tag{2.242}$$

where k_0 is the rate constant when $\gamma \to 1$(ideal solution). The dependence of γ on the solvent type is best represented by the Scatchard-Hildebrand equation:

$$RT \ln \gamma_i = V_i \left(\delta_i - \delta_s\right)^2 \tag{2.243}$$

where V_i is the molar volume of solute (A, B, or $AB^\dagger$). $\delta_i^2 = \dfrac{\Delta E_i}{V_i}$ is called the "internal pressure" or the "cohesive energy density." δ_s is the "solvent internal pressure." If δ is the same for both solvent and solute, solubility maximum is observed. Values for δ are listed elsewhere (e.g., Moore and Pearson, 1981). Utilizing this, we can write

$$\ln k_{sol} = \ln k_0 + \frac{V_A}{RT}\left(\delta_s - \delta_A\right)^2 + \frac{V_B}{RT}\left(\delta_s - \delta_B\right)^2 + \frac{V_{AB^\dagger}}{RT}\left(\delta_s - \delta_{AB^\dagger}\right)^2 \tag{2.244}$$

Since V_A, V_B, and $V_{AB^\dagger}$ are similar, it is the δ term that determines the value of $\ln k_{sol}$. If the internal pressures of A and B are similar to that of the solvent s, but different from that of $AB^\dagger$, then the last term dominates and $\ln k_{sol}$ will be lower than the ideal value.

The value of ln k_{sol} will be high if either the internal pressure for A or for B differs considerably from that for s, but is similar to that for $AB^\dagger$.

2.3.4.3 Linear Free Energy Relationships

There exists a relationship between equilibrium constant K_{eq} ($= \exp -\Delta G^0/RT$) and the rate constant k ($= \exp -\Delta G^\dagger/RT$) for several reactions. The correlation is linear and is evident for reactions in solution of the type R-X + A $\rightarrow$ Products, where R is the reactive site and X is a substituent that does not directly participate in the reaction. When ln k is plotted against ln K_{eq} for different reactions that involve only a change in X, a linear correlation is evident. The linearity signifies that as the reaction becomes thermodynamically more favorable, the rate constant also increases. These correlations are called "linear free energy relationships" (LFERs) and are similar to the LFERs between equilibrium constants that we encountered in Chapter 4 (e.g., K_{oc} versus K_{ow}, C_i^* versus K_{ow}). LFERs are particularly useful in estimating k values when no such data are available for a particular reaction. The correlations are of the type

$$\left(\frac{k'}{k}\right) = \left(\frac{K'_{eq}}{K_{eq}}\right)^{\alpha} \qquad\qquad (2.245)$$

where k' and K' are for the reaction with the substituent X'. The relationship between free energies is therefore

$$\Delta G^\dagger = \alpha \ \Delta G^0 + \beta \qquad\qquad (2.246)$$

In environmental chemistry, several LFERs have been obtained for the reactions of organic molecules in the water environment. These relationships are particularly valuable in environmental engineering since they afford a predictive tool for rate constants of several compounds for which experimental data are unavailable. Brezonik (1990), Wolfe et al. (1980a,b), Betterton et al. (1988), and Schwarzenbach et al. (1988) have reported LFERs for the alkaline hydrolysis of several aromatic and aliphatic compounds. The hydrolysis of triarylphosphate esters follows the Hammett LFER between log k and log k_o most readily. The organophosphates and organophosphorothionates appear to obey the LFER between log k and log K_a. The primary amides and diphthalate esters obey the Taft LFER. The rates of photooxidation of substituted phenols by single oxygen ($^1O_2$) also have been shown to obey an LFER. Table 2.18 is a compilation of some of the LFERs reported for pollutants in the aquatic environment. Though several hundred LFERs exist in the organic chemistry literature, few have direct applicability in environmental engineering since (1) the compounds are not of relevance to environmental engineers, (2) the solvents used to develop these are not representative of environmental matrices, or (3) considerable uncertainty exists in these predictions due to the fact that they are based on limited data. More work is certainly warranted in this area. Nonetheless, at least order of magnitude estimates of reaction rates are possible using these relationships as starting points.

TABLE 2.18
Different Types of Acid-Base Hydrolysis Mechanisms in Environmental Chemistry

Type	Reaction	Rate Expression	Examples
I	$S + H^+ \rightleftharpoons SH^+$ $SH^+ + W \xrightarrow{\text{slow}} P$	$kK_{eq}[S][H^+][R]$	Ester, amide, and ether hydrolysis
II	$SH + H^+ \rightleftharpoons HSH^+$ $HSH^+ + B \xrightarrow{\text{slow}} BH^+ + SH$	$kK_{eq}K_a[BH^+][HA]$	Hydrolysis of alkyl-benzoimides, keto-enol changes
III	$HS + HA \rightleftharpoons HS \cdot HA$ $HS \cdot HA + B \xrightarrow{\text{slow}} P$	$kK_{eq}[HS][HA][B]$	Mutarotation of glucose
IV	$S + HA \rightleftharpoons S \cdot HA$ $S \cdot HA + R \rightarrow P$	$kK_{eq}[S][HA][R]$	General acid catalysis, hydration of aldehydes
V	$S^- + HA \xrightarrow{\text{slow}} SH + A^-$ $SH \xrightarrow{\text{fast}} P$	$k[S^-][HA]$	Decomposition of diazo-acetate
VI	$HS + B \rightleftharpoons S^- + BH^+$ $S^- + R \xrightarrow{\text{slow}} P$	$kK_{eq}[SH][R][OH^-]/K_b$	Claisen condensation
VII	$HS + B \xrightarrow{\text{slow}} S^- + BH^+$ $S^- \xrightarrow{\text{fast}} \text{Products}$	$k[HS][B]$	General base catalysis
VIII	$R + S \rightleftharpoons T$	$[T][\sum_i k_i[B_i] + \sum_j k_j[HA_j]]$	Aromatic substitutions
IX	$HS + B \rightleftharpoons B \cdot HS$ $B \cdot HS \xrightarrow{\text{slow}} P$	$kK_{eq}[B][HS][R]$	General base catalysis, ester hydrolysis

Source: Moore, J.W. and Pearson, R.G., *Kinetics and Mechanism*, 3rd edn., Wiley-Interscience, New York, 1981.

Note: S represents reactant, B is the general base, HA is the general acid, R is a reactant whether acid or base.

2.3.5 Reaction Mechanisms

Reactions in the environment are complex in nature, consisting of several steps. Empirical rate laws determined through experiments can give us ideas about the underlying mechanisms of these complex reactions (Moore and Pearson, 1981). It is best illustrated using a reaction, which admittedly has only limited significance in environmental engineering. The reaction is the formation of hydrogen halides from its elements. The hydrogen halides do play a central role in the destruction of ozone in the stratosphere through their involvement in the reaction of ozone with chlorofluorocarbons. Several introductory physical chemistry textbooks use this reaction as an illustration of complex reactions. Hence, only a cursory look at the essential concept is intended, namely, that of the deduction of reaction mechanism from knowledge of the rate law for the formation of a hydrogen halide.

2.3.5.1   Chain Reactions

In the early part of the twentieth century, Bodenstein and Lind (1907) studied in great detail the reaction

$$H_2 + Br_2 \rightleftharpoons 2HBr \tag{2.247}$$

It was observed that the rate expression at the initial stages of the reaction where $[HBr] \ll [H_2]$ or $[Br_2]$ was

$$r = \frac{d[HBr]}{dt} = k'[H_2][Br_2]^{1/2} \tag{2.248}$$

At later times, the rate was

$$r = \frac{d[HBr]}{dt} = \frac{k''[H_2][Br_2]^{1/2}}{1 + \frac{k'''[HBr]}{[Br_2]}} \tag{2.249}$$

Thus, as time progressed, the rate was inhibited by HBr formed during the reaction. The fractional orders in the rate expression invariably indicate a "chain reaction" involving "free radicals." A chain reaction is one in which an intermediate compound is generated and consumed that initiates a series of several other reactions leading to the final product. A chain reaction involves an "initiation" step, a "propagation" step, and a "termination" step.

The "initiation" step will involve a decomposition of one of the reactants. Usually, this will occur for the reactant with the smallest bond energy. For example, between H_2 and Br_2, with dissociation energies of 430 and 190 kJ mol^{-1}, respectively, Br_2 will dissociate easily as

$$Br_2 \xrightarrow{k_1} 2Br \tag{2.250}$$

The propagation step consists of the following reactions:

$$\begin{aligned} H_2 + Br &\xrightarrow{k_2} HBr + H \\ H + Br_2 &\xrightarrow{k_3} HBr + Br \end{aligned} \tag{2.251}$$

where it should be noted the Br radical is consumed and regenerated. In complex reactions of this type, HBr can also react with H to give H_2 and Br:

$$H + HBr \xrightarrow{k_4} H_2 + Br \tag{2.252}$$

This is called an "inhibition" step since the H radical is consumed by a reaction other than by chain termination. The reason why HBr reacts with H and not Br is that the former is an exothermic reaction (-84.5 kJ mol^{-1}) while the latter is an endothermic reaction (126.4 kJ mol^{-1}). The "termination" step occurs by the reaction

$$2Br + M \xrightarrow{k_5} Br_2 + M \tag{2.253}$$

where M is a third body that absorbs the energy of recombination and thereby helps to terminate the chain.

In order to derive the overall rate expression, we must first note that the concentrations of the intermediate H and Br radicals are at steady state, and hence,

$$\frac{d[H]}{dt} = 0 = k_2 [Br][H_2] - k_3 [H][Br_2] - k_4 [H][HBr] \tag{2.254}$$

and

$$\frac{d[Br]}{dt} = 0 = 2k_1 [Br_2] - k_2 [Br][H_2] + k_3 [H][Br_2] + k_4 [H][HBr] - 2k_5 [Br]^2 \tag{2.255}$$

Upon solving these equations simultaneously, we get

$$[Br] = \left(\frac{k_1 [Br_2]}{k_5} \right)^{1/2} \tag{2.256}$$

and

$$[H] = \frac{k_2 \left(\dfrac{k_1}{k_5} \right)^{1/2} [H_2][Br_2]^{1/2}}{k_3 [Br_2] + k_4 [HBr]} \tag{2.257}$$

The overall rate of production of HBr is

$$\frac{d[HBr]}{dt} = \frac{2k_2 \left(\dfrac{k_1}{k_5} \right)^{1/2} [H_2][Br_2]^{1/2}}{1 + \left(\dfrac{k_4}{k_3} \cdot \dfrac{[HBr]}{[Br_2]} \right)} \tag{2.258}$$

Note that this expression gives the correct dependencies at both initial times and later times. The constants k'' and k''' in the earlier equation can be identified as related to the individual rate constants.

Such chain reactions are of significance in environmental engineering and will be frequently encountered in atmospheric (gas phase) reaction chemistry (Seinfeld, 1986), in catalytic reactions in wastewater treatment, heterogeneous catalysis of atmospheric solution chemistry (Hoffmann, 1990), and in combustion engineering. An interesting example of a free radical chain reaction discussed by Hoffmann (1990) is the oxidation of S(IV) by Fe(III), which is a prevalent metal in atmospheric particles.

$$Fe^{3+} + SO_3^{2-} \rightarrow Fe^{2+} + SO_3^{\bullet-} \tag{2.259}$$

Example 2.39   Chain Reaction for the Oxidation of Organic Compounds in Natural Waters

An example of a chain reaction is the oxidation of organic compounds by peroxides in water, sediment, and atmospheric environments (Ernestova et al., 1992). Let R-H represent an organic compound and AB an initiator (e.g., hydrogen peroxide, metal salts, or organic azo compounds). H_2O_2 is an excellent oxidant in natural water. There are at least three possible initiation steps in this case:

$$
\begin{aligned}
AB &\xrightarrow{h\nu,T} A^\bullet + B^\bullet \\
RH + A^\bullet &\xrightarrow{fast} R^\bullet + AH \\
RH + O_2 &\rightarrow R^\bullet + HOO^\bullet
\end{aligned}
\tag{2.260}
$$

The initiation can be induced by sunlight, ionizing radiation (cosmic rays for example), acoustic waves, and temperature fluctuations. Dissolved gases such as ozone can also initiate free radicals. In shallow water bodies, the principal factor is sunlight. H_2O_2 has been recognized as an important component in the self-purification of contaminated natural waters. It also provides $OH^\bullet$ radical in the atmosphere that reacts with most other organics, which leads to increased oxidation, aqueous solubility, and scavenging of organics. The hydroxyl radical is appropriately termed the "atmosphere's detergent." The last reaction in Equation 2.260 is an initiation of oxygen in the absence of any AB.

The propagation steps are

$$
\begin{aligned}
R^\bullet + O_2 &\xrightarrow{k_1} ROO^\bullet \\
ROO^\bullet + RH &\xrightarrow{k_2} ROOH + R^\bullet
\end{aligned}
\tag{2.261}
$$

Thus, the radical $R^\bullet$ is reformed in the last propagation step. These steps can be repeated several times depending on the light or thermal energies available for initiation. The only termination steps are radical recombinations, such as

$$
\begin{aligned}
ROO^\bullet + ROO^\bullet &\xrightarrow{k_t} Pr \\
ROO^\bullet + R^\bullet &\rightarrow Pr \\
R^\bullet + R^\bullet &\rightarrow Pr
\end{aligned}
\tag{2.262}
$$

If the pollutant concentration in the water column is low, the most likely termination step is the first expression. The rate of oxidation of the organic compound is given by Equation 2.262:

$$r = \frac{d[RH]}{dt} = -k_2 [ROO^\bullet][RH] \tag{2.263}$$

In natural waters, $ROO^\bullet$ concentration may reach steady state concentrations of $\sim 10^{-9}$ mol L^{-1}. Applying the pseudo-steady state approximation for the peroxide radical, we obtain

$$[ROO^\bullet] = \left(\frac{k_1 [R^\bullet][O_2]}{k_t} \right)^{1/2} \tag{2.264}$$

If we denote the rate of peroxide formation as $r_{PER} = k_1 [R^\bullet][O_2]$, then we can write

$$r = -k_2 \left(\frac{r_{PER}}{k_t} \right)^{1/2} [RH] = k[RH] \tag{2.265}$$

In most natural waters, the rate of peroxide formation, r_{PER}, is constant. Hence, k is a pseudo-first-order rate constant. There are a few classes of compounds (e.g., polyaromatic hydrocarbons, nitrosoamines) that are subject to free radical oxidation in both natural waters and atmospheric moisture (Schnoor, 1992).

2.3.6 REACTIONS IN SOLUTIONS

2.3.6.1 Effects of Ionic Strength on Rate Constants

Reactions in the hydrosphere occur in the presence of many different ions. Hence, the ionic strength influences the rate of a reaction. The following equation has been derived to relate the ionic strength (I) effect on a reaction in solution.

$$\ln \left(\frac{k}{k^0} \right) = 2 z_A z_B I^{1/2} \tag{2.266}$$

where k^0 denotes the rate constant in zero ionic strength solution ($I = 0$). If the ionic charges, z_A and z_B, are similar, increasing I increases the rate constant, whereas for ions of opposite charge, the rate constant decreases with increasing I. If either of the species in uncharged, I will have no effect on the rate constant. These effects are called "kinetic salt effects." Using the value of $A = 0.51$ derived in Section 2.2.4.4, the approximate dependence of I on the rate constant can be readily estimated using the following equation:

$$\ln \left(\frac{k}{k^0} \right) = 1.02 z_A z_B I^{1/2} \tag{2.267}$$

Hence, a slope of the plot of ln k versus $I^{1/2}$ should give a slope of $1.02 z_A z_B$. Note that one can also substitute other relationships for γ, as given in Table 2.10, and obtain the appropriate relationship between rate constant and ionic strength that are applicable at higher values of I.

The ionic strength effect on rate constants will become significant only for I > 0.001 M. For rainwater, I is small, whereas for lakes and rivers, it is close to 0.001 M. For atmospheric moisture (fog, cloud) and for seawater, I exceeds 0.001 M. Wastewater also has values of I > 0.001 M. Only for these latter systems does the dependence of I on k become significant.

Example 2.40   Effect of Ionic Strength on Rate Constant

The ionic strength affects a bimolecular reaction rate constant as follows:

I (mol kg^{-1})	k (mol L^{-1} s^{-1})
2.5×10^{-3}	1.05
3.7×10^{-3}	1.12
4.5×10^{-3}	1.16
6.5×10^{-3}	1.18
8.5×10^{-3}	1.26

A plot of the rate constant versus I gives as intercept $k^0 = 0.992$ L mol^{-1} s^{-1} with $r^2 = 0.936$ and slope = 31.5. A plot of ln (k/k^0) versus $I^{1/2}$ gives a straight line with slope 2.2 and a correlation coefficient $r^2 = 0.756$. The slope, $2.2 = 1.02 z_A z_B$. If one of the ions A has $z_A = -1$ (e.g., OH), the other ion B must have a charge $z_B = -2$. Thus, we can infer the charge of the ions involved in the activated complex.

2.3.7   ENVIRONMENTAL CATALYSIS

The rates of environmental reactions are influenced by many organic or inorganic species, solid particles, and liquid surfaces. The resulting change in reaction rates is called "catalysis" and the entities responsible for the charge are called "catalysts." Catalysis is prevalent in the natural environment (water, air, and soil) and also in waste treatment and pollution prevention processes. If the process occurs such that the catalysts are in the same phase as the reactants, it is termed "homogeneous catalysis." Some reactions are, however, effected by the presence of a separate phase (e.g., solid particles in water); these are called "heterogeneous catalysis." Most surface reactions, in one way or another, belong to the latter category. A list of typical examples in environmental engineering is given in Table 2.19.

Catalysts participate in a reaction but are regenerated in the system such that there is no net concentration change. The equilibrium constant $K_{eq} = k_f/k_b$ remains unchanged. Therefore, it must be that both k_f and k_b are influenced to the same extent. For most environmental catalysis reactions, the general rate expression will be

$$r = \left[f\left([A], [B]... \right) \right] \cdot [X] + f'\left([A], [B]... \right) \qquad (2.268)$$

TABLE 2.19

Examples of Homogeneous and Heterogeneous Catalysis in Environmental Engineering

Catalysis Type	Reaction	Applications
Homogeneous		Oxidation of S(IV) by H_2O_2 atmospheric chemistry, aquatic chemistry, waste treatment.
Homogeneous and heterogeneous	Acid and base hydrolysis of pesticides and esters	Aquatic, soil, and sediment chemistry
Homogeneous	Enzyme catalyzed biodegradation	Aquatic, soil chemistry, waste treatment
Homogeneous and heterogeneous	NO_x formation in combustion reactors	Atmospheric chemistry, hazardous waste incinerators
Homogeneous	Ozone destruction in gas phase	Stratospheric ozone chemistry
Heterogeneous	Production of hydrochlorofluoro-carbons	Manufacture of CFC replacement chemicals
Heterogeneous	Hydroxylation of N_2O over zeolite catalysts	Oxidative removal of N_2O, which contributes to greenhouse effect
Heterogeneous	Oxidation of organics in water on TiO_2	Removing pollutants from oil slicks
Heterogeneous	Dehalogenation of pesticides using membrane catalysts and bacteria	Hazardous waste treatment of soils

where [X] represents the catalyst concentration. $f(A, B, \ldots)$ and $f'(A, B, \ldots)$ denote the functional dependence on the substrate concentration. As $[X] \to 0$, $r \to f'(A, B, \ldots)$, and in some cases, since $f'(A, B, \ldots) = 0$, $r \to 0$.

2.3.7.1 Mechanisms and Rate Expressions for Environmental Catalysis

For environmental catalysis, the primary step is the formation of a complex Z between the catalyst X and substrate A.

$$X + A \underset{k_b}{\overset{k_f}{\rightleftharpoons}} Z + Y \tag{2.269}$$

The complex Z further reacts with another reactant W (e.g., the solvent) to give the desired products.

$$Z + W \overset{k'}{\longrightarrow} P + X \tag{2.270}$$

The products consist of the regenerated catalyst X and other reaction products P. Note that the second step is a nonequilibrium reaction. As we will see in the next section, Z is a surface adsorbed complex for a heterogeneous catalysis. Y and W do not exist in that case. Acids or bases catalyze many reactions in the environment. In these cases, X can be H^+ or OH^-. If $X \equiv H^+$, then Z is the conjugate base of A, and the reaction is "acid catalyzed," whereas if $X \equiv OH^-$, then Z is the conjugate acid of A and the reaction is said to be "base catalyzed."

The equilibrium constant for the formation of complex Z is

$$K_{eq} = \frac{k_f}{k_b} = \frac{[Z][Y]}{[X][A]} \qquad (2.271)$$

If we start with initial concentrations $[X]_0$ and $[A]_0$ and $[Z]$ is the concentration of the intermediate complex, then $[X] = [X]_0 - [Z]$ and $[A] = [A]_0 - [Z]$. We can now obtain an expression that is only dependent on $[Z]$ and $[Y]$.

$$K_{eq} = \frac{[Z][Y]}{\left([X]_0 - [Z]\right) \cdot \left([A]_0 - [Z]\right)} \qquad (2.272)$$

If, for example, $[A]_0 \gg [X]_0$, then

$$[Z] = \frac{K_{eq}[X]_0[A]_0}{K_{eq}[A]_0 + [Y]} \qquad (2.273)$$

and the rate of product formation is

$$r = k'[Z][W] = k' \frac{K_{eq}[X]_0[A]_0}{K_{eq}[A]_0 + [Y]}[W] \qquad (2.274)$$

Atmospheric chemical reactions follow this rate expression when Y is absent and W is either O_2 or N_2 (see Example 2.36). In cases where $K_{eq}[A]_0 \gg [Y]$, $r = k'[X]_0[W]$ and the rate is linear in $[X]_0$ and independent of $[A]_0$. In many environmental systems where acid or base catalysis prevails, the condition of interest is $K_{eq}[A]_0 \ll [Y]$. The rate is then proportional to the first power in $[A]_0$.

A different reaction rate will ensue if we start with a high catalyst concentration, $[X]_0 \gg [A]_0$. We have then $[X] = [X]_0$, $[A] = [A]_0 - [Z]$, and hence, we have the following rate:

$$r = k' \frac{K_{eq}[X]_0[A]_0}{K_{eq}[X]_0 + [Y]} \qquad (2.275)$$

The catalyst concentration enters the rate expression in a distinctly nonlinear fashion.

If the second reaction (dissipation of the complex Z) is extremely fast, then the rate of dissipation can be handled using a pseudo-steady state approximation. Thus,

$$\frac{d[Z]}{dt} = 0 = k_f[X][A] - k_b[Z][Y] - k'[Z][W] \qquad (2.276)$$

Since $[X] = [X]_0 - [Z]$, $[A] = [A]_0 - [Z]$, $[Z]$ is small, and $[Z]^2$ is even smaller,

$$[Z] = \frac{k_f[X]_0[A]_0}{k_f([X]_0 + [A]_0) + k_b[Y] + k'[W]} \qquad (2.277)$$

and the rate is

$$r = \frac{k_f[X]_0[A]_0}{k_f([X]_0 + [A]_0) + k_b[Y] + k'[W]} \cdot k'[W] \qquad (2.278)$$

When either $[X]_0$ or $[A]_0$ is low, the rate is linear in both. At high concentrations, the rate is independent of both $[X]$ and $[A]$.

Now that we have seen how a catalyst affects the reaction rate, it is easy to understand how it affects the energy of the reaction. For instance, consider the decomposition of hydrogen peroxide in the aqueous phase. Under normal conditions, the activation energy is ~76 kJ mol^{-1}. This reduces to ~57 kJ mol^{-1} in the presence of a little bromide in the aqueous phase. The enhancement in rate is $\exp[(-57 + 76)/RT] = 2140$. Sometimes, $k' \gg k_b$, and hence the first energy barrier E_1 is rate controlling; this is termed an "Arrhenius complex" mechanism. In the second case, $k_b \gg k'$, and a second barrier controls the rate, which is termed the "van 't Hoff complex" mechanism. In either case, the catalyst provides an alternate route of low energy for the reaction to occur.

2.3.7.2 Homogeneous Catalysis

Homogeneous-catalyzed gas-phase reactions that do not involve chain reactions have no general mechanisms. An example of such a reaction in atmospheric chemistry is the combination of NO and Cl_2 catalyzed by Br_2 molecules. The reaction mechanism is

$$2NO + Br_2 \underset{k_b}{\overset{k_f}{\rightleftharpoons}} 2NOBr$$

$$2NOBr + Cl_2 \xrightarrow{k'} 2NOCl + Br_2 \qquad (2.279)$$

Bromine is regenerated in the process and is therefore the catalyst in this reaction. If the second reaction is rate controlling, the rate is given by

$$r = k'\frac{k_f}{k_b}[NO]^2[Cl_2][Br_2] \qquad (2.280)$$

The experimentally determined rate expression is in agreement with this equation.

Another gas phase homogeneous catalysis of significance in atmospheric chemistry is the decomposition of ozone in the upper atmosphere catalyzed by oxides of nitrogen and other chlorine-containing compounds such as Freon.

$$2O_3 \xrightarrow{\text{catalyst}} 3O_2 \qquad (2.281)$$

This reaction has serious consequences since ozone plays a significant role in moderating the UV light that reaches the earth. We shall discuss this reaction in detail in Chapter 4.

In the aqueous environment, and particularly in surface waters and soil/sediment pore waters, an important reaction is the hydrolysis of organic pollutants such as alkyl halide, ester, aromatic acid ester, amide, carbamate, etc. Many pesticides and herbicides are also hydrolysable. Hydrolysis plays an important role in deciding how nature tends to cleanse itself.

Mabey and Mill (1978) reviewed the environmental hydrolysis of several organic compounds. pH had the most profound effect on the hydrolysis rate. Both acids and bases were found to catalyze the reaction rates. The following discussion is on the specific acid-base catalysis where the acid is H^+ and the base is OH^-. If a catalyzed reaction is carried out at a high $[H^+]$ such that $[OH^-]$ is negligible, the rate of the reaction will be directly proportional to $[H^+]$ and $[A]$.

$$r = k'_H \left[H^+ \right] \left[A \right] \qquad (2.282)$$

For a constant pH, $[H^+]$ is constant, and therefore we have a pseudo-first-order reaction in $[A]$, the rate of which is given by

$$r = k_H \left[A \right] \qquad (2.283)$$

Similarly the rate of a base-catalyzed reaction is given by

$$r = k_{OH} \left[A \right] \qquad (2.284)$$

For an uncatalyzed reaction the rate is

$$r = k_0 \left[A \right] \qquad (2.285)$$

The overall rate of an acid-base catalyzed reaction is therefore

$$r = \left(k_0 + k_H + k_{OH} \right) \left[A \right] = k \left[A \right] \qquad (2.286)$$

The overall rate constant can also be written in terms of [H$^+$] and [OH$^-$] as follows:

$$k = k_0 + k'_H\left[H^+\right] + k'_{OH}\frac{K_w}{\left[H^+\right]} \qquad (2.287)$$

For general acid-base catalysis, we have

$$k = k_0 + k'_H\left[H^+\right] + k'_{OH}\frac{K_w}{\left[H^+\right]} + k'_{Acid}\left[Acid\right] + k'_{Base}\left[Base\right] \qquad (2.288)$$

The shape of the log k versus [H$^+$] curve will be as shown in Figure 2.36. If $k'_H\left[H^+\right]$ is large, then $k = k'_H\left[H^+\right]$ and the slope of the curve is -1. This is the left limb of the curve in Figure 2.36. The right limb has a slope of $+1$. In some cases, the transition region where k is independent of pH is not clear. This happens if $k_0 \ll \left(K_w k'_H k'_{OH}\right)$. The term I$_{AB}$ in the figure denotes the pH at which acid and base catalysis rates are equal. Hence, $I_{AB} = \left(1/2\right)\log\left(k'_H k'_{OH}/K_w\right)$. The rates of both acid- and base-catalyzed reactions are dependent on the substituent and active moieties on the reactant (Schwarzenbach et al., 1993).

The mechanisms of acid or base catalysis will depend upon the moiety W, as defined in the general reaction scheme earlier. W can be a solvent molecule, another base, or an acid in the aqueous phase. The catalyst X can be either an acid or a base. Hence, there are several permutations that should be contemplated in deriving rate equations for acid-base catalysis. Moore and Pearson (1981) enumerate nine such mechanisms. Table 2.18 lists these reaction mechanisms, the rate expressions, and some examples from environmental engineering.

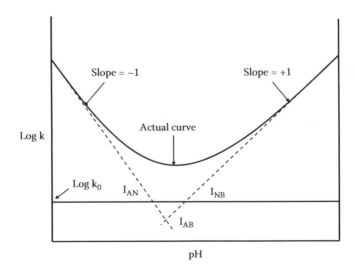

FIGURE 2.36 Variation in acid-base hydrolysis rate with pH for organic compounds in the environment.

An aspect of homogeneous catalysis that we have not contemplated, yet is the action of enzymes (X = enzyme). This is an important aspect of environmental bioengineering and is relegated to Chapter 4 where rates of enzyme reactions are considered.

Example 2.41   Obtaining k_0, k_A, and k_B from Rate Data

The reaction α glucose $\rightarrow \beta$ glucose is called mutarotation. Acids and bases catalyze it. At 291 K, the following first-order rate constants were obtained for the process using acetic acid in an aqueous solution containing 0.02 M sodium acetate.

[Acetid acid] (mol L^{-1})	0.02	0.105	0.199
k (min^{-1})	1.36×10^{-4}	1.40×10^{-4}	1.46×10^{-4}

In the general expression for k, both k_H and k_{OH} are negligible under these conditions. The term k_B is also negligible under these conditions. Hence, $k = k_0 + k_{Acid}[\text{Acid}]$. A plot of k versus [Acid] gives as intercept $k_0 = 1.35 \times 10^{-4}$ min^{-1}, and slope $k_{Acid} = 5.6 \times 10^{-5}$ L (mol min)$^{-1}$. The correlation coefficient is 0.992. This is a general method of obtaining k. For catalysis by different species, each k value can be isolated as shown here.

Example 2.42   Effect of Suspended Sediment (Soil) on the Homogeneously Catalyzed Hydrolysis of Organic Compounds

Since atmospheric moisture (e.g., fog water, rain) and the lake and river water contain suspended solids, the rate of hydrolysis of organics is likely to be influenced by them. This example will illustrate the effect for some organic compounds.

Many organic compounds that are common pollutants (e.g., pesticides—malathion, DDT) have low reactivity and are known to associate with colloidal matter that has high organic carbon content (see Chapter 4). These organic compounds generally do not enter chemical reactions with the colloid, which acts as an inert sink for organic pollutants. The concentration of a compound *truly* dissolved in water will be largely small if organic compounds adsorb to colloids (particulates). If the total amount of an organic in water is W_i and W_{ci} is the mass adsorbed from a solution containing ρ_s mass of a sorbent per unit volume of solution, then from the equations developed in Section 3.4.2, we have $\dfrac{W_{ci}}{W_i} = \dfrac{1}{1 + \rho_s K_{sw}}$.

There are two competing rates to consider: the organic adsorption rate to the sorbent and its hydrolysis rate in water. Usually, the rate at which the organic is adsorbed to the sorbent is relatively fast compared to the rate of hydrolysis. Hence, hydrolysis is the rate-limiting step. The rate constant for hydrolysis will be reduced by $(1 + \rho_s K_{sw})^{-1}$ since the concentration of a *truly* dissolved organic that can hydrolyze will be reduced to the same extent. Hence, the modified rate constant is $k^* = k(1 + \rho_s K_{sw})^{-1}$, where k is the hydrolysis rate constant without solids. Consider the hydrolysis of DDT. At 298 K, $k = 2.95 \times 10^{-8}$ s^{-1} at a pH of 8 (Wolfe et al., 1977). If K_{sw} for DDT on a sediment is 10^4 L kg^{-1}, then for a ρ_s of 10^{-5} g cm^{-3}, $k^* = 2.3 \times 10^{-8}$ s^{-1}, whereas for a ρ_s of 10^{-3} g cm^{-3}, $k^* = 2.3 \times 10^{-9}$ s^{-1}. The hydrolysis of DDT in slightly alkaline solutions is dramatically reduced in the presence of solid sorbents. These calculations are valid only if the sorbents do not participate as heterogeneous catalysts, as discussed in the next section.

2.3.7.3 Heterogeneous Catalysis

In the atmosphere, hydrosphere, and lithosphere, there are numerous reactions that occur at the surfaces of solids or liquids. These gas/solid, gas/liquid, and solid/liquid interface reactions are influenced by the nature and property of the surface. For example, the hydrolysis rates of organic esters and ethers are accelerated in the presence of sediment particles. The particle-mediated catalysis plays a large role in many atmospheric photochemical reactions. Many waste treatment processes also rely on reactions at surfaces. Examples are catalysts for air pollution control. Removal of volatile organics from automobile exhaust involves the use of sophisticated catalysts. Reactions such as the scrubbing of stack gases using solvents can be accelerated if the gaseous species reacts at the gas/liquid interface. Thus, the removal of ammonia by gas scrubbing is enhanced if the solution is slightly acidic.

2.3.7.4 General Mechanisms of Surface Catalysis

Reactions on surfaces differ greatly from those in bulk phases. A surface reaction involves a series of successive steps. These are shown schematically in Figure 2.37. The first step, viz., bulk phase diffusion, is generally fast. Hence, it is unlikely to be a rate-determining step, except for diffusion in solutions. It is difficult to differentiate between steps 2, 3, and 4. For example, a molecule can simultaneously react with the surface as it adsorbs. Hence, the entire process of adsorption, reaction, and desorption are usually considered a single rate-limiting step.

A heterogeneous surface reaction mechanism involves postulating a molecule (A) become adsorbed on a surface (X), which further becomes an activated complex (Z) that then breaks down to give the product (P). This is the basis of the "Langmuir-Hinshelwood" mechanism for heterogeneous surface reactions.

$$A + X \rightleftharpoons Z \rightarrow X + P \qquad (2.289)$$

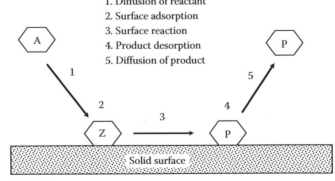

1. Diffusion of reactant
2. Surface adsorption
3. Surface reaction
4. Product desorption
5. Diffusion of product

FIGURE 2.37   Schematic of the steps in a heterogeneous reaction on the surface of a solid.

If two species A and B are involved, we have the following scheme:

$$A + X \rightleftharpoons Z_1$$
$$B + X \rightleftharpoons Z_2 \qquad (2.290)$$
$$Z_1 + Z_2 \rightarrow X + P$$

This scheme requires that two species be adsorbed on adjacent surface sites. In some cases, only one (say B) gets adsorbed, which then reacts with a gaseous species (say A) to give the products. This is the "Langmuir-Rideal" mechanism.

For the Langmuir-Hinshelwood mechanism, the rate of the reaction depends on the surface concentration of A. The Langmuir isotherm for adsorption from the gas phase gives the surface coverage of A.

$$\theta_A = \frac{K_{Lang,A} P_A}{1 + K_{Lang,A} P_A} \qquad (2.291)$$

where θ_A is given by the ratio S/S_0; S_0 being the total number of binding sites available on the surface. The rate of conversion of the adsorbed complex to products is

$$r = k\theta_A = k \frac{K_{Lang,A} P_A}{1 + K_{Lang,A} P_A} \qquad (2.292)$$

At high pressures, $K_{Lang,A} P_A \gg 1$, $r \rightarrow k$ and is independent of the concentration of A. At low pressures, $K_{Lang,A} P_A \ll 1$ and $r \rightarrow k K_{Lang,A} P_A$ and the rate is first order in A.

This formalism also allows the elucidation of reaction rates when more than one species is involved in the reaction. The Langmuir isotherm for two competing species A and B gives

$$\theta_A = \frac{K_{Lang,A} P_A}{1 + K_{Lang,A} P_A + K_{Lang,B} P_B}$$
$$\theta_B = \frac{K_{Lang,B} P_B}{1 + K_{Lang,A} P_A + K_{Lang,B} P_B} \qquad (2.293)$$

If the heterogeneous catalysis is bimolecular involving both adsorbed species A and B (Langmuir-Hinshelwood mechanism), we have the following overall rate:

$$r = k\theta_A\theta_B = k \cdot \frac{K_{Lang,A} K_{Lang,B} P_A P_B}{\left(1 + K_{Lang,A} P_A + K_{Lang,B} P_B\right)^2} \qquad (2.294)$$

Note that this equation indicates the competition for surface sites between A and B. As a consequence, if P_A is held constant, the rate will go through a maximum as P_B is varied.

If the heterogeneous catalysis is bimolecular involving adsorbed B reacting with the gas phase species A (Langmuir-Rideal mechanism), we have the following rate:

$$r = k\theta_B P_A = k\,\frac{K_{Lang,B}P_A P_B}{\left(1 + K_{Lang,A}P_A + K_{Lang,B}P_B\right)} \qquad (2.295)$$

By replacing pressure P_A with the aqueous concentration $[A]$, we obtain the rate of heterogeneous surface catalysis in solutions.

Let us now consider the energetics of a heterogeneous catalysis. Consider the rate of the unimolecular reaction at low pressures:

$$r = kK_{Lang,A}P_A = k'P_A \qquad (2.296)$$

where k' is the overall first-order rate constant. Note that the first-order rate constant k varies with T according to the Arrhenius expression:

$$\frac{d\ln k}{dT} = \frac{E_a}{RT^2} \qquad (2.297)$$

Similarly, as we discussed in Section 2.2.5.3, the Langmuir adsorption constant $K_{Lang,A}$ also has a relationship with T:

$$\frac{d\ln K_{Lang,A}}{dT} = -\frac{q_{ads}}{RT^2} \qquad (2.298)$$

where q_{ads} is the heat evolved during adsorption. Hence,

$$\frac{d\ln k'}{dT} = \frac{\left(E_a - q_{ads}\right)}{RT^2} = \frac{E_a'}{RT^2} \qquad (2.299)$$

The "true activation energy" E_a' is smaller than E_a by the quantity q_{ads}, as shown schematically in Figure 2.38. For the case of high pressures, E_a is the same as E_a'. At low P_A, only few A are adsorbed and these need an energy $E_a - q_{ads}$ to cross the barrier. At high P_A, most of A is on the surface and the system needs to overcome a larger barrier to form products. Most environmental reactions in the natural environment fall in the category of low pressure or concentration reactions and have potential energy diagrams, such as those depicted in Figure 2.38.

Example 2.43   Heterogeneous Catalysis of Ester Hydrolysis in Water

An example of heterogeneous catalysis in the environment is the metal oxide (mineral surface) catalyzed hydrolysis of esters. Stone (1989) discussed an environmentally important reaction, namely, the influence of alumina on the base hydrolysis of monophenyl terephthalate (MPT) in aqueous solution. Phthalate esters are ever-present pollutants in wastewater and atmospheric moisture.

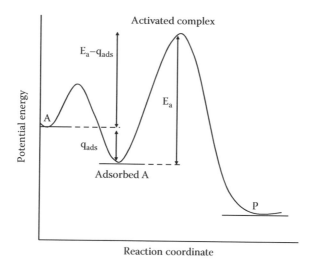

FIGURE 2.38 Potential energy surface for a heterogeneous reaction where the pressure of the reactant in the gas phase is low.

The hydrolysis in a homogeneous system without alumina follows the reaction: $MPT^- + OH^- \rightarrow PhT + PhOH$, where PhT is phenyl phthalate and PhOh is phenol. The rate is given by

$$r = -\frac{d\left[MPT^-\right]}{dt} = k_{base}\left[OH^-\right]\left[MPT^-\right]$$

with the base hydrolysis rate constant $k_{base} = 0.241$ L (mol s)$^{-1}$ at 298 K in buffered solutions (pH 7.6–9.4). At pH > $pK_a = 3.4$, MPT exists mostly as MPT^-, which is the species of interest throughout the pH range of hydrolysis. The addition of a small concentration of a heterogeneous surface (alumina) increases the rate of hydrolysis substantially (Figure 2.39). The rate increased with increasing pH and $[Al_2O_3]$, whereas ionic strength adversely affected the rate. It was shown that the adsorption of MPT^- on alumina surface enhanced the hydrolysis rate, thus providing an additional pathway for the reaction. The overall reaction rate can be written as

$$r = -\frac{d\left[MPT^-\right]}{dt} = k_{base}\left[OH^-\right]\left[MPT^-\right] + k_r K_{ads}\Gamma_{OH^-}\Gamma_{MPT^-} \qquad (2.300)$$

where
  K_{ads} is the adsorbed complex formation constant for MPT^- on alumina
  Γ_{OH^-} is the OH^- concentration in the diffuse layer
  Γ_{MPT^-} is the adsorbed MPT^- concentration (obtained from the Poisson-Boltzmann equation)

$$\Gamma_{OH^-}\Gamma_{MPT^-} = \left[OH^-\right]\left[MPT^-\right] \cdot e^{\left(\frac{F(\psi_s + \psi_d)}{RT}\right)} \qquad (2.301)$$

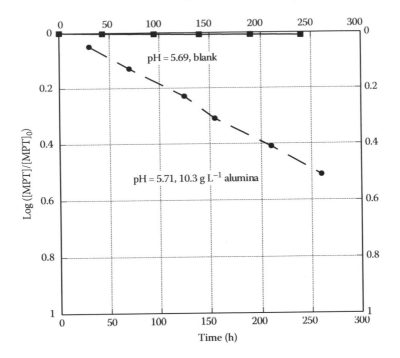

FIGURE 2.39 Heterogeneous catalysis of monophenyl terephthalate (MPT) by alumina in the aqueous phase. Square symbol represents particle-free solution and circular symbol represents particle-laden solution. The reaction was conducted in 0.003 M acetate buffer. (Data from Stone, A.T., *J. Colloid Interface Sci.*, 127, 429, 1989.)

Utilizing this equation, we have the overall rate constant for hydrolysis k_H^*.

$$k_H^* = \left[k_\Gamma K_{ads} e^{\left(\frac{F(\psi_s + \psi_d)}{RT} \right)} + k_{base} \right] \left[OH^- \right] \qquad (2.302)$$

Thus, k_H^* is substantially larger than k_{base}. At a constant pH, the exponential term decreases with ionic strength. Similarly, due to competition from other ions in solution, K_{ads} for MPT^- also decreases. Thus, k_H^* will decrease with increasing I. This equation also predicts a maximum in the rate with pH. With increasing pH, Γ_{OH^-} increases and Γ_{MPT^-} decreases. These opposing effects should cancel each other at some pH where the maximum in rate is observed.

2.3.7.5 Autocatalysis in Environmental Systems

In some environmental chemical reactions, some of the products of the reaction act as catalysts. A bimolecular autocatalysis reaction is represented as A + B → 2B.

The rate expression is $r = -d[A]/dt = k[A][B]$. Let us define the progress of the reaction by ξî. If at any time $[A]_0 - [A] = [B] - [B]_0$ such that $[B] = [A]_0 + [B]_0 - [A] = [A]_0 + [B]_0 - \xi$,

$$-\frac{d\xi}{dt} = k\xi \left(\left[A \right]_0 + \left[B \right]_0 - \xi \right) \qquad (2.303)$$

This equation can be integrated to obtain

$$\left[B \right] = \frac{\left[A \right]_0 + \left[B \right]_0}{1 + \dfrac{\left[A \right]_0}{\left[B \right]_0} \cdot e^{-k\left(\left[A \right]_0 + \left[B \right]_0 \right)t}} \qquad (2.304)$$

Note from this that at $t = 0$, $[B] = [B]_0$. For $[B]_0$, the only condition is that $[B] = 0$ for all t. However, if $[B]_0 \neq 0$ at $t = 0$, $[B]$ will slowly increase; this is termed the "induction period." The value of $[B]$ increases continuously and reaches its maximum value of $[A]_0 + [B]_0$ as $t \to \infty$. The characteristic S-shaped curve shown in Figure 2.40 is characteristic of autocatalytic reactions in the environment. Oscillatory reactions such as the oxidation of malonic acid by bromate and catalyzed by cerium ions (otherwise called the "Belousov-Zhabotinsky reaction") are classic examples of autocatalysis.

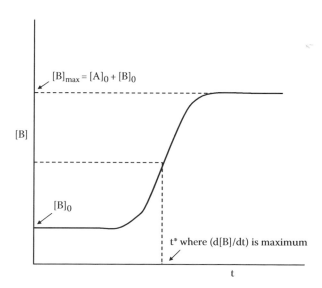

FIGURE 2.40 General shape of a concentration profile for an autocatalytic reaction.

Example 2.44 Autocatalysis in Natural Waters–Mn (II) Oxidation

The oxidation of Mn(II) in aqueous solutions is known to be base catalyzed

$$Mn(II) \xrightarrow[O_2]{OH^-} MnO_x(s) \tag{2.305}$$

with the rate law

$$r_1 = -\frac{d[Mn(II)]}{dt} = k'[OH^-]^2 P_{O_2}[Mn(II)] \tag{2.306}$$

At constant pH and P_{O_2}, the reaction follows pseudo-first-order kinetics with a rate constant, $k = k'[OH^-]^2 P_{O_2}$. Thus,

$$r_1 = k[Mn(II)] \tag{2.307}$$

The oxidation of Mn(II) gives MnO_x in the solution, which acts as an auto-catalyzing agent.

$$MnO_x(s) + Mn(II) \xrightarrow{O_2} 2MnO_x(s) \tag{2.308}$$

The rate of this heterogeneous base catalysis is

$$r_2 = -\frac{d[Mn(II)]}{dt} = k'' P_{O_2} \frac{K_{ads}}{[H^+]}[Mn(II)][MnO_x] \tag{2.309}$$

where $[MnO_x] = [Mn(II)]_0 - [Mn(II)]$, and K_{ads} is the adsorption constant for Mn(II) on the MnO_x catalyst. At constant pH and $[O_2]$, the rate is

$$r_2 = k^*[Mn(II)][MnO_x] \tag{2.310}$$

The overall rate of the auto-catalyzed reaction is

$$r = -\frac{d[Mn(II)]}{dt} = (k + k^*[MnO_x])[Mn(II)] \tag{2.311}$$

Upon integration, using the boundary condition $[MnO_x]_0 = 0$, we get

$$\frac{[MnO_x]}{([Mn(II)]_0 - [MnO_x])} = \frac{k}{(k^*[Mn(II)]_0 + k)} \cdot \left[e^{\left\{ (k + k^*[Mn(II)]_0)t \right\}} - 1 \right] \tag{2.312}$$

If $k^*[Mn(II)]_0 > k$, we can approximate this equation and rearrange to obtain

$$\ln\left(\frac{[MnO_x]}{[Mn(II)]_0 - [MnO_x]} \right) = \ln\left(\frac{k}{k^*[Mn(II)]_0} \right) + \left[k + k^*[Mn(II)]_0 \right] \cdot t \tag{2.313}$$

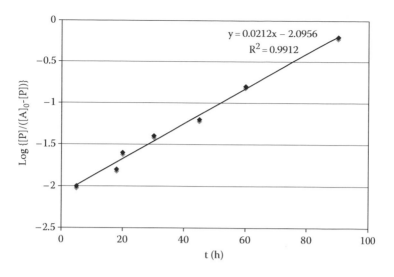

FIGURE 2.41   Kinetics of autooxidation of Mn(II) in alkaline solutions at pH = 9.8 at 298 K. $[A]_0 = 8 \times 10^{-5}$ M; $P_{O_2} = 1$ atm. (Data from Stumm, W. and Morgan, J.J., *Aquatic Chemistry*, 4th edn, New York: John Wiley and Sons, Inc, 1996.)

Thus, a plot of $\ln\left[\dfrac{[MnO_x]}{[Mn(II)]_0 - [MnO_x]}\right]$ versus t will give, at sufficiently large t values, a straight line slope of k ı k*[Mn(II)]_0 and an intercept of ln (k/k*[Mn(II)]_0), from which the values of k and k* can be ascertained. The data for Mn(II) oxidation at a constant partial pressure of oxygen and different pH values were reported by Morgan and Stumm (1964), and are plotted in Figure 2.41. The linear fit to the data shows the appropriateness of the rate mechanism given earlier.

2.3.8   REDOX REACTIONS IN ENVIRONMENTAL SYSTEMS

The transfer of electrons (e^-) between compounds is an important aspect of several reactions in the environment. If a compound accepts an e^- it is said to undergo reduction, and the process of donation of an e^- is called "oxidation." These come under the umbrella term "redox reactions." The study of redox systems is the central aim of the discipline called "electrochemistry," which is a specialized branch of "physical chemistry." Most photo-assisted and biochemical degradation of organic compounds in the natural environment and hazardous waste treatment processes include e^- mediation. Redox reactions are also prevalent in the dark and hence can occur in the subsurface environment as well. As a result of redox processes, remarkable differences in the properties of surficial and deep sediment layers are observed. Since the surface sediment is in the aerobic (oxygen-rich) zone, it is easily oxidized. This layer contains several important species such as O_2, CO_2, SO_4^{2-}, NO_3^-, and oxides of iron. In the aerobic zone, most substances are rapidly oxidized. The redox potential (as shall be discussed in the subsequent paragraphs) is always positive in the aerobic zone and ranges from +0.4 to +0.1 V. The deeper sediments are anaerobic and have redox potentials between −1 and −2.5 V. It contains predominantly reduced species such as H_2S, NH_3, CH_4, and several organic compounds. The "redox potential discontinuity" (RPD) zone separates the aerobic and anaerobic zones. The RPD zone, where there is a decrease in redox potential from +1 to −1 V, has a very short depth. Processes occur at different rates in these zones and are mediated by the biota that inhabits the area. The microbes that are present in both aerobic and anaerobic zones can act as intermediaries in e^- exchange between compounds.

Redox processes are generally slower than most other chemical reactions, and systems involving them are in disequilibrium. Microbial-mediated e^- exchange plays a large role in the bioremediation of contaminated groundwater aquifers.

The transfer of e^- can be understood via an inventory of the so-called "oxidation states" of reactants and products. The oxidation state of an atom in a molecule is the charge associated with that atom if the ion or molecule were to be dissociated. The oxidation state of a monoatomic species is its electron charge. The sum of oxidation states is zero for a molecule, but for an ion it is equal to its charge. In the natural environment, compounds that undergo redox reactions are comprised of those that have C, N, or S atoms. For redox reactions to occur in the environment, there has to be a source and a sink for e^- in the system. It has been shown through both laboratory and filed observations that even the most recalcitrant (refractory) organic compounds can undergo redox reactions. Let us take a simple example, viz., chloroform that undergoes a transformation to methylene chloride as follows:

$$CHCl_3 + H^+ + 2e^- \rightarrow CH_2Cl_2 + Cl^- \qquad (2.314)$$

In the reactant $CHCl_3$, the oxidation state of C is +2. The additions of H^+ and the removal of Cl^- reduce the oxidation state of C to 0 in the product (CH_2Cl_2). Since a reduction in oxidation state and the release of a Cl species have been made simultaneously it is called a "reductive dechlorination" process. In complex environmental matrices, it is often difficult to clearly delineate the exact source or sink for e^-.

Formally there is an analogy between the transfer of e^- in a redox process and the transfer of H^+ in an acid-base reaction. It is useful to first understand how the "electron activity" in solutions is represented. A "redox reaction" can be generally represented as the sum of two "half-cell reactions," one where the e^- is accepted (reduction) and the other where it is donated (oxidation). For example, a redox reaction involving the oxidation of Zn by Cu^{2+} can be represented as

$$
\begin{aligned}
Cu^{2+}(aq) + 2e^- &\rightarrow Cu(s) \quad \{Reduction\} \\
Zn(s) &\rightarrow Zn^{2+}(aq) + 2e^- \quad \{Oxidation\} \\
Cu^{2+}(aq) + Zn(s) &\rightarrow Cu(s) + Zn^{2+}(aq) \quad \{Redox\}
\end{aligned}
\qquad (2.315)
$$

It is useful to represent each as a reduction reaction and then the overall process is the difference between the two:

$$
\begin{aligned}
Cu^{2+}(aq) + 2e^- &\rightarrow Cu(s) \quad \{Reduction\} \\
Zn^{2+}(aq) + 2e^- &\rightarrow Zn(s) \quad \{Reduction\} \\
Cu^{2+}(aq) + Zn(s) &\rightarrow Cu(s) + Zn^{2+}(aq) \quad \{Redox\}
\end{aligned}
\qquad (2.316)
$$

Each half reaction is denoted as Ox/Red and is represented by $Ox + ne^- \rightarrow Red$. The equilibrium quotient for the reaction is $K = \dfrac{\left[Red\right]}{\left[Ox\right]\left[e^-\right]^n}$.

Analogous to the definition of H^+ activity, $pH = -\log [H^+]$, we can define the "electron activity" $[e^-]$ using $pe = -\log [e^-]$. Thus, for this redox reaction,

$$pe = pe^0 - \frac{1}{n} \cdot \log\left(\frac{\left[Red\right]}{\left[Ox\right]}\right) = pe^0 + \frac{1}{n} \cdot \log\left(\frac{\left[Ox\right]}{\left[Red\right]}\right) \tag{2.317}$$

where $pe^0 = (1/n) \log K$ is the electron activity at unit activities of [Red] and [Ox] species. This definition is generally applicable to any reaction involving e^-, such as $\sum_i \nu_i A_i + ne^- = 0$, where ν_i is the stoichiometric coefficient (positive for reactants, negative for products), and for which we have

$$pe = pe^0 + \frac{1}{n} \cdot \log\left(\prod_i (A_i)^{\nu_i}\right) \tag{2.318}$$

Example 2.45   pe of Rainwater

Neglecting all other ions, we have the following equation driving the pe of rainwater: $\frac{1}{2}O_2(g) + 2H^+ + 2e^- \rightarrow H_2O(l)$. Since the activity of pure water is 1, we can write $pe = pe^0 + \frac{1}{2}\log\left[P_{O_2}^{0.5}\left[H^+\right]^2\right]$, where $pe^0 = (1/2)\log K$ with $K = \dfrac{1}{P_{O_2}^{0.5}\left[H^+\right]^2\left[e^-\right]^2} = 10^{41}$. If atmospheric partial pressure of O_2 (= 0.21 atm) is used and a pH of −5.6 in rainwater is considered, pe = 14.7.

Pe value can be obtained experimentally from the *electrode potential of a redox reaction*, E_H (in volts). The subscript H denotes that it is measured on a hydrogen scale. Consider the redox reaction, $Ox + ne^- \rightarrow Red$. The Gibbs function for the reaction is given by $\Delta G = -nFE_H$, which is the work required to move n electrons from the anode to the cathode of an electrochemical cell. E_H is the electrode potential (Volts) on a hydrogen scale, that is, assuming the reduction of H^+ to H_2 has a zero reduction potential. F is the Faraday constant (96,485 Coulombs mol^{-1}). The standard Gibbs energy ΔG^0 is defined as equal to $-nFE_H^0$. Since we know that $pe = pe^0 + (1/n) \log ([Ox]/[Red])$, and $pe^0 = (1/n) \log K = -\Delta G^0/2.303RT$, we have

$$-\Delta G = -\Delta G^0 + \frac{2.303RT}{n} \cdot \log \frac{\left[Ox\right]}{\left[Red\right]} \tag{2.319}$$

TABLE 2.20
Equilibrium and Rate Constants for SiO$_2$ Dissolution Reactions at 298 K

Type	ln K$_{eq}$	ln k$_d$ (s^{-1})	ln k$_p$ (s^{-1})
Quartz	−9.11	−30.81	−21.70
α-Cristobalite	−7.71	−29.41	−21.70
β-Cristobalite	−6.72	−28.44	−21.70
Amorphous silica	−6.25	−27.61	−21.40

Sources: Stumm, W. and Morgan, J.J., *Aquatic Chemistry*, 4th edn., John Wiley & Sons, Inc., New York, 1981; Brezonik, 1994.

or

$$E_H = E_H^0 + \frac{2.303RT}{n} \cdot \log\frac{\left[Ox\right]}{\left[Red\right]} \tag{2.320}$$

which is called the "Nernst equation" for electrochemical cells (Table 2.20). It is also apparent from these expressions that

$$pe = \frac{FE_H}{2.303RT} \tag{2.321}$$

Thus, pe is obtained from E_H, which is easily measured using electrochemical cells. Some examples of E_H values for redox reactions of environmental interest are given in Table 2.21.

For redox equilibria in natural waters, it is convenient to assign activities of H$^+$ and OH$^-$ values that are applicable to neutral water. The pe^0 values relative to E_H^0 are now designated pe^0(w) relative to $E_H^0\left(w\right)$. The two are related through the ion product of water, $pe^0\left(w\right) = pe^0 + \frac{n_H}{2}\log K_w$, where n_H denotes the number of protons exchanged per mole of electrons. This type of characterization of pe^0 allows one to grade the oxidizing capacity of ions at a specified pH (viz., that of neutral water, 7). Thus, a compound of higher pe^0(w) will oxidize one with a lower pe^0(w).

Example 2.46   Calculation of E_H^0 and pe^0(w) for a Half-Cell Reaction

Consider the half-cell reaction, $SO_4^{2-} + 9H^+ + 8e^- \rightleftharpoons HS^- + 4H_2O$. The standard free energy of the reaction is $\Delta G^0 = \Delta G_H^0 = G_f^0\left(HS^-, aq\right) + 4G_f^0\left(H_2O\right) - G_f^0\left(SO_4^{2-}, aq\right)$, since both H$^+$ and e$^-$ have zero G_f^0 by convention. Therefore, $\Delta G_H^0 = 12 + 4\left(-237\right) - \left(-744\right) = -192$ kJ mol^{-1}. Hence, $E_H^0 = \Delta G_H^0/nF = \left(192 \times 1,000\right)/\left(8 \times 96,485\right) = +0.24$ V. To obtain E_H at any other pH (say 7), we can use the Nernst equation to obtain E_H at a pH of 7, $E_H = E_H^0 + \frac{0.059}{8}\log\frac{\left[SO_4^{2-}\right]\left[H^+\right]^9}{\left[HS^-\right]\left[H_2O\right]^6}$.

If all species are at their standard states of unit activities, except [H$^+$] = 10^{-7} M, we have $E_H^0\left(w\right) = +0.24 + \left(0.0074\right)\log\left[10^{-7}\right]^9 = -0.22$ V. Hence, $pe^\sigma\left(w\right) = E_H^0\left(w\right)/0.059 = -3.73$.

TABLE 2.21
E_H^0 and pe^0 Values for Selected Reactions of Environmental Significance

Reaction	E_H^0 (V)	pe^0
$O_2(g)+4H^++4e^- \rightleftharpoons 2H_2O$	+1.22	+20.62
$Fe^{3+}+e^- \rightleftharpoons Fe^{2+}$	+0.77	+13.01
$2NO_3^-+12H^++10e^- \rightleftharpoons N_2(g)+6H_2O$	+1.24	+20.96
$CHCl_3+H^++2e^- \rightleftharpoons CH_2Cl_2+Cl^-$	+0.97	+16.44
$CH_3OH+2H^++2e^- \rightleftharpoons CH_4(g)+H_2O$	+0.58	+9.88
$S(s)+2H^++2e^- \rightleftharpoons H_2S(aq)$	+0.17	+2.89
$\alpha\text{-}FeOOH\ (s)+HCO_3^-+2H^++e^- \rightleftharpoons FeCO_3(s)+2H_2O$	−0.04	−0.80
$SO_4^{2-}+10H^++4e^- \rightleftharpoons H_2S(g)+H_2O$	+0.31	+5.25
$SO_4^{2-}+9H^++8e^- \rightleftharpoons HS^-+4H_2O$	+0.25	+4.25
$N_2(g)+8H^++6e^- \rightleftharpoons 2NH_4^+$	+0.27	+4.68
$2H^++2e^- \rightleftharpoons H_2(g)$	0.00	0.00
$6CO_2(g)+24H^++24e^- \rightleftharpoons C_6H_{12}O_6+6H_2O$	−0.01	−0.17
$CO_2(g)+4H^++4e^- \rightleftharpoons CH_2O+H_2O$	−0.07	−1.20
$CO_2(g)+H^++2e^- \rightleftharpoons HCOO^-$	−0.28	−4.83

Since the transfer of protons (acid-base reactions) and electrons (redox reactions) in environmental processes is closely linked, it is convenient to relate the pe and pH of the reactions. It can provide information when either of these reactions predominates. The following discussion is a brief annotation of the salient aspects of pe-pH diagrams. The reader is referred to Stumm and Morgan (1996) for further details.

2.3.8.1 Rates of Redox Reactions

The rate law for a general redox reaction of the form $A_{ox}+B_{red} \rightarrow A_{red}+B_{ox}$ is given by $r = k_{AB}[A_{ox}][B_{red}]$. As an example, let us consider the homogeneous oxidation of Fe(II) in water at a given partial pressure of O_2 above the solution. The rate law is

$$r = -\frac{d\left[Fe(II)\right]}{dt} = k\left[Fe(II)\right]P_{O_2} \tag{2.322}$$

For fixed P_{O_2}, the rate is $r = k_{obs}[Fe(II)]$, where k_{obs} is a pseudo-first-order rate constant. It has been shown that the rate derives contributions from three species of Fe (II), namely, Fe^{2+}, $FeOH^+$, and $Fe(OH)_2$ such that

$$k_{obs}\left[Fe(II)\right] = k_0\left[Fe^{2+}\right]+k_1\left[FeOH^+\right]+k_2\left[Fe(OH)_2\right] \tag{2.323}$$

For a general metal-ligand addition reaction proceeding as follows

$$M+L \rightleftharpoons ML+L \rightleftharpoons ML_2 \cdots \rightleftharpoons ML_n \tag{2.324}$$

we can define an "equilibrium constant" for each step

$$K_{eq,i} = \frac{[ML_i]}{[ML_{i-1}][L]} \tag{2.325}$$

and a "stability constant" for each complex

$$\beta_i = \frac{[ML_i]}{[M][L]^i} \tag{2.326}$$

Using these definitions one can write for the Fe(II) system discussed here (see Morel and Hering, 1993 for details).

$$k_{obs} = \frac{k_0 + \left(k_1 \dfrac{K_1 K_w}{[H^+]} \right) + \left(k_2 \dfrac{\beta_2 K_w^2}{[H^+]^2} \right)}{\left(1 + \dfrac{K_1 K_w}{[H^+]} + \dfrac{\beta_2 K_w^2}{[H^+]^2} \right)} \tag{2.327}$$

Typical values of k_0, k_1, and k_2 at 298 K are 1×10^{-8} s^{-1}, 3.2×10^{-2} s^{-1}, and 1×10^4 s^{-1}, respectively. Since $k_0 \ll k_1 \ll k_2$, we can see why predominantly hydroxo species of Fe(II) are formed. Competition for ligands (e.g., Cl$^-$, SO$_4^{2-}$) will significantly lower the rate of oxidation. Therefore, iron oxidation is far more favorable in freshwaters than in open marine systems. Reported rate constant in marine systems is a hundred times lower than in freshwater systems.

The redox reaction, such as discussed here, proceeds with a transfer of electrons between molecules. There are two known mechanisms of electron transfer between the oxidant and the reductant. Both involve the formation of an activated complex. The distinguishing feature is the type of activated complex. In the first variety, the hydration shells of the two ions interpenetrate each other sharing a common solvent molecule. This is called the "inner sphere (IS) complex." In the second type, the two hydration shells are separated by one or more solvent molecules, this being termed the "outer sphere (OS) complex." In the OS complex, the solvent will mediate the electron transfer. The two mechanisms can be distinguished from one another since, in the IS case, the rates will depend on the type of ligand forming the bridged complex. Electron transfer reactions are generally slower than other reactions because of the rearrangements and orientations of the solvent molecule required for the reaction to proceed.

A formal theory of e$^-$ transfer in reactions is provided by the well-known "Marcus theory." The basic tenet of the theory is that the overall free energy of activation for electron transfer is made up of three components (Marcus, 1963). The first one involves the electrostatic potential ($\Delta G_{elec}^\dagger$) required to bring two ions together. The second one is the free energy for restructuring the solvent around each ion ($\Delta G_{solv}^\dagger$).

The last and final term arises from the distortions in bonds between ligands in products and reactants ($\Delta G^{\dagger}_{lig}$).

$$\Delta G^{\dagger}_{AB} = \Delta G^{\dagger}_{elec} + \Delta G^{\dagger}_{solv} + \Delta G^{\dagger}_{lig} \qquad (2.328)$$

Marcus derived expressions for the free energy terms and combined them to give the overall free energy. This resulted in the following equation for the rate of electron transfer:

$$k_{AB} = \kappa \frac{kT}{h} \cdot e^{-\frac{\Delta G^{\dagger}_{AB}}{RT}} \qquad (2.329)$$

Marcus also derived the following relationship between individual ionic interactions relating k_{AB} to k_{AA} and k_{BB}:

$$k_{AB} = \left(k_{AA} k_{BB} K_{AB} f \right)^{1/2} \qquad (2.330)$$

where $\ln f = (1/4) (\ln K_{AB})^2 / \ln (k_{AA} k_{BB} / Z^2)$. Z is the collision frequency between uncharged A and B ($\approx 10^{12}$ L (mol s)$^{-1}$). From the expression for k_{AB}, we can obtain

$$\Delta G^{\dagger}_{AB} = \frac{1}{2} \cdot \left(\Delta G^{\dagger}_{AA} + \Delta G^{\dagger}_{BB} + \Delta G^{0}_{AB} \right) - \frac{1}{2} \cdot RT \ln f \qquad (2.331)$$

For most reactions, $\Delta G^{0}_{AB} \sim 0$. Note that k_{AA} and k_{BB} are so defined that they represent the following redox reactions:

$$\begin{aligned} A_{ox} + A^{*}_{red} &\rightleftharpoons A^{*}_{red} + A_{ox} \\ B_{ox} + B^{*}_{red} &\rightleftharpoons B^{*}_{red} + B_{ox} \end{aligned} \qquad (2.332)$$

These equations represent "self-exchange reactions." The overall redox process has K_{AB} as the equilibrium constant.

$$K_{AB} = \frac{\left[A_{red} \right]\left[B_{ox} \right]}{\left[A \right]_{ox}\left[B_{red} \right]} \qquad (2.333)$$

Since for one-electron exchange reactions we have already seen that

$$\ln K_{AB} = \frac{\left(E^{0}_{HA} - E^{0}_{HB} \right)}{0.059} = pe^{0}_{A} - pe^{0}_{B} \qquad (2.334)$$

we can compare the equilibrium constant calculated using Marcus theory to that determined experimentally. Good agreement is generally observed between the observed and predicted values, lending validity to the Marcus relationship.

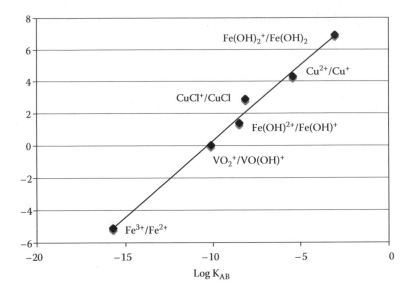

FIGURE 2.42 Marcus free energy relationships for the oxidation of several ions of environmental interest. (Data taken from Wehrli, B., Redox reactions of metal ions at mineral surfaces, in: Stumm, W. (ed.), *Aquatic Chemical Kinetics*, John Wiley & Sons, Inc., New York, 1990, pp. 311–336.)

The limiting case of the Marcus relationship leads to a simple LFER between k_{AB} and K_{AB} for outer sphere electron transfer reactions of the type $A_{ox}L + B_{red} \rightarrow A_{red}L + B_{ox}$. If a plot of $\ln k_{AB}$ versus $\ln K_{AB}$ is made, a linear relationship with a slope of 0.5 is observed, provided $f \approx 1$ and ΔG^0_{AB} is small, representing near-equilibrium conditions. For endergonic reactions, the slope is ≈ 1. Several examples of such relationships in environmental reactions have been established (Wehrli, 1990). Several autooxidation reactions of interest in environmental science were considered by Wehrli (1990) and data tabulated for both k_{AB} and K_{AB} for reactions of the type $A_{red} + O_2 \rightarrow A_{ox} + O_2^-$. Figure 2.42 is a plot showing the unit slope for the LFER involving different redox couples.

It is appropriate at this stage to summarize the various LFERs that we have discussed thus far for the prediction of rate constants and equilibrium constants in a variety of contexts. These are compiled in Table 2.22. Armed with a knowledge of these categories of LFERs, one should be able to predict the rates of many common environmental reactions and/or the equilibrium constants for various partitioning processes. This can be especially useful if only few values are available within a group.

2.3.9 ENVIRONMENTAL PHOTOCHEMICAL REACTIONS

Solar radiation is the most abundant form of energy on earth. All regions of the sun's spectrum (UV, visible, and infrared) reach the earth's atmosphere. However, only a small fraction of it is absorbed by water or land, while the rest

TABLE 2.22
Summary of Linear Free Energy Relationships in Environmental Engineering

Type	Application	Special Feature
Kinetic rate constants:		
Brönsted	Acid and base catalysis, hydrolysis, association and dissociation reactions	Log k_a or k_b related to log K_a or K_b
Hammet σ	p- and m-substituted aromatic hydrolysis, enzyme catalysis	Substituent effects on organic reactions
Taft σ^*	Hydrolysis and other reactions for aliphatic compounds	Steric effects on substituents
Marcus	Electron transfer (outer sphere), metal ion auto oxidations	Log k_{AB} related to log K_{AB}
Equilibrium partition constants:		
$C_i^* - K_{ow}$	Aqueous solubility	Log C_i^* related to log K_{ow}
$K_{oc} - K_{ow}$	Soil-water partition constant	Log K_{oc} related to log K_{ow}
$K_{BW} - K_{ow}$	Bioconcentration factor	Log K_{BW} related to log K_{ow}
$K_{mic} - K_{ow}$	Solute solubility in surfactant micelles	Log K_{mic} related to log K_{ow}

is reflected, absorbed, or dissipated as heat. Several reactions are initiated in the natural environment because of the absorbed radiation. The term "photochemistry" refers to these transformations. Photochemistry is basic to the world we live in. Plants depend on solar energy for photosynthesis. Photolytic bacteria derive solar energy for conversion of organic molecules to other products. Absorption of photoenergy by organic carbon in natural waters leads to the development of color in lakes and rivers. Redox reactions in the aquatic environment are initiated by absorption of light energy. Reactive free radicals ($OH^\cdot$, $NO_3^\cdot$, $HO_2^\cdot$) are formed by photochemical reactions in the atmosphere where they react with other species.

There are two fundamental laws in photochemistry:

1. *Grotthuss-Draper law*: Only light absorbed by a system can cause chemical transformations.
2. *Stark-Einstein law*: Only one quantum of light is absorbed per molecule of absorbing and reacting species that disappear.

There are two important laws that are derived from the Grotthuss-Draper law. The first one is called "Lambert's law," which states that the fraction of incident radiation absorbed by a transparent medium is independent of the intensity of the incident light, and that successive layers in the medium absorb the same fraction of incident light. The second is called "Beer's law," which states that absorption of incident light is directly proportional to the concentration of the absorbing species in the medium. By combining the two laws we can express the ratio of change in absorption to the total incident radiation as follows:

$$\frac{dI}{I} = \alpha_v C_i dz \qquad (2.335)$$

This equation is called the "Beer-Lambert law." α_v is proportionality constant (m^2 mol^{-1}). dz is the increment in thickness of the medium perpendicular to the incident radiation. v is the frequency of the incident light. If $I = I_o$ at $z = 0$, we can obtain upon integration

$$\log \frac{I_0}{I} = \varepsilon_v C_i Z \tag{2.336}$$

where $\varepsilon_v = \alpha_v/2.303$ is called the "molar extinction coefficient" (m^2 mol^{-1}). It is specific to a specific frequency, v, or wavelength, $\lambda = 1/v$. C_i is expressed in mol m^{-3}, and Z is expressed in m. If C_i is in mol dm^{-3}, then Z is expressed in mm. The expression $\log (I_0/I)$ is called the absorbance A_λ. I is expressed in J $(m^2$ $s)^{-1}$. The amount of light absorbed by the medium is

$$I_{abs} = I_0 - I = I\left(1 - 10^{-\varepsilon_v C_i Z}\right) \tag{2.337}$$

If several components are absorbing simultaneously in the sample, then

$$I_{abs} = I_0 - I = I\left(1 - 10^{-\sum_i \varepsilon_v C_i Z}\right) \tag{2.338}$$

The Stark-Einstein law explicitly identifies that only molecules that are electronically excited take part in photochemical reactions, that is, those that are chemically "active" must be distinguished from those that are "excited" by photons. Excited species can lose energy by nonchemical pathways and via thermal reactions. Einstein defined the efficiency of a photochemical reaction as "quantum efficiency," ϕ,

$$\phi = \frac{\text{Number of molecules formed}}{\text{Number of quanta absorbed}} \tag{2.339}$$

This concept can be extended to all physical and chemical processes following the absorption of light. It can therefore be identified as a means of keeping tabs on the partitioning of absorbed quanta into the various modes.

Consider a molecule B that received a quantum of light energy to form the excited species B^*:

$$B + h v \rightarrow B^* \tag{2.340}$$

The absorption spectrum for the molecule is the plot of absorbance (A) versus wavelength (λ) or frequency (v). Since the absorbance depends on the nature of the functional groups in the molecules that are photoexcited, the absorption spectrum of each compound can be considered to be its fingerprint. These functional groups are called "chromophores." The wavelength at which maximum absorption is possible (designated λ_{max}) and the corresponding ε are listed in standard handbooks (Lide and Frederikse, 1994). Organic compounds with fused aromatic rings or unsaturated heteroatom functionalities have generally high absorbances.

A molecule in its excited state can undergo four main types of primary photochemical processes:

$$\text{Fluorescence: } B^* \rightarrow B + h\nu'$$
$$\text{Dissociation: } B^* \rightarrow P_1 + P_2$$
$$\text{Quenching: } B^* + M \rightarrow B + M$$
$$\text{Energy transfer: } B^* + C \rightarrow B + C^*$$
$$\text{Reaction: } B^* + C \rightarrow S_1 + S_2$$

Each of the processes given here has a ϕ_i associated with it. The value of the quantum efficiency is a function of the wavelength and is denoted by ϕ_λ. In the atmosphere (troposphere), the range of wavelengths of interest is between 280 to 730 nm, since most of the UV wavelengths < 280 nm are blocked by the stratospheric ozone layer. Beyond 730 nm no reactions of interest take place. In the aquatic environment (lakes, rivers, oceans), the upper range is rarely above 400 nm.

The general form of the photolysis reaction is represented as

$$B + h\nu \rightarrow P_1 + S_1 \tag{2.341}$$

The rate of the photochemical reaction is

$$-\frac{d[B]}{dt} = J[B] \tag{2.342}$$

where J is the first-order photochemical rate constant. Since the reaction occurs over the entire wavelength range $\lambda_1 - \lambda_2$, we have to obtain an average rate constant.

The photolysis rate constant J can be expressed in terms of the absorption cross section $\sigma_\lambda(T)$, which is a function of both λ and T. The rate constant is

$$J = \int_{\lambda_1}^{\lambda_2} \sigma(\lambda) \cdot \phi(\lambda) \cdot I_\lambda \, d\lambda \tag{2.343}$$

or

$$J = \sum_{\lambda_1}^{\lambda_2} \overline{\sigma_{\lambda_i}} \cdot \overline{\phi_{\lambda_i}} \cdot I_\lambda \, \Delta\lambda \tag{2.344}$$

where the over bar denotes the values at midrange centered at λ_i in the interval $\Delta\lambda$. In using this expression, one should separately evaluate $\sigma_{\lambda i}$ and I_λ. Normally, the shortest wavelength for photochemistry in the atmosphere is 290 nm. Finlayson-Pitts and Pitts (1986) list the values of the absorption cross section $\sigma_{\lambda i}$ and the actinide flux I_λ for many atmospheric reactions. Values of J for a number of reactions are listed in Table 2.23.

The following two examples will illustrate the use of these equations for determining the photolytic rate constants in water and air environments.

TABLE 2.23
Photolytic Rate Constants for Compounds in Air and Water

Air Environment

Reaction	J (s^{-1})
$O_3 \rightarrow O_2 + O(^1D)$	10^{-5} at 10 km, 10^{-3} at 40 km
$NO_2 \xrightarrow{hv} NO + O$	0.008 (surface), 0.01 (30 km)
$NO_3 \xrightarrow{hv} NO + O_2$	0.016
$NO_3 \xrightarrow{hv} NO_2 + O$	0.19
$CH_3COCH_3 \xrightarrow{hv} CH_3 + CH_3CO$	12.4×10^{-6}
$HCHO \xrightarrow{hv} HCO + O$	10.1×10^{-6}
$CO_2 \xrightarrow{hv} CO + O(^3P)$	2.2×10^{-8}

Water Environment

Compound	J (s^{-1})
PAHs	
Naphthalene	2.7×10^{-6}
Pyrene	2.8×10^{-4}
Anthracene	2.6×10^{-4}
Chrysene	4.4×10^{-5}
Pesticides	
Malathion	1.3×10^{-5}
Sevin	7.3×10^{-7}
Trifluralin	2.0×10^{-4}
Mirex	2.2×10^{-8}
Parathion	8.0×10^{-7}

Sources: Seinfeld and Pandis, 1998; Finlayson-Pitts and Pitts, 1986; Lyman et al., 1990; Warneck, 1986.

Example 2.47 Rate Constant for Photolysis in the Atmosphere

For the case of nitrogen dioxide photolysis in air, we have the following data:

λ (nm)	$\overline{\sigma_{\lambda_i}}$ (cm²)	$\overline{\phi_{\lambda_i}}$ (photons (cm² s)⁻¹)	$I_\lambda \Delta\lambda$	$\sigma_{\lambda_i}\phi_{\lambda_i}I_\lambda\Delta\lambda$
300	2E–19	0.98	1.6E+13	3.14E–06
310	3E–19	0.972	2.81E+14	8.19E–05
320	3.5E–19	0.964	7.19E+14	0.000243
330	4E–19	0.956	1.29E+15	0.000494
340	4E–19	0.948	1.42E+15	0.000537
350	5E–19	0.94	1.59E+15	0.000748
360	5.5E–19	0.932	1.66E+15	0.000852
370	6E–19	0.924	2.07E+15	0.001146
380	6E–19	0.916	2.02E+15	0.001110
390	6E–19	0.908	2.06E+15	0.001122
400	6E–19	0.8	2.81E+15	0.001349
410	6E–19	0.17	3.56E+15	0.000363
420	6E–19	0.03	3.7E+15	6.65E–05
			Sum:	0.008114

Hence, $J = 0.0081$ s⁻¹ and $t_{1/2} = 85$ s. In this case, since $I_{\lambda i}\Delta\lambda$ changes with the zenith angle, the rate constant varies diurnally, and hence, NO_2 decomposition by photolysis also varies diurnally.

2.3.10 ENZYME CATALYSIS

2.3.10.1 Michaelis-Menton and Monod Kinetics

Knowledge of the biochemical reaction kinetics will allow us to predict not only the reaction rates, but also the present and future concentrations of a pollutant involved in a natural biochemical process. It will also give us information how one can influence a reaction to procure a desired product in exclusion of other undesired products. Biochemical reactions can be thermodynamically feasible if the overall free energy change is negative. However, even if ΔG is negative, not all the reactions will occur at appreciable rates to be of any use in a living cell, unless other criteria are satisfied. The answer to this is catalysis caused by the enzymes present in all living organisms. "Enzymes" are a class of proteins that are polymeric chains of amino acids held together by peptide bonds. Chemical forces arrange these amino acids in a special three-dimensional pattern. The enzymes are designed to bind a substrate to a particular site where the conversion to the product species occurs. This active site is designed to accommodate specific amino acids in a given arrangement for optimal catalysis. "Enzyme catalysis" is, therefore, very specific, but quite fast. Generally, there is a complimentary structure relationship between the active site of the enzyme and the molecule upon which the enzyme acts. Two models represent this. The first one is the "lock-and-key model," where the substrate and enzyme fit together like a jigsaw puzzle (Figure 2.43a). In the second model called the "induced fit model," the active site wraps around the substrate to adopt the required conformation to cause catalysis (Figure 2.43b).

The rate of an enzyme-catalyzed reaction can be obtained by following the conversion of the substrate into products. The basic mechanism is as follows:

$$E + S \underset{k_{-1}}{\overset{k_1}{\rightleftharpoons}} E-S \xrightarrow{k_2} E + P \tag{2.345}$$

where
   E is the enzyme
   S is the substrate
   P is the product

Note that this mechanism is a special case of the general catalysis mechanism described in Section 2.3.7. The enzyme-substrate complex, E–S, is assumed to be at pseudo-steady state. Conservation of enzyme concentration is evident from this mechanism, and we have $[E]_0 = [E] + [E-S]$. For the substrate, we have $[S]_0 = [S] + [P]$, with the caveat that $[E] \approx [E-S] \ll [S]$. The rate of formation of the product is given by

$$r = \frac{d[P]}{dt} = -\frac{d[S]}{dt} = k_2[E-S] \tag{2.346}$$

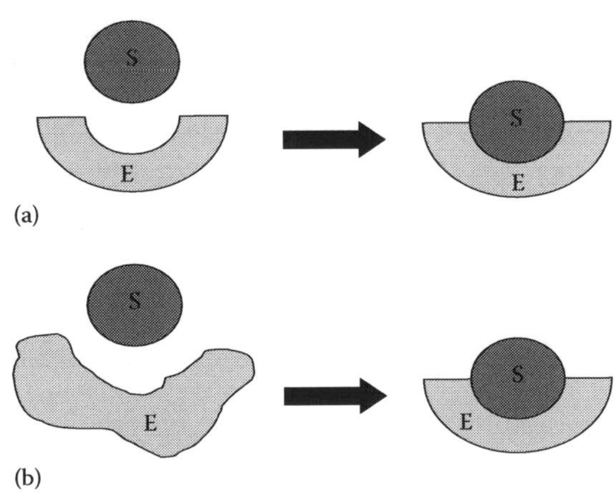

(a)

(b)

FIGURE 2.43   Different models of enzyme-substrate interactions. (a) Lock-and-key model and (b) induced fit model.

The rate of formation of [E-S] is given by

$$\frac{d[E-S]}{dt} = k_1\left([E]_0 - [E-S]\right)[S] - \left(k_{-1} + k_2\right)[E-S] \qquad (2.347)$$

Applying the pseudo-steady state approximation,

$$[E-S]_{ss} = \frac{k_1[E]_0[S]}{k_1[S] + k_{-1} + k_2} \qquad (2.348)$$

The expression for the rate can now be written as

$$r = \frac{V_{max}[S]}{K_m + [S]} \qquad (2.349)$$

with $V_{max} = k_2[E]_0$ and $K_m = (k_{-1} + k_2)/k_1$. This equation is called the "Michaelis-Menten equation." Note the similarity of the equation to that derived for heterogeneous catalysis obeying the Langmuir isotherm. Figure 2.44 represents the equation. The value of r used is generally the initial reaction rate. The limiting conditions of the equation are of special interest. When $[S] \gg K_m$, that is, at high substrate concentration $r \to V_{max}$, the maximum rate is achieved. At low substrate concentration, $[S] \ll K_m$, $r \to (V_{max}/K_m)[S]$, that is, the rate is directly proportional to the substrate concentration. It is seen from the rate equation that when $r \to V_{max}/2$, $K_m \to [S]_{1/2}$, which is the substrate concentration at which the enzyme reaction rate is one half of the maximum rate. The parameters, V_{max} and K_m, therefore characterize the kinetics

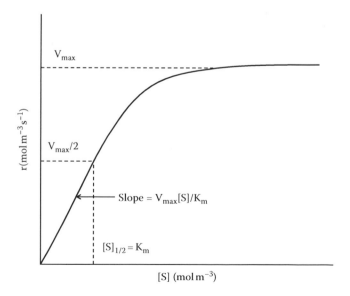

FIGURE 2.44 The variation in an enzyme-catalyzed reaction rate with substrate concentration as per the Michaelis-Menten equation.

of a reaction catalyzed by enzymes. The following example is an illustration of how these constants can be obtained from experimental data on enzyme-catalyzed substrate conversions.

Example 2.48   Estimation of the Michaelis-Menten Kinetic Parameters from Experimental Data

The enzymatic degradation of a pollutant to CO_2 and H_2O was measured over a range of pollutant concentration [S] (mol L^{-1}). The initial rates are given here:

[S] (mol L^{-1})	r (mM min^{-1})
0.002	3.3
0.005	6.6
0.010	10.1
0.017	12.4
0.050	16.6

Obtain the Michaelis-Menten constants.

The Michaelis-Menten equation can be recast in three different forms to obtain linear fits of the experimental data whereby the values of K_m and V_{max} can be determined from the slopes and intercepts of the plots. These are as follows:

1. *Lineweaver-Burk plot*: A plot of 1/r versus 1/[S]

$$\frac{1}{r} = \frac{K_m}{V_{max}} \cdot \frac{1}{[S]} + \frac{1}{V_{max}} \tag{2.350}$$

The slope is K_m/V_{max} and the intercept is $1/V_{max}$.

2. *Langmuir plot*: A plot of [S]/r versus [S]

$$\frac{[S]}{r} = \frac{1}{V_{max}} \cdot [S] + \frac{K_m}{V_{max}} \tag{2.351}$$

The slope is $1/V_{max}$ and the intercept is K_m/V_{max}.

3. *Eadie-Hofstee plot*: A plot of r versus r/[S]

$$r = -K_m \cdot \frac{r}{[S]} + V_{max} \tag{2.352}$$

The slope is $-K_m$ and the intercept is V_{max}.

The three plots for the data are shown in Figure 2.45. The values of K_m and V_{max} obtained are given here:

Method	r^2	K_m (mM)	V_{max} (mM min^{-1})
Lineweaver-Burk	0.9999	10.1	20.0
Langmuir	0.9998	10.0	19.9
Eadie-Hofstee	0.9986	10.0	19.9

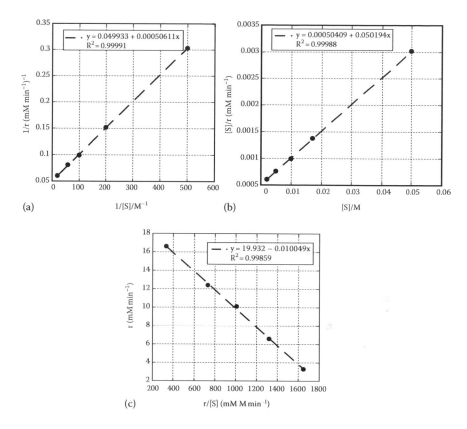

FIGURE 2.45 Estimation of Michaelis-Menten parameters using different methods. (a) Lineweaver-Burk plot, (b) Langmuir plot, (c) Eadie-Hofstee plot.

Often the Lineweaver-Burk plot is preferred since it gives a direct relationship between the independent variable [S] and the dependent variable r. There is one shortcoming, that is, as [S] → 0, 1/r → ∞, and hence is inappropriate at low [S]. The value of the Eadie-Hofstee plot lies in the fact that it gives equal weight to all points unlike the Lineweaver-Burk plot. In the present case, the Lineweaver-Burk plot appears to be the best.

Michaelis-Menten kinetics considers the case where living cells producing the enzymes are so large that little or no increase in cell number occurs. Michaelis-Menten law is applicable for a no-growth situation with a given fixed enzyme concentration. We know that food metabolism in biological species leads to growth and reproduction. As cells grow rapidly and multiply in number, the concentration of the enzyme and the substrate degradation rate will vary. These aspects come under "growth kinetics" or "cell kinetics."

Generally, the growth of microorganisms occurs in phases. Figure 2.46 is a typical growth curve. A measure of the microbial growth is the cell number density (number of cells per unit volume) denoted [X]. During the initial stage, the change in cell density is

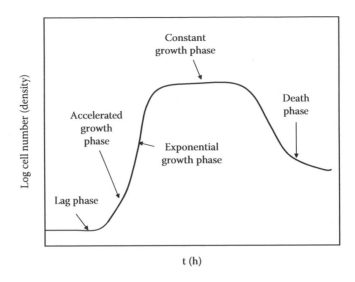

FIGURE 2.46 Typical growth curve for microorganisms.

zero for some time. This is the "lag phase." Subsequent to the lag phase is the phase of growth and cell division. This is called the "accelerated growth phase." This is followed by the "exponential growth phase" in which the growth rate increases exponentially. Further, the growth rate starts to plateau. This is called the "decelerated growth phase." This leads to a constant maximum cell density in what is called the "stationary phase." As time progresses, the cells deplete the nutrient source and begin to die. The cell density rapidly decreases during the final stage called the "death phase."

The rate of cell growth per unit time is represented by the following equation:

$$r_g = \frac{d[X]}{dt} = \zeta[X] \tag{2.353}$$

where ζ is the "specific growth rate" (h^{-1}). The effect of substrate concentration [S] on ζ is described by the "Monod equation":

$$\zeta = \frac{\zeta_{max}[S]}{K_s + [S]} \tag{2.354}$$

Note that when $[S] \ll K_s$, $\zeta \propto [S]$ and when $[S] \gg K_s$, $\zeta = \zeta_{max}$. This equation is an empirical expression and states that if the nutrient concentration [S] reaches a large value, the specific growth rate plateaus out. However, this is not always the case. In spite of this, the Monod kinetics is a simplified expression for growth kinetics at low substrate concentrations.

Introducing the Monod equation in the overall expression we get

$$r_g = \frac{d[X]}{dt} = \frac{\zeta_{max}[S]}{K_s + [S]} \cdot [X] \tag{2.355}$$

When $[S] \ll K_s$, we have

$$r_g = \frac{\zeta_{max}}{K_s} \cdot [S][X] = k''[S][X] \tag{2.356}$$

where k'' is a pseudo-second-order rate constant. This expression is an adequate representation for the growth rate of microorganisms in the soil and groundwater environments (McCarty et al., 1992). k'' has units of L $(mol\ h)^{-1}$ (or $dm^3\ (mol\ h)^{-1}$). Note that although Monod kinetics for cell growth and the Michaelis-Menten kinetics for enzyme catalysis give rise to similar expressions, they do differ in the added term $[X]$ in the expression for Monod kinetics.

The Michaelis-Menten expression gives the rate at which a chemical transformation occurs if the enzyme were separated from the organisms, and the reaction conducted in solution. If we represent the rate r as the *change in moles of substrate per unit organism in a given volume per time*, then $R = r[X]_0$ is the *moles of chemical transformed per unit volume per unit time*. $[X]_0$ is the initial number of organisms per volume of solution. If r is given by the Michaelis-Menten expression with V_{max} in moles per organism per time, we have

$$R = -\frac{d[S]}{dt} = \frac{V_{max}[S]}{K_m + [S]} \cdot [X]_0 \tag{2.357}$$

The cell yield y_c is defined as the ratio of mass of new organisms produced to the mass of chemical degraded.

$$y_c = \frac{(d[X]/dt)}{-(d[S]/dt)} \tag{2.358}$$

Note that the maximum specific growth rate ζ_{max} is the product of the maximum rate of degradation (V_{max}) and y_c.

$$\zeta_{max} = V_{max} \cdot y_c \tag{2.359}$$

Hence we can write the Monod equation as

$$-\frac{d[S]}{dt} = \frac{V_{max}[S]}{K_s + [S]} \cdot [X] \tag{2.360}$$

If we identify K_m with K_s and $[X]$ with $[X]_0$, one can infer the Monod kinetics expression directly from the Michaelis-Menten kinetics expression. This corroborates the earlier statement that the Monod no-growth kinetics and Michaelis-Menten kinetics are identical.

In order to take into consideration the fact that as organisms grow they also decay at a rate k_{dec} (h^{-1}), one rewrites the Monod equation as

$$\frac{d[X]}{dt} = \left(\frac{\zeta_{max}[S]}{K_s + [S]} \right) \cdot [X] - k_{dec}[X] \qquad (2.361)$$

At low values of [S], there is no net increase in organism density since its growth is balanced by its decay. Thus, $d[X]/dt = 0$, and

$$[S]_{min} = \frac{K_s k_{dec}}{\zeta_{max} - k_{dec}} \qquad (2.362)$$

represents the minimum concentration of substrate that can maintain a given microbial population. However, the organism may find another compound with a different $[S]_{min}$ that can sustain another level of population. This compound is called a "cometabolite." In the natural environment, this type of biodegradation kinetics is often observed.

2.4 CHEMICAL KINETICS AND REACTOR DESIGN

2.4.1 Types of Reactors

Chemical reactors are classified into batch and flow systems (Figure 2.47). In a "batch reactor," a fixed volume (mass) of material is treated for a specified length of time. If material flows into and out of the reactor in a continuous manner, it is called a "continuous flow reactor."

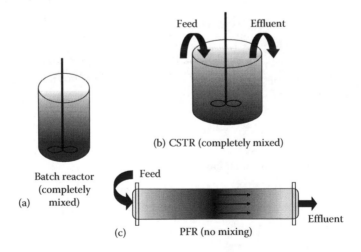

FIGURE 2.47 Different types of ideal reactors used in environmental engineering processes: (a) A batch reactor where there is uniform concentration in the reactor, (b) A CSTR where there is uniform mixing and composition everywhere within the reactor and at the exit, (c) A PFR where there is no mixing between the earlier and later entering fluid packets.

2.4.1.1 Ideal Reactors

To analyze a reactor, we use the basic concept of "mass balance" or "conservation of mass." This principle dates from 1789 when the eminent French scientist Antöine Lavoisier, considered to be the father of modern chemistry, first enunciated it in his *Traité élémentaire de chimie*. His observations translated into English (Holmes, 1994) read as follows:

> For nothing is created, either in the operations of art, or in those of nature, and one can state as a principle that in every operation there is an equal quantity of material before and after operation; that the quality and quantity of the [simple] principles are the same, and that there are nothing but changes, modifications.

We establish a control volume as shown in Figure 2.48, where the general case of a continuous system is represented. The feed rate of solute j is F_{j0} (moles per time). The effluent rate from the system is F_j (moles per time). If r represents the rate of the reaction, then Vv_jr represents the rate of generation of j (moles per time). Note that v_j is the stoichiometric coefficient of j for the reaction, as shown in Section 2.3. A total mass (mole) balance within the reactor is then written as (Schmidt, 2005)

$$\left[\text{Accumulation}\right] = \left[\text{Input}\right] - \left[\text{Output}\right] + \left[\text{Generation by reaction}\right]$$

$$\frac{dN_j}{dt} = F_{j0} - F_j + Vv_jr \tag{2.363}$$

Note that N_j is the total moles of j in the reactor. This equation is the basis for reactor design. Note that if there are additional source terms within the system they are included in the input term. Specifically, we can determine either the time or the reactor volume required for a specified conversion of the reactants to products.

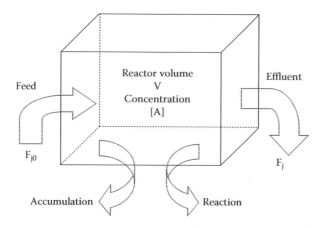

FIGURE 2.48 Mass balance for a reactor.

2.4.1.1.1 Batch Reactor

Since both F_{j0} and F_j are zero in this case, we have

$$\frac{dN_j}{dt} = V\nu_j r \tag{2.364}$$

Consider a first-order reaction $A \rightarrow B$, for which $r = kC_A$ and $\nu_A = -1$.
    Hence,

$$\frac{dN_A}{dt} = -kC_A V \tag{2.365}$$

where V is the total volume of the reactor. Further, since $N_A = C_A V$ in a constant-volume batch reactor, we can write

$$\frac{dC_A}{dt} = -kC_A \tag{2.366}$$

Thus, for this case, the rate of change in concentration of A in the reactor, $C_A(t)$, is given by the rate of the reaction. We can substitute the appropriate expression for the rate (r_A), and solve the resulting differential equation for $C_A(t)$. A batch reactor is an unsteady state operation where the concentration changes with time but is the same throughout the reactor.

2.4.1.1.2 Continuous-Flow Stirred Tank Reactor

A common type of reactor encountered in environmental engineering is the "continuous-flow stirred tank reactor" (CSTR), in which the effluent concentration is the same as the concentration everywhere within the reactor at all times (Figure 2.47). In other words, the feed stream is instantaneously mixed in the reactor. The mole balance for a pollutant j in the CSTR is

$$\frac{dN_j}{dt} = F_{j0} - F_j + V\nu_j r \tag{2.367}$$

In many cases, this equation can be further simplified by assuming a "steady state" such that $dN_j/dt = 0$, that is, the rate of accumulation is zero. If we also stipulate that the rate of the reaction is uniform throughout the reactor, we get the following equation:

$$F_{j0} = F_j + V\nu_j r \tag{2.368}$$

Thus, the volume of a CSTR is given by

$$V = \frac{\left(F_j - F_{j0}\right)}{\nu_j r} \tag{2.369}$$

Let us consider the special case where we have a "constant" volumetric feed flow rate Q_0 and an effluent volumetric flow rate of Q_e. If C_{A0} and C_A represent the concentration of A in the feed and the effluent, respectively, we can write $F_{j0} = Q_0C_{j0}$, and $F_j = QC_j$. Then, we have the following equation for the reactor volume:

$$V = \frac{(QC_j - Q_0C_{j0})}{v_j r} \tag{2.370}$$

2.4.1.1.3 Plug Flow or Tubular Reactor

Another type of a continuous flow reactor is the "tubular reactor," where the reaction occurs such that radial concentration gradient is nonexistent, but a gradient exists in the axial direction. This is called "plug flow." The flow of the fluid is one in which no fluid element is mixed with any other either behind or ahead of it. The reactants are continually consumed as they flow along the length of the reactor, and an axial concentration gradient develops. Therefore, the reaction rate depends on the axial direction. A polluted river that flows as a narrow channel can be considered to be in plug flow.

In order to analyze the tubular reactor, we consider a section of a cylindrical pipe of length Δz, as shown in Figure 2.49. If the rate is constant within the small volume element $\Delta V = A_c\Delta z$ and we assume steady state, the following mass balance on compound j for the volume element results:

$$F_j(z) - F_j(z + \Delta z) + A_c dz v_j r = 0 \tag{2.371}$$

Since $\Delta V = A_c\Delta z$, where A_c is the cross-sectional area, we can write this equation as

$$\frac{F_j(z + \Delta z) - F_j(z)}{A_c\Delta z} = v_j r \tag{2.372}$$

Taking the limit as $\Delta z \to 0$ and rewriting, we get

$$\frac{1}{A_c} \frac{dF_A}{dz} = v_j r \tag{2.373}$$

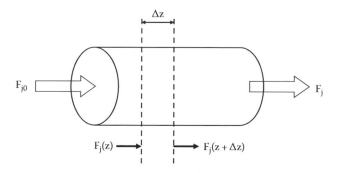

FIGURE 2.49 Schematic of a tubular or plug-flow reactor.

or

$$\frac{dF_A}{dV} = \nu_j r \tag{2.374}$$

It can be shown that this equation also applies to the case where the cross-sectional area varies along the length of the reactor. This is the case of a river going through several narrow channels along its path.

If the reaction is first order, viz., A $\rightarrow$ B, the rate of disappearance of A is $r = kC_A$. The stoichiometry of the reaction is such that $\nu_A = -1$. If we consider a constant-density reactor, and note that $F_A = Q_0 C_A$ with Q_0 being the volumetric flow rate, we have

$$-\frac{dC_A}{dV} = \frac{k}{Q_0} \cdot C_A \tag{2.375}$$

Integrating with the initial condition, $C_A = C_{A0}$ at $t = 0$, we get

$$V = \frac{Q_0}{k} \cdot \ln\left(\frac{C_{A0}}{C_A}\right) \tag{2.376}$$

If 50% of C_{A0} has to be converted into products, then $V = 0.693 Q_0/k$ is the reactor volume required.

2.4.1.1.4 Design Equations for CSTR and PFR

For any general stoichiometric equation of the form $aA + bB \rightarrow cC + dD$, we can write the following equation on a per mole of A basis:

$$A + \frac{b}{a} B \rightarrow \frac{c}{a} C + \frac{d}{a} D \tag{2.377}$$

For continuous systems (CSTR and PFR) it is convenient to express the change in terms of conversion of A, χ which is defined as the fractional conversion of A at steady state. Then the mass balance can be written as

Molar rate of A leaving the reactor, F_A = Molar rate of A fed into the reactor, F_{A0}

    – Molar rate of consumption of A in the reactor, $F_{A0}\chi$.

Therefore,

$$F_A = F_{A0}(1 - \chi) \tag{2.378}$$

For a CSTR, the reactor volume is given by

$$V = \frac{F_{A0}\chi}{r} \tag{2.379}$$

For a PFR, we have

$$\frac{d(F_{A0}\chi)}{dV} = r \tag{2.380}$$

Since F_{A0} is a constant, we can integrate this expression to obtain V.

Example 2.49   Comparison of Reactor Volumes for a CSTR and a PFR

In both a CSTR and a PFR, r_A can be expressed as a function of the conversion χ. A most frequently observed relationship between r_A and χ (for a first-order reaction) in environmental engineering is $r_A = kC_{A0}(1-\chi)$. Using this relationship for a CSTR, we have

$$\frac{V_{CSTR}}{F_{A0}} = \frac{\chi}{kC_{A0}(1-\chi)}$$

For a PFR we get the following:

$$\frac{V_{PFR}}{F_{A0}} = \frac{1}{kC_{A0}} \int_0^\chi \frac{\chi}{(1-\chi)}$$

If the molar feed rate F_{A0} and initial feed concentration C_{A0} are the same in both cases, we have

$$\frac{V_{CSTR}}{V_{PFR}} = \frac{\chi}{(1-\chi) \ln\left(\frac{1}{1-\chi}\right)}$$

If the desired conversion is $\chi = 0.6$, then the ratio of volumes is 1.63. Thus a 63% larger volume of the CSTR than that of a PFR is required to achieve a 60% conversion.

Example 2.50   CSTR Model for a Surface Impoundment

Surface impoundments are used at many industrial sites for the treatment of wastewater. It is generally a holding tank where water is continuously fed for treatment. Enhanced aeration (using surface agitation) can volatilize organic compounds from surface impoundments. This process also facilitates the growth of bioorganisms that are capable of destroying the waste. Surface impoundments also perform the function of a settling pond in which particulates and other sludge

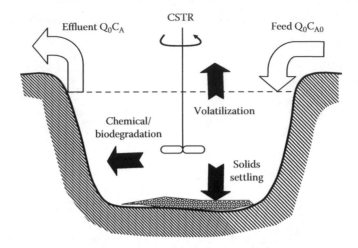

FIGURE 2.50 Primary loss mechanisms in a surface impoundment modeled as a CSTR.

solids are separated from the waste stream. This is called "activated sludge process." Figure 2.50 depicts the three primary pathways by which a pollutant is removed from a waste stream entering an impoundment.

A surface impoundment can be considered to be a CSTR with complete mixing within the aqueous phase, and the performance characteristics analyzed. If the reactor is unmixed (i.e., no back mixing) such that a distinct solute concentration gradient exists across the reactor, the impoundment will have to be modeled as a PFR. The three primary loss mechanisms for a chemical entering the impoundment are (1) volatilization across the air-water interface, (2) adsorption to settle able solids (biomass), and (3) chemical or biochemical reaction. Each of these loss processes within the reactor can be considered to be a first-order process. Chemical reactions degrade the pollutant. Exchange between phases (e.g., volatilization, settling) within the reactor tends to deplete the chemical concentration; it does not, however, change the chemical. The overall mass balance for a chemical A in the impoundment is

$$\frac{dN_A}{dt} = F_{A0} - F_A + V\nu_A r_{tot}$$

where r_{tot} is the total rate of loss of A by all processes within the reactor.

The rate of volatilization of A is

$$r_{voln} = k_v C_A$$

The rate of settling of A on solids in the water column is given by

$$r_{set} = k_s C_A$$

The chemical reaction rate is given by

$$r_{react} = k_r C_A$$

The total rate of loss of A from the surface impoundment is given by

$$r_{tot} = r_{react} + r_{voln} + r_{set} = (k_r + k_v + k_s)C_A = k^*C_A$$

The overall first-order rate constant is k^*. For a first-order loss process, $\nu_A = -1$. For a constant density reactor, $N_A = VC_A$. If the influent and effluent feed rates are the same ($= Q_0$), we have the following mass balance equation:

$$V\frac{dC_A}{dt} = Q_0C_{A0} - Q_0C_A - k^*C_A V$$

Assuming steady state,

$$\frac{C_A}{C_{A0}} = \frac{1}{1 + \left(k^* \dfrac{V}{Q_0}\right)}$$

where C_{A0} is the concentration of A in the influent. The quantity V/Q_0 has units of time and is called the "mean detention time," "residence time," or "contact time" in the reactor for the aqueous phase and is designated τ_d.

Blackburn (1987) reported the determination of the three transport rates for 2,4-dichlorophenol in an activated sludge process in an impoundment. The impoundment had an aggregate volume of 10^6 L with a solid (biomass) concentration of 2000 mg L^{-1}. The mean detention time was 1 d. The depth of the impoundment was 1.5 m. The average biodegradation rate k_r was 0.05 h^{-1}. The rate constant for settling of the biomass was estimated to be 2.2×10^{-6} h^{-1}. The rate constant for volatilization was 2×10^{-5} h^{-1}. The overall rate constant $k^* = 0.05$ h^{-1}. Therefore, $C_A/C_{A0} = (1 + 0.05\tau_{rd})^{-1}$. This is shown in Figure 2.51. The contribution

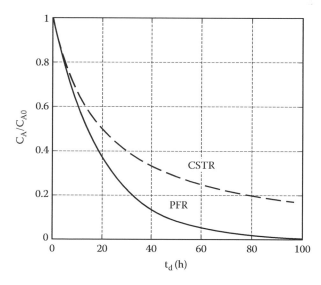

FIGURE 2.51 Comparison of exit to inlet concentration ratio in a surface impoundment modeled as a CSTR and a PFR.

of each mechanism toward the loss of DCP can also be ascertained. For example, the fractional loss from biodegradation is $f_{bio} = k_r \tau_d / (1 + k_r \tau_d)$.

If the surface impoundment behaves as a PFR, then the steady state mass balance will be given by $-\dfrac{dC_A}{dV} = \dfrac{k^*}{Q_0} C_A$, which gives $C_A / C_{A0} = \exp(-k^* \tau_d) = \exp(-0.05 \tau_d)$. This is also shown in Figure 2.51. Note that the decay of DCP is faster in the PFR than the CSTR. It is clear that an ideal PFR can achieve the same removal efficiency as a CSTR, but utilizing a much smaller volume of the reactor.

Example 2.51   Combustion Incinerator as a PFR

Combustion incinerators are used to incinerate municipal and industrial waste. Generally, they require high temperatures and an oxygen-rich environment within the chamber. Consider a waste-containing benzene being incinerated using an air stream at a velocity of 5 m s⁻¹. A typical inlet temperature is 900°C (1173 K). Due to heat transfer, the temperature changes along the chamber length from inlet to outlet. Assume a gradient of 10°C m⁻¹. Therefore, the rate constant will also vary linearly with length.

For benzene, the Arrhenius parameters are activation energy E_a = 225 kJ mol⁻¹ and pre-exponential factor A = 9 × 10¹⁰ s⁻¹. A typical length of the incinerator chamber is 10 m. Hence, the exit temperature will be 900 − 10 × 10 = 800°C (1073 K). The first-order rate constant for benzene will be $k_{1173\ K}$ = 8 s⁻¹ and $k_{1073\ K}$ = 1 s⁻¹. As an approximation, let us consider an average first-order rate constant k = 4 s⁻¹. The concentration in the exit stream can be obtained using the equation for a plug-flow reactor: $\dfrac{C_A}{C_{A0}} = e^{-k\tau}$. Since τ = 10 m/5 (m s⁻¹) = 2 s, C_A / C_{A0} = 3 × 10⁻⁴. Hence, the destruction efficiency is 99.96%.

Frequently, in environmental reactors, the reactor (e.g., a CSTR) may not be at steady state always. Examples of these include a waste impoundment and chemical spills into waterways. In these cases, we need to obtain the solution to the mass balance equation under "nonsteady state" conditions. For illustrative purposes, we will use a first-order reaction.

$$V \frac{dC_A}{dt} = Q_0 \left(C_{A0} - C_A \right) - k C_A V \qquad (2.381)$$

Rearranging,

$$\frac{dC_A}{dt} + \left(k + \frac{Q_0}{V} \right) C_A = \frac{Q_0}{V} C_{A0} \qquad (2.382)$$

Integrating this equation with the initial condition (at t = 0) $C_A = C_{A0}$, we get

$$C_A(t) = \frac{1}{\left(1 + k\tau_d \right)} C_{A0} \left[1 - e^{-\left(k + \frac{1}{\tau_d} \right)t} \right] + C_{A0} e^{-\left(k + \frac{1}{\tau_d} \right)t} \qquad (2.383)$$

where τ_d is as defined earlier. Note that as $t \to \infty$, this equation reduces to the steady state solution obtained earlier. Depending on the functionality of C_{A0}, the response of the CSTR will vary. Levenspiel (1999) considered several such input functions and derived the corresponding response functions.

2.4.1.1.5 Relationship between Steady State and Equilibrium

As we discussed earlier, "equilibrium" for closed systems is the state of minimum free energy (or zero entropy production), which is a "time-invariant state." For open systems, the equivalent time-invariant state is called the "steady state." How closely does a steady state resemble the equilibrium state? This question is important in assessing the applicability of steady state models in environmental engineering to depict time-invariant behavior of systems.

As an example, consider the reversible reaction

$$A \underset{k_b}{\overset{k_f}{\rightleftharpoons}} B \tag{2.384}$$

Consider the reaction in an open system where both species A and B are entering and leaving the system at a feed rate Q_0 with feed concentrations C_{A0} and C_{B0}, respectively. The mass balance relations are

$$V\frac{dC_A}{dt} = Q_0\left(C_{A0} - C_A\right) - k_f C_A + k_b C_B$$
$$V\frac{dC_B}{dt} = Q_0\left(C_{B0} - C_B\right) + k_f C_A - k_b C_B \tag{2.385}$$

Applying the steady state approximation to both species and solving the resulting equations, we obtain

$$\left(\frac{C_A}{C_B}\right)_{ss} = \frac{1}{K_{eq}} + \frac{1}{k_f \tau_d} \tag{2.386}$$

where $K_{eq} = \dfrac{k_f}{k_b}$ and $\tau_d = \dfrac{V}{Q_0}$. Recalling that $k_f = 0.693/t_{1/2}$ for a first-order reaction, we have

$$\left(\frac{C_A}{C_B}\right)_{ss} = \frac{1}{K_{eq}} + \frac{t_{1/2}}{0.693 \cdot \tau_d} \tag{2.387}$$

Note that if τ_d is very large and $t_{1/2}$ is very small, $(C_A/C_B)_{ss} \to 1/K_{eq}$. Thus, if the mean residence time is very large and the reaction is fast, the steady state ratio approaches that at equilibrium. For many natural environmental conditions even though τ_d may be quite large, $t_{1/2}$ is also relatively large (since k_f is small), and hence steady state and equilibrium are not identical.

2.4.1.2 Nonideal Reactors

The reactor designs we have discussed thus far assumed ideal flow patterns—plug flow (no mixing) as one extreme and CSTR (complete mixing) as the other extreme. Although many reactors show ideal behavior, natural environmental reactors fall somewhere in between the two ideal reactors. Fluid channeling, recycling, or stagnation points in the reactor cause the deviations. There are two models in the chemical engineering literature that are used to explain nonideal flows in reactors. These are called "dispersion" and "tanks-in-series" models.

2.4.1.2.1 Dispersion Model

Ideal plug-flow involves no intermixing between fluid packets. However, usually molecular and turbulent diffusion can skew the profile as shown in Figure 2.52. As the disturbances increase, the reactor will approach the characteristic of complete mixing as in a CSTR. The change in concentration is represented by the Fick's equation for molecular diffusion, written as follows:

$$\frac{\partial C}{\partial t} = D_{ax} \frac{\partial^2 C}{\partial x^2} \tag{2.388}$$

where D_{ax} (m^2 s^{-1}) is called the axial dispersion coefficient. If $D_{ax} = 0$, we have plug flow (PFR) and as $D_{ax} \to \infty$, the flow is completely mixed (CSTR). In order to nondimensionalize this equation, we express $\zeta = (ut + x)/L$ and $\theta = ut/L$, where u represents the fluid velocity and L is the length of the reactor. We have then (Levenspiel, 1999)

$$\frac{\partial C}{\partial \theta} = \frac{D_{ax}}{uL} \frac{\partial^2 C}{\partial \zeta^2} - \frac{\partial C}{\partial \zeta} \tag{2.389}$$

The term D_{ax}/uL is the "dispersion number."

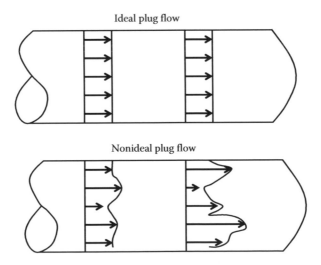

FIGURE 2.52   Schematic of the plug flow and dispersed plug flow models.

If a "perfect" pulse is introduced in a flowing fluid, the solution to this equation gives the exit concentration (Hill, 1977).

$$\frac{C(L,t)}{\int_0^\infty C(L,t)d\left(\frac{t}{\tau}\right)} = \frac{1}{2V_R\sqrt{\frac{\pi D_{ax}}{uL}\cdot\frac{t}{\tau}}}\cdot e^{-\left[\frac{\left(1-\frac{t}{\tau}\right)^2}{\frac{4D_{ax}}{uL}\cdot\frac{t}{\tau}}\right]}$$ (2.390)

A plot of the right-hand side of this equation versus t/τ is given in Figure 2.53. For small values of D_{ax}/uL, the curve approaches that of a normal Gaussian error curve. However, with increasing D_{ax}/uL (>0.01), the shape changes significantly over time. For small values of dispersion ($D_{ax}/uL < 0.01$), it is possible to calculate D_{ax} by plotting log $C(L, t)/C(0)$ versus t and obtaining the standard deviation σ from the data. $C(0)$ is the influent pulse input concentration. The equation is

$$D_{ax} = \frac{1}{2}\cdot\sigma^2\cdot\frac{u^3}{L}$$ (2.391)

2.4.1.2.2 Tanks-in-Series Model

The second model for dispersion is a series of CSTRs. The actual reactor is then composed of n CSTRs, the total volume is $V_R = nV_{CSTR}$. The average residence time in the actual reactor is $\tau = V_R/Q_0 = nV_{CSTR}/Q_0$, where Q_0 is the volumetric flow rate

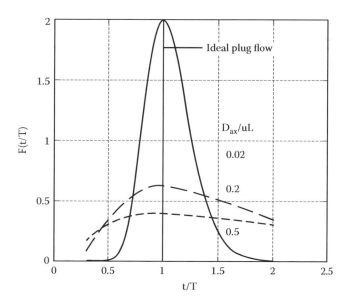

FIGURE 2.53 Relationship between $F(t/\tau)$ and dimensionless time for different values of D_{ax}/uL. True plug flow model is represented by the vertical line at dimensionless time of 1.0.

into the reactor. For any reactor n in the series, the following material balance holds:

$$C_{n-1}Q_0 = C_n Q_0 + V_{CSTR} \cdot \frac{dC_n}{dt} \tag{2.392}$$

The concentration leaving the jth reactor is given by

$$\frac{C_j}{C(0)} = 1 - e^{-n\frac{t}{\tau}} \left\{ 1 + n\frac{t}{\tau} + \frac{1}{2!}\left(n\frac{t}{\tau}\right)^2 + \cdots + \frac{1}{(j-1)!}\left(n\frac{t}{\tau}\right)^{j-1} \right\} \tag{2.393}$$

Example 2.52　Dispersion in a Soil Column

Chloride is a tracer used to determine the dispersivity in a soil column. It is conservative and does not react with soil particles. A pulse input is introduced at the bottom of a soil column of height 6.35 cm at an interstitial pore water velocity of 5.37×10^{-4} cm s^{-1}. The soil has a porosity of 0.574. The chloride concentration at the exit (top) as a function of time is given here:

t (min)	C(L, t)/C(0)
125	0.05
150	0.15
180	0.30
200	0.40
250	0.90

Determine the dispersivity in the column.

First, plot log C(L, t)/C(0) as a function of t on a probability plot, as shown in Figure 2.54. The standard deviation in the data is $\sigma_t = 46$ min, which is the difference between the 84th and 50th percentile points, $t_{84\%} - t_{50\%}$. Hence, $D_{ax} = (1/2)$ $\sigma_t^2 u^3/L = 9.3 \times 10^{-5}$ cm^2 s^{-1}.

Example 2.53　Pulse Input in a River

A conservative (nonreactive, nonvolatile, water soluble) pollutant is suddenly discharged into a river flowing at an average speed of 0.5 m s^{-1}. A monitoring station is located 100 miles downstream from the spill. The dispersion number is 0.1. Determine the time when the concentration at the station is 50% of the input.

$D_{ax}/uL = 0.1$ (intermediate dispersion). $\tau = L/u = (100\text{ miles})(1600\text{ m mile}^{-1})/0.5$ (m s^{-1}) $= 3.2 \times 10^5$ s $= 3.7$ days. For C(L, t) $= 0.5$ C(0),

$$0.5 = \frac{1}{\sqrt{3.14 \times 0.1 \cdot \frac{t}{\tau}}} \cdot \exp\left(-\frac{\left(1 - \frac{t}{\tau}\right)^2}{0.4 \cdot \frac{t}{\tau}} \right)$$

By trial and error $t_{0.5}/\tau = 1.9$. Hence, $t_{1/2} = 7$ days.

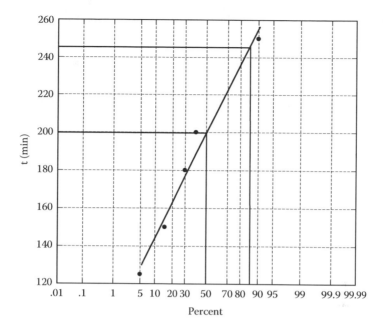

FIGURE 2.54 Probability plot of dimensionless concentration versus time to obtain the dispersion coefficient from experimental data.

2.4.1.3 Dispersion and Reaction

Let us now consider the case where the flow is nonideal, and the compound entering the reactor is undergoing reaction as it flows through the reactor. Let the reaction rate be r, with the stoichiometric coefficient for A being ν_A.

Figure 2.55 shows the reactor configuration in which we perform a material balance over the volume between x and x + Δx. There are two separate inputs and outputs,

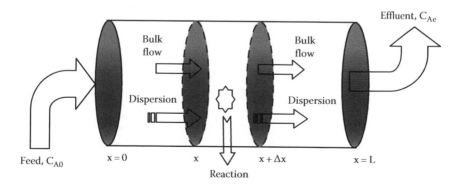

FIGURE 2.55 Material balance in a reactor with dispersion and reaction.

one due to bulk flow ($uC_A A_c$) and the other due to axial dispersion ($D_{ax} A_c \partial C_A / \partial x$). Within the reactor the compound reacts at a rate r. The overall material balance is

$$[\text{Accumulation}] = [\text{Input}(\text{bulk flow} + \text{dispersion})]$$

$$- [\text{Output}(\text{bulk flow} + \text{dispersion})] + [\text{Reaction}]$$

At steady state, accumulation is zero. Hence,

$$0 = \left(uA_c C_A \big|_x - D_{ax} A_c \frac{dC_A}{dx} \bigg|_x \right) - \left(uA_c C_A \big|_{x+\Delta x} - D_{ax} A_c \frac{dC_A}{dx} \bigg|_{x+\Delta x} \right) + v_A r A_c \Delta x \quad (2.394)$$

Rearranging the terms and dividing by the volume $A_c \Delta x$ gives

$$u \cdot \left(\frac{C_A \big|_x - C_A \big|_{x+\Delta x}}{\Delta x} \right) - D_{ax} \left(\frac{\frac{dC_A}{dx}\big|_x - \frac{dC_A}{dx}\big|_{x+\Delta x}}{\Delta x} \right) + v_A r = 0 \quad (2.395)$$

Taking limit as $\Delta x \to 0$, we get

$$-u \frac{\partial C_A}{\partial x} + D_{ax} \frac{\partial^2 C_A}{\partial x^2} + v_A r = 0 \quad (2.396)$$

This equation and its variations appear in many cases in environmental engineering. The first term on the left-hand side is called the "advection" or "convection" term, the second is the "dispersion" term, and the last is the "reaction" term. For a first-order reaction A → B, the equation can be written in the following form:

$$D_{ax} \frac{d^2 C_A}{dx^2} - u \frac{dC_A}{dx} - k C_A = 0 \quad (2.397)$$

To make this equation dimensionless, we use the following transformations: z = x/L and dz = dx/L, with L being the length of the tube. Hence, we have

$$\frac{d^2 C_A}{dz^2} - Pe_L \frac{dC_A}{dz} - Da_L C_A = 0 \quad (2.398)$$

where $Pe_L = uL/D_{ax}$ is called the "Peclet number," and $Da_L = kL^2/D_{ax}$ is called the "Damköhler number." If mixing is rapid, that is, D_{ax}/uL is very small or uL/D_{ax} is large, the axial dispersion term will be negligible and the system will approach plug flow behavior. This is a good approximation for analyzing contaminant dynamics in surface waters. In the atmosphere, since the dispersion is large, we have to consider the entire advective-dispersion equation.

2.4.1.4   Reaction in Heterogeneous Medium

Often, the reactant is transported from one medium to another where it reacts. A good example is gaseous NH_3 that dissolves and reacts with an aqueous acidic solution to give NH_4OH. Another example is the reaction and transport of gases (CO_2, O_2) and volatile species in soil. As a general example, consider a compound A in air that diffuses in air and reacts transforming to a compound B. There are three steps involved:

$$A_g \xrightarrow{\text{gas diffusion}} A_\sigma$$
$$A_\sigma \xrightarrow{k_A} B_\sigma \qquad (2.399)$$
$$B_\sigma \xrightarrow{\text{liquid diffusion}} B_{aq}$$

Depending on which of these reactions is slow the overall rate will vary.

For the heterogeneous reaction scheme considered earlier, the bulk gas phase will be well mixed with a concentration C_A^g, whereas the solute concentration at the interface is C_A^σ. The change occurs across a very thin film of thickness δ near the interface. The diffusive flux of A to the interface is given by

$$n_{A,\text{diff}} = \frac{D_{AL}}{\delta} \cdot \left(C_A^g - C_A^\sigma \right) \qquad (2.400)$$

where D_{AL} is the diffusion constant of A in solution. D_{AL}/δ is the film transfer coefficient for A. More appropriately, we can define a general mass transfer coefficient k_c to represent this term.

The first-order surface reaction rate is given by

$$r_\sigma = k_\sigma C_A^\sigma \qquad (2.401)$$

At steady state, the flux to the interface must be the same as the reactive loss at the surface. This gives

$$C_A^\sigma = \frac{k_c C_A^g}{k_c + k_\sigma} \qquad (2.402)$$

The rate of the reaction at steady state is

$$r_\sigma = \frac{k_\sigma k_c C_A^g}{k_\sigma + k_c} \qquad (2.403)$$

There are two limiting conditions for this rate: (1) if diffusion is controlling, $k_\sigma \gg k_c$, $r \rightarrow k_c C_A^g$, and (2) if reaction is controlling, $k_\sigma \ll k_c$, $r_\sigma \rightarrow k_\sigma C_A^g$.

Example 2.54   Reaction and Diffusion Limited Regions

The transport limited mass transfer coefficient can be related to other system parameters such as the flow velocity, particle size, and fluid properties using correlations, for example, the *Fröessling correlation*:

$$Sh = \frac{k_c D_p}{D_A} = 2 + 0.6 \, Re^{1/2} \, Sc^{1/3} \tag{2.404}$$

where
   Sh is the Sherwood number
   Re is Reynold's number $(= u D_p / \nu)$
   Sc is the Schmidt number $(= \nu / D_A)$

For Re > 25, this equation can be simplified to get

$$Sh = 0.6 \cdot Re^{1/2} \, Sc^{1/3} \tag{2.405}$$

Thus, we have the following equation for k_c:

$$k_c = 0.6 \cdot \left(\frac{D_A^{2/3}}{\nu^{1/6}} \right) \cdot \left(\frac{u}{D_p} \right)^{1/2} \tag{2.406}$$

For aqueous solutions, $D_A \approx 10^{-10} \, m^2 \, s^{-1}$, $\nu = 1 \times 10^{-6} \, m^2 \, s^{-1}$, and $k_c \approx 1.2 \times 10^{-6}$ $(u/D_p)^{1/2}$. Using a typical environmental reaction condition, $C_A^g \approx 10^{-5} \, mol \, m^{-3}$, and $k_A^\sigma \approx 0.1 \, s^{-1}$, we obtain

$$r = \frac{\left(1 \times 10^{-6} \right)}{1 + \dfrac{8.3 \times 10^4}{\left(u / D_p \right)^{1/2}}} \tag{2.407}$$

This rate is plotted in Figure 2.56. There are two distinct regions. In the first, the rate is proportional to $(u/D_p)^{1/2}$ and is called the "diffusion-limited region." In this region, r increases with increasing D_p and decreasing u. In the second region, r is independent of $(u/D_p)^{1/2}$ and is called the "reaction-limited region." Diffusion limited reactions are important in several important environmental processes.

2.4.1.4.1   Kinetics and Transport at Fluid/Fluid Interfaces

Consider a gas and a liquid in contact (Figure 2.57). Let us consider a component i from the gas phase that exchanges with the liquid phase. This component can either be reactive in the liquid phase or not. Let us first consider the base case where there is no reaction in either phase. The two bulk fluids are completely mixed so that the concentration of species is C_{Gi}^∞ and C_{Li}^∞. Note that we can also represent the gas phase composition by a partial pressure P_i. Mixing and turbulence in the bulk phase quickly disperse the species in the solution. However, near the interface on both sides there is insignificant mixing (low turbulence). Hence, diffusion through the interfacial films

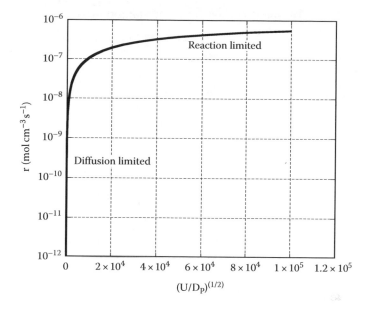

FIGURE 2.56 Diffusion and reaction-limited regions for a heterogeneous reaction involving transport of reactants to the surface.

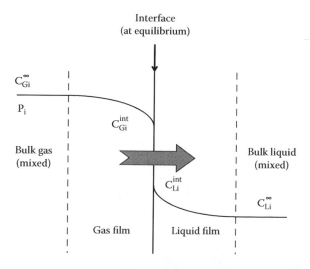

FIGURE 2.57 Schematic of the two-film theory of mass transfer for transfer of a solute from the gas to the liquid, namely absorption. Note that equilibrium exists only at the arbitrary dividing plane called the interface. Mixing is complete in both bulk phases at distances away from the interface.

limits mass transfer. The interface, however, is at equilibrium and air-water partitioning equilibrium applies so that

$$C_{Gi}^{int} = K_{AW} C_{Li}^{int} \qquad (2.408)$$

The rate of mass transfer from gas to liquid through the gas film is given by

$$r_1 = k_G \left(C_{Gi}^{\infty} - C_{Gi}^{int} \right) \qquad (2.409)$$

and that through the liquid film

$$r_2 = k_L \left(C_{Li}^{int} - C_{Li}^{\infty} \right) \qquad (2.410)$$

At steady state, the rate of at which the mass reaching the interface from the gas-side should equal that leaving through the liquid film, that is, $r_1 = r_2$,

$$k_G \left(C_{Gi}^{\infty} - C_{Gi}^{int} \right) = k_L \left(C_{Li}^{int} - C_{Li}^{\infty} \right) \qquad (2.411)$$

Since C_{Gi}^{int} and C_{Li}^{int} are not known or obtainable from experiments, we need to eliminate these using the air-water partition equilibrium to get

$$C_{Gi}^{int} = \frac{\left(k_G C_{Gi}^{\infty} + k_L C_{Li}^{\infty} \right)}{\left(k_G + \dfrac{k_L}{K_{AW}} \right)} \qquad (2.412)$$

Hence,

$$r_1 = \frac{1}{\left(\dfrac{1}{k_L} + \dfrac{1}{k_G K_{AW}} \right)} \cdot \left(\frac{C_{Gi}^{\infty}}{K_{AW}} - C_{Li}^{\infty} \right) \qquad (2.413)$$

is the rate of transfer from gas to liquid ("absorption"). If the transfer is from liquid to gas ("stripping"), the rate is

$$r_1 = \frac{1}{\left(\dfrac{1}{k_L} + \dfrac{1}{k_G K_{AW}} \right)} \cdot \left(C_{Li}^{\infty} - \frac{C_{Gi}^{\infty}}{K_{AW}} \right) \qquad (2.414)$$

If the rate of absorption is expressed in the form

$$r_1 = K_L \cdot \left(\frac{C_{Gi}^{\infty}}{K_{AW}} - C_{Li}^{\infty} \right) \qquad (2.415)$$

where K_L is the overall mass transfer coefficient based on the bulk phase concentrations, we recognize that

$$\frac{1}{K_L} = \frac{1}{k_L} + \frac{1}{k_G K_{AW}}$$ (2.416)

where each term represents a resistance to mass transfer.

$$R_T = R_L + R_G$$ (2.417)

The rate can also be expressed as follows:

$$r_l = K_G \cdot \left(C_{Gi}^{\infty} - K_{AW} C_{Li}^{\infty} \right)$$ (2.418)

where

$$\frac{1}{K_G} = \frac{1}{k_G} + \frac{K_{AW}}{k_L}$$ (2.419)

The general equation for the rate of absorption in the case of an enhancement in the liquid phase due to reaction is given by

$$r_l = \frac{1}{\left(\dfrac{1}{k_L} + \dfrac{K_{AW}}{k_G \cdot E} \right)} \cdot C_{Gi}^{\infty}$$ (2.420)

where E is the enhancement in mass transfer due to reaction

$$E \equiv \frac{\text{Rate of uptake with reaction}}{\text{Rate of uptake without reaction}}$$ (2.421)

For example, consider an instantaneous reaction given by A (g) + B (aq) → products, the enhancement factor E is $1 + \dfrac{D_{B,aq}}{D_{A,aq}} \cdot \dfrac{C_{B,aq}}{C_A^{int}}$.

The derivation of the two-film mass transfer rate at fluid-fluid interfaces can be generalized to a number of other cases in environmental engineering such as soil-water and sediment-water interfaces.

2.4.1.5 Diffusion and Reaction in a Porous Medium

The lithosphere (e.g., soils, sediments, aerosols, activated carbon) is characterized by an important property, namely, its porous structure. Therefore, all of the accessible area around a particle is not exposed to the pollutants in the bulk fluid (air or water). The diffusion of pollutants within the pores will lead to a concentration gradient from the particle surface to the pore. The overall resistance to mass transfer from the bulk fluid to the pore will be composed of the following: (1) diffusion resistance within the thin-film boundary layer surrounding the particle, (2) diffusion resistance within the pore fluid, and (3) the final resistance from that due to the reaction at the solid-liquid boundary within the pores. This is shown schematically in Figure 2.58a and b.

If the external film diffusion controls mass transfer, the concentration gradient is $C_A^\infty - C_A^\sigma$, and the diffusivity of A is the molecular diffusivity in the bulk liquid phase, D_A. If internal resistance controls the mass transfer, the gradient is $C_A^\infty - C_A(r)$, but the diffusivity is different from D_A. The different diffusivity results from the fact that the solute has to diffuse through the tortuous porous space within the particle. A tortuosity factor τ is defined, which is the ratio of the actual path length between two points to the shortest distance between the same two points. Since only a portion of the solid particle is available for diffusion, we have to consider the porosity ε of the medium. ε is defined as the ratio of the void volume to the total volume. Thus, the bulk phase diffusivity D_A is corrected for internal diffusion by incorporating ε and τ.

$$D_A^{eff} = D_A \cdot \frac{\varepsilon}{\tau} \tag{2.422}$$

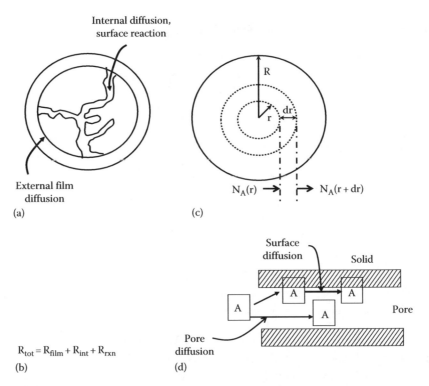

FIGURE 2.58 Schematic of diffusion and reaction/sorption in a porous medium. (a) Internal and external resistance to mass transfer and reaction/sorption within a spherical porous particle. (b) Various resistance to mass transfer and reaction/sorption. (c) Material balance on a spherical shell. (d) Simultaneous bulk diffusion and surface diffusion within a pore.

For most environmental applications, ε/τ is represented by the "Millington-Quirk approximation," which gives $\theta^{10/3}/\varepsilon^2$, where θ is the volumetric content of the fluid (air or water).

The process of reaction occurring within the pore space can be either a surface transformation of $A \to B$, or simply a change from the pore water to the adsorbed state. We shall represent this by a general first-order surface reaction such that

$$r_\sigma = k_\sigma C_{A\sigma} \qquad (2.423)$$

where

k_σ has units of length per time so that r_σ can be expressed in moles per area per time
$C_{A\sigma}$ is expressed in moles per volume

Consider Figure 2.58c. A mass balance on the spherical shell of thickness Δr can be made. The diffusion of solute into the center of the sphere dictates that the flux expression should have a negative sign so that $N_A(r)$ is in the direction of increasing r. The overall balance is Flux of A in at r – Flux of A out at $(r + \Delta r)$ + Rate of generation of A by reaction = Rate of accumulation of A in the solid. The rate of accumulation of A in the solid is given by $\varepsilon \partial C_A/\partial t$, with ε being the porosity and C_A the concentration of A per unit volume of the void space in the solid.

$$N_A(r) \cdot 4\pi r^2 \big|_r - N_A(r) \cdot 4\pi r^2 \big|_{r+\Delta r} + r_\sigma A_s \cdot 4\pi r^2 \cdot \Delta r = \varepsilon \cdot \frac{\partial C_A}{\partial t} \qquad (2.424)$$

where A_s is the internal surface area per unit volume and $4\pi r^2 \Delta r$ is the volume of the shell. At steady state, $\partial C_A/\partial t$ will be zero. Then, by dividing the expression into $4\pi\Delta r$, and letting $\Delta r \to 0$, we can obtain

$$\frac{d}{dr}\left(r^2 N_A(r)\right) - r^2 r_\sigma A_s = 0 \qquad (2.425)$$

The expression for $N_A(r)$ is given by Fick's law of diffusion, $N_A(r) = -D_A^{eff} \frac{dC_A}{dr}$. The surface reaction rate $r_\sigma = k_\sigma C_A(r)$.

Hence,

$$\frac{1}{r^2}\frac{d}{dr}\left(r^2 \frac{dC_A}{dr}\right) - \frac{k_\sigma A_s}{D_A^{eff}} C_A = 0 \qquad (2.426)$$

This differential equation has to be solved with appropriate boundary conditions to get $C_A(r)$.

Let the outer surface of the sphere be at a constant concentration $C_{A\sigma}$. This implies that external film diffusion is not important. Hence the first boundary condition is $C_A = C_{A\sigma}$ at $r = R$. A second boundary condition is that the concentration at the center of the spherical particle is finite, that is, $dC_A/dr|_{r=0} = 0$. At this point, we can introduce a set of dimensionless variables that will simplify the form of the differential equation. These are $\Psi = C_A(r)/C_{A\sigma}$ and $\Lambda = r/R$. This gives

$$\frac{1}{\Lambda^2}\frac{d}{d\Lambda}\left(\Lambda^2 \frac{d\Psi}{d\Lambda}\right) - \Phi^2 \Psi = 0 \qquad (2.427)$$

where $\Phi = \left(\dfrac{k_\sigma A_s}{D_A^{eff}}\right)^{1/2}$ is called the "Thiele modulus." Upon inspection of the equation for Φ, we note that it is the ratio of the surface reaction rate to the rate of diffusion. Thus, the Thiele modulus gives the relative importance of reaction and diffusion rates. For large Φ, the surface reaction rate is large and diffusion is rate limiting, whereas for small Φ, surface reaction is rate limiting. The transformed boundary conditions are $\Phi = 1$ at $\Lambda = 1$, and $d\Psi/d\Lambda|_{\Lambda = 0} = 0$. The solution to this differential equation is (Smith, 1970; Fogler, 2006)

$$\frac{C_A}{C_{A\sigma}} = \frac{1}{\Lambda} \cdot \frac{\sinh(\Phi\Lambda)}{\sinh(\Phi)} \qquad (2.428)$$

Figure 2.59a is a plot of C_A/C_{As} versus Λ (i.e., r) for three values of the Thiele modulus, Φ.

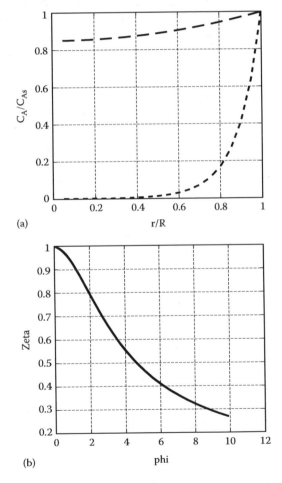

(a)

(b)

FIGURE 2.59 (a) Variation in pollutant concentration with radial position for various Thiele modulus. (b) Variation of overall effectiveness factor with Thiele modules.

In the field of catalysis, it is conventional to define a related term called the "over-all effectiveness factor" ξ, which is an indication of how far a molecule can diffuse within a solid before it disappears via reaction. It is defined as the ratio of the actual overall reaction rate to the rate that would result if the entire interior surface were exposed to the surface concentration C_{As}. It is given by (Fogler, 2006)

$$\xi = \frac{3}{\Phi^2}\left(\Phi \coth \Phi - 1\right) \tag{2.429}$$

This function is plotted in Figure 2.59b, which shows that with increasing Φ, the effectiveness factor decreases. With increasing Thiele modulus, the accessibility of the reactant to the interior surface sites is reduced, and consequently the process is diffusion limited within the particle. The overall reaction rate is therefore $r_{A\sigma} = \xi k_1 C_{A\sigma}$ if the surface reaction is first order.

Let us now include the external film resistance to mass transfer. This can be introduced to make the problem general and also to evaluate the relative importance of processes external and internal to the particle. At steady state, the net rate of transfer of mass to the surface of the particle should equal the net reaction (on exterior surface and interior surface). Thus,

$$k_{mt}\left(C_A^\infty - C_{A\sigma}\right)A_{ext} = r_{A\sigma}A_{int} \tag{2.430}$$

A_{ext} is the external surface area of the particle and the internal surface area is A_{int}. Utilizing the expression for $r_{A\sigma}$, we have upon rearranging,

$$C_{A\sigma} = \frac{k_{mt}A_{ext}C_A^\infty}{k_{mt}A_{ext} + \xi k_\sigma A_{int}} \tag{2.431}$$

The steady state rate of mass transport to the surface is then given by

$$r_{mt}^{ss} = r_{A\sigma}A_{int} = \xi k_\sigma A_{int}C_{A\sigma} = \omega k_\sigma A_{int}C_A^\infty \tag{2.432}$$

where $\omega = \dfrac{\xi}{1 + \left(\xi \cdot \dfrac{k_1 A_{int}}{k_\sigma A_{ext}}\right)}$ denotes the change in the effectiveness factor as the exter-nal mass transfer becomes significant. The transfer rate is dependent only on the bulk phase (air or water) concentration. As k_σ becomes smaller ω decreases and the external mass transfer resistance becomes important.

Most reactions in environmental systems occur in assemblages of porous particles. This gives rise to both macropores and micropores within the medium. The diffusion and kinetics in such a porous bed medium can be modeled as a catalytic reactor with the aim of obtaining the degree of conversion at any defined position within the reactor. The analysis can be used to model a waste treatment unit operation such as activated carbon, ion-exchange, or other physicochemical treatment processes. It can also be applied to the modeling of fate and transport of pollutants in a porous medium such as a soil or sediment.

PROBLEMS

2.1$_1$ Classify the following properties as intensive and extensive. Give appropriate explanations: Temperature, entropy, pressure, volume, number of moles, density, internal energy, enthalpy, molar volume, mass, chemical potential, Helmholtz free energy.

2.2$_1$ Given the ideal gas law, $pV = nRT$, find q for an isothermal reversible expansion. What is the work required to compress 1 mol of an ideal gas from 1 to 100 atm at room temperature? Express the work in Joules, ergs, calories, and liter-atm. units. Notice the latter three are not SI units.

2.3$_2$ For an ideal gas show that for an adiabatic expansion TV^r is a constant. r is given by R/C_v.

2.4$_2$ For an ideal gas show that the difference between C_p and C_v is the gas constant R.

2.5$_1$ Calculate the rate of change in temperature per height for the rise of 1 g dry air parcel in the lower atmosphere. The actual rate is $6.5°C \ km^{-1}$. Suggest reasons for the difference. $C_{p,m} = 1.005 \ kJ \ (kg \ °C)^{-1}$.

2.6$_2$ Calculate the heat of vaporization of water at 298 K given its value at 393 K is $2.26 \times 10^6 \ J \ kg^{-1}$. Specific heat of water is $4.184 \times 10^3 \ J \ K^{-1} \ kg^{-1}$ and its heat capacity at constant pressure is $33.47 \ J \ K^{-1} \ mol^{-1}$. Assume that the heat capacity is constant within the range of temperature.

2.7$_2$ Find the heat capacity at constant pressure in $J \ K^{-1} \ mol^{-1}$ for $CHCl_3$ at 400 K. Use data from Table 2.1. Obtain the heat capacity at constant volume at 400 K.

2.8$_2$ $0.5 \ m^3$ of nitrogen at 400 kPa and 300 K is contained in a vessel insulated from the atmosphere. A heater within the device is turned on and allowed to pass a current of 2 A for 5 min from a 120 V source. The electrical work is done on the system to heat the nitrogen gas. Find the final temperature of the gas in the tank. The C_p for nitrogen is $1.039 \ kJ \ (kg \ K)^{-1}$.

2.9$_2$ A 5 L tank contains 2 mol of methane gas. The pressure of the gas is increased from 2 to 4 atm while maintaining a constant volume of the tank. What is the total internal energy change during the process? Heat capacity at constant pressure for methane is $34 \ J \ (mol \ K)^{-1}$ and is constant.

2.10$_2$ A 5 L plastic coca-cola bottle contains air at 300 K and 12.5 bar gauge pressure. How much work the gas would do if you could expand the gas so that the final pressure is 1 bar isothermally and reversibly?

2.11$_2$ The Melvin Equation of state was found to be $(P - b)V = nR(T - a)^2$ where a and b are constants that adjust for the phase of the system. Derive an expression for Coefficient of Thermal Expansion: $\alpha = -\dfrac{1}{V}\left(\dfrac{\partial V}{\partial T}\right)_P$ using the Melvin Equation of State. NOTE: The Melvin Equation of State is not based on any experimental data or system. Do not use this equation of state in real world applications.

2.12$_2$ A closed container containing a 20–80 mol mixture of water and carbon dioxide at 2 atm and 25°C is set in direct sunlight on a hot day. The vapor pressures of water and carbon dioxide at 25°C are 0.031 and 63.56 atm. The container's temperature is raised to 40°C and the pressure inside the can is raised to 2.25 atm.

1. Determine the vapor and liquid compositions of CO_2 and water after the can is heated if the standard heat of vaporization of water and carbon dioxide is 40.86 kJ/mol and 15.325 kJ/mol. The constant-pressure heat capacity of water is 75.327 J $(mol*K)^{-1}$. The constant-pressure heat capacity of carbon dioxide is 36.94 J $(mol*K)^{-1}$.

2. If the can is suddenly open to the atmosphere, but the temperature does not change, what is the vapor (just above the liquid assuming no air is in the mix) and liquid compositions of CO_2 and water assuming that the temperature does not change? What is the Henry's constant of CO_2?

2.13_1 Supercooled water at $-3°C$ is frozen at atmospheric pressure. Calculate the maximum work for this process. Density of water is 0.999 g cm^{-3} and that of ice is 0.917 g cm^{-3} at $-3°C$.

2.14_1 An ideal gas is subjected to a change in pressure from 2 to 20 atm at a constant temperature of 323 K. Calculate the change in chemical potential for the gas.

2.15_2 A reaction that is of considerable importance in nature is the transformation of methane utilizing ozone, $3CH_4(g) + 4O_3(g) \rightarrow 3CO_2(g) + 6H_2O(l)$. This reaction is the main source for water in the stratosphere and is augmented by photochemical processes.

1. Let us carry out the reaction in a laboratory at 298 K. If the standard heat of formation of ozone is 142 kJ mol^{-1}, and that of carbon dioxide, water and, methane are $-393, -285$, and -75 kJ mol^{-1}, respectively, at 298 K, determine the standard heat of the overall reaction at 298 K. Compare this enthalpy change with the reaction of methane with oxygen in the previous problem.

2. If the standard Gibbs free energy of formation of ozone, carbon dioxide, water, and methane are $163, -394, -237$, and -50 kJ mol^{-1}, respectively, at 298 K, calculate the standard free energy change of the overall reaction at 298 K.

3. What will be the free energy change if the temperature is decreased to 278 K? Is the reaction more or less favorable at this temperature?

2.16_3 Capillary rise is an important phenomenon in enhanced oil recovery, recovery of nonaqueous phase liquids from contaminated aquifers, rise of sap in trees, and also in the determination of surface tension of liquids. This is based on the balance of forces between the surface tension and hydrostatic forces due to gravity.

1. Consider a small capillary partially immersed in water. The water rises to an equilibrium height given by h. Equate the pressure given by the Young-Laplace equation for the hemispherical interface with the hydrostatic pressure and derive the equation $\sigma = \dfrac{rgh}{2} \Delta\rho$, where r is the radius of the capillary, g is gravitational constant, and $\Delta\rho$ is the difference in density between the water and air.

2. Soils are considered to have capillary size pores. Calculate the capillary rise of water in soil pores of diameter 100, 1,000, and 10,000 μm. This gives the shape of the boundary between air and water in soil pores. Repeat the calculation for a nonaqueous phase contaminant such as chloroform in contact with groundwater. $\sigma = 20$ mN m^{-1} and $\Delta\rho = 150$ kg m^{-3}. What conclusions can you draw about the shape of the boundary in this case?

2.17$_1$ Calculate the minimum work necessary to increase the area of the surface of water in a beaker from 0.001 to 0.005 m^2 at 298 K.

2.18$_2$ A parcel of water 1 kg in mass is located 1 km above a large body of water. The temperature of the air, lake, and the water parcel is uniform and remains at 22°C. What will be the change in entropy if the water parcel descends and mixes with the lake water and reaches equilibrium with it?

2.19$_2$ A gas container has compressed air at 5×10^5 Pa at 300 K. What is the maximum useful work available from the system per kilogram of air? The atmospheric pressure is 1×10^5 Pa at 300 K?

2.20$_2$ Hydroxyl radical (OH•) plays an important role in atmospheric chemistry.
  1. It is produced mainly via the reaction of photoexcited oxygen atom (O^*) with moisture: $O^* + H_2O \rightarrow 2OH•$. Using Appendix B and the fact that the heat of formation of O^* in the gaseous state is 440 kJ mol^{-1}, obtain the enthalpy of this reaction at standard state.
  2. OH• is an effective scavenger of other molecules, for example, OH• + $NO_2 \rightarrow HNO_3$. Obtain the standard enthalpy of the reaction.

2.21$_1$ Oxides of sulfur are an important class of inorganic pollutants in the atmosphere resulting mainly from coal burning. For the oxidation of SO_2 in air by the reaction $SO_2(g) + ½O_2(g) \rightarrow SO_3(g)$, what is the free energy at 298 K? Use Appendix B.

2.22$_2$ Sun's heat on a lake surface leads to evaporation of water. If the heat flux from the sun is assumed to be 5 J min^{-1} cm^{-2}, how much water will be evaporated on a clear summer day from a square meter of the lake surface in 5 h? What volume of the atmosphere is required to hold the evaporated water? Assume that both air and water remain at a constant temperature of 308 K. Heat of vaporization of liquid water at 308 K is 2425 kJ kg^{-1}.

2.23$_3$ The predominant form of $CaCO_3$ in nature is calcite. Marine shells are made up of this form of calcium carbonate. However, there also exists another form of $CaCO_3$, called aragonite. They differ in properties at 298 K as follows:

Property	Calcite	Aragonite
Standard free energy of formation (kJ mol^{-1})	−1128.7	−1127.7
Density (g cm^{-3})	2.71	2.93

  1. Is the conversion of calcite to aragonite spontaneous at 298 K?
  2. At what pressure will the conversion be spontaneous at 298 K?

2.24$_3$ Municipal landfills contain refuse that are continually decomposed by indigenous bacteria. This produces large quantities of methane gas. Let us consider a large landfill that contains approximately 10^{10} m^3 of methane. Is it possible that this landfill gas by being allowed to expand and cool to ambient conditions (without combustion) can be used to produce power? In other words, you are asked to determine the useful work that can be obtained. Assume the following: The heat capacity at constant pressure is 36 J (mol K)$^{-1}$, pressure in the landfill is 8 atm and temperature in the landfill is 240°C. The molecular weight of methane is 0.016 kg mol^{-1}.

2.25$_2$ Methyl mercury $(CH_3)_2Hg$ is the most easily assimilable form of mercury by humans. It is present in both air and water in many parts of the United States and the world. Consider the following process by which it converts to elemental Hg: $(CH_3)_2Hg \rightarrow C_2H_6 + Hg$. Is the reaction favorable under ambient conditions? Look up ΔG_f^σ for the compounds in appropriate references.

2.26$_2$ For the reaction H_2S (g) $\rightarrow H_2S$ (aq), the enthalpy of dissolution is -19.2 kJ mol^{-1} at 298 K. If the temperature is raised to 318 K, by how much will the equilibrium constant of the reaction increase?

2.27$_1$ State whether the following statements are true or false:
1. The surface temperature in Baton Rouge was 94°F and 500 m above the temperature was 70°F on 9/15/97. The atmosphere was stable.
2. Surface tension can be expressed in Pa m^{-2}.
3. Entropy of an ideal gas is only a function of temperature.
4. The first law requires that the total energy of a system can only be conserved *within* the system.
5. The entropy of an isolated system must be a constant.
6. C_p of an ideal gas is independent of P.
7. Heat is *always* given by the integral of TdS.
8. Chemical potential is an extensive property.
9. For an adiabatic process, dU and δw are not equal.
10. Heat of reaction can be obtained from the heats of formation of reactants and products.
11. If $\Delta H > 0$, the reaction is always possible.
12. Although S_{sys} may increase, decrease, or remain constant, S_{univ} cannot decrease.

2.28$_2$ The molar Gibbs free energy for CO_2 in the gas phase is -394.4 kJ mol^{-1} and that in water is -386.2 kJ mol^{-1}. Is dissolution of CO_2 a spontaneous process at 298 K and 1 atm total pressure?

2.29$_1$ What pressure is required to boil water at 393 K? The heat of vaporization of water is 40.6 kJ mol^{-1}.

2.30$_2$ What is the pH of pure water at 37°C? For the reaction, $H_2O \Leftrightarrow H^+ + OH^-$, K_{eq} is 1×10^{-14} M^2 at 25°C and enthalpy of dissociation is 55.8 kJ mol^{-1}.

2.31$_3$ A human being is an open system and maintains a constant body temperature of 310 K by removing excess heat via evaporation of water through the skin. Consider a person weighing 175 lb capable of generating heat by digesting 100 mol of glucose through the reaction: $C_6H_{12}O_6$ (s) $+ 6O_2$ (g) $\rightarrow 6CO_2$ (g) $+ 6H_2O$ (l). How much water will have to be evaporated to maintain the body temperature at 310 K? C_p for water is 33.6 J (K mol)$^{-1}$. For water, $\Delta H_v^o = 40.5$ kJ mol^{-1}. Ambient temperature is 298 K.

2.32$_3$ A fundamental energy giving reaction for the creation of the primordial living cell is proposed to be the formation of iron pyrites by the reaction: FeS (s) $+$ H_2S (aq) $\rightarrow FeS_2$ (s) $+ 2H^+$ (aq) $+ 2e^-$ (aq). The energy released by this reaction is used to break up CO_2, and form the carbon compounds essential to life. Estimate the maximum work obtainable from this reaction under standard conditions.

2.33_2 For a hurricane (Figure 2.2), follow Example 2.2 to derive the maximum wind velocity predicted from theory. The specific enthalpy of air under moist conditions is given by $C_{p,a}t + (C_{p,w}\,t + h_{w,e})\,x_w$, where $C_{p,a} = 1.006$ kJ (kg °C)$^{-1}$ and $C_{p,w} = 1.84$ kJ (kg °C)$^{-1}$ are the specific heat capacities of air and water, respectively. The water content of moist air is $x_w = 0.0203$ kg kg^{-1}. Temperature t is given in °C and enthalpy of water evaporation $h_{w,e} = 2520$ kJ kg^{-1}. The surface temperature of air in contact with warm, moist water can be taken to be 27°C, while that of ambient dry air is 23°C. Give the answer for velocity in miles per hour.

2.34_1 Calculate the mole fraction, molarity, and molality of each compound in an aqueous mixture containing 4 g of ethanol, 0.6 g of chloroform, 0.1 g of benzene, and 5×10^{-7} g of hexachlorobenzene in 200 mL of water.

2.35_1 A solution of benzene in water contains 0.002 mole fraction of benzene. The total volume of the solution is 100 mL and has a density of 0.95 g cm^{-3}. What is the molality of benzene in solution?

2.36_2 Given the partial pressure of a solution of KCl (4.8 molal) is 20.22 Torr at 298 K and that of pure water at 298 K is 23.76 Torr, calculate the activity and activity coefficient of water in solution.

2.37_2 Give that the $\mu^0 = -386$ kJ mol^{-1} for carbon dioxide on the molality scale, calculate the standard chemical potential for carbon dioxide on the mol fraction and molarity scale.

2.38_1 Estimate the mean ionic activity coefficients of the following aqueous solutions: (1) 0.001 molal KCl, (2) 0.01 molal CaCl$_2$, and (3) 0.1 molal Na$_3$PO$_4$.

2.39_2 The problem of oxygen solubility in water is important in water quality engineering. Reaeration of streams and lakes is important for maintaining an acceptable oxygen level for aquatic organisms to thrive.
1. The solubility of oxygen in water at 1 bar pressure and 298 K is given as 1.13×10^{-3} molal. Calculate the free energy of solution of oxygen in water.
2. It is known that atmospheric air has about 78% by volume of nitrogen and 21% by volume of oxygen. If Henry's constant (K_H) for oxygen and nitrogen are 2.9×10^7 and 5.7×10^7 Torr, respectively, at 293 K, calculate the mass of oxygen and nitrogen in 1 kg of pristine water at atmospheric pressure.
3. If Raoult's law is obeyed by the solvent (water), what will be the partial pressure of water above the solution at 298 K. The vapor pressure of pure water is 23.756 Torr at 298 K.

2.40_2 Henry's constants for oxygen at three different temperatures are given here:

T (K)	K_H (Torr)
293	2.9×10^7
298	3.2×10^7
303	3.5×10^7

1. Calculate the heat of solution of oxygen in water assuming no dependence on temperature between 293 and 303 K.
2. Estimate the standard molar entropy of solution at 298 K.

2.41_3 Henry's constants for ethylbenzene at different temperatures are given here:

T (K)	K'_{AW} (atm m³ mol⁻¹)
283	0.00326
288	0.00451
293	0.00601
298	0.00788
303	0.01050

1. Express Henry's constant in various forms (K_H, K_{AW}, K''_{AW}, and K_x) at 298 K.
2. Calculate the enthalpy of solution for ethylbenzene.

2.42_3 For naphthalene the following vapor pressure data was obtained from the literature (Sato et al., 1986; Schwarzenbach et al., 1993)

T (K)	Vapor Pressure (Pa)
298.5	10.7
301.5	14.9
304.5	19.2
307.3	25.9
314.8	52.8
318.5	71.1
322.1	97.5
322.6	101
325.0	124
330.9	200
358.0	1275
393.0	5078
423.0	12756

1. Plot the log of vapor pressure as a function of 1/T and explain the behavior.
2. Estimate from the data the heat of sublimation of naphthalene and the heat of melting (fusion) of solid naphthalene to subcooled liquid.
3. Estimate the vapor pressure of naphthalene over the temperature range given using the equations from the text and compare with the actual values.

2.43_2 The enthalpy of vaporization of chloroform at its boiling point 313.5 K is 28.2 kJ mol⁻¹ at 25.3 kPa (atmospheric pressure).
1. Estimate the rate of change of vapor pressure with temperature at the boiling point.
2. Determine the boiling point of chloroform at a pressure of 10 kPa.
3. What will be its vapor pressure at 316 K?

2.44_3 1. The normal boiling point of benzene is 353.1 K. Estimate the boiling point of benzene in a vacuum still at 2.6 kPa pressure. Use Trouton's rule.
2. The vapor pressure of solid benzene is 0.3 kPa at 243 K and 3.2 kPa at 273 K. The vapor pressure of liquid benzene is 6.2 kPa at 283 K and 15.8 kPa at 303 K. Estimate the heat of fusion of benzene.

2.45$_3$ 0.5 m^3 of nitrogen at 400 kPa and 27°C is contained in a vessel kept insulated from the atmosphere. A heater within the device is turned on and allowed to pass current of 2 A for 5 min from a 120 V source. Thus, electrical work is done to heat the nitrogen gas. Find the final temperature for nitrogen in the tank. C$_P$ for nitrogen gas is 1.039 kJ kg K^{-1}.

2.46$_2$ What pressure is required to boil water at 393 K? The enthalpy of vaporization of water is 40.6 kJ mol^{-1}.

2.47$_2$ The equilibrium constant for the dissociation reaction A$_2$ ↔ 2A is 6 × 10^{-12} atm at 600 K and 1 × 10^{-7} atm at 800 K. Calculate the standard enthalpy change for this reaction assuming that the enthalpy is constant. What is the entropy change at 600 K?

2.48$_2$ In a chemodynamics laboratory, an experiment involving the activity of sediment dwelling worms in a small aquarium was found to be seriously affected for a few days. The problem was traced to bacterial growth arising from dirty glassware. Bacteria can survive even in boiling water at atmospheric pressure. However, literature data suggested that they can be effectively destroyed at 393 K. What pressure is required to boil water at 393 K to clean the glassware? Heat of vaporization of water is 40.6 kJ mol^{-1}.

2.49$_2$ Obtain the enthalpy of vaporization of methanol. The vapor pressures at different temperatures are given here:

Temperature (°C)	Vapor Pressure (mmHg)
15	74.1
20	97.6
25	127.2
30	164.2

2.50$_1$ Can an assemblage of liquid droplets be in true equilibrium with its bulk liquid? Explain your answer.

2.51$_2$ The formation of fog is accompanied by a simultaneous decrease in temperature and an increase in relative humidity. If water in the atmosphere is cooled rapidly to 298 K, a degree of supersaturation is reached when water drops begin to nucleate. It has been observed that for this to happen, the vapor pressure of water has to be at least 12.7 kPa. What will be the radius of a water droplet formed under supersaturation? How many water molecules will be in the droplet? Conventional wisdom says that the number of water molecules on a drop should be less than the total number. Calculate the ratio of water molecules on the surface to the number in the droplet. Comment on your result. Molecular area of water = 33.5 Å^2.

2.52$_1$ The value of K$_{ow}$ as a ratio of molar concentrations is always smaller than that expressed as a ratio of mol fractions. Explain why.

2.53$_3$ Using the values from Appendix C and the rules given in Lyman et al. (1990), obtain the log K$_{ow}$ for the following compounds. Obtain the correct structures for the compounds from standard organic chemistry texts:
1. Pentachlorophenol
2. Hexachlorobenzene

3. p,p'-DDT
4. Lindane
5. Phenanthrene
6. Isopropyl alcohol

2.54_2 The melting point of naphthalene is 353.6 K. Its solubility in water at room temperature has been measured to be 32 mg L^{-1}. (1) Calculate the activity coefficient of naphthalene in water. (2) What is its ideal solubility in water?

2.55_2 The solubility of anthracene in water at different temperatures has been reported by May et al. (1983) as follows:

t (°C)	Solubility (μg kg^{-1})
5.2	12.7
10.0	17.5
14.1	22.2
18.3	29.1
22.4	37.2
24.6	43.4

Determine the excess thermodynamic functions for the solution of anthracene in water at 25°C. ΔH_m is 28.8 kJ mol^{-1}.

2.56_1 The cavity surface area for accommodating benzene in water is 240.7 Å^2, whereas the molecular area of benzene is only 110.5 Å^2. Can you explain the difference?

2.57_2 Calculate the aqueous solubility of hexachlorobenzene in water using appropriate correlations and calculations involving (1) molecular area, (2) molecular volume, and (3) molar volume.

2.58_2 The compound 2,2',3,3'-tetrachlorobiphenyl is a component of Aroclor, a major contaminant found in sedimentary porewater of New Bedford harbor, Massachusetts. In order to perform fate and transport estimates on this compound, we need first and foremost its aqueous solubility in seawater.
1. Use the fragment constant approach to obtain K_{ow}.
2. Obtain the aqueous solubility in ordinary water from the appropriate K_{ow} correlation.
3. Check the uncertainty in the prediction using alternate methods such as molecular area (225.6 Å^2) and molar volume (268 cm^3 mol^{-1}).

2.59_2 The mean activity coefficients of KCl in water at 25°C are given here:

Molality of KCl	γ_γ
0.1	0.77
0.2	0.72
0.3	0.69
0.5	0.65
0.6	0.64
0.8	0.62
1	0.60

Test whether this data is in compliance with the Debye-Huckel limiting law.

2.60_2 State whether the following statements are true or false:
1. Octanol-water partition constant is a measure of the aqueous activity of a species.
2. Excess free energy of dissolution for a large hydrophobic molecule is positive since the excess enthalpy is always positive.
3. Methanol does not increase the solubility of dichlorobenzene in water.
4. At equilibrium the standard state chemical potential for a species i in air and water is the same.
5. Fugacity is a measure of the degree to which equilibrium is established for a species in a given environmental compartment.
6. If two aqueous solutions containing different nonvolatile solutes exhibit the same vapor pressure at the same temperature, the activities of water are identical in both solutions.
7. If two liquids (toluene and water) are not completely miscible with one another, then a mixture of the two can never be at equilibrium.

2.61_2 Chlorpyrifos is an insecticide. It is a chlorinated heterocyclic compound and is only sparingly soluble in water. Its properties are $t_m = 42°C$, $M = 350.6$, log $K_{ow} = 5.11$, and $a = 4.16 \times 10^{-3}$ atm mol^{-1}. Estimate its vapor pressure using *only* the given data.

2.62_2 Calculate the infinite dilution activity coefficient of pentachlorobenzene (C_6HCl_5) in water. It is a solid at room temperature and has a melting point of 86°C and a molecular weight of 250.3. It has a log K_{ow} of 4.65.

2.63_1 Match equation on the right with the appropriate statement on the left. Only one equation applies to each statement.

Raoult's law	(A) $G_c + G_t$
Henry's constant	(B) $RT \ln \gamma_i$
Octanol-water partition constant	(C) $(d \ln P/dT) = \Delta H/RT^2$
Ionic strength	(D) $K\, C^{1/m}$
Gibbs excess free energy	(E) $f_i^w = f_i^a$
Freundlich isotherm	(F) $0.5 \sum m_i z_i^2$
Clausius-Clapeyron equation	(G) P_i^*/C_i^*
Free energy of solution	(H) P_i/P_i^*
	(I) $K\, C$
	(J) $H + TS$

2.64_2 A mixture of polystyrene (2) in toluene (1) is said to follow the Flory-Huggins model for activity coefficients. Given that $v_2 = 1000 v_1$ and $\chi = 0.6$, obtain γ_1 and γ_2 as functions of x_2 at 298 K.

2.65_2 A mixture of methyl acetate (1) and methanol (2) has a Margules constant $A = 1.06RT$. Obtain a plot of the activity coefficient of methanol as function of its mole fraction at 298 K. Use the one constant Margules equation.

2.66_2 If sea water at 293 K has the following composition, what is the total ionic strength of sea water?

Electrolyte	Molality
NaCl	0.46
$MgSO_4$	0.019
$MgCl_2$	0.034
$CaSO_4$	0.009

2.67_3 One gram of a soil (total surface area 10 m^2 g^{-1}) from a hazardous waste site in North Baton Rouge, LA, was determined to contain 1.5 mg of chloroform (molecular surface area 24 Å^2) at 298 K. The soil sample was placed in a closed vessel and the partial pressure of chloroform was measured to be 10 kPa at 298 K. The heat of adsorption of chloroform on the soil from the vapor phase was estimated to be -5 kJ mol^{-1}. If Langmuir adsorption isotherm is assumed, what is the adsorption constant at 298 K?

2.68_3 The following data was reported by Roy et al. (1992) for the adsorption of herbicide (2,4-dichlorophenoxy acetic acid designated 2,4D) on a typical Louisiana soil. 5 g of soil was equilibrated with 200 mL of aqueous solution containing known concentrations of 2,4D and kept vigorously shaken for 48 h. The final equilibrium concentrations of 2,4D in water were measured using a liquid chromatograph.

Final Equilibrium Aqueous Concn. (mg L^{-1})	Initial Spike Aqueous Concn. (mg L^{-1})
1.0	1.1
2.9	3.1
6.1	6.2
14.8	15.0
45.3	48.9
101.9	105.3
148.0	152.8
239.5	255.0

Based on appropriate statistics, suggest which of the three isotherms— linear, Langmuir, or Freundlich—best represents the data.

2.69_2 The following adsorption data was reported by Thoma (1992) for the adsorption of a polyaromatic hydrocarbon (pyrene) on a local sediment (University lake, Baton Rouge, LA).

Equilibrium Aqueous Concentration (µg L^{-1})	Adsorbed Concentration (ng g^{-1})
5.1	35.6
7.0	51.8
9.7	51.9
13.4	67.2
13.0	70.9

The molecular weight of pyrene is 202 and its aqueous solubility is 130 $\mu g \, L^{-1}$. Obtain the linear partition constant (in $L \, kg^{-1}$) for pyrene between the sediment and water. What is the adsorption free energy for pyrene on this sediment?

2.70_2 The following data was obtained by Poe et al. (1988) for the adsorption of an aromatic pollutant (benzene) on Arkansas soil (Weller soil, Fayetteville, AR). Plot the data as a BET isotherm and obtain the BET constants. From the BET constants, obtain the surface area of the soil and the adsorption energy for benzene on this soil.

Relative Vapor Pressure	Amount Adsorbed (mg g^{-1})
0.107	7.07
0.139	10.68
0.253	14.18
0.392	18.32
0.466	20.89
0.499	22.15

2.71_2 The following data was reported by Hartkopf and Karger (1973) for the adsorption of hydrocarbon (*n*-nonane) vapor on a water surface at 12.5°C.

Relative Partial Pressure	Amount Adsorbed (mol cm^{-2})
0.07	0.20×10^{-11}
0.12	0.36
0.17	0.48
0.22	0.60
0.25	0.78
0.30	0.94
0.34	1.12
0.38	1.26
0.43	1.48
0.47	1.68
0.52	1.88
0.56	2.10

What type of an isotherm does this data represent? Obtain the isotherm constants and the free energy of adsorption. The reported free energy of adsorption of 12.5°C is $-4.7 \, kcal \, mol^{-1}$.

2.72_1 The following two observations have the same explanation. River water is often muddy, whereas seawater is not. At the confluence of a river with the sea, a large silt deposit is always observed. Based on the principles of charged interfaces, provide a qualitative explanation for these observations.

2.73$_2$ Determine the ionic strength and corresponding Debye lengths for the following solutions: 0.005 M sodium sulfate, 0.001 M calcium sulfate, and 0.001 M ferrous ammonium sulfate.

2.74$_2$ Typical concentrations of ions in natural waters are given here. Calculate the ionic strength and Debye lengths in these solutions:

Medium	Concentration (mmol kg^{-1})							
	Na$^+$	K$^+$	Ca^{2+}	Mg^{2+}	Cl$^-$	SO$_4$$^{2-}$	NO$_3$$^-$	F$^-$
Seawater	468	10	10.3	53	546	28	0.05	0.07
Rainwater	0.27	0.06	0.37	0.17	0.22	0.12	0.02	0.005
Fog water	0.08	—	0.2	0.08	0.2	0.3	—	1

2.75$_2$ Trichloroethylene is used as a dry cleaning fluid and is an environmentally significant air and water pollutant. Obtain the enthalpy of vaporization from the following vapor pressure data:

Temperature (°C)	Vapor Pressure (mm of Hg)
20	56.8
25	72.6
30	91.5

2.76$_2$ Estimate the vapor pressure of (1) chlorobenzene (a liquid) and (2) 1,3,5-trichlorobenzene (a solid) at a temperature of 25°C. The melting point of trichlorobenzene is 63°C. Compare with the reported values in Appendix A.

2.77$_2$ Alkylethoxylates (AE) are nonionic surfactants that find widespread use in cleaning, industrial, and personal care products. As such their environmental mobility is predicated upon a knowledge of their adsorption to sediments. Consider n-alkylethers of poly (ethylene glycol) of general formula CH_3-$(CH_2)_{n-1}$-$(OCH_2CH_2)_xOH$, i.e., A_nE_x. If $n = 13$ and $x = 3$, it has a molecular weight of 332. The following data were obtained for adsorption of $A_{13}E_3$ to a sediment (EPA 12):

Mass in Aqueous Solution (nM)	Mass Adsorbed (μmol kg^{-1})
175	25
270	45
350	80
450	100

Determine whether Langmuir or Freundlich isotherm best fits this data.

2.78_3 DDT is a pollutant that can be transported through the atmosphere to remote locations. The subcooled liquid vapor pressure (below 110°C) and liquid vapor pressure of p,p'-DDT at different temperatures are given here:

t (°C)	P_i^* (Torr)
25	2.58×10^{-6}
90	2.25×10^{-3}
100	5.18×10^{-3}
110	0.0114
120	0.0243
130	0.0497
140	0.0984
150	0.189

The melting point of DDT is 109°C. (1) Obtain from this data the enthalpy of vaporization of DDT. (2) Estimate the vapor pressure of solid DDT at 25°C.

2.79_3 An aqueous solution containing 2.9×10^{-4} mole fraction of CO_2 is in equilibrium with pure CO_2 vapor at 101.325 kPa total pressure and 333 K. Calculate the mole fraction of CO_2 in water if the total pressure is 10132.5 kPa at 333 K. Neglect the effect of pressure on Henry's constant. Also, assume that the vapor phase behaves ideally, whereas the aqueous phase does not.

2.80_2 Given the following data on the aqueous solubility of SO_2 at 298 K, calculate Henry's constant in various forms:

x_i^w	P_i (atm)
3.3×10^{-4}	0.0066
5.7×10^{-4}	0.013
2.3×10^{-3}	0.065
4.2×10^{-3}	0.131
7.4×10^{-3}	0.263
1.0×10^{-2}	0.395

2.81_3 Oil field waste, considered nonhazardous, is transported in a tanker truck of volume 16 m³ and disposed in land farms. The total waste is composed of 8 m³ of water over 1 m³ of waste oil. The waste oil is sour and there is the possibility of hydrogen sulfide generation. Use the fugacity level I model to determine the percent hydrogen sulfide in the vapor phase inside the truck. Hydrogen sulfide properties: $P_a = 560$ atm, waste oil-water partition constant = 10, waste density = 1.1 g cm⁻³.

2.82_3 Freshwater lakes receive water from the oceans through evaporation by sunlight. Consider the ocean to be made entirely of NaCl of 0.4 M in ionic strength and the lake to be 0.001 M in $MgCl_2$. Compute the free energy required to transfer 1 mol of pure water from ocean to lake. Assume a constant temperature of 298 K.

2.83$_2$ Henry's constant for phenanthrene at 20°C was measured as 2.9 Pa m^3 mol^{-1} and its enthalpy of volatilization was determined to be 41 kJ mol^{-1}. What is Henry's constant at 25°C? The melting point of phenanthrene is 100.5°C.

2.84$_2$ A solution of benzene and toluene behaves ideally at 25°C. The pure component vapor pressures of benzene and toluene are 12 and 3.8 kPa, respectively, at 298 K. Consider a spill of the mixture in a poorly ventilated laboratory. What is the gas phase mole fraction of benzene in the room? Use Raoult's law. The mixture is 60% by weight of benzene.

2.85$_2$ 5 g of soil contaminated with 1,4-dichlorobenzene from a local Superfund site was placed in contact with 40 mL of clean distilled water in a closed vessel for 24 h at 298 K. The solution was stirred and after 72 h the soil was separated by centrifugation at 12,000 rpm. The aqueous concentration was determined to be 35 μg L^{-1}. If the partition constant was estimated to be 13 mL kg^{-1}, what is the original contaminant loading on the soil?

2.86$_2$ Estimate the activity coefficient of oxygen in the following solutions at 298 K: (1) 0.01 M NaCl, (2) seawater (see Problem 2.95 for composition), (3) wastewater containing 0.001 M Na$_2$SO$_4$ and 0.001 M CaSO$_4$. Use the McDevit-Long equation from the text. φ = 0.0025. Oxygen solubility in pure water is 8.4 mg L^{-1} at 298 K and 1 atm total pressure.

2.87$_2$ Henry's constant for methyl mercury at 0°C is 0.15 and at 25°C is 0.31. Estimate the enthalpy of dissolution of gaseous methyl mercury in water.

2.88$_3$ In pure water the solubility of benzo-[a]-pyrene varies as follows:

t (°C)	C$_i^*$ (nmol L^{-1})
8.0	2.64
12.4	3.00
16.7	3.74
20.9	4.61
25.0	6.09

The melting point of the compound is 179°C. Calculate the excess entropy of solution at 298 K.

2.89$_2$ Atmospheric reactions in hydrometeors (rain, fog, mist, dew) produce aldehydes and ketones. Consider the conversion of pyruvic acid to acetaldehyde in fog droplets. The reaction is CH$_3$–C(O)–COOH (g) → CH$_3$CHO (g) + CO$_2$ (g). (1) Obtain the standard free energy change for the reaction at 298 K. (2) What is the equilibrium constant for the reaction at 298 K?

2.90$_2$ It is known that when deep sea divers come up rapidly from their dives, they get what is called the "bends." This is due to the formation of bubbles of nitrogen in their blood serum (approximated as water). If Henry's constant for nitrogen is 8.6 × 10^3 atm, what is the concentration of nitrogen in the blood serum of a diver who came up from 200 ft below sea level?

2.91_2 Acid rain occurs by dissolution of SO_3 in aqueous droplets. SO_2 from coal burning power plants is the culprit. The reaction is SO_2 (g) + 0.5O_2 (g) → SO_3 (g). What is the equilibrium constant for this reaction at 298 K? If the partial pressure of O_2 is 0.21 atm, what is the equilibrium ratio of partial pressures of SO_3 to SO_2 at 298 K?

2.92_3 Consider a 0.1 mole fraction of ammonia solution in water at 298 K. Use the ideal solution approximation to obtain the partial pressure of water in this system. If the actual measured vapor pressure is 3.0 kPa, obtain the activity and activity coefficient of water in the solution.

2.93_2 Find the fugacity of a liquid A in water. The concentration of A in water is 1 mg L^{-1}. Molecular weight of A is 78. The vapor pressure of A is 0.02 atm. Assume that A behaves ideally in water.

2.94_2 For the reaction H_2S (g) → H_2S (aq), the enthalpy of dissolution is −19.2 kJ mol^{-1} at 298 K. If the temperature is raised to 318 K, by how much will the equilibrium constant of the reaction change?

2.95_2 The equilibrium constant K_{eq} for the dissociation of bromine into atoms, the reaction being $Br_2 \rightleftharpoons Br^{\bullet} + Br^{\bullet}$, is 6×10^{-12} at 600 K and 1×10^{-7} at 800 K. Calculate the standard enthalpy change for this reaction assuming that it is constant in the temperature range. What is the entropy change at 600 K?

2.96_2 If the standard enthalpy and entropy of dissolution of naphthalene in water are 9.9 kJ mol^{-1} and 59 J mol^{-1} K^{-1}, estimate the equilibrium partial pressure of naphthalene over an aqueous solution containing 1×10^{-4} mol L^{-1} of naphthalene.

2.97_2 The rate of formation of a pollutant NO (g) by the reaction $2NO_2$ → 2NO (g) + O_2 (g) is 1×10^{-5} mol L^{-1} s^{-1}. Obtain the rate of the reaction and the rate of dissipation of NO_2. If the ΔG^0 for the given reaction is 35 kJ mol^{-1}, what is the concentration of NO_2 in ppmv in air that has 21% oxygen at 1 atm. The total concentration of NO_2 is 10 ppmv.

2.98_1 For the reaction A + 2B → 3C, assume that it follows the stoichiometric relation given. What is the order of the reaction in B? Write down the expression for the rate of disappearance of B and the rate of appearance of A.

2.99_1 For the following hypothetical reaction, $A_2 + B_2$ → 2AB, the reciprocal of the plot of [A] (in mol per 22.4 L) versus time has a slope of −5 when both A_2 and B_2 are at unit concentrations. Determine the rate constant for the disappearance of B_2 in units of m^3 mol^{-1} s^{-1} and m^3 s^{-1}.

2.100_2 Derive the equation for the half-life of a nth order reaction, nA → C. Would you be able to determine the order of a reaction if the only data available are different initial concentrations of A and the corresponding $t_{1/2}$ values? If so, how?

2.101_1 The reaction 2A → C proceeds with a rate constant of 5×10^{-4} mol L^{-1} s^{-1}: (1) Calculate the half-life of the reaction when C_A^0 is 2 mol L^{-1}. (2) How long will it take for C_A to change from 5 to 1 mol L^{-1}?

2.102_2 The decomposition of a species A in air occurs via the following reaction: A + A → C. The pressure of C as a function of t was monitored in a reactor.

t (min)	p_C (atm)
0	0
20	0.01
40	0.05
60	0.10
90	0.12
120	0.14
∞	0.16

Obtain (1) the order of the reaction, (2) the rate constant, and (3) the half-life of A.

2.103$_2$ For the following auto-catalytic reaction scheme for the conversion of A to C, determine the rate of formation of C using the pseudo-steady state approximation:

$$A \xrightarrow{k_1} B$$
$$B + C \xrightarrow{k_2} C + A$$

2.104$_3$ Sulfur in the +4 oxidation state participates in a number of atmospheric reactions. S(IV) represents the total concentration in the +4 oxidation state: $[S(IV)] = [SO_2] + [SO_2 \cdot H_2O] + [HSO_3^-]$. The second-order reaction of S(IV) with a component A is given by A + S(IV) → products. (1) What will be the rate expression for the process? (2) Rewrite this expression in terms of the partial pressures of the species.

2.105$_1$ A first-order reaction is 30% complete in 45 min. What is the rate constant in s^{-1}? How long will it take for the reaction to be 90% complete?

2.106$_3$ An important reaction in both atmospheric chemistry and wastewater treatment is the conversion of ozone to oxygen. In the gas phase, the reaction is $2O_3$ (g) ⇌ $3O_2$ (g). This proceeds in steps as follows. Step 1: $O_3 + M \underset{k_a'}{\overset{k_a}{\rightleftharpoons}} O_2 + O + M$, and Step 2: $O + O_3 \xrightarrow{k_b} 2O_2$. M is a third species.

1. Using the steady state approximation, obtain an expression for the decomposition rate of ozone.
2. The following data were obtained for the thermal decomposition of ozone in aqueous acidic solutions (Sehestad et al., 1991) at 304 K in the presence of $[HClO_4] = 0.01$ M:

$[O_2]$ (μM)	$[O_3]$ (μM)	Initial Rate Constant (s^{-1})
10	200	1.3×10^{-4}
400	200	7.0×10^{-5}
1000	200	4.7×10^{-5}
1000	95	2.4×10^{-5}
1000	380	8.0×10^{-5}

What can you infer regarding the mechanism of decomposition of ozone in solution?

2.107_2 The oxidation of polyaromatic hydrocarbons is an important reaction in atmospheric chemistry. The reaction between benzene and oxygen was studied by Atkinson and Pitts (1975) at various temperatures.

T (K)	k (L mol^{-1} s^{-1})
300	1.44×10^{-7}
341	3.03×10^{-7}
392	6.9×10^{-7}

Obtain E_a, for this reaction.

2.108_2 The atmospheric reaction between methane (CH_4) and hydroxyl radical ($OH^\bullet$) is of importance in understanding the fate of several hydrocarbon pollutants in the air environment. An activation energy of 15 kJ mol^{-1} and a rate constant of 3×10^{-12} cm^3 molecule^{-1} s^{-1} were determined in laboratory experiments for the reaction. Using the temperature profiles in the atmosphere (Chapter 2), determine the rate constants for this reaction at various altitudes. The approximate distances of the various layers from ground level are (1) atmospheric boundary layer ($= 0.1$ km), (2) lower troposphere ($= 1$ km), (3) middle troposphere ($= 6$ km), and (4) the lower stratosphere ($= 20$ km).

2.109_3 The thermal decomposition of alkane (e.g., ethane) is of importance in environmental engineering (e.g., combustion kinetics of petroleum hydrocarbons). A chain reaction mechanism is proposed. Identify the initiation, propagation, and termination steps in the reaction. Using the steady state approximation obtain the overall rate of decomposition of ethane.

$$C_2H_6 \xrightarrow{k_1} 2CH_3^\bullet$$
$$CH_3^\bullet + C_2H_6 \xrightarrow{k_2} C_2H_5^\bullet + CH_4$$
$$C_2H_5^\bullet \xrightarrow{k_3} C_2H_4 + H^\bullet$$
$$H^\bullet + C_2H_6 \xrightarrow{k_4} C_2H_5^\bullet + H_2$$
$$2C_2H_5^\bullet \xrightarrow{k_5} n\text{-}C_4H_{10}$$
$$2C_2H_5^\bullet \xrightarrow{k_6} C_2H_4 + C_2H_6$$

2.110_3 The concentrations of some major ions in natural waters are given here:

Medium	Concentration (mmol kg^{-1})							
	Na$^+$	K$^+$	Ca^{2+}	Mg^{2+}	Cl$^-$	SO$_4^{2-}$	NO$_3^-$	F$^-$
Seawater	468	10	10.3	53	546	28	0.05	0.07
River water	0.27	0.06	0.37	0.17	0.22	0.12	0.02	0.005
Fog water	0.08	—	0.2	0.08	0.2	0.3	—	1

The following two reactions are of importance in both atmospheric moisture and seawater:

1. Oxidation of H_2S by peroxymonosulfate (an important oxidant in clouds), $HSO_5^- + SO_5^{2-} \rightarrow 2SO_4^{2-} + O_2 + H^+$, with $k = 0.1$ L mol^{-1} s^{-1} at 298 K in distilled water

2. Oxidation of nitrite by ozone to form nitrate, $NO_2^- + O_3 \rightarrow NO_3^- + O_2$, with $k = 1.5 \times 10^{-5}$ L mol^{-1} s^{-1} at 298 K in distilled water

   Estimate the rate constants of these reactions in the natural waters given in the table.

2.111$_2$ The following rate data were reported by Sung and Morgan (1980) for the homogeneous oxidation of Fe(II) to Fe(III) in aqueous solution:

I (M)	k (L² mol⁻² atm min)
0.009	4.0×10^{13}
0.012	3.1×10^{13}
0.020	2.9×10^{13}
0.040	2.2×10^{13}
0.060	1.8×10^{13}
0.110	1.2×10^{13}

Obtain the product of the charges of the species involved in the reaction. Tamura et al. (1976) proposed the following rate-limiting step for the reaction: $FeOH^+ + O_2OH^- \rightarrow Fe(OH)_2^+ + O_2^-$. Does your result support this contention?

2.112$_3$ Corrosion in wastewater collection systems by sulfuric acid generated from H_2S is a nuisance. Oxidation of H_2S by O_2 is therefore an important issue. The rate of this reaction is accelerated by catalysts. Kotranarou and Hoffmann (1991) discussed the use of Co^{II}TSP (a tetrasulfophthalocyanine) as a possible catalyst for autooxidation of H_2S in wastewater. The rate constant for the reaction was of the form $k = \dfrac{k'[O_2]}{K + [O_2]}$. The following reaction scheme was suggested:

$$Co^{II}(TSP)_2^{\,4-} \rightleftharpoons 2Co^{II}(TSP)^{2-}$$

$$Co^{II}(TSP)^{2-} + HS^- \rightleftharpoons HSCo^{II}(TSP)^{3-}$$

$$HSCo^{II}(TSP)^{3-} + O_2 \rightleftharpoons HSCo^{II}(TSP)(O_2{}^{\cdot-})^{3-}$$

$$HSCo^{II}(TSP)(O_2{}^{\cdot-})^{3-} + HS^- \rightleftharpoons HSCo^{II}(TSP)(O_2{}^{\cdot-})(HS^-)^{4-}$$

$$HSCo^{II}(TSP)(O_2{}^{\cdot-})(HS^-)^{4-} \xrightarrow{\text{slow}} HSCo^{II}(TSP)(O_2{}^{2-})^{5-} S^0 + H$$

$$HSCo^{II}(TSP)(O_2{}^{2-})^{5-} + 2H^+ \xrightarrow{\text{fast}} HSCo^{II}(TSP)^{3-} + H_2O_2$$

    Obtain the rate expression for $-d[S(-II)]/dt$, where $S(-II)$ is the reduced form of sulfur in this reaction scheme. Simplify the expression for the pseudo-first-order kinetics to show that the rate constant has the general form given here.

2.113_3 Different isomers of hexachlorocyclohexane (HCH) are abundant pesticides in the world's oceans and lakes. They are also known to be present in atmospheric particles and in the sedimentary environment. Ngabe et al. (1993) studied the hydrolysis rate of the α-isomer using buffered distilled water where all other processes, including biotransformations and photo-assisted processes, were suppressed. The data obtained were as follows:

T (K)	pH	k (min^{-1})
278	9.04	5.01×10^{-7}
288	9.01	4.19×10^{-6}
303	9.01	9.74×10^{-5}
318	9.01	6.45×10^{-4}
318	7.87	4.86×10^{-5}
318	7.01	7.52×10^{-6}

    Given that α-HCH is resistant to acid attack, use this data to obtain (1) the specific neutral and base-catalyzed hydrolysis rate constants, (2) the relative contribution of each mechanism, and (3) the activation energy for the base-catalyzed reaction.

    If a local lake in Baton Rouge, LA, has an average pH of 7.2, and a temperature of 14°C, what will be the time required to reduce the α-HCH concentration in the lake to 75% of its original value. Neglect any sedimentation or biotransformation in the lake.

2.114_3 Consider a body of water containing 10^{-4} M of a pollutant A that is in contact with air that contains hydrogen sulfide. The dissolved H_2S converts compound A to its reduced form according to the equation $A_{ox} + H_2S \rightarrow A_{red}$ with an E_H^σ of 0.4 V. Given that the total gas + aqueous concentration of hydrogen sulfide is $[H_2S]_T \cdot 10^{-2}$ M, and that the pK_a of H_2S is 7 at 298 K, obtain the following: (1) The redox potential at a pH of 8.5 (*Note*: $E_H^\sigma (H_2S, aq)) = +0.14$ V at 298 K) and (2) the fraction of A_{ox} converted at pH 8.5.

2.115_3 Chlorine dioxide (ClO_2) is used as an oxidant in wastewater treatment and drinking water disinfection processes. It is preferred over other oxidants since it produces no chloroform and very few other chloroorganics. It also has its limitations since one of its reduced forms (chlorite) is a blood poison at high concentrations. The reaction of ClO_2 with a pollutant A follows the stoichiometric equation $ClO_2 + \alpha A \rightarrow$ products, with $\alpha = (\Delta A)/(\Delta ClO_2)$, and follows the rate law $-d[ClO_2]/dt = k_{tot}[ClO_2]^n[A_o]^n$. In a series of experiments, Hoigne and Bader (1994) used $[A_o] \gg 5[ClO_2]$ to determine the rate constants for this reaction with several pollutants. The reaction with H_2O_2 is of particular importance. Since H_2O_2 dissociates with pH changes, $k_{tot} = k_{HA} + \beta(k_{A^-} - k_{HA})$, with HA denoting the undissociated H_2O_2 and A^- is its dissociated form, HO_2^-.

β is the fraction in the dissociated form. pK_a of H_2O_2 is 11.7. $k_{HA} = 0.1$ L $(mol\ s)^{-1}$, $k_A = 1.3 \times 10^{-5}$ L $(mol\ s)^{-1}$ at 298 K, with $[H_2O_2]$ in the range 0.3–20 mM and pH between 2 to 11. Determine the half-life of a 10 μM H_2O_2 solution at a pH of 8 if the water is disinfected with 1 μM of ClO_2.

2.116_3 Barry et al. (1994) studied the oxidation kinetics of Fe(II) (a prevalent metal in most natural waters) in an acidic alpine lake. The total rate of oxidation was found to be a combination of the following: (1) homogeneous oxidation governed by

$$r = -\frac{d\left[Fe\left(II\right)\right]}{dt} = \left(k_o + \frac{k_1 K_1}{\left[H^+\right]} + \frac{k_2 \beta_2}{\left[H^+\right]^2}\right)\left[Fe^{2+}\right]P_{O_2}$$

where K_1 and β_2 are the equilibrium and stability constants of the two major hydroxo species, Fe(OH)$^+$ and Fe(OH)$_2$, respectively, (2) an abiotic oxidation given by $r = k_3' A\left[Fe\left(II\right)\right]\left[OH^-\right]^2 P_{O_2}$, where A is the area of the surface to which iron is attached, and (3) a biotic oxidation process given by $r = k_4[Bac][Fe(II)][OH^-]^2 P_{O_2}$, with [Bac] measured as grams volatile solids per liter. The rate constants at 298 K were determined to be $k_o = 1 \times 10^{-8}$ s^{-1}, $k_1 = 3.2 \times 10^{-2}$ s^{-1}, $k_2 = 1 \times 10^4$ s^{-1}, $k_3' = 4.9 \times 10^{14}\ (L\ mol)^{-2}$ atm^{-1} m^{-2} s^{-1}, $k_4 = 8.1 \times 10^{16}$ (L mol)$^{-2}$ atm^{-1} s^{-1} (g volatile solids L^{-1})$^{-1}$, $K_1 = 10^{-9.5}$, $\beta_2 = 10^{-20.6}$, and A = 4.2 m^2 g^{-1}. Calculate the half-life for Fe(II) in the acidic lake if the pH is 5.2. The sediment showed a 7% loss of weight upon combustion. The suspended sediment concentration in the lake water was 1 g L^{-1}.

2.117_2 Determine the equilibrium constant for the following reactions at 298 K. Assume unit activities of all species except $[H^+] = 10^{-6}$ M.

1. $PbO\ (s) + 2H^+\ (aq) \rightleftharpoons Pb^{2+}\ (aq) + H_2O\ (l)$

2. $\frac{1}{4}CH_2O + Fe(OH)_3\ (s) + 2H^+ \rightleftharpoons Fe^{2+} + \frac{1}{4}CO_2 + \frac{11}{4}H_2O$

   The first reaction is of importance in corrosion problems, whereas the second reaction is an important oxidation reaction occurring in soils and sediments.

2.118_2 Nitrogen fixation in the environment is mediated by bacteria called "nitrobacter." The conversion of N_2 to NH_4^+ in nature is the primary reduction reaction at pH 7. This is accompanied by the oxidation of natural organic matter (designated CH_2O) to CO_2. What are the free energy changes for the two half reactions? What is the total free energy of the redox process? How is the reaction effected in a slightly acidic lake water (pH = 6)?

2.119_2 Assume a linear adsorption constant K_d for a nonionic organic pollutant with suspended particles of concentration ρ_p. If the pollutant has an average rate constant of k_{ads} in the sorbed state and k_{aq} in the aqueous phase for transformation to a product B, obtain the ratio of the overall reaction rate k_{obs} to k_{aq}. For the following conditions, how is the rate affected: (1) lake waters of $\rho_p = 10$ mg L^{-1}, (2) bottom sediments and soils of $\rho_p - 10^6$ mg L^{-1}. Discuss your results with respect to the two compounds 1,2-dichloroethane and p,p'-DDT.

2.120_2 The oxygen transfer coefficient ($k_l a$) into a biological fermenter can be obtained using the sodium sulfite method where the following reaction is supposed to occur: $Na_2SO_3 + \dfrac{1}{2}O_2 \xrightarrow{Co^{2+}} Na_2SO_4$. Co^{2+} catalyzes the reaction. In a batch fermenter filled with 0.01 m³ of 0.5 M sodium sulfite containing 0.002 M Co^{2+} ion, fresh air is bubbled in at a constant rate. After 10 min of bubbling, an aqueous sample (10×10^{-6} m³) was withdrawn and analyzed for sodium sulfite. The final concentration of sodium sulfite was 0.2 M. Calculate $k_l a$ for oxygen from this data. Given that the temperature of the experiment was 298 K and the pressure was 1 atm. Obtain Henry's constant for oxygen from the text.

2.121_3 Sedlak and Andren (1994) showed that the aqueous phase oxidation of polychlorinated biphenyls (PCBs) by $OH^\bullet$ is influenced by the presence of particles such as diatomaceous earth (represented P). The mechanism representing this is given here, where A–OH is the hydroxylated PCB.

$$A - P \underset{k_{-1}}{\overset{k_1}{\rightleftharpoons}} A + P$$
$$A + OH^\bullet \xrightarrow{k_2} A - OH$$

Using appropriate assumptions derive the following equation for the concentration of A in the aqueous phase:

$$\frac{[A]}{[A]_0} = \frac{k_1}{(\beta - \alpha)} \cdot \left(e^{-\alpha t} - e^{-\beta t} \right)$$

where $\alpha\beta = k_1 k_2 [OH^\bullet]$ and $\alpha + \beta = k_1 + k_{-1}[P] + k_2[OH^\bullet]$. For 2,2',4,5,5'-pentachlorobiphenyl, $k_1 = 4.53 \times 10^{-4}$ s⁻¹, $k_{-1} = 1.41$ L (g s)⁻¹ and $k_2 = 4.6 \times 10^{-9}$ cm³ (mol s)⁻¹. At an average concentration of $[OH^\bullet] = 1 \times 10^{-14}$ mol L⁻¹, and $[P] \equiv [Fe] = 10$ μM, sketch the progress of hydroxylation of PCB. Repeat your calculation for $[P] = 50$ μM. What conclusions can your draw from your plots?

2.122_2 The photochemical reaction in the upper atmosphere that leads to the formation and dissipation of O_2 follows the following pathway:

$$O_2 \xrightarrow{h\nu} O\left({}^1D\right) + O$$
$$O\left({}^1D\right) + M \xrightarrow{k_2} O + M$$

$O(^1D)$ is a transient state (singlet) for O atom and M is a third body such as N_2. The rate of the first reaction is represented R_1. If $[M] = 10^{14}$ molecules cm⁻³ and $k_2 = 1 \times 10^{-25}$ cm⁶(molecule)⁻² s⁻¹, determine the steady state concentration of singlet oxygen as a function of R_1.

2.123_2 Atmospheric photooxidation of propane to acetone is said to occur by the following pathway:

$$CH_3CH_2CH_3 + OH^{\cdot} \rightarrow CH_3CH^{\cdot}CH_3 + H_2O$$

$$CH_3CH^{\cdot}CH_3 + O_2 \xrightarrow{\text{fast}} CH_3CH_2\left(O_2^{\cdot}\right)CH_3$$

$$CH_3CH_2\left(O_2^{\cdot}\right)CH_3 + NO \rightarrow NO_2 + CH_3CH\left(O^{\cdot}\right)CH_3$$

$$CH_3CH\left(O^{\cdot}\right)CH_3 + O_2 \xrightarrow{\text{fast}} CH_3COCH_3 + HO_2^{\cdot}$$

Obtain an expression for the rate of production of NO_2 due to the dissipation of propane. Make appropriate assumptions.

2.124_3 Atmospheric chemists consider the oxidation of CH_4 by $OH^{\cdot}$ to be a convenient way to estimate the concentration of hydroxyl ion in the atmosphere: $CH_4 + OH^{\cdot} \xrightarrow{k} CH_3^{\cdot} + H_2O$. In so doing, it is assumed that the hydroxyl radical concentration is constant, since it is much larger than the methane concentration. Given that $k = 107$ m^3 (mol s)$^{-1}$ at 300 K and that methane has an atmospheric residence time of 10 years, obtain the mean $OH^{\cdot}$ concentration in the atmosphere.

2.125_2 The acid-catalyzed decomposition of urea has a rate constant of 4.3×10^{-4} h^{-1} at 334 K and 1.6×10^{-3} h^{-1} at 344 K. Estimate the activation energy and preexponential factor for the reaction. What will be the rate at a room temperature of 300 K?

2.126_2 A first-order process has the following rate constants at different ionic strengths of a 1:1 electrolyte:

I (mol L^{-1})	0.007	0.01
	0.015	0.020
k (h^{-1})	2.06	3
	5.1	7.02

If one of the ions involved in the reaction has a positive charge, what is the charge on the other ion?

2.127_1 The rate constant for the decay of natural ^{14}C radio isotope is 1.2×10^{-4} y^{-1}. What is the half-life of ^{14}C?

2.128_3 The hydrolysis of $ArOCOCH_3$ to Ar OH is base catalyzed. At values of pH > 2 acid catalysis is unimportant. The uncatalyzed reaction rate k_0 is 4×10^{-5} s^{-1} at 20°C. At pH = 8, what is the rate constant of the reaction? How much time will it take to reduce the concentration of the compound to 95% of the original value in a lake of pH 8 at 20°C?

2.129_3 Pentachloroethane (PCA) is spilled into the Mississippi river at a mean pH of 7.2 and a temperature of 20°C. Haag and Mill (1988) report the following parameters for the uncatalyzed and base-catalyzed reactions of PCA:

$$k_0\left(h^{-1}\right) = 4.9 \times 10^{-8}, \quad E_a = 95 \, kJ \, mol^{-1}$$

$$k_b \left(h^{-1} \right) = 1320, \quad E_a = 81 \, kJ \, mol^{-1}.$$

How long will it take to reduce the concentration of PCA in the river to 95% of its original concentration?

2.130$_3$ Loftabad et al. (1986) proposed that polyaromatic hydrocarbons ($\equiv$A) in contact with dry clay surfaces underwent oligomerization. This was inhibited by moisture present in the soil. The reaction scheme proposed was as follows:

$$S + A \rightleftharpoons S - A$$
$$S + W \rightleftharpoons S - W$$
$$S - A \rightarrow S + \text{products}$$

W represents water and S represents a surface site for compound A. Based on this, derive the following rate expression for the reaction of A: $-r_A = k_o C_A /$ $(1 + K\theta)$, where K is the equilibrium constant for water adsorption and θ is the water content (g kg^{-1} of soil). K_o is the apparent rate constant in dry soil ($\theta = 0$). State the assumptions made in deriving the rate expression.

2.131$_2$ The decomposition of N_2O_5 to NO_2 in the gas phase occur via the following reaction scheme:

$$N_2O_5 \underset{k_{b1}}{\overset{k_{f1}}{\rightleftharpoons}} NO_2 + NO_3$$
$$NO_2 + NO_3 \overset{k_2}{\longrightarrow} NO + O_2 + NO_2$$
$$NO + NO_3 \overset{k_3}{\longrightarrow} 2NO_2$$

Derive the following equation for the overall decomposition rate of N_2O_5 by using the PSSA to show that $-\dfrac{d\left[N_2O_5 \right]}{dt} = k_{tot} \left[N_2O_5 \right]$. Identify the expression for k_{tot}.

2.132$_3$ A complete and general mechanism for enzyme catalysis should include a reversible decomposition of the E-S complex:

$$E + S \underset{k_{-1}}{\overset{k_1}{\rightleftharpoons}} E - S \underset{k_{-2}}{\overset{k_2}{\rightleftharpoons}} E + P$$

Derive an equation for the rate of product formation. Contrast it with that derived in the text where reversible decomposition of E-S was not considered.

2.133$_2$ Some substrates can bind irreversibly either to an enzyme or the enzyme-substrate complex and interfere with enzymatic reactions. These are called "inhibitors" and are of two types:
1. Derive the rate expressions for product formation in each case and sketch the behavior as Michaelis-Menten plots.

2. The hydrolysis of sucrose is said to be inhibited by urea. The rate data for the same is given here:

[S] (mol L⁻¹)	r (mol (L⁻¹ s)⁻¹)	r (mol (L s)⁻¹) with [I] = 2 M
0.03	0.18	0.08
0.06	0.27	0.11
0.10	0.33	0.18
0.18	0.37	0.19
0.24	0.37	0.18

What type of an inhibitor is urea?

2.134[1] The uptake of CO_2 in water and subsequent dissociation into bicarbonate ion is shown to be accelerated by an enzyme called "bovine carbonic anhydrase." The following data is provided on the initial rate of the reaction

$$CO_2\ (aq) + H_2O\ (aq) \xrightarrow{\ E\ } H^+\ (aq) + HCO_3^-\ (aq)$$

At pH 7 for different CO_2 (aq) concentration:

$[CO_2]_0$ (mol dm⁻³)	r_0 (mol (dm³ s)⁻¹)
1.25×10^{-3}	3×10^{-5}
2.5×10^{-3}	5×10^{-5}
5×10^{-3}	8×10^{-5}
2×10^{-2}	1.5×10^{-4}

Obtain the Michaelis-Menten parameters for this reaction.

2.135[3] The atmospheric reaction of chlorine monoxide (ClO) with NO_2 proceeds as follows:

$$ClO + NO_2 + M \underset{k_b}{\overset{k_f}{\rightleftharpoons}} ClONO_2 + M$$

with $k_b = 10^{-6.16} \exp(-90.7\ kJ\ mol^{-1}/RT)$ in units of cm³ molecule⁻¹ s⁻¹. The high and low pressure limiting rate constants at 298 K were obtained from experiments as follows:

$$k_0 = 1.8 \times 10^{-31}\ cm^6\ molecule^{-2}\ s^{-1}$$

$$k_\infty = 1.5 \times 10^{-11}\ cm^3\ molecule^{-1}\ s^{-1}$$

Obtain the effective bimolecular rate constant k_f at 298 K and 1 atm.

2.136[2] A recent article (Li et al., 2008) showed that apart from the well known ozone dissociation that produces OH in the atmosphere, there also exist another pathway for OH production. This involves the reaction of electronically excited NO_2 with H_2O as follows:

$$NO_2 \xrightarrow[hv]{J_1} NO_2^*$$
$$NO_2^* + M \xrightarrow{k_2} NO_2 + M$$
$$NO_2^* + H_2O \xrightarrow{k_3} OH + HONO$$
$$HONO \xrightarrow[hv]{J_4} OH + NO$$

Apply a steady state approximation to NO_2^* to obtain the following equation:

$$r_{OH} = \frac{2J_1 k_3 [NO_2][H_2O]}{k_3 [H_2O] + k_2 [M]}.$$

2.137$_2$ An autocatalysis reaction is one in which the reactant is reformed as the reaction proceeds:

$$A \xrightarrow{k_1} B$$
$$B + A \xrightarrow{k_2} C + A$$

1. Using the pseudo-steady state approximation, derive the rate equation for the formation of the product C.
2. If the overall rate constant is 2 h^{-1}, and the initial concentration of A is 10 mol m^{-3}, what is the initial rate of formation of C?

2.138$_2$ For the reaction: $2A \xrightarrow{k} B$,
   1. Write the differential rate equation for the dissipation of A.
   2. Integrate this expression with the initial condition that at $t = 0$, $[A] = [A]_0$, and obtain the equation for A as a function of time.
   3. If the rate constant is 2 h^{-1}, what is the half-life of the reaction?

2.139$_3$ The decomposition of di-2-methylpropan-2-yl into proponanone and ethane goes through free radical mechanisms given as follows:

$$(CH_3)COOC(CH_3)_3 \xrightarrow{k_1} 2(CH_3)_3 CO^\bullet$$
$$(CH_3)_3 CO^\bullet \xrightarrow{k_2} CH_3COCH_3 + CH_3^\bullet$$
$$CH_3^\bullet + CH_3^\bullet \xrightarrow{k_3} C_2H_6$$

Formulate the expression for the rate of production of ethane and show that the reaction is first order in the reactant, di-2-methylpropan-2-yl. Use the pseudo-steady state approximation and perform algebra. Remember that two $(CH_3)_3CO^\bullet$ radicals are formed for every one molecule of the reactant.

2.140$_2$ For the reaction: $A + B \underset{k_{-1}}{\overset{k_1}{\rightleftharpoons}} C \xrightarrow{k_2} D$, obtain the expression for the rate of production of D in terms of the reactant concentrations. What is the order of the overall reaction?

2.141_3 The reaction of gas phase ozone with dimethyl sulfide (DMS) in a glass reactor occurs via the following two-step mechanism:

$$O_3 \xrightarrow{\ k_1\ } \text{Wall decay}$$
$$O_3 + DMS \xrightarrow{\ k_2\ } \text{Products}$$

Experiments were conducted under a large excess of DMS concentration, $[DMS]_0$ to obtain the value of $-\dfrac{d\ln[O_3]}{dt}$ (Du et al., 2007):

$[DMS]_0$ (10^{14} molecules cm^{-3})	$-\dfrac{d\ln[O_3]}{dt}$ (10^5 s^{-1})
0	0.609
3.63	6.00
4.84	7.29
6.05	7.83

Derive the rate equation for the conditions stated, and from an appropriate plot of the data obtain k_1 and k_2.

2.142_3 Consider the troposphere (total volume V) to be a well-mixed CSTR. A man-made chemical A is released from surface sources at a constant rate of N_A (Tg y^{-1}). Assume that the northern and southern hemispheres show uniform concentrations in A. The initial concentration is $[A]_0 = 0$ Tg m^{-3}. If the only removal process in the troposphere is a first-order reaction with a rate constant k_A (y^{-1}), what will be the concentration of A in the troposphere at any given time? (*Note:* $\dfrac{1}{[A]} \cdot \dfrac{d[A]}{dt} = \dfrac{1}{\tau}$, where τ is the characteristic time of A in the reservoir. $k_A = \dfrac{1}{\tau_r}$, where τ_r is the characteristic time for reaction of A and $\dfrac{N_A}{V[A]} = \dfrac{1}{\tau_d}$, where τ_d is the characteristic time for inflow of A. Express $[A]/[A]_0$ as a function of τ_r and τ_d alone.)

2.143_2 Constructed wetlands have been proposed as wastewater treatment systems. Plants produce enzymes (extracellular and intracellular) that breakdown organic pollutants. The system involves a gentle down-flow of water over sloping vegetation with a slow accompanying water evaporation rate (20 mol (m^2 min)$^{-1}$). Consider a 1 km $\times$ 0.1 km $\times$ 0.2 km deep wetland. A nonvolatile pesticide run-off (concentration 10^{-2} mol m^{-3}) is introduced into the wetland at a flow rate of 0.05 m^3 min^{-1} and undergoes a first-order biochemical reaction with a rate constant of 10^{-6} min^{-1}. What is the steady state concentration of the pesticide at the outflow from the wetland? Note that the water can be considered to be in plug flow.

2.144$_2$ A type of waste combustion incinerator is called a "rotary kiln." It is a cylindrical reactor inclined to the horizontal and undergoing slow axial rotation to achieve adequate mixing of the reactants. Consider a pesticide (e.g., DDT) containing waste incinerated in the kiln at a temperature of 300°C. If its decomposition is first order with a rate constant k of 15 s^{-1} at 300°C, determine the reactor volume required for 99.99% destruction of DDT from the waste stream flowing into the kiln at a rate of 0.01 m^3 s^{-1}. Consider the kiln to be first a CSTR and then repeat the calculation for a PFR.

2.145$_2$ Consider a beaker of water in equilibrium with CO_2 in the air (partial pressure P_{CO_2}). The pH of the water is made slightly alkaline so that the dissolution of CO_2 in water is enhanced. The solution process is controlled by diffusion and reaction in the stagnant boundary layer of thickness δ. CO_2 undergoes two competing reactions in water:

1. Dissolution in water: $CO_2 + H_2O \overset{k_1}{\rightleftharpoons} HCO_3^- + H^+$

2. Reaction with OH^-: $CO_2 + OH^- \overset{k_2}{\rightleftharpoons} HCO_3^-$

    a. Obtain first the amount of CO_2 in the boundary layer at any time t and the flux to the surface. The ratio of the two will give the time constant for diffusion in the boundary layer.

    b. Assume that only the two forward reactions are of importance and write down the reaction rate equation for the dissolution/reaction of CO_2 in water. The reciprocal of the first-order rate constant (at constant pH) is the time constant for reaction in the boundary layer.

    c. Equate the two rate constants and obtain the value of δ if the pH of the solution is 5.5. What does this δ represent?

2.146$_3$ Assume a global CO_2 increase in air at the rate of 0.41% per year. The current P_{CO_2} is 358 ppmv. Predict what will be the average pH of rainwater in year 2050? Is your estimate realistic? Justify your conclusions.

2.147$_2$ An enzyme catalysis is carried out in a batch reactor. The enzyme concentration is held constant at 0.003 mol dm^{-3}. The decomposition of the pollutant A $\overset{E}{\longrightarrow}$ P was studied separately in the laboratory. It obeys the Michaelis-Menten kinetics with K_M = 5 mM and V_{max} = 10 mM min^{-1}, but at an enzyme concentration of 0.001 mol dm^{-3}. How long will it take in the batch reactor to achieve 90% conversion of A to P?

2.148$_2$ An organism (*Zoogloea ramigera*) is to be used in a CSTR for wastewater treatment. It obeys Monod kinetics with μ_{max} = 5.5 d^{-1} and K_s = 0.03 mg dm^{-3}. A total flow rate of 100 mL min^{-1} is being envisaged at an initial pollutant concentration of 100 mg dm^{-3}. C_X = 10 mg L^{-1}: (1) If the desired yield factor is 0.5, what reactor volume is required? (2) Design a CSTR in series with the first one that can reduce the pollutant concentration to 0.1 mg dm^{-3}. (3) If the second reactor is a PFR, what size reactor will be required in part (2)? (4) Which combination do you recommend and why?

2.149$_2$ A solution containing a denitrifying bacteria *Micrococcus denitrificans* at a concentration of 1×10^{-6} kg dm^{-3} is introduced into a semibatch reactor where the growth rate is given by $r = k\dfrac{[X][O_2][S]}{K_s + [S]}$ where $[O_2]$ is the oxygen concentration (mol dm^{-3}). An excess of substrate [S] was used. If after 3 h of operation the concentration of the bacteria is 3×10^{-5} kg dm^{-3}, what is the specific growth rate of the bacteria?

2.150$_2$ The rate of growth of *E. coli* obeys Monod kinetics with $\mu_{max} = 0.9$ h^{-1} and $K_s = 0.7$ kg m^{-3}. A CSTR is used to grow the bacteria at 1 kg m^{-3} with inlet substrate pollutant feed rate of 1 m^3 h^{-1} and [S]$_{in} = 100$ kg m^{-3}. If the cell yield is 0.7, calculate (1) the reactor volume required for the maximum rate of production of *E. coli*?. (2) If the effluent from the reactor is fed to a series reactor, what will be the volume of the second reactor if the desired pollutant concentration in the exit stream is 0.1 kg m^{-3}.

REFERENCES

Adamson, A. (1990) *Physical Chemistry of Surfaces*, 4th edn. New York: John Wiley & Sons, Inc.

Adamson, A. and Gast, A.P. (1997) *Physical Chemistry of Surfaces*, 6th edn. New York: John Wiley & Sons, Inc.

Aquan-Yeun, M., Mackay, D., and Shiu, W.Y. (1979) Solubility of hexane, phenanthrene, chlorobenzene, and *p*-dichlorobenzene in aqueous electrolyte solutions. *Journal of Chemical and Engineering Data* **24**, 30–34.

Arbuckle, W.B. (1983) Estimating activity coefficients for use in calculating environmental parameters. *Environmental Science and Technology* **17**, 537–542.

Atkins, P.W. and de Paula, J. (2006) *Physical Chemistry*, 8th edn. New York: W. H. Freeman.

Atkinson, R. and Pitts, J.N. (1975) Temperature measurements of the absolute rate constants for the reaction of oxygen (^{3}P) atoms with a series of aromatic hydrocarbons over the temperature range 299-292K. *The Journal of Physical Chemistry* **79**, 295.

Banerjee, S. and Howard, P.H. (1988) Improved estimation of solubility and partitioning through correction of UNIFAC-derived activity coefficients. *Environmental Science and Technology* **22**, 839–848.

Barry, R.C., Schnoor, J.L., Sulzberger, B., Sigg, L., and Stumm, W. (1994) Iron oxidation kinetics in an acidic alpine lake. *Water Research* **28**, 323–333.

Ben Naim, A. (1980) *Hydrophobic Interactions*. New York: Plenum Press.

Betterton, E.A., Erel, Y., and Hoffmann, M.R. (1988) Aldehyde-bisulfite adducts: Prediction of some of their thermodynamic and kinetic properties. *Environmental Science and Technology* **22**, 92–97.

Biggar, J.W. and Riggs, R.I. (1974) Aqueous solubility of heptachlor. *Hilgardia* **42**, 383–391.

Blackburn, J.W. (1987) Prediction of organic chemical fates in biological treatment systems. *Environmental Progress* **6**, 217–223.

Bockris, J.O. and Reddy, A.K.N. (1970) *Modern Electrochemistry*. New York: Plenum Press.

Bodenstein, M. and Lind, S.C. (1907) Geschwindigkeit der Bilding der Bromwasserstoffes aus sienen Elementen. *Zeitschrift für Physikalische Chemie* **57**, 168–192.

Brezonik, P.L. (1990) Principles of linear free-energy and structure-activity relationships and their applications to the fate of chemicals in aquatic systems. In: Stumm, W. (ed.), *Aquatic Chemical Kinetics*, pp. 113–144. New York: John Wiley & Sons, Inc.

Brezonik, P.L. (1994) *Chemical Kinetics and Process Dynamics in Aquatic Systems*. Boca Raton, FL: Lewis Publishers/CRC Press, Inc.

Butler, J.N. (1982) *Carbondioxide Equilibria and Its Applications*, 2nd edn. New York: Addison Wesley Publishing Co.

Chiou, C.T. (1981) Partition coefficient and water solubility in environmental chemistry. In: Saxena, J. and Fisher, F. (eds.), *Hazard Assessment of Chemicals*, Vol. 1. New York: Academic Press.

Chiou, C.T., Peters, L.J., and Freed, V.H. (1979) A physical concept of soil-water equilibria for nonionic organic compounds. *Science* **206**, 831–832.

Chiou, C.T. and Shoup, T.D. (1985) Soil sorption of organic vapors and effects of humidity on sorptive mechanisms and capacity. *Environmental Science and Technology* **19**, 1196–1200.

Choi, D.S., Jhon, M.S., and Eyring, H. (1970) Curvature dependence of the surface tension and the theory of solubility. *Journal of Chemical Physics* **53**, 2608–2614.

Collander, R. (1951) The partition of organic compounds between higher alcohols and water. *ACTA Chemica Scandinaca* **5**, 774–780.

Denbigh, K. (1981) *The Principles of Chemical Equilibrium*, 4th edn. New York: Cambridge University Press.

Dobbs, R.A. and Cohen, J.M. (1980) Carbon adsorption isotherms for toxic organics. Report No.: EPA-600/8-80-023. Cincinnati, OH: ORD, U.S. Environmental Protection Agency.

Du, L. et al. (2007). Rate constant of the gas phase reaction of dimethyl sulfide (CH_3SCH_3) with ozone. *Chemical Physics Letter* **436**, 36–40.

Einstein, A. (1949) Autobiographical notes. In: Schlipp, P.A. (ed.), *Albert Einstein: Philosopher-Scientist*. Evanston, IL: Library of Living Philosophers.

Eley, D.D. (1939) On the solubility of gases. Part I. The inert gases in water. *Transactions of the Faraday Society* **35**, 1281–1293.

Emanuel, K. (2003) Tropical cyclones. *Annual Review of Earth and Planetary Sciences* **31**, 75–104.

Emanuel, K.E. (2005) *Divine Wind: The History and Science of Hurricanes*. New York: Oxford University Press.

Ernestova, L.S., Shtamm, Y.Y., Semenyak, L.V., and Skurlatov, Y.I. (1992) Importance of free radicals in the transformation of pollutants. In: Schnoor, J.L. (ed.), *Fate of Pesticides and Chemicals in the Environment*, pp. 115–126. New York: John Wiley & Sons, Inc.

Finlayson-Pitts, B.J. and Pitts, J.N. (1986) *Atmospheric Chemistry*. New York: John Wiley & Sons, Inc.

Fogler, H.S. (2006) *Elements of Chemical Reaction Engineering*, 4th edn. Englewood Cliffs, NJ: Prentice-Hall Inc.

Frank, H.S. and Evans, W.F. (1945) Free volume and entropy in condensed systems. *Journal of Chemical Physics* **13**, 507–552.

Franks, F. (1983) *Water*. London, England: The Royal Society of Chemistry.

Haag, W.R. and Mill, T. (1988) Effect of a subsurface sediment on the hydrolysis of haloalkanes and epoxides. *Environmental Science and Technology* **22**, 658–663.

Hansch, C. and Leo, A.J. (1979) *Substituent Constants for Correlation Analysis in Chemistry and Biology*. New York: John Wiley & Sons, Inc.

Hartkopf, A. and Karger, B.L. (1973) Study of the interfacial properties of water by gas chromatography. *Accounts of Chemical Research* **6**, 209–216.

Hermann, R.B. (1972) Theory of hydrophobic bonding. II. Correlation of hydrocarbon solubility in water with solvent cavity surface area. *Journal of Physical Chemistry* **76**, 2754–2759.

Hill, C.G. (1977) *An Introduction to Chemical Engineering Kinetics and Reactor Design*. New York: John Wiley & Sons, Inc.

Hoffmann, M.R. (1990) Catalysis in aquatic environments. In: Stumm, W. (ed.), *Aquatic Chemical Kinetics*, pp. 71–112. New York: John Wiley & Sons, Inc.

Hoigne, J. and Bader, H. (1994) Kinetics of reactions of OClO in water I. Rate constants for inorganic and organic compounds. *Water Research* **28**, 45–55.

Holmes, F. (1994) Antione Lavoisier & The Conservation of Matter: Delving deeper than the thumbnail sketches often found in chemistry textbooks into the way this seminal 18th-century French chemist designed and conducted his experiments reveals a scientist very recognizable to practicing today. *Chemical & Engineering News* **72**, 38–45.

Israelchvili, J.N. (1992) *Intermolecular and Surface Forces*, 2nd edn. New York: Academic Press.

IUPAC. (1988) *Quantities, Units and Symbols in Physical Chemistry*. Oxford, U.K.: Blackwell Scientific Publications.

Jasper, J.J. (1972) The surface tension of pure liquid compounds. *Journal of Physical and Chemical Reference Data* **1**, 841–1009.

Kipling, J.J. (1965) *Adsorption from Solutions of Non-Electrolytes*. New York: Academic Press.

Kistiakowsky, W. (1923) Uber verdampfungswarme und einige fleichungen walche die eigendraften der unassorzierten flussigkerten bestimmen. *Zeitschrift fur Physikalishe Chemie* **62**, 1334.

Kondepudi, D. (2008) *Introduction to Modern Thermodynamics*. New York: John Wiley & Sons, Inc.

Kotranarou, A. and Hoffmann, M.R. (1991) Catalytic autooxidation of hydrogen sulfide in wastewater. *Environmental Science and Technology* **25**, 1153–1160.

Laidler, K.J. (1965) *Chemical Kinetics*. New York: McGraw-Hill, Inc.

Laidler, K.J. (1993) *The World of Physical Chemistry*. New York: Oxford University Press.

Langmuir, I. (1925) The distribution and orientation of molecules. *ACS Colloid Symposium Monograph* **3**, 48–75.

Leo, A., Hansch, C., and Elkins, D. (1971) Partition coefficients and their uses. *Chemical Reviews* **71**, 525–613.

Levenspiel, O. (1999) *Chemical Reaction Engineering*, 3rd edn. New York: John Wiley & Sons, Inc.

Lewis, G.N. and Randall, M. (1961) *Thermodynamics*, 2nd end. New York: McGraw-Hill Book Co.

Li, S. et al. (2008). Atmospheric hydroxyl radical production from electonically excited NO_2 and H_2O. *Science* **319**, 1657.

Lide, D.R. and Frederikse, H.P.R. (1994) *CRC Handbook of Chemistry and Physics*. Boca Raton, FL: CRC.

Loftabad, S.K., Pickard, M.A., and Gray, M.R. (1986) Reactions of polynuclear aromatic hydrocarbons on soil. *Environmental Science and Technology* **30**, 1145–1151.

Lyman, W.J., Reehl, W.F., and Rosenblatt, D.H. (1990) *Handbook of Chemical Property Estimation Methods*. Washington, DC: American Chemical Society.

Mabey, W. and Mill, T. (1978) Critical review of hydrolysis of organic compounds in water under environmental conditions. *Journal of Physical and Chemical Reference Data* **7**, 383–415.

Mackay, D. (1982) Volatilization of organic pollutants from water. EPA Report No.: 600/3-82-019, NTIS No.: PB 82-230939. Springfield, VA: National Technical Information Service.

Mackay, D. (1991) *Multimedia Environmental Models*. Chelsea, MI: Lewis Publishers, Inc.

Marcus, R.A. (1963) On the theory of oxidation-reduction reactions involving electron transfer V. Comparison and properties of electrochemical and chemical rate constants. *Journal of Physical Chemistry* **67**, 853–857.

May, W.E., Wasik, S.P., Miller, M.M., Tewari, Y.B., Brown-Thomas, J.M., and Goldberg, R.N. (1983) Solution thermodynamics of some slightly soluble hydrocarbons in water. *Journal of Chemical and Engineering Data* **28**, 197–200.

McAuliffe, C. (1966) Solubility in water of paraffin, cycloparaffin, olefin, acetylene, cycloolefin, and aromatic hydrocarbons. *Journal of Physical Chemistry* **70**, 1267–1275.

McCarty, P.L., Roberts, P.V., Reinhard, M., and Hopkins, G. (1992) Movement and transformation of halogenated aliphatic compounds in natural systems. In: Schnoor, J.L. (ed.), *Fate of Pesticides and Chemicals in the Environment*, pp. 191–210. New York: John Wiley & Sons, Inc.

Moore, J.W. and Pearson, R.G. (1981) *Kinetics and Mechanism*, 3rd edn. New York: Wiley-Interscience.

Morel, F.M.M. and Hering, J.G. (1993) *Principles and Applications of Aquatic Chemistry*. New York: John Wiley & Sons, Inc.

Morgan, J.J. and Stumm, W. (1964) The role of multivalent metal oxides in limnological transformations, as exemplified by iron and manganese. *Proceedings of the Second International Water Pollution Research Conference*, Tokyo, Japan. Elmsford, NY: Pergamon Press.

Ngabe, B., Bidleman, T.F., and Falconer, R.L. (1993) Base hydrolysis of α- and Γ-hexachlorocyclohexanes. *Environmental Science and Technology* **27**, 1930–1933.

Nemethy, G. and Scheraga, H.A. (1962) Structure of water and hydrophobic bonding in proteins. II. Model for the thermodynamic properties of aqueous solutions of hydrocarbons. *Journal of Chemical Physics* **36**, 3401.

Pankow, J.F. (1992) *Aquatic Chemistry Concepts*. Chelsea, MI: Lewis Publishers, Inc.

Pankow, J.F. and Morgan, J.J. (1981) Kinetics for the aquatic environment. *Environmental Science and Technology* **15**, 1155–1164.

Pennel, K.D., Rhue, R.D., Rao, P.S.C., and Johnston, C.T. (1992) Vapor phase sorption of *p*-xylene and water on soils and clay minerals. *Environmental Science and Technology* **26**, 756–763.

Pierotti, G.J., Deal, C.H., and Derr, E.L. (1959) Activity coefficients and molecular structure. *Industrial and Engineering Chemistry* **51**, 95–102.

Poe, S.H., Valsaraj, K.T., Thibodeaux, L.J., and Springer, C. (1988) Equilibrium vapor phase adsorption of volatile organic chemicals on dry soils. *Journal of Hazardous Materials* **19**, 17–32.

Prausnitz, J.M., Lichtenthaler, R.N., and de Azevedo, E.G. (1999) *Molecular Thermodynamics of Fluid-Phase Equilibria*, 3rd edn. Englewood Cliffs, NJ: Prentice-Hall Inc.

Roy, D., Tamayo, A., and Valsaraj, K.T. (1992) Comparison of soil washing using conventional surfactant solutions and colloidal gas aphron suspensions. *Separation Science and Technology* **27**, 1555–1568.

Sandler, S.I. (1999) *Chemical and Engineering Thermodynamics*, 2nd edn. New York: John Wiley & Sons, Inc.

Sato, N., Inomata, H., Arai, K., and Saito, S. (1986) Measurement of vapor pressures for coal-related aromatic compounds by gas saturation method. *Journal of Chemical Engineering of Japan* **19**, 145–147.

Saylor, J.H. and Battino, R. (1958). *Journal of Physical Chemistry* **62**, 1334.

Schmidt, L.D. (2005) *The Engineering of Chemical Reactions*, 2nd edn. New York: Oxford University Press.

Schnoor, J.L. (1992) Chemical fate and transport in the environment. In: Schnoor, J.L. (ed.), *Fate of Pesticides and Chemicals in the Environment*, pp. 1–25. New York: John Wiley & Sons, Inc.

Schwarzenbach, R.P., Gschwend, P.M., and Imboden, D.M. (1993) *Environmental Organic Chemistry*. New York: John Wiley & Sons, Inc.

Schwarzenbach, R.P., Stierli, R., Folsom, B.R., and Zeyer, J. (1988) Compound properties relevant for assessing the environmental partitioning of nitrophenols. *Environmental Science and Technology* **22**, 83–92.

Sedlak, D.L. and Andren, A.W. (1994) The effect of sorption on the oxidation of polychlorinated biphenyls by hydroxyl radical. *Water Research* **28**, 1207–1215.

Sehested, K., Corfitzen, H., Holdman, J., Fischer, C.H., and Hart, E.J. (1991) The primary reaction in the decomposition of ozone in acidic aqueous solutions. *Environmental Science and Technology*. **25**, 1589–1596.

Seinfeld, J.H. (1986) *Atmospheric Physics and Chemistry of Air Pollution*. New York: John Wiley & Sons, Inc.

Seinfeld, J.H. and Pandis, S.N. (1998) *Atmospheric Chemistry and Physics*, 2nd edn. New York: John Wiley & Sons, Inc.

Sinanoglu, O. (1981) What size cluster is like a surface? *Chemical Physics Letters* **81**, 188–190.

Smith, J.M. (1970) *Chemical Engineering Kinetics*, 2nd edn. New York: John Wiley & Sons, Inc.

Sposito, G. (1984) *The Surface Chemistry of Soils*. New York: Oxford University Press.

Stillinger, F.H. and Rahman, A. (1974). Improved simulation of liquid water by molecular dynamics. *Journal of Chemical Physics* **60**, 1545–1557.

Stone, A.T. (1989) Enhanced rates of monophenyl terephthalate hydrolysis in aluminum oxide suspensions. *Journal of Colloid and Interface Science* **127**, 429–441.

Stumm, W. (1993) *Chemistry of the Solid-Water Interface*. New York: John Wiley & Sons, Inc.

Stumm, W. and Morgan, J.M. (1981) *Aquatic Chemistry*, 2nd edn. New York: John Wiley & Sons, Inc.

Stumm, W. and Morgan, J.J. (1996) *Aquatic Chemistry*, 4th edn. New York: John Wiley and Sons, Inc.

Sung, W. and Morgan, J.J. (1980) Kinetics and product of ferrous iron oxygenation in aqueous systems. *Environmental Science and Technology* **14**, 561–568.

Tamura, H., Goto, K., and Nagayama, M. (1976) Oxygenation of ferrous iron in neutral solutions. *Corrosion Science* **16**, 197–207.

Thibodeaux, L.J., Nadler, K.C., Valsaraj, K.T., and Reible, D.D. (1991) The effect of moisture on volatile organic chemical gas-to-particle partitioning with atmospheric aerosols—Competitive adsorption theory predictions. *Atmospheric Environment* **25**, 1649–1656.

Thoma, G.J. (1992) Studies on the diffusive transport of hydrophobic organic chemicals in bed sediments, PhD dissertation, Louisiana State University, Baton Rouge, LA.

Tolman, R.C. (1927) *Statistical Mechanics with Applications to Physics and Chemistry*. New York: American Chemical Society.

Tse, G., Orbey, H., and Sandler, S.I. (1992) Infinite dilution activity coefficients and Henry's law coefficients of some priority water pollutants determined by a relative gas chromatographic method. *Environmental Science and Technology* **26**, 2017–2022.

Tsonopoulos, C. and Prausnitz, J.M. (1971) Activity coefficients of aromatic solutes in dilute aqueous solutions. *Industrial and Engineering Chemistry Fundamentals* **10**, 593–600.

Tucker, E.E. and Christian, S.D. (1979) A prototype hydrophobic interaction: The dimerization of benzene in water. *Journal of Physical Chemistry* **83**, 426–427.

Uhlig, H.H. (1937) The solubility of gases and surface tension. *Journal of Physical Chemistry* **41**, 1215–1225.

Valsaraj, K.T. (1988) On the physico-chemical aspects of partitioning of hydrophobic non-polar organics at the air-water interface. *Chemosphere* **17**, 875–887.

Valsaraj, K.T. and Thibodeaux, L.J. (1988) Equilibrium adsorption of chemical vapors on surface soils, landfills and landfarms—A review. *Journal of Hazardous Materials* **19**, 79–100.

Wakita, K., Yoshimoto, M., Miyamoto, S., and Wtanabe, H. (1986) A method of calculation of the aqueous solubility of organic compounds by using the new fragment solubility constants. *Chemical and Pharmacuetical Bulletin (Tokyo)* **34**, 4663–4681.

Warneck, P.A. (1986) *Chemistry of the Natural Atmosphere*. New York: Academic Press.

Wehrli, B. (1990) Redox reactions of metal ions at mineral surfaces. In: Stumm, W. (ed.), *Aquatic Chemical Kinetics*, pp. 311–336. New York: John Wiley & Sons, Inc.

Wolfe, N.L., Paris, D.F., Steen, W.C., and Baughman, G.L. (1980a) Correlation of microbial degradation rates with chemical structure. *Environmental Science and Technology* **14**, 1143–1144.

Wolfe, N.L., Steen, W.C., and Burns, L.A. (1980b) Phthalate ester hydrolysis: Linear free energy relationships. *Chemosphere* **9**, 403–408.

Wolfe, N.L., Zepp, R.G., Paris, D.F., Baughman, G.L., and Hollis, R.C. (1977) Methoxychlor and DDT degradation in water: Rates and products. *Environmental Science and Technology* **11**, 1077–1081.

Yalkowsky, S.H. and Banerjee, S. (1992) *Aqueous Solubilities—Methods of Estimation for Organic Compounds*. New York: Marcel Dekker Inc.

Young, T. (1805) *Miscellaneous Works*. London, England: J Murray.

3 Applications of Equilibrium Thermodynamics

Chemical thermodynamics has numerous applications in environmental engineering. We will discuss specific examples that are relevant in understanding (1) the fate, transport, and transformations of chemicals in air, water, sediment, and soil environments and (2) the design of waste treatment and control operations. The discussion will primarily involve the equilibrium partitioning of chemicals between different phases. In general, the distribution of species i between any two phases A and B at a given temperature is given by

$$K_{AB}(T) = \frac{C_{iA}}{C_{iB}} \tag{3.1}$$

where C_{iB} and C_{iA} are concentrations of component i in phases B and A, respectively, at equilibrium.

3.1 FUGACITY AND ENVIRONMENTAL MODELS

In Chapter 2, the concept of "fugacity" as it relates to chemical potential was introduced. In environmental application, fugacity is used to estimate the tendency of molecules to partition into the various environmental compartments (Mackay, 1979). Since fugacity is identical to partial pressure in ideal gases, it is also related to the vapor pressures of liquids and solids. Fugacity is directly measurable (e.g., at low pressures, fugacity and partial pressure of an ideal gas are the same), and since it is linearly related to partial pressure or concentration, it is a better criterion for equilibrium than the elusive chemical potential. In fact, the criterion of equal chemical potential for equilibrium can be replaced without the loss of generality with the criterion of equal fugacity between phases. If an aqueous solution of a compound i is brought into contact with a given volume of air, the species i will transfer from water into air until the following criterion is established:

$$\mu_i^{w0} + RT\ln\left(\frac{f_i^w}{f_i^{w0}}\right) = \mu_i^{a0} + RT\ln\left(\frac{f_i^a}{f_i^{a0}}\right) \tag{3.2}$$

If we choose the same standard states for both phases, we have the following criterion for equilibrium:

$$f_i^w = f_i^a \tag{3.3}$$

This equality will hold even when the standard states are so chosen that they be at the same temperature, but at different pressures and compositions. We then have an exact relation between the standard states, that is,

$$\mu_i^{w0} - \mu_i^{a0} = RT \ln\left(\frac{f_i^{w0}}{f_i^{a0}}\right) \tag{3.4}$$

The equality of fugacity can be generalized for many different phases in equilibrium with one another and containing multicomponents. We now have three equivalent criteria for equilibrium between phases, as described in Table 3.1.

Mackay (1991) proposed a term "fugacity capacity," Z that related fugacity (of any phase expressed in Pa) to concentration (expressed in $mol\,m^{-3}$). Thus,

$$Z\left(mol\ m^{-3}\ Pa^{-1}\right) = \frac{C\left(mol\ m^{-3}\right)}{f\left(Pa\right)} \tag{3.5}$$

The value of Z depends on a number of factors such as identity of the solute, nature of the environmental compartment, temperature T, and pressure P. A fugacity capacity can be defined for each environmental compartment (Table 3.2). In order to obtain the value of Z, knowledge of other equilibrium relationships between phases ("partition coefficients") is required. These relationships will be described in detail later in this chapter. Suffice it to say, at this point the partition coefficients are to be either experimentally determined or estimated from correlations.

TABLE 3.1
Criteria for Equilibrium between Two Phases a and b

Property	Criteria
Gibbs free energy	$G_i^a = G_i^b$
Chemical potential	$\mu_i^a = \mu_i^b$
Fugacity	$f_i^a = f_i^b$

Notes: In terms of chemical potentials, at constant temperature, and either at constant V or P, the criteria for irreversible and reversible processes are as follows:
$\Delta\mu_i < 0$ (irreversible, spontaneous process)
$\Delta\mu_i = 0$ (reversible, equilibrium process)
$\Delta\mu_i > 0$ (nonspontaneous process).

TABLE 3.2

Definition of Fugacity Capacities for Environmental Compartments

Compartment	Definition of Z ($mol\,m^{-3}\,Pa^{-1}$)
Air	$1/RT$
Water	$1/K''_{AW}$
Soil or sediment	$K_{sw}\rho_s/K''_{AW}$
Biota	$K_{bw}\rho_b/K''_{AW}$

Source: Mackay, D., *Multimedia Environmental Models*, Lewis Publishers, Chelsea, MI, 1991.

Note: R is the gas constant ($= 8.314\,Pa\,m^3\,mol^{-1}\,K^{-1}$), K_{aw} is a form of Henry's constant for species ($Pa\,m^3\,mol^{-1}$), K_{sw} is the partition constant for species between the soil or sediment and water ($dm^3\,kg^{-1}$), K_{bw} is the bioconcentration factor ($dm^3\,kg^{-1}$), ρ_s and ρ_b are the densities ($kg\,dm^{-3}$) of soil/sediment and biota, respectively.

Once the fugacity capacities are known for individual compartments, the mean Z value can be determined by multiplying the Z_j value of each compartment with the volume of that compartment (V_j) and summing the calculated value of each compartment. If the total mass of the compound or chemical input or inventory in all the compartments is known, then the fugacity of the compound is given by

$$f = \frac{\sum_j n_j}{\sum_j V_j Z_j} \tag{3.6}$$

where the numerator is the sum of all the moles of the chemical in all the compartments. This fugacity is common to all phases at equilibrium. Hence, the concentration in each compartment is given by the following equation:

$$C_j = f \cdot Z_j \tag{3.7}$$

More sophisticated calculations suitable for realistic situations involving time-dependent inflow and outflow of chemicals into various compartments have been proposed and discussed in detail by Mackay (1991). The student is referred to this excellent reference source for more details. The purpose of this section has been to impress upon the student how the concept of fugacity can be applied to environmental modeling. The following example should illustrate the concept.

Example 3.1   Fugacity Model (Level I) for Environmental Partitioning

Problem statement: Consider an evaluative environment consisting of air, water, soil, and sediment. The volumes of the phases are as follows: air $= 6 \times 10^9\,m^3$, water $= 7 \times 10^6\,m^3$, soil $= 4.5 \times 10^4\,m^3$, and sediment $= 2.1 \times 10^4\,m^3$. Determine the equilibrium distribution of a hydrophobic pollutant such as pyrene in this four-compartment model. The properties for pyrene are as follows: Henry's constant = 0.9 Pa m^3 mol^{-1}, K_d (soil) = 1.23 × 10^3 L kg^{-1}, ρ_s (for sediment and soil) = 1.5 × 10^{-2} kg L^{-1}, K_d (sed) = 2.05 × 10^3 L kg^{-1}. Let the temperature be 300 K and the total inventory of pyrene be 1000 mol.

Solution: First step is to calculate the fugacity capacity Z (mol m^{-3} Pa^{-1}) as follows:

Air: Z_1 = 1/RT = 1/(8.314 × 300) = 4 × 10^{-4}
Water: Z_2 = 1/H = 1/0.89 = 1.1
Soil: Z_3 = $K_d\rho_s$/H = (1.23 × 10^3) (1.5 × 10^{-2})/0.89 = 20.7
Sediment: Z_4 = (2.05 × 10^3) (1.5 × 10^{-2})/0.89 = 34.5

The second step is to calculate the fugacity, f:

$$f = 1000/1.2 \times 10^7 = 8.5 \times 10^{-5} \text{ Pa}$$

The last step is to calculate the concentrations in each phase, C_j:

$$C_{air} = fZ_1 = 3.4 \times 10^{-8} \text{ mol m}^{-3}$$

$$C_{water} - fZ_2 - 9.4 \times 10^{-5} \text{ mol m}^{-3}$$

$$C_{soil} = fZ_3 = 1.7 \times 10^{-3} \text{ mol m}^{-3}$$

$$C_{sediment} = fZ_4 = 2.9 \times 10^{-3} \text{ mol m}^{-3}$$

We can now calculate the total mol of pyrene in each compartment, m_j:

$$m_{air} = C_{air}V_{air} = 205 \text{ mol}$$

$$m_{water} = C_{water}V_{water} = 658 \text{ mol}$$

$$m_{soil} = C_{soil}V_{soil} = 76 \text{ mol}$$

$$m_{sediment} = C_{sediment}V_{sediment} = 61 \text{ mol}$$

Thus, the largest fraction (65.8%) of pyrene is in water. The next largest fraction (20.5%) resides in the air. Both sediment and soil environments contain less than 10% of pyrene each.

If there is inflow and outflow from the evaluative environment, and steady state exists, we have to use a Level II fugacity calculation. If G_j (m^3 h^{-1}) represents both the inflow and outflow rates from compartment j, then the total influx rate I (mol h^{-1}) is related to the fugacity:

$$f = \frac{I}{\sum_j G_j Z_j} \tag{3.8}$$

The concentrations are then given by $C_j = fZ_j$ and mass by $m_j = C_jV_j$. The total influx rate is given by $I = E + \Sigma G_jC_j^0$, where E is the total emission rate (mol h^{-1}) from the environment and C_j^0 is the influent concentration (mol m^{-3}) in the stream.

Example 3.2   Level II Fugacity Calculation

For the environment consisting of 10^4 m^3 air, 1000 m^3 of water, and 1 m^3 sediment, an air flow rate of 100 m^3 h^{-1}, a water inflow of 1 m^3 h^{-1}, and an overall emission rate of 2 mol h^{-1} are known for pyrene. The influent concentration in air is 0.1 mol m^{-3} and in water is 1 mol m^{-3}. Given Z values of 10^{-4} for air, 0.1 for water, and 1 for sediment, calculate the concentration in each compartment.

Total influx $I = 2$ (mol h^{-1}) + 100 (m^3 h^{-1}) 0.1 (mol m^{-3}) + 1 (m^3 h^{-1}) 1 (mol m^{-3}) = 13 mol h^{-1}. Fugacity $f = 13/[(100) (10^{-4}) + (1) (1)] = 13$ Pa. $C_{air} = (13) (0.0001) = 0.0013$ mol m^{-3}, $C_{water} = (13) (0.1) = 1.3$ mol m^{-3}, and $C_{sed} = (13) (1) = 13$ mol m^{-3}.

3.2   AIR-WATER PHASE EQUILIBRIUM

In a number of situations in the environment, the two phases, air and water, coexist. The largest area of contact is between these two phases in the natural environment. Exchange of mass and heat across this interface is extensive. In some cases, one phase will be dispersed in the other. A large number of waste treatment operations involve comminuting either of the phases to maximize the surface area and mass transfer. First, we shall discuss the area of fate and transport modeling (chemodynamics). Second, we will discuss an area within the realm of separation processes for waste treatment.

Consider the exchange of a compound at the air-water interface (Figure 3.1). Compounds transfer between air and water through volatilization and absorption. Gas bubbles transport materials from the ocean floor to the atmosphere. Ejections of particulates attached to the bubbles occur upon bubble bursting at the interface. Dissolved compounds and particulates in air can be deposited to sea or land via attachment or dissolution in fog, mist, and rain. As discussed in Chapter 2, Henry's law describes the equilibrium at this interface. The "air-water equilibrium constant," K_{AW}, at any temperature is given by

$$K_{AW}(T) = \frac{C_{ia}}{C_{iw}} \tag{3.9}$$

Example 3.3   Use of Air-Water Partition Constant

Water from an oil-field waste pond contains hydrogen sulfide at a concentration of 500 mg L^{-1} at an ambient temperature of 298 K. What will be the maximum equilibrium concentration of H$_2$S above the solution?

At 298 K, $K_{AW}^{''}$ for H$_2$S is 0.41 (Appendix A). $C_{iw} = 500$ mg L$^{-1} = 0.5/34 = 0.0147$ mol L^{-1}. Hence, $C_{ia} = (0.41) (0.0147) = 0.006$ mol L^{-1}. Partial pressure in air, $P_i = C_{ia}RT = 0.147$ atm.

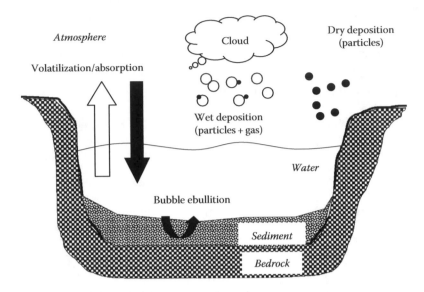

FIGURE 3.1 Exchange of chemicals between water and atmosphere. Equilibrium is important in many cases such as volatilization/absorption, wet deposition, dry deposition, and gas bubble transport.

Many "waste treatment separation processes" also involve air-water exchange of chemicals (Figure 3.2). Examples include wastewater aeration ponds and surface impoundments in which surface and submerged aerators are used to transfer oxygen to the water and to remove volatile organic and inorganic compounds from water using flotation. Volatile organics are removed using a tower where air and water are passed countercurrent to one another through a packed bed. The operations are non-equilibrium processes, but where, the tendency of the system to attain equilibrium provides the driving force for separation.

The transfer of material (say i) from one phase to the other (e.g., water to air) is due to a gradient in concentration of i between the phases (Figure 3.3). This flow of material i from water to air is termed "flux" and is defined as the moles of i passing a unit area of interface per unit time. The equation for flux is

$$N_i = K_w \left(C_{iw} - C_{iw}^* \right) \tag{3.10}$$

where C_{iw}^* is the concentration of i in water that would be in equilibrium with air. This is given by the air-water distribution constant:

$$C_{iw}^* = \frac{C_{ia}}{K_{AW}} \tag{3.11}$$

The maximum overall flux from water to air will be attained when $C_{ia} = 0$. Thus, $N_i^{max} = K_w C_{iw}$. If $C_{iw} = 0$, then the flux will be from air to water. Thus, if the sign is

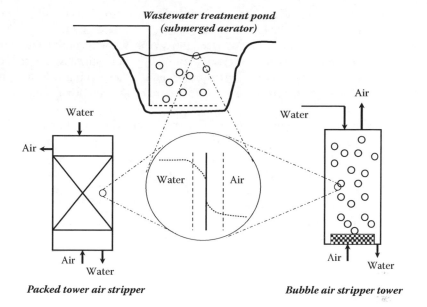

FIGURE 3.2 Examples of air-water contact and equilibrium in separation processes. In each of the three cases, contact between air and water is important.

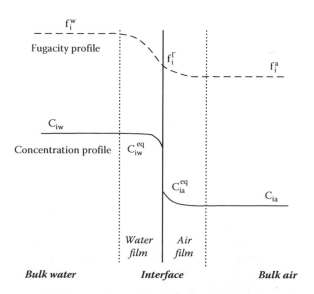

FIGURE 3.3 Film theory of mass transfer of solutes between air and water. This diagram is for the volatilization of a compound from water. The concentration profile shows a discontinuity at the interface, whereas the fugacity profile does not. The overall resistance to mass transfer is composed of four individual resistances to diffusion of chemical from water to air. For more details, see also Section 2.4.1.4.1.

positive, the flux is from water to air, and if negative, it is from air to water. If the bulk air and water phases are in equilibrium, then $C_{iw} = C_{ia}/K_{AW}$ and the net overall flux $N_i = 0$. Thus, the flow of material between the two phases will continue until the concentrations of the two bulk phases reach their equilibrium values as determined by the air-water equilibrium constant. At equilibrium the "net overall flux" is zero, and the water-to-air and the air-to-water diffusion rates are equal but opposite in direction.

The flux expression given in Equation 3.3 is predicated upon the assumption that compound i diffuses through a stagnant water film to a stagnant air film across an interface of zero volume (Figure 3.3). Equilibrium is assumed to exist at the interface, but the two bulk phases are not in equilibrium and have uniform concentrations beyond the film thickness, δ. The term $(C_{iw} - C_{ia}/K_{AW})$ is called the "concentration driving force" for mass transfer. The "overall mass transfer coefficient" K_w includes the value of the equilibrium constant in the following form (see Section 2.4.1.4)

$$\frac{1}{K_w} = \frac{1}{k_w} + \frac{1}{k_a K_{AW}} \tag{3.12}$$

where k_w and k_a are the individual film mass transfer coefficients on the water and air side, respectively. Each term in this equation corresponds to a resistance to mass transfer. $1/K_w$ is the total resistance, whereas $1/k_w$ is the resistance to diffusion through the water film (δ_w). $1/k_a K_{AW}$ is the resistance provided by the air film (δ_a).

Example 3.4   Overall and Individual Mass Transfer Coefficients

For evaporation of benzene from water, $k_w = 7.5 \times 10^{-5}$ m s^{-1} and $k_a = 6 \times 10^{-3}$ m s^{-1}. Estimate the overall mass transfer coefficient K_w. What is the percent resistance in the air side of the interface?

At 298 K, $K_{AW} = 0.23$ (from Appendix A). Hence $1/K_w = 1/k_w + 1/(k_a K_{AW}) = 1.4 \times 10^4$ and $K_w = 7.1 \times 10^{-5}$ m s^{-1}. Percent resistance in the air film $= (K_w/k_a K_{AW}) \times 100 = 5.1\%$. Benzene volatilization is therefore water-phase controlled. Note that if K'_{AW} is small, percent resistance air side will increase.

Quantitative determination of the mass transfer rates thus requires estimation of the equilibrium partition constant between air and water phases. Note that it appears both in the driving force and the overall mass transfer coefficient expressions.

Notice from Figure 3.3 that there exists a discontinuity in concentrations at the interface. However, fugacity changes gradually without discontinuity. Since the fugacity in air is different from that in water, a concentration difference between the two phases exists and the system moves toward equilibrium where $f_i^w = f_i^\Gamma = f_i^a$. From the discussion in Chapter 3 on fugacity capacity, it can be shown (Mackay, 1991) that the mass transfer flux can also be written in terms of fugacity as

$$N_i = K_{w,f} \left(f_i^w - f_i^a \right)$$

where N_i is in $mol\,m^{-2}s^{-1}$, as defined earlier, and $K_{w,f}$ is given in terms of fugacity capacities ($Z_w = 1/K'_{AW}$ and $Z_a = 1/RT$) as

$$\frac{1}{K_{w,f}} = \frac{1}{k_w Z_w} + \frac{1}{k_a Z_a} \tag{3.13}$$

Note that $K_{w,f} = K_w/K_{AW}$ with K_w as in Equation 3.5. The definition of K_{AW} is given in Chapter 2. The net flux is identified as the algebraic sum of the net volatilization from water, $K_{w,f}f_i^w$ and the net absorption rate from air, $K_{w,f}f_i^a$. When these two rates are equal the system is at equilibrium. The term $\left(f_i^w - f_i^a\right)$ is appropriately called the "departure from equilibrium."

A variety of methods can be employed to obtain Henry's constants for both organic and inorganic compounds in air-water systems. Following are the descriptions of some of the more common methods.

3.2.1 ESTIMATION OF HENRY'S CONSTANT FROM GROUP CONTRIBUTIONS

Hine and Mookerjee (1975) suggested one method of estimating the air-water partition constant by summing bond or group contributions in a molecule. Meylan and Howard (1992) recently refined the method. The method is analogous to the one discussed in Chapter 3 for the determination of $\log K_{ow}$. It relies on the assumption that the free energy of transfer of a molecule between air and water is an additive function of the various groups or bonds present in the molecule. The correlation was developed to obtain the reciprocal of the air-water partition constant defined in this book. Thus, the method gives $1/K''_{AW}$. An extensive listing of the bond and group contributions is given in Hine and Mookerjee (1975). It is best to illustrate this method through examples.

Example 3.5 K_{AW} from a Bond Contribution Scheme

Given the following bond contributions to $\log K''_{AW}$, calculate K_{AW} for (1) benzene, (2) hexachlorobenzene, and (3) chloroform.

Bond	Log K''_{AW}
$H-C_{ar}$	−0.154
$C_{ar}-C_{ar}$	0.264
$H-C_{al}$	−0.119
$Cl-C_{ar}$	−0.024
$Cl-C_{al}$	0.333

These symbols are self-explanatory. The subscript ar denotes aromatic while al refers to aliphatic.

1. In benzene there are six $C_{ar}-C_{ar}$ bonds and six $H-C_{ar}$ bonds. Thus, $-\log K''_{AW} = 6(-0.154) + 6(0.264) = 0.66$. Hence, $K''_{AW} = 0.219$. The experimental value is 0.231.

2. In hexachlorobenzene, there are six C_{ar}—C_{ar} bonds and six Cl—C_{ar} bonds. Hence, $-\log K''_{AW} = 6(0.264) + 6(-0.024) = 1.46$. $K''_{AW} = 0.038$. The experimental value is 0.002.

3. In chloroform, we have one H—C_{al} and three Cl—C_{al} bonds. Hence, $-\log K''_{AW} = -0.119 + 3(0.333) = 0.87$. $K''_{AW} = 0.134$. The experimental value is 0.175.

Bond contribution scheme suffers from a major disadvantage in that interactions between bonds are usually not considered. This will be a serious drawback when polar bonds are involved.

Example 3.6 K_{AW} from a Group Contribution Scheme

Estimate Henry's constant for benzene and hexachlorobenzene using the following group contributions:

Group	$-Log\ K_{AW}$
C_{ar}—H$(C_{ar})_2$	0.11
C_{ar}—(C)—$(C_{ar})_2$	0.70
C_{ar}—Cl$(C_{ar})_2$	0.18

For benzene we have six C_{ar}–H$(C_{ar})_2$ groups. Hence, $-\log K_{AW} = 6(0.11) = 0.66$. $K_{AW} = 0.218$. The experimental value is 0.231. The prediction is only marginally better than the bond contribution scheme.

For hexachlorobenzene we have six C_{ar}–Cl$(C_{ar})_2$. Hence, $-\log K_{AW} = 6(0.18) = 1.08$. $K_{AW} = 0.083$. The experimental value is 0.002. The agreement is poor and is no better than that for the bond contribution scheme.

Since the group contribution scheme of Hine and Mookerjee is of limited applicability for compounds with multiple polar groups, Meylan and Howard (1992) proposed an alternative but a more reliable group contribution scheme whereby

$$\text{Log } K_{AW} = \sum_i a_i q_i + \sum_j b_j Q_j \tag{3.14}$$

Appendix F lists q_i and Q_j values for several bonds.

3.2.2   Experimental Determination of Henry's Constants

In Chapter 3, we noted that

$$K_{AW} = \left(\frac{V_w}{RT}\right) \cdot \left(\frac{P_i^*}{x_i^*}\right) \tag{3.15}$$

which means that K_{AW} can be obtained from a ratio of the saturated vapor pressure and aqueous solubility of a sparingly soluble compound.

Based on this premise, there are several methods for K_{AW} measurements described in the literature. The different methods fall generally into three categories: (1) from a ratio of vapor pressure and aqueous solubility that are independently measured, (2) static methods, and (3) mechanical recirculation methods. In the first category of methods, the errors in independent measurements of vapor pressure and aqueous solubility lead to additive errors in K_{AW}. In the second category of methods, because of difficulties associated with the simultaneous precise measurements of concentrations in both air and water, they are restricted to compounds with large vapor pressures and aqueous solubilities. Such methods are ideally suited for soluble solutes such as CO_2, SO_2, and several volatile organics. The EPICS method of Gossett (1987) and the direct measurement technique of Leighton and Calo (1981) fall under this category. In reality, methods from both categories 1 and 2 should be superior to the third since they require the measurement in both phases, which would allow mass balance closure if the initial mass of organic solute introduced into the system is accurately known. The errors in these methods arise directly from the concentration measurements. The third class of methods (e.g., batch air stripping, wetted wall column, and fog chamber) suffer from a serious deficiency and that is the necessity to ascertain the approach to equilibrium between phases.

Several investigators have compiled Henry's constants for a variety of compounds of environmental interest (e.g., Mackay and Shiu, 1981; Ashworth et al., 1988). The values reported show considerable scatter for a single compound. Table 3.3 is an example of the degree of agreement reported by different workers. It is not

TABLE 3.3
Comparison of K_{AW} Values from the Literature

Technique	Reference[a]	Chloroform	Carbon Tetrachloride	Benzene	Lindane
Equilibrium	1	0.175	1.229	0.22	—
Batch stripping	2	0.141	1.258	0.231	—
Multiple equilibration	3	0.125	0.975	—	—
Vapor pressure and solubility	4	0.153	0.807	0.222	1.33×10^{-2}
Direct measurement	5	0.153	1.14	0.227	—
Wetted wall	6	—	—	—	3.43×10^{-3}
Fog chamber	7	—	—	—	3.54×10^{-3}
Vapor pressure and solubility	8	—	—	—	8.41×10^{-4} to 2.18×10^{-3}
K_{AW} (mean)		0.149	1.082	0.225	7.17×10^{-3}
Std. devn.		0.018	0.189	0.005	4.82×10^{-3}

[a] (1) Ashworth et al. (1988)—equilibrium partitioning in closed systems; (2) Warner et al. (1987)—bubble air stripping; (3) Munz and Roberts (1986)—multiple equilibration; (4) Mackay and Shiu (1981)—vapor pressure-solubility relationship; (5) Leighton and Calo (1981)—experimental vapor pressure and solubility; (6) Fendinger and Glotfelty (1988)—wetted wall column; (7) Fendinger et al. (1989)—fog chamber experiment; (8) Suntio et al. (1987)—vapor pressure–solubility relationship.

unreasonable to expect a great deal of scatter, especially for those compounds that have low vapor pressures and aqueous solubilities. The standard deviation in H_c is only 2% of the mean for benzene, whereas it is 12% of the mean for chloroform and 17% of the mean for carbon tetrachloride. The standard deviation is 67% of the mean for lindane, which has a low H_c value. For most environmental purposes, it has been suggested that a reasonable standard error in H_c is about 5%–10% (Mackay and Shiu, 1981).

3.2.3 EFFECTS OF ENVIRONMENTAL VARIABLES ON K_{AW}

The air–water partition constants estimated at room temperature using binary aqueous systems (solute + water) are not particularly applicable to natural systems (wastewater, atmospheric moisture, seawater, etc.). Several environmental variables affect K_{AW} in different ways. In this section, we briefly review four important variables and their effects on air–water partition constants. These are (1) temperature, (2) presence of other miscible solvents, (3) presence of colloids, and (4) variations in pH.

As noted here, since K_{AW} is a ratio of P_i^* to x_i^*, the effects of temperature on K_{AW} are directly related to the variations in vapor pressure and aqueous solubility with temperature. As per the Clausius–Clapeyron equation, both P_i^* and x_i^* show logarithmic dependence on temperature. Hence, the variation in K_{AW} with temperature is given by the equation

$$\ln K_{AW}(T) = A - \frac{B}{T} \qquad (3.16)$$

where A and B are constants for a given compound. The values for A and B for a variety of compounds have been tabulated (Table 3.4). In reality, Equation 3.16 is a method whereby single values of air-water partition constants at a standard temperature T_0 are extrapolated over a modest range using the solute enthalpy of solution. A thermodynamics expression associated with van't Hoff relates the air-water partition constants at two temperatures (Smith and Harvey, 2007). The implicit assumption is that the enthalpy of solution is constant over the temperature range and this is valid only over a narrow range of temperatures (≈ 20 K).

Over a wide range of temperature, K_{AW} is correlated to T using the following polynomial, $K_{AW}(T) = a + bT + cT^2$, where a, b, and c for several compounds are listed in Table 3.4.

Example 3.7 K_{AW} and Temperature

Determine K_{AW} for benzene at 30°C using Table 3.4.

Using $K_{AW} = (1/(8.205 \times 10^{-5} \times 303)) * \exp(5.534 - (3194/303)) = 0.267$.
Using $K_{AW} = 0.0763 + 0.00211 * 30 + 0.000162 * 900 = 0.285$.

Wastewaters and contaminated waters have varied composition and often contain other organic solvents that are miscible with water. These solvent + water mixtures

TABLE 3.4

Temperature Dependence of K_{AW}

$$K_{AW} = \frac{1}{RT} \cdot \exp\left(A - \frac{B}{T}\right)$$

Compound	Temp. Range (°C)	A	B
Benzene	10–30	5.534	3194
Toluene	10–30	5.133	3024
Ethylbenzene	10–30	11.92	4994
o-Xylene	10–30	5.541	3220
Chloroform	10–30	11.41	5030
1,1,1-Trichloroethane	10–30	7.351	3399
Trichloroethylene	10–30	7.845	3702

$$K_{AW} = a + bt + ct^2$$

Compound	Temp. Range (°C)	a	b	c
Benzene	0–50	0.0763	0.00211	0.000162
Toluene	0–50	0.115	−0.00474	0.000466
Ethylbenzene	5–45	0.05	0.00487	0.000250
o-Xylene	15–45	0.0353	0.00444	0.000131
Chloroform	0–60	0.0394	0.00486	—
1,1,1-Tichloroethane	0–35	0.204	0.0182	0.000173
Trichloroethylene	5–45	0.151	−0.00597	0.000680

Sources: Ashworth, R.A. et al., *J. Hazard. Mater.*, 18, 25, 1988; Turner et al., 1996.
Note: $R = 8.205 \times 10^{-5}$ atm m^3 (K mol)$^{-1}$, T is in K, t is in °C.

affect the overall solubility of organic solutes, thus varying x_i^* in the equation for K_{AW}. There is evidence in the literature that the effects do not become predominant unless significant concentrations (>10 weight percent) of a very soluble cosolvent are present in natural systems. An important paper in the literature in this regard is the one by Munz and Roberts (1987), where they evaluated Henry's constants of three chloromethanes (chloroform, carbon tetrachloride, and hexachloroethane), often encountered in wastewaters and atmospheric water. Munz and Roberts (1987) observed no significant effects on Henry's constants in the presence of cosolvents such as methanol and isopropanol up to concentrations as large as 10 g L^{-1} (i.e., cosolvent mol fraction of $\approx 5 \times 10^{-3}$). It was observed that the effects were pronounced for very hydrophobic solutes. A compound that interacts better with the cosolvent exhibits the effect at low cosolvent concentrations than does a compound that is more hydrophilic. In general, for any hydrophobic solute, the presence of a cosolvent increases the aqueous solubility and lowers K_{AW}. These conclusions have been substantiated by others who determined Henry's constants of volatile solutes such as toluene and chloroform in distilled, tap, and natural lake waters. Table 3.5 is a summary of these observations.

TABLE 3.5
Measured Air-Water Partition Constants for Selected Compounds in Natural Waters

Compound	Matrix Composition	K_{AW}	Reference
Toluene	Distilled water	0.244	Yurteri et al. (1987)
	Tap water	0.231	
	"Creek" water	0.251	
Chloroform	Distilled water	0.125	Nicholson et al. (1984)
	Natural lake water	0.121	
CH_3CCl_3	Distilled water	0.53	Hunter-Smith et al. (1983)
	Sea water	0.94	
$CHClBr_2$	Distilled water	0.036	Nicholson et al. (1984)
	Natural lake water	0.034	
$CHBr_3$	Distilled water	0.018	Nicholson et al. (1984)
	Natural lake water	0.017	
Hexachlorobenzene	Distilled water	0.054	Brownawell (1986)
	Sea water	0.070	
2,4,4'-Trichlorobiphenyl	Distilled water	0.00595	Brownawell (1986)
	Sea water	0.00885	

The effects of inorganic salts on air-water partition constants are manifest through their effects on the infinite dilution activity coefficients that are described in Chapter 3. Most hydrophobic organic compounds show the "salting-out" behavior, whereby the mole fraction solubility in water is decreased (or activity coefficient is increased). Consequently, the air-water partition constant decreases in the presence of salts. A number of investigators have found that, for example, the partition constant for chlorinated volatile organic compounds (VOCs) such as chlorofluorocarbons and chlorobiphenyls in seawater is 20%–80% greater than the values determined in distilled or deionized water (Table 3.5).

Example 3.8 K_{AW} and Ionic Strength

Determine K_{AW} for phenanthrene in an industrial wastewater with 1 M total ion concentration at 292 K.

In water, without added salts, $K_{AW}^o = C_i^a/C_i^w$. C_i^a is not effected by ionic strength in water, but C_i^w is modified using the McDevit-Long theory (Section 2.2.4.3). Let C_i^{ww} represent the new wastewater concentration. Then new $K_{AW} = C_i^a/C_i^{ww}$. By McDevit-Long theory, $\gamma_i^{ww} = \gamma_i^w \exp(\varphi V_H C_s)$. Since $\gamma_i^w \infty 1/x_i^w \infty 1/C_w^i$, we can write $C_i^w = C_i^{ww} \exp(\varphi V_H C_s)$. Therefore, $K_{AW} = K_{AW}^o \exp(\varphi V_H C_s)$. From Table 2.11, for phenanthrene, $\varphi V_H = (0.00213)(182) = 0.38$. From Appendix A, $K_{AW}^{"o} = 3.16$ kPa dm^3 mol^{-1}. Hence, $K_{AW} = 3.16 \exp(0.38 \times 1) = 4.62$.

TABLE 3.6
Sizes of Organic and Inorganic Particulates
in Natural and Wastewaters

Type of Particle	Size (nm)
Organic macromolecules (humics)	1–10
Virus	10–100
Oxides (iron and aluminum)	10–1000
Clays	10–1000
Bacteria	10^3–10^4
Soil particles	10^3–10^6
Calcium carbonate, silica	10^4–10^6

Note: 1 nm = 10^{-3} μm.

Colloids and particulates are ubiquitous in both natural waters and wastewaters. They are characterized by their size, which can range from a few nanometers to thousands of nanometers. Examples of colloidal material encountered in natural and wastewaters are given in Table 3.6. Organic colloids are characterized by sizes 10^3 nm or less. They are dispersed phases composed of high-molecular-weight (>1000) macromolecules of plant origin. They are composed of C, H, and O with traces of N and S. They possess ionizable groups (OH or COOH) and are macro ions. These are classified as dissolved organic compounds (DOCs) and range in concentration from a few mg L^{-1} in oceans to as large as 200 mg L^{-1} in peaty catchments or swamps, and are known to give a distinct brownish tinge to water. DOCs are also observed at concentrations ranging from 10 to 200 mg L^{-1} in atmospheric water (fog and rainwater). DOCs are known to complex with inorganic metal ions and bind organic pollutants. The concentration of metals and organic compounds bound to a single macromolecule can be large (Wijayaratne and Means, 1984). The effective solubilities of both inorganic and organic species can, therefore, be several times larger in the presence of DOCs.

Hydrophobic pollutants will preferentially associate with DOCs since they provide an organic medium shielding the pollutants from interactions with water. It has been observed that the association of organic compounds is correlated to their hydrophobicity (K_{ow}). The more hydrophobic the compound, the greater is its sorption to DOCs. Chiou et al. (1986, 1987) determined the effect of natural humic acids (extracted from soil) upon the aqueous solubility of several chlorinated organics (1,2,4,-trichlorobenzene, lindane, and several polychlorinated biphenyls). Haas and Kaplan (1985) reported a similar study on the solubility of toluene in water containing humic acids. Carter and Suffet (1982) determined the effect of humic acids on the aqueous solubility of several polyaromatic hydrocarbons. Boehm and Quinn (1973) reported the enhanced solubility of several hydrocarbons in seawater brought about by the presence of DOCs. The effects were observed to be specific to the type of DOCs. Commercial humic acids behaved differently from marine or soil humic acids.

In general, a larger solubility was always noted in the presence of DOCs in water. Chiou et al. (1986) showed that if C_i^* represents the solubility of the organic compound in pure water and C_i^{app} is the apparent solubility in DOC-rich water, then

$$\frac{C_i^{app}}{C_i^*} = 1 + C_C K_{cw} \qquad (3.17)$$

where

C_C is the DOC concentration (g mL^{-1}) in water

K_{cw} is the equilibrium partition constant for the organic solute between DOC and water (expressed in mL g^{-1})

The values of K_{cw} for several compounds are given in Table 3.7.

Since K_{AW} is inversely proportional to the aqueous solubility, the air-water partition constant in the presence of DOCs will be given by

$$K_{AW}^* = \frac{K_{AW}}{1 + C_C K_{cw}} \qquad (3.18)$$

It can therefore be seen that the effects on air-water partition constants for a compound of relatively high aqueous solubility will be less than that for one with a very low aqueous solubility.

Mackay et al. (1982) noted that when commercially available humic acid or fulvic acid was used in the aqueous solutions, up to concentrations of 54 mg L^{-1}, the air-water partition constant for naphthalene showed a reduction of about 7×10^{-5} for every 1 mg L^{-1} of the colloid added. It was concluded that although natural organic colloids reduce the value of air-water partition constant, it is probably negligible for most of the soluble organics at typical environmental concentrations of DOCs. Callaway et al. (1984) reported a reduction in K_{AW} for two low-molecular-weight

---

TABLE 3.7

Partition Constants for Pollutants between DOC and Water

Compound	Type of DOC	Log K_C	Reference
p,p'-DDT	Soil humic	5.06	Chiou et al. (1986)
p,p'-DDT	Aldrich humic acid	5.56	Chiou et al. (1987)
1,2,3-Trichlorobenzene	Soil humic	3.00	Chiou et al. (1987)
Toluene	Aldrich humic acid	3.3	Haas and Kaplan (1985)
Lindane	Soil humic	2.7	Chiou et al. (1986)
Pyrene	Soil humic	4.9–5.5	Gauthier et al. (1986)
2,4,4'-PCB	Soil humic	4.24	Chiou et al. (1987)
Anthracene	Soil humic	4.92	Carter and Suffet (1982)
Fluoranthene	Soil humic	5.32	Acha and Rebhun (1992)

organic compounds, namely, chloroform and trichloroethene (TCE) in the presence of humic acid. The experiments involved measuring the concentrations of the VOCs in closed vessels after equilibration of a known volume of the aqueous phase containing humic acid (10 wt%). It was then compared with a system without humic acid. The ratio K^*_{AW}/K_{AW} was 0.39 and 0.79 for chloroform, the different values corresponding to the pretreatment procedure used for preparing the humic acid solution. The ratio for TCE was 0.28. Yurteri et al. (1987) observed that the ratio of air-water partition constants decreased to 0.94 for toluene in the presence of 10 mg L^{-1} humic acid in water whereas it was 0.99 for 5 mg L^{-1} humic acid in water. A significant observation was that the combined effects of inorganic salts, surfactants, and humic acids on the partition constant were far more significant than the effect of any single component in wastewater samples.

Municipal wastewaters in many regions of the world contain high concentrations of detergents (e.g., see Takada and Ishiwatari, 1987). Linear alkyl benzene sulfonates (LABs) and linear alkyl sulfates (LAS) are common constituents of synthetic detergents. These and other surfactants behave in a manner similar to the DOCs described earlier. They tend to increase the aqueous solubility of many organic and inorganic species.

Valsaraj (1988) studied the effects of surfactants (both anionic and cationic) on the air-water partition constants of low-molecular-weight hydrocarbons of environmental interest. The effects of two anionic surfactants (sodium dodecylbenzene sulfate, DDS; and sodium dodecylbenzenesulfonate, DDBS) and one cationic surfactant (hexadecyltrimethyl ammonium bromide, HTAB) were studied. The reduction in air-water partition constant from that in pure water was deduced. It was observed that the partition constant for chloroform decreased to 0.77 times its value in pure water in the presence of 42 mM of DDBS and it further decreased to 0.66 times its pure water value in the presence of 82 mM of DDBS. There was no decrease till the critical micellar concentration (CMC) of DDBS was exceeded. The results with DDS and HTAB micelles were similar to that with DDBS.

Example 3.9　K_{AW} and DOC

For anthracene, estimate what will be the concentration in water in the presence of 1000 mg L^{-1} of a DOC.

From Table 3.7, K_{cw} for anthracene is $10^{4.92}$ L kg^{-1}. Given C_C is 1000 mg L$^{-1} = 10^{-3}$ kg L^{-1}, the increase in aqueous phase concentration is $C^{app}_i/C^*_i = 1 + C_C K_{cw} = 16.8$. Since $C^*_i = 3.3 \times 10^{-5}$ mol L^{-1}, $C^{app}_i = 5.6 \times 10^{-4}$ mol L^{-1}. Also note that since $C^{app}_i/C^*_i = x^{app}_i/x^*_i = \gamma^*_i/\gamma^{app}_i$, the activity coefficient is 16.8 times lower in the presence of DOC.

Many inorganic gases (e.g., carbon dioxide, sulfur dioxide, and ammonia) that dissolve in water undergo chemical reactions depending on the pH of the solution. For example, SO_2 dissolving in rainwater gives rise to sulfuric acid, which is the cause of "acid rain." Many organic species also undergo reactions with water depending on the pH. For example, phenol is converted into phenolate ion under alkaline conditions.

The ionization of water gives rise to hydrogen and hydroxide ions:

$$H_2O \rightleftharpoons H^+ + OH^- \tag{3.19}$$

At equilibrium, in "pure water," the concentrations of both ions are equal and have a value 1×10^{-7} M. The equilibrium constant for the reaction $K_w = [H^+][OH^-] = 1 \times 10^{-14}$ M^2. The pH of pure water is given by $-\log[H^+] = 7.0$. "Pure" rainwater is slightly acidic with a pH of 5.6. The dissolution of gases and organic compounds that ionize in solution can change the pH of water. This process can also change the equilibrium constant for dissolution of gases in water, given by Henry's law.

Let us consider SO_2 dissolution in water. The transfer of SO_2 from gas to water is given by the following equilibrium reaction:

$$SO_2(g) + H_2O(l) \rightleftharpoons SO_2 \cdot H_2O \ (l) \tag{3.20}$$

The "reaction equilibrium" is discussed in Chapter 2 (see Section 2.3.3.3.1). For this reaction, the equilibrium constant is given by

$$K_{eq} = \frac{\left[SO_2 \cdot H_2O\right]_l}{P_{SO_2}\left[H_2O\right]_l} \tag{3.21}$$

Since the concentration of water is a constant, it can be incorporated into the equilibrium constant. If we express the $[SO_2 \cdot H_2O]_l$ in molar concentration, we have a new equilibrium constant

$$K_{eq}^* = \frac{\left[SO_2 \cdot H_2O\right]_l}{P_{SO_2}} \tag{3.22}$$

expressed in units of $mol \, L^{-1} atm^{-1}$. Note that this is the inverse of the conventional definition of the air-water partition constant.

$$K_{AW}' = \frac{P_{SO_2}}{\left[SO_2 \cdot H_2O\right]_l} = \frac{1}{K_{eq}^*} \tag{3.23}$$

The complex species $SO_2 \cdot H_2O$ (l) can undergo subsequent ionization to produce bisulfite, HSO_3^- and sulfite, SO_3^{2-} ions according to the following reactions:

$$SO_2 \cdot H_2O \ (l) \rightleftharpoons HSO_3^-(l) + H^+(l) \tag{3.24}$$

$$HSO_3^- \ (l) \rightleftharpoons SO_3^{2-}(l) + H^+(l) \tag{3.25}$$

The respective equilibrium constants for the reactions can be written in terms of Henry's constants for SO_2 as follows:

$$K_{s1} = \frac{\left[HSO_3^-\right]_l \left[H^+\right]_l}{\left[SO_2 \cdot H_2O\right]_l} = \left[HSO_3^-\right]_l \left[H^+\right]_l \cdot \frac{K_{AW}'}{P_{SO_2}} \tag{3.26}$$

$$K_{s2} = \left[SO_3^{2-} \right]_l \left[H^+ \right]_l^2 \cdot \frac{K'_{AW}}{K_{s1} P_{SO_2}} \tag{3.27}$$

The total concentration of SO_2 that exists in the various forms, $[SO_2 \cdot H_2O]_l$, $[HSO_3^-]_l$, and $[SO_3^{2-}]_l$, is given by

$$\left[SO_2 \right]_T = \frac{P_{SO_2}}{K'_{AW}} \cdot \left[1 + \frac{K_{s1}}{\left[H^+ \right]_l} + \frac{K_{s1} K_{s2}}{\left[H^+ \right]_l^2} \right] \tag{3.28}$$

The apparent air-water partition constant K^*_{AW} defined byt $P_{SO_2} / [SO_2]_T$ is given by

$$K^*_{AW} = \frac{K'_{AW}}{\left[1 + \frac{K_{s1}}{\left[H^+ \right]_l} + \frac{K_{s1} K_{s2}}{\left[H^+ \right]_l^2} \right]} \tag{3.29}$$

Thus, as $[H^+]_l$ increases (or pH decreases), more and more of SO_2 appears in solution in the form of bisulfite and sulfite ions. Therefore, the apparent partition constant decreases. For the SO_2 dissolution, $K_{s1} = 0.0129$ M and $K_{s2} = 6.014 \times 10^{-8}$ M (Seinfeld, 1986).

The dissolution of CO_2 in water is similar to that of SO_2 in that three different reaction equilibria are possible. These and the respective equilibrium constants are given here:

$$CO_2(g) + H_2O(l) \rightleftharpoons CO_2 \cdot H_2O(l); \quad K'_{AW} = \frac{P_{CO_2}}{\left[CO_2 \cdot H_2O \right]_l} \tag{3.30}$$

$$CO_2 \cdot H_2O(l) \rightleftharpoons H^+(l) + HCO_3^-(l); \quad K_{c1} = \frac{\left[HCO_3^- \right]_l \left[H^+ \right]_l}{\left[CO_2 \cdot H_2O \right]_l} \tag{3.31}$$

$$HCO_3^-(l) \rightleftharpoons H^+(l) + CO_3^{2-}(l); \quad K_{c2} = \frac{\left[CO_3^{2-} \right]_l \left[H^+ \right]_l}{\left[HCO_3^- \right]_l} \tag{3.32}$$

The apparent air-water partition constant is given by

$$K^*_{AW} = \frac{K'_{AW}}{\left[1 + \frac{K_{c1}}{\left[H^+ \right]_l} + \frac{K_{c1} K_{c2}}{\left[H^+ \right]_l^2} \right]} \tag{3.33}$$

Just as for the SO_2 example, increasing pH decreases the air-water partition constant for the CO_2 case also. The values of the dissociation constants are $K_{c1} = 4.28 \times 10^{-7}$ M and $K_{c2} = 4.687 \times 10^{-11}$ M (Seinfeld and Pandis, 1998).

The effect of pH on the dissolution of ammonia in water is opposite to the ones for CO_2 and SO_2. NH_3 dissolves in water to form ammonium hydroxide, which further dissociates to give NH_4^+ and OH^- ions. Thus the solution is made more alkaline by the presence of ammonia. As pH increases and the solution becomes more alkaline, more of ammonia remains as the gaseous species and hence its solubility in water decreases. Therefore, the effective Henry's constant will increase with increasing pH. The reactions of relevance here are the following:

$$NH_3(g) + H_2O(l) \rightleftharpoons NH_3 \cdot H_2O(l); \quad K'_{AW} = \frac{P_{NH_3}}{\left[NH_3 \cdot H_2O\right]_l} \quad (3.34)$$

$$NH_3 \cdot H_2O(l) \rightleftharpoons NH_4^+(l) + OH^-(l); \quad K_{a1} = \frac{\left[NH_4^+\right]_l\left[OH^-\right]_l}{\left[NH_3 \cdot H_2O\right]_l} \quad (3.35)$$

Substituting for $[OH^-] = K_w/[H^+]$ in the second of these equations, one obtains the following expression for total ammonia:

$$\left[NH_3\right]_T = \frac{P_{NH_3}}{K'_{AW}} \cdot \left[1 + \frac{K_{a1}}{K_w}\left[H^+\right]_l\right] \quad (3.36)$$

and hence the apparent air-water partition constant is

$$K^*_{AW} = \frac{K'_{AW}}{\left[1 + \frac{K_{a1}}{K_w}\left[H^+\right]_l\right]} \quad (3.37)$$

From this equation it is clear that as $[H^+]_l$ decreases (or solution becomes more alkaline), the apparent air-water partition constant increases. The value of K_{a1} is 1.709×10^{-5} M (Seinfeld and Pandis, 1998).

The effect of pH on the ratio of air-water partition constants for the three species CO_2, SO_2, and NH_3 is shown in Figure 3.4.

Example 3.10 CO_2 Equilibrium between Air and Water in a Closed System

Consider a 1 mM solution of sodium bicarbonate in an aqueous phase in a closed vessel of total volume 1 L at 298 K. Let the ratio of gas to aqueous volume be 1:100. If the pH of the solution is 8.5, what will be the equilibrium concentration of CO_2 in the gas phase?

In order to solve this problem, the primary relationship we need is the equilibrium between air and water for CO_2, given by Henry's law. $K''_{AW} = 31.6$ atm mol^{-1} L^{-1} at 298 K. Hence,

$$\frac{\left[CO_2\right]_g}{\left[CO_2 \cdot H_2O\right]_w} = \frac{K'_{AW}}{RT} = 1.29 \quad (3.38)$$

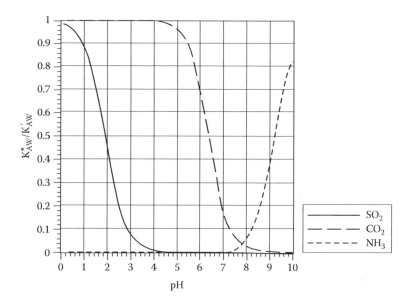

FIGURE 3.4   The effect of pH on the air-water partition constant of CO_2, SO_2, and NH_3.

CO_2 in the gas phase is generated from the sodium bicarbonate in solution and follows the reaction equilibria mentioned earlier in this chapter. The total CO_2 in the aqueous phase is given by

$$[CO_2]_{T,w} = [CO_2 \cdot H_2O]_w + [HCO_3^-]_w + [CO_3^{2-}]_w \qquad (3.39)$$

Using the expressions for K_{c1} and K_{c2} given in the text we can show that $[CO_2 \cdot H_2O]_w = \alpha_0 [CO_2]_{T,w}$ where

$$\alpha_0 = \cfrac{1}{1 + \cfrac{K_{c1}}{[H^+]_l} + \cfrac{K_{c1}K_{c2}}{[H^+]_l^2}} \qquad (3.40)$$

is the aqueous phase mole fraction of $[CO_2 \cdot H_2O]_w$ species. Similar expressions can be derived for the other species $[HCO_3^-]_w$ and $[CO_3^{2-}]_w$ also (see Stumm and Morgan, 1996). Since pH = 8.5, $[H^+] = 3.2 \times 10^{-9}$ M. Using $K_{c1} = 4.3 \times 10^{-7}$ M and $K_{c2} = 4.7 \times 10^{-11}$ M, we then have $\alpha_0 = 7.3 \times 10^{-3}$. Hence we have $[CO_2 \cdot H_2O]_w = 7.3 \times 10^{-3} [CO_2]_{T,w}$. A total mass balance requires that in the given closed system, the total CO_2 mass arising from the initial concentration of $NaHCO_3$ should equal $[CO_2]_{T,w}$ in the aqueous phase at equilibrium plus the $[CO_2]_g$ in the gas phase. Thus,

$$V_w (1 \times 10^{-3}) = V_w [CO_2]_{T,w} + V_G [CO_2]_g \qquad (3.41)$$

$$1 \times 10^{-3} = \frac{[CO_2 \cdot H_2O]_w}{7.3 \times 10^{-3}} + \frac{1}{100} \cdot (1.29 \times [CO_2 \cdot H_2]_w) \qquad (3.42)$$

Therefore, $[CO_2 \cdot H_2O]_w = 7.3 \times 10^{-6}$ M, and $[CO_2]_g = 9.4 \times 10^{-6}$ M.

Example 3.11   Benzene Concentration near a Wastewater Lagoon

The atmosphere above a wastewater lagoon contained 0.01 µg P^{-1} of benzene. The average surface water concentration in the lagoon was measured as 10 µg P^{-1}. If the temperature of air and surface water was 20°C, in what direction is the flux of benzene?

$C_i^a = 0.01$ µg L^{-1}, $C_i^w = 10$ µg L^{-1}. K_{AW}'' at 298 K is 0.23 (from Appendix A). Using the temperature variation for benzene (Table 3.4), we obtain at 293 K, $K_{AW} = 0.23$. The driving force for flux is, therefore, (0.01 − 0.01 × 10^{-9}/0.23) = 0.01 g cm^{-3}. Since the flux is positive, we conclude that the impoundment acts as a source of benzene to the air.

3.3   AIR-WATER EQUILIBRIUM IN ATMOSPHERIC CHEMISTRY

Atmospheric moisture is present in the form of hydrometeors (fog, rain, cloud droplets, aerosols, and hydrosols). An "aerosol" is a stable suspension of solid or liquid particles or both in air, whereas a "hydrosol" is a stable suspension of particles in water. Aerosols provide the medium for a variety of atmospheric reactions. In the atmosphere, solar radiation is absorbed and scattered by both particles and gas molecules that comprise aerosols and hydrosols. Atmospheric aerosol mass is rarely lower than 1 µg m^{-3} close to the earth's surface and occasionally increases to 100 µg m^{-3} in urban areas. The concentration of most trace gases in the atmosphere is in the vicinity of 10^3–10^4 µg m^{-3}. Thus, except for some rare cases (such as rural areas) the trace gases in the atmosphere far exceed the aerosol concentrations. Aerosols, however, play a larger role in the hydrologic cycle by providing condensation nuclei for both fog and cloud formation. Sea spray from the ocean surface provides a rich source of liquid droplets that upon evaporation of water form sea-salt crystals. Some of the liquid droplets arise from condensation of organic vapors when the vapor pressure exceeds the saturation point. An example of this type of atmospheric droplets is the aerosol generated from incomplete combustion of wood or agricultural residues.

The primary focus of interest for us in these cases is the equilibrium between the dispersed phase and the continuous phase that contain trace gases (organic and inorganic). We wish to analyze the extent of removal of molecules from the gas phase by various processes.

Table 3.8 lists the typical properties of atmospheric droplets and particles. Typical aerosol and fog droplets are smaller than 10 microns in diameter. Cloud drops, raindrops, and snowflakes have larger diameters. The surface areas of most particles are in the range of 10^{-4}–10^{-1} m^2 m^{-3}, except for snowflakes that have large surface areas. Small particles (e.g., aerosols and fog) have relatively large atmospheric lifetimes compared to large particles. Large particles settle out faster through sedimentation and gravity settling. Small particles (colloids) are kept in suspension by frequent collisions resulting from their Brownian motion. Large particles (raindrops, cloud drops, and snowflakes) have high liquid water content than aerosols and fog. The water content of aerosols depends largely on the relative humidity of air.

Many organic compounds (volatile or semivolatile) as well as inorganic compounds are transported over long distances and appear even in the remote arctic regions via dispersion through the troposphere. Compounds volatilize from their

TABLE 3.8
Properties of Atmospheric Particles and Droplets

Nature of Droplet	Size (μm)	Surface Area, S_T (m^2 m^{-3})	Liquid Water Content, L (m^2 m^{-3} of Air)	Typical Atmospheric Lifetime, τ
Aerosols	10^{-2}–10	-1×10^{-3}	10^{-11}–10^{-10}	4–7 days
Fog droplets	1–10	-8×10^{-4}	5×10^{-8}–5×10^{-7}	3 h
Cloud drops	10–10^2	-2×10^{-1}	10^{-7}–10^{-6}	7 h
Raindrops	10^2–10^3	-5×10^{-4}	10^{-7}–10^{-6}	3–15 min
Snowflakes	10^3–10^5	0.3		15–50 min

Sources: Seinfeld, J.H., *Atmospheric Chemistry and Physics of Air Pollution*, John Wiley & Sons, Inc., New York, 1986; Gill, P.S. et al., *Rev. Geophys. Space Phys.*, 21, 903, 1983; Graedel, T.E. and Crutzen, P.C., *Atmospheric Change: An Earth System Perspective*, W.H. Freeman Co., New York, 1993.

sources in the temperate and tropical regions and are transported through the atmosphere to the oceans and polar caps. Pollutants exist both in the gaseous form ($\equiv$G) and bound to particulates ($\equiv$P). A large fraction of the material is transported in the gaseous form, which exchanges directly with the earth and oceans via "dry deposition," the flux being $J_{Dry(G)}$. Gaseous materials are also solubilized and absorbed by hydrometeors and scavenged to the earth. This is called "wet deposition." Materials bound to aerosols in the atmosphere are also scavenged by hydrometeors, which is also called wet deposition. The total flux due to wet deposition is $J_{Wet(G+P)}$. The aerosols grow in size and settle to the earth by dry deposition and the flux is $J_{Dry(P)}$. A schematic of these exchange processes is given in Figure 3.5. The total flux is

$$J_{Total} = J_{Wet} + J_{Dry} = J_{Wet(G+P)} + J_{Dry(P)} + J_{Dry(G)} \tag{3.43}$$

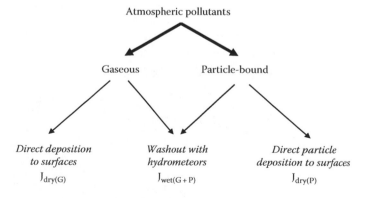

FIGURE 3.5 Mechanisms of deposition of pollutants from the atmosphere.

3.3.1 WET DEPOSITION OF VAPOR SPECIES

The concentration of a gaseous species i scavenged by a hydrometeor at equilibrium is given by the ratio of concentration per m^3 of water to the concentration per m^3 of air. This is called the "washout ratio," W_g, from the gas phase. It is easy to recognize that for equilibrium scavenging $W_g = 1/K_{AW}$.

For closed systems, the total concentration of a species is fixed. If C_{iT} represents the total concentration of species i per unit volume of air, it is distributed between the air and water content θ_L (g m^{-3}), and hence,

$$C_{iT} = C_{ia} + \theta_L \cdot C_{iw} \tag{3.44}$$

Applying Henry's law to the aqueous concentration, $C_{iw} = C_{ia}/K_{AW}$, we get the following equation for the fraction of species i in the aqueous phase:

$$\phi_{iw} = \frac{\theta_L}{K_{AW} + \theta_L} \tag{3.45}$$

For compounds with low Henry's constant, K_{AW}, a significant fraction of species i will exist in the aqueous phase. With increasing θ_L, a greater amount of water becomes available to solubilize species i and hence the value of ϕ_i^w increases. The concentration of species i in the aqueous phase is then given by

$$C_{iw} = \frac{C_{iT}}{\theta_L + K_{AW}} \tag{3.46}$$

3.3.2 WET DEPOSITION OF AEROSOL-BOUND FRACTION

Compounds that exist in the gaseous form in air are in equilibrium with those that are "loosely bound" to atmospheric aerosols; this is called the "exchangeable" fraction. The remaining fraction is incorporated within the particle matrix and is "tightly bound" or "nonexchangeable" and does not participate in the equilibrium with the gas phase. The extent to which the exchangeable molecules are associated with the aerosol is directly related to the partial pressure of the compound. Junge (1977) provided an equation relating the adsorbed fraction (ϕ_{ip}) to the total surface area of the aerosol (S_T) and the vapor pressure of the compound (P_i^*).

Assume that adsorption on aerosols is governed by a BET isotherm:

$$\frac{\Gamma_i}{\Gamma_i^m} = \frac{K_{BET}}{(1 - y_i)\left[1 + (K_{BET} - 1)y_i\right]} \tag{3.47}$$

Since $y_i \ll 1$,

$$\Gamma_i = \Gamma_i^m K_{BET} \frac{P_i}{P_i^*} \tag{3.48}$$

Using the ideal gas law and M_i the molar mass of compound i, we get

$$\Gamma_i = \Gamma_i^m K_{BET} \frac{C_{ia}RT}{M_i P_i^*} \tag{3.49}$$

Consider a total suspended particle concentration in air, C_s, then the fraction of i in the solids is

$$\phi_{iP} = \frac{\Gamma_i C_s}{\Gamma_i C_s + C_{ia}} \tag{3.50}$$

Now, $C_s = A_v/A_m$, where A_v is the total surface area per unit volume of air, and A_m is the surface area of one particle. Hence,

$$\phi_{iP} = \frac{\rho_i A_v}{\rho_i A_v + P_i^*} \tag{3.51}$$

where ρ_i is given by $\dfrac{\Gamma_i^m K_B RT}{M_i A_m}$. It is characteristic of the compound and depends on the compound molar mass, and the surface concentration for monolayer coverage of the compound (Pankow, 1987). P_i^* is the subcooled liquid vapor pressure for compounds that are solids at room temperature. Compounds that have large vapor pressures tend to have small values of φ_i^p, whereas compounds of small vapor pressures are predominantly associated with the aerosols. φ_i^p is small for $P_i^* > 10^{-6}$ mm of Hg. In a clean atmosphere, therefore, many of the hydrophobic compounds such as PCB congeners, DDT, and Hg should exist in the vapor phase. A range of vapor pressures from 10^{-4} to 10^{-8} mm of Hg are observed for environmentally significant compounds and those with $P_i^* > 10^{-4}$ mm Hg must be predominantly in the vapor phase while those with $P_i^* < 10^{-8}$ mm Hg must exist almost entirely in the particulate phase. The fraction occurring in the aerosol phase generally increases with increasing molecular weight and decreasing P_i^* for a homologous series of compounds (e.g., n-alkanes).

For a compound that is removed by both gas and aerosol scavenging by rain, cloud, or fog droplets, the following is the overall expression for washout ratio:

$$W_T = W_g \left(1 - \phi_{iP}\right) + W_P \phi_{iP} \tag{3.52}$$

W_p is the aerosol washout ratio, which is a function of meteorological factors and particle size. A raindrop reaches equilibrium with the surrounding atmosphere within a 10 m fall, and hence the washout of vapors may be viewed as an equilibrium partitioning process. Because of their small sizes and long atmospheric residence times, fog droplets can also be considered to attain equilibrium with the atmosphere rather quickly.

Using Equation 3.52 for washout ratio, the overall flux of pollutants to the surface (in $mol\ m^{-2}\ s^{-1}$) by wet deposition can be obtained as

$$J_{wet(G+P)} = R_I \cdot W_T \cdot C_{ia} \tag{3.53}$$

where
   R_I is the rainfall intensity (cm h^{-1})
   C_{ia} is the measured concentration of i in air

If K_{AW} is large, vapor uptake into droplets is negligible and only the aerosol particulate fraction is removed by wet deposition. A number of environmentally significant hydrophobic organic compounds fall in this category. Examples are n-alkanes, polychlorinated biphenyls, chlorinated pesticides, and polyaromatic hydrocarbons. Compounds with small K_{AW} values are removed mostly via gas scavenging by droplets. Examples of this category include phenols, low-molecular-weight chlorobenzenes, and phthalate esters. A number of investigators have reported experimentally determined values of both W_g and W_p for a variety of metals and organics. The range of values observed for typical pollutants in some regions of the world are summarized in Table 3.9. Aerosols have overall washout ratios approximately 10^6. Aerosols with washout ratios less that 10^5 are generally considered to be insoluble and relatively young in that their dimensions are of the order of 0.1 μm. It has been shown that for most nonreactive organic compounds such as PAHs in Table 3.9, the vapor scavenging washout ratios indicated that equilibrium was attained between the falling raindrops and the gas phase within about a 10 m fall through the atmosphere (Ligocki et al., 1985). Hence, these values can be satisfactorily predicted from the temperature-corrected Henry's constants for the compounds. The aerosol scavenging washout ratios of organic compounds were observed to be similar to the washout ratios of fine-particle elements such as Pb, Zn, As, and V.

TABLE 3.9
Range of Washout Ratios for Typical Pollutants

Compound	W_g	W_p	
n-Alkanes		$1.3–2.2 \times 10^4$	Portland, OR
		$3.5–5.8 \times 10^5$	College Stn., TX
		$4.5–1.6 \times 10^6$	Norfolk, VA
PAHs	$1.9 \times 10^2–1.8 \times 10^4$	$2 \times 10^3–1.1 \times 10^4$	Portland, OR
		$1.4–2.5 \times 10^5$	Isle Royale
Phenanthrene	64 ± 46	1.1×10^6	Chesapeake Bay, MD
Pyrene	390 ± 280	8.5×10^5	Chesapeake Bay, MD
Metals		$2.0 \times 10^5–1 \times 10^6$	Ontario, Canada
		$1 \times 10^5–5 \times 10^5$	Northeast USA
Mercury, Hg^0	$10–100$		
Radionuclides		$2 \times 10^5–2 \times 10^6$	
PCDDs/PSDFs		$1 \times 10^4–1 \times 10^5$	
PCBs		$1 \times 10^2–2 \times 10^6$ (rain)	
		$5 \times 10^4–4 \times 10^6$ (snow)	
Sulfate, SO_4^{2-}		$6 \times 10^4–5 \times 10^6$	

Sources: Bidleman, T.F., *Environ. Sci. Technol.*, 22, 361, 1988; Ligocki, M.P. et al., *Atmos. Environ.*, 19, 1609, 1985; Baker, J.T., *Atmospheric Deposition of Contaminants to the Great Lakes and Coastal Waters*, SETAC Press, Pensacola, FL, 1997.

Example 3.12   Typical Calculation of Washout Ratios

Given typical values of particle washout ratios, $W_p = 2 \times 10^5$, aerosol surface area, $A_v = 3.5 \times 10^{-6}$ cm^2 cm^{-3} of air, and the constant $\rho_i = 1.7 \times 10^{-4}$ atm cm. Let us estimate the total washout ratios and the relative contributions of gas and aerosol scavenging for benzene, benzo-[a]-pyrene, and γ-hexachlorocyclohexane (HCH). Let us also assume that ρ_i is approximately a constant between these compounds. The necessary properties are as follows:

Compound	P_i^* (atm)	K_{AW} (–)
Benzene	0.125	0.228
γ-HCH	3.1×10^{-7}	2.0×10^{-5}
Benzo-[a]-pyrene	1.2×10^{-10}	2.3×10^{-6}

The calculated values of φ_i^p and washout ratios are as follows:

Compound	φ_i^p	$W_g\left(1 - \varphi_i^p\right)$	$W_p \varphi_i^p$	W_T
Benzene	4.7×10^{-9}	4	9×10^{-4}	4
γ-HCH	1.6×10^{-3}	5×10^4	3×10^3	5.3×10^4
Benzo-[a]-pyrene	0.83	7×10^4	2×10^5	2.7×10^5

For benzene, the deposition is wholly via gas scavenging. For γ-HCH, although vapor scavenging dominates, there is a small contribution from the aerosol fraction. However, for benzo-[a]-pyrene, the contribution is almost completely from the particle (aerosol) scavenging process.

Fog forms close to the ground where most of the gases and aerosols are concentrated. It is formed as a result of a decrease in air temperature and an increase in relative humidity, whereby water vapor from the atmosphere condenses on tiny aerosol particles. Fog droplets have diameters generally between 1 and 10 μm. Since a fog droplet is approximately 1/100th of a raindrop, it is more concentrated than rain. In addition to its importance as a site for chemical reactions, fog exerts a significant influence as a scavenger on both organic and inorganic materials from the near-surface atmosphere. Hence, it can have a significant effect on human health and vegetation. Since fog forms close to the earth's surface, its dissipation leaves particles in the near-surface atmosphere, which are highly concentrated with pollutants. This observation has significant implications for the long-range transport and deposition of environmental pollutants (Munger et al., 1983). Knowledge of fogwater chemistry also has relevance to the ambient acid formation and acidic precipitation problems. SO_2 oxidation reactions increase the acidity of fog. Fog droplets have greater deposition rates than dry gases and aerosols. Most of the liquid water content in fog is lost to the surface. Thus, in regions of heavy fog events, calculation of dry deposition and rain fluxes may not account for the actual deposition of pollutants. Historically, fog has been a cause for severe human health effects such as those of the infamous London fog of 1952, which caused close to 10,000 deaths in 4 months.

Example 3.13   Total Deposition of a Pesticide by Fog

The concentration of a pesticide (chlorpyrifos) in air from a foggy atmosphere in Parlier, CA was 17.2 ng m^{-3} at 25°C. The aerosol particle concentration in air was negligible. If the moisture deposition rate from fog is 0.2 mm h^{-1} and it lasts for 4 h, how much chlorpyrifos will be deposited over one acre of land surface covered by fog?

   Henry's constant for chlorpyrifos is $K_{AW} = 1.7 \times 10^{-4}$ at 25°C. $C_i^\beta = 17.2$ ng m^{-3}. Particle deposition is zero, only gaseous species need be considered. $W_i = 1/K_{AW}$. Flux to surface is $J_i = R_i C_{ia}/K_{AW} = 2 \times 10^{-12}$ g cm^{-2} h. Amount deposited = (2×10^{-12}) (4) $(4.04 \times 10^7) = 3.27 \times 10^{-4}$ g.

Droplets larger than about 10 μm contribute the bulk of liquid water in fogs. However, droplets smaller than this will be more concentrated in pollutants. Large concentrations of colloidal organic matter (between 10 and 200 mg L^{-1}) are sometimes observed in fogwater. These colloids arise from dissolved humics and fulvics and are surface active (Capel et al., 1991). The fogwater collected in California were observed to contain a variety of inorganic ions (NO_3^-, SO_4^{2-}, H^+, and NH_4^+), which were also the major components of aerosols collected from the same region (Munger et al., 1983). This showed that highly concentrated fogwater appeared to result from highly concentrated aerosols. Likewise, dissipation of highly concentrated fog resulted in highly reactive and concentrated aerosols. This points to the fact that fog is formed via condensation on aerosol nuclei. The chemical composition of fog is not only determined by the composition of their condensation nuclei (e.g., ammonium sulfate or trace-metal-rich dust) but also by the gases absorbed into fog droplets. High concentrations of formaldehyde (~0.5 mM) were also observed in California fog. Many other inorganic species such as Na^+, K^+, Ca^{2+}, and Mg^{2+} have also been identified in fog collected from different regions of the world. Fog is an efficient scavenger of both gaseous and particulate forms of S(IV) and N(III). The high millimolar concentrations of metal ions in fogwater and the resultant metal ion catalysis for the transformation of S(IV) to S(VI) in a foggy atmosphere is an interesting research area. A number of investigators have shown that fogwater contains high concentrations of polyaromatic hydrocarbons and chlorinated pesticides (Glotfelty et al., 1987; Capel et al., 1991).

3.3.3   Dry Deposition of Aerosol-Bound Pollutants

The transfer of aerosol-bound pollutants directly from air to water is called dry deposition. The deposition flux is parameterized as

$$J_{Dry(P)} = V_d \cdot C_{ia} \qquad (3.54)$$

where V_d is the deposition velocity (cm s^{-1}). Typical deposition velocities are given in Table 3.10. V_d is composed of three resistance terms:

$$V_d = \frac{1}{r_{al}} + \frac{1}{r_{bl}} + \frac{1}{r_s}$$

TABLE 3.10
Typical Dry Deposition Velocities for Gases and Aerosols

Compound	Range of V_d (cm s^{-1})
SO_2 (g)	0.3–1.6
N_{ox} (g)	0.01–0.5
HCl (g)	0.6–0.8
O_3 (g)	0.01–1.5
SO_4^{2-}	0.01–1.2
NH_4^+	0.05–2.0
Cl^-	1–5
Pb	0.1–1.0
Hg^0	0.02–0.11
Ca, Mg, Fe, Mn	0.3–3.0
Al, Mn, V	0.03–3.0
PCB, DDT	0.19–1.0
Fine particles (1 μm)	0.1–1.2

Sources: Davidson, C.I. and Wu, Y.L., *Acidic Precipitation—Sources, Deposition and Canopy Interactions,* Lindberg, S.E. et al., eds., New York: Springer-Verlag, Vol. 3, pp. 103–216, 1992; Eisenreich, S.J. et al., *Environ. Sci. Technol.,* 5, 30, 1981.

where
r_{al} is the aerodynamic layer resistance
r_{bl} is the boundary layer resistance
r_s is the surface resistance

V_d is affected by environmental factors, such as type of aerosol, particle size, wind velocity, surface roughness, atmospheric stability, and temperature. An excellent summary is provided by Davidson and Wu (1992).

3.3.4 DRY DEPOSITION FLUX OF GASES FROM THE ATMOSPHERE

Gaseous molecules can exchange with the aqueous phase via dissolution and/or volatilization. This dissolution is driven by the concentration gradient between the gas and water phases. At the beginning of this chapter (Section 3.2), the concept of a two-film theory of mass transfer across the air-water interface was introduced. The flux of solute i (mass transferred per unit area per unit time) across the interface is given by

$$J_{Dry(G)} = K_w \cdot \left(\frac{C_{ia}}{K_{AW}} - C_{iw} \right) \tag{3.55}$$

where the overall mass transfer coefficient K_w is given by

$$\frac{1}{K_w} = \frac{1}{k_w} + \frac{1}{k_a K_{AW}} \qquad (3.56)$$

The equilibrium air-water partition constant appears both in the concentration driving force and mass transfer coefficient terms. The overall mass transfer coefficient is composed of two terms, an aqueous phase coefficient k_w and an air phase coefficient k_a. The reciprocal of the transfer coefficient k_w is the resistance to mass transfer in the aqueous film, whereas the reciprocal of $k_a K_{AW}$ is the mass transfer resistance in the gaseous film. If $1/k_w \gg 1/k_a K_{AW}$, the solute transfer is said to be aqueous phase controlled and $K_w \approx k_w$. On the other hand, if $1/k_w \ll 1/k_a K_{AW}$, then the solute transfer is said to be gas phase controlled and $K_w \approx k_a K_{AW}$. A number of gaseous compounds of interest (e.g., oxygen, chlorofluorocarbons, methane, carbon monoxide) are said to be liquid phase controlled. So also are many of the VOCs (e.g., chloroform, carbon tetrachloride, perchloroethylene, 1,2-dichloroethane). Some highly chlorinated compounds (e.g., dieldrin, pentachlorophenol, lindane) are predominantly gas phase controlled. A general rule of thumb is that compounds with $K_{AW} \geq 0.2$ are liquid (aqueous) phase controlled, whereas those with $K_{AW} \leq 2 \times 10^{-4}$ are gas phase controlled. Table 3.11 lists the values of liquid and gas phase mass transfer coefficients for some compounds of environmental significance. For water crossing the interface from the bulk phase into air, there exists no resistance in the liquid phase, and hence is completely gas phase controlled. A mean value of k_a for water was estimated as 8.3×10^{-3} m s^{-1} (Liss and Slater, 1974). For all other gas-phase-controlled chemicals, one can obtain k_a values by multiplying this value for water with the ratio of the gas phase diffusivity of water and the specific gas. Generally, the mass transfer coefficient is a function of the wind velocity over the water body (atmospheric turbulence). In the case of those species that dissociate in aqueous solution, associate with, or react with other species (such as colloids, ions, and other macromolecules), a more complex dependence of the transfer rate upon the aqueous chemistry has been observed. For these compounds only the "truly" dissolved, unassociated fraction in water will equilibrate with the vapor in the air. Therefore, the effects of the various parameters on the equilibrium partitioning (Henry's law) for compounds that we studied in Section 3.2.1 are of direct relevance in this context.

Example 3.14   Flux of SO$_2$ in the Environment

The mean concentration of SO$_2$ in our atmosphere has been determined to be approximately 3 μg m^{-3}. SO$_2$ is fairly rapidly oxidized in the slightly alkaline environment of seawater and hence its concentration in surface ocean waters is approximately zero. Using the data from Table 3.12 for k_w, k_a, and Henry's law constant K_{AW}, we have $K_w = \{(1/k_w) + (1/k_a K_{AW})\}^{-1} = 9.0 \times 10^{-5}$ m s^{-1}. Since $C_i^w - 0$, we have for the flux $J_{Dry(G)} = 9.0 \times 10^{-5} \{3/0.02\} = 0.013$ μg m^{-2} s^{-1}. The flux is directed from the atmosphere to the ocean. If the total surface area of all oceans is 3.6×10^{14} m^2, the total annual flux would be 1.5×10^{14} g, which is the same order of magnitude as the estimated emissions from fossil fuel burning (Liss and Slater, 1974).

TABLE 3.11

Air/Water Partition Constants and Air/Water Mass Transfer Coefficients for Selected Pollutants in the Natural Environment

Compound	K_{AW} (–)	k_w (m s^{-1})	k_a (m s^{-1})
Inorganic			
Sulfur dioxide	0.02	9.6×10^{-2}	4.4×10^{-3}
Carbon monoxide	50	5.5×10^{-5}	8.8×10^{-3}
Nitrous oxide	1.6	5.5×10^{-5}	5.3×10^{-3}
Ozone	3	4.9×10^{-4}	5.1×10^{-3}
Ammonia	7.3×10^{-4}	6.7×10^{-5}	8.5×10^{-3}
Oxygen	25.6	5.5×10^{-5}	—
Water	—	—	8.4×10^{-3}
Organic			
Methane	42	5.5×10^{-5}	8.8×10^{-3}
Trichlorofluoromethane	5	3.1×10^{-5}	3.0×10^{-3}
Chloroform	0.183	4.2×10^{-5}	3.2×10^{-3}
Carbon tetrachloride	0.96	5.2×10^{-5}	4.1×10^{-3}
Benzene	0.23	7.5×10^{-5}	6.0×10^{-3}
Dichlorobenzene	0.069	3.5×10^{-5}	4.5×10^{-3}
Lindane	2×10^{-5}	3.8×10^{-5}	3.1×10^{-3}
DDT	1.6×10^{-3}	3.6×10^{-5}	2.8×10^{-3}
Naphthalene	0.021	3.6×10^{-5}	4.6×10^{-3}
Phenanthrene	1.6×10^{-3}	5.0×10^{-5}	4.0×10^{-3}
Pentachlorophenol	1.1×10^{-5}	3.3×10^{-5}	3.2×10^{-3}
PCB (Aroclor 1254)	4.4×10^{-4}	1.7×10^{-2}	1.9

Sources: Warneck, P.A., *Chemistry of the Natural Atmosphere*, New York: Academic Press, 1986; Eisenreich, S.J. et al., *Environ. Sci. Technol.*, 5, 30, 1981; Mackay, D. and Leinonen, P.J., *Environ. Sci. Technol.*, 9, 1178, 1975; Liss, P.S. and Slater, P.G., *Nature*, 247, 181, 1974.

Note: All values are at 298 K.

Example 3.15   Flux of Hg^0 in the Environment

In the equatorial Pacific Ocean, Kim and Fitzgerald (1986) reported that the principal species of mercury in the atmosphere is elemental Hg^0. It was shown that a significant fraction of the dissolved mercury in the surface water was elemental Hg^0. The concentration in the surface water, C_{wi} was 30 fM (= 30×10^{-15} mol L^{-1} = 6 ng m^{-3}). The mean atmospheric concentration was 1 ng m^{-3}. Henry's constant at 25°C for Hg^0 is 0.37. The overall mass transfer coefficient K_w = 5.1 m d^{-1}. The flux of Hg^0 is therefore given by $J_{Dry(G)}$ = 5.1[1/0.37 – 6] = –16.8 ng m^{-2} d^{-1}. The annual flux of Hg^0 is given by –16.8 × 365 = –6.1 × 10^3 ng m^{-2} = –6.1 μg m^{-2}. The flux is negative and hence is directed from the ocean to the atmosphere. The total flux from 1.1 × 10^{13} m^2 surface area of the equatorial Pacific ocean is 0.67 × 10^8 g year^{-1}. The total flux of Hg^0 to air from all regions of the world combined is 8 × 10^9 g year^{-1}. Hence, the flux from the equatorial Pacific ocean is only 0.08% of the total.

TABLE 3.12
Typical Louisiana Sediment/Soil Composition

Component/Property	Sediment	Soil
Bulk density	1.54 g cm^{-3}	1.5 g cm^{-3}
Particle density	2.63 g cm^{-3}	
Bed porosity	0.59 cm^3 cm^{-3}	0.4 cm^3 cm^{-3}
Organic carbon, ϕ_{oc} (humics, fulvics, etc.)	2.1%	2.3%
Sand	27%	57.8%
Silt	37%	25.5%
Clay	34%	16.7%
Iron	13 g kg^{-1}	23 g kg^{-1}
Aluminum	12 g kg^{-1}	25 g kg^{-1}
Calcium	3 g kg^{-1}	1 g kg^{-1}
Magnesium	2 g kg^{-1}	0.4 g kg^{-1}
Manganese	206 mg kg^{-1}	100 mg kg^{-1}
Phosphorus	527 mg kg^{-1}	166 mg kg^{-1}
Zinc	86 mg kg^{-1}	54 mg kg^{-1}
Sodium	77 mg kg^{-1}	29 mg kg^{-1}
Lead	56 mg kg^{-1}	35 mg kg^{-1}
Chromium	20 mg kg^{-1}	23 mg kg^{-1}
Copper	17 mg kg^{-1}	20 mg kg^{-1}
Nickel	12 mg kg^{-1}	15 mg kg^{-1}

Notes: Sediment sample obtained from Bayou Manchac, LA. Soil sample obtained from North Baton Rouge, LA.

Example 3.16   Flux of CO_2 between Atmosphere and Seawater

Broeker and Peng (1974) estimated that the mean K_w for CO_2 is 11 cm h^{-1} in seawater at 20°C. This is representative of the world's oceans. The maximum rate of CO_2 transfer can be obtained by assuming that $C_{iw} \approx 0$, that is, CO_2 is rapidly assimilated in the surface waters of the oceans. Since P_i for CO_2 is 0.003 atm, the value of $C_{ia} = 0.003/RT = 1.25 \times 10^{-7}$ mol cm^{-3}. K_{AW} for CO_2 is 1.29. Hence $C_{ia}/K_{AW} = 9.7 \times 10^{-8}$ mol cm^{-3}. The flux is therefore $J_{Dry(G)} = 11(9.7 \times 10^{-8}) = -1 \times 10^{-6}$ mol (cm^2 h)$^{-1}$. The flux is from the atmosphere to seawater.

3.3.5   THERMODYNAMICS OF AQUEOUS DROPLETS IN THE ATMOSPHERE

In Section 2.2.3.4, it was noted that for pure liquids, the vapor pressure above a curved air-water interface is larger than that over a flat interface. This was termed the "Kelvin effect." The Kelvin equation representing this effect was derived as

$$\ln\left(\frac{P}{P^*}\right) = \frac{2\sigma_{aw}}{r} \cdot \frac{V_w}{RT} \tag{3.57}$$

where
  r is the radius of the droplet
  P and P* are the vapor pressures over the curved interface and the flat surface, respectively
  V_w is the molar volume of water

Let us now consider the aqueous droplet to contain a nonvolatile species i with molar volume V_i. If the number of moles of i is n_i and the number of moles of water is n_w, the total volume of the drop $(4/3)\pi r^3 = n_i V_i + n_w V_w$. The mole fraction of water in the droplet is given by $x_w = n_w/(n_i + n_w)$. Using these relations, one can write

$$x_w = \cfrac{1}{\left[1 + \left(\cfrac{n_i V_w}{\cfrac{4}{3}\pi r^3 - n_i V_i}\right)\right]} \qquad (3.58)$$

If the Raoult's law convention is applied to the case of the flat air-water interface (refer to Section 2.2.2.3), the vapor pressure of water above the solution will be given by

$$P_w = \gamma_w x_w P_w^* \qquad (3.59)$$

where P_w^* is the vapor pressure of water over the flat interface. The Kelvin equation now takes the form (Seinfeld and Pandis, 1998)

$$\ln\left(\frac{P_w}{\gamma_w x_w P_w^*}\right) = \frac{2\sigma_{aw}}{r} \cdot \frac{V_w}{RT} \qquad (3.60)$$

Substituting for x_w, rearranging this equation, and using the dilute solution definition, that is, $(4/3)\pi r^3 \gg n_i V_i$. Also, for a dilute solution according to the Raoult's law convention, the activity coefficient $\gamma_w = 1$. Therefore the final equation is

$$\ln\left(\frac{P_w}{P_w^*}\right) = \frac{B_1}{r} - \frac{B_2}{r^3}$$

where $B_1 = 2\sigma_{aw}V_w/RT$ and $B_2 = 3n_i V_w/4\pi$. Note that B_1 (in μm) $\approx 0.66/T$ and B_2 (in μm³) $\approx 3.44 \times 10^{13}\ \nu\ m_s/M_s$, where ν is the number of ions resulting from solute dissociation, m_s is the mass (g) per particle of solute, and M_s is the molecular weight (g mol⁻¹) (Seinfeld and Pandis, 1998).

Note the difference between the equation for a pure water droplet and that for an aqueous solution droplet. Whereas for a pure water droplet there is a gradual approach of $P_w \rightarrow P_w^*$ as r is increased, in the case of an aqueous solution droplet P_w can either increase or decrease with r depending on the magnitude of the second term B_2/r^3, which results solely from the solute effects and is a function of its mole number m_s/M_s.

When the two terms on the right-hand side become equal, $P_w = P_w^*$, the radius at which this is achieved is called the "potential radius" and is given by

$$r^* = \left(\frac{3}{8} \cdot \frac{n_i}{\pi} \cdot \frac{RT}{\sigma_{aw}} \right)^{1/2} \tag{3.61}$$

The maximum value of $\ln\left(P_w/P_w^*\right)$ will be reached when the derivative with respect to the radius goes to zero, which gives a "critical radius" given by

$$r_c = \sqrt{3 \cdot \frac{B_1}{B_2}} = \sqrt{3} \cdot r^* \tag{3.62}$$

A typical plot of P_w/P_w^* for both pure water droplets and solution droplets (solute being nonvolatile, e.g., an inorganic salt or an organic compound) is shown in Figure 3.6. We shall assume that the concentration of i is low enough that σ_{aw} remains unaffected. If the compound is surface active, σ_{aw} will be lower and hence the Kelvin term will be even smaller. This can be the case especially for many of the hydrophobic compounds of environmental interest (Perona, 1992). Although electrolytes also affect the surface tension of water, their influence is mostly less than 20% within the observed ranges of concentrations.

For pure water droplets, the curve shows a steep decrease with increasing r. For the solution droplet, there is an initial steeply rising portion till r_c is reached. This results from the solute effects. Beyond this point, the Kelvin term dominates and the slow decrease in P_w/P_w^* is apparent. Curves such as these are called "Kohler curves." They are useful in estimating the size of an aqueous droplet given the relative humidity $\left(P_w/P_w^*\right)$ and the concentration of the solute.

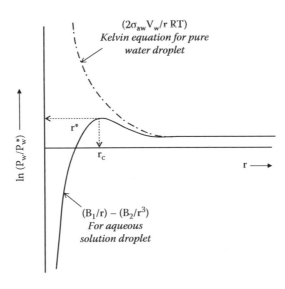

FIGURE 3.6 Kohler curves for pure water droplet and aqueous solution droplet in air. Notice the differences in the shape of the curve when water contains a solute.

Consider a fixed ratio of P_w/P_w^*. For this fixed value there are two possible radii, one less than r_c and one greater than r_c. For $r < r_c$, the drop is at its equilibrium state. If it adds any water its equilibrium vapor pressure will be larger than the ambient value and it therefore quickly loses that water through evaporation and reverts to its original equilibrium. If $r > r_c$, the drop will have a lower than ambient vapor pressure upon gaining more water molecules. Therefore it will continue to accumulate water and grow in size. If the value of P_w/P_w^* is greater than the maximum value, whatever be the size of the drop, it always has lower than ambient vapor pressure and hence it continues to grow in size. A drop that has crossed this threshold is said to be "activated." With increasing radius of the drop the height of the maximum decreases, and hence the larger particles are activated preferentially. In fact, this is how fog and cloud droplets that are $-10\ \mu m$ in diameter are formed from aerosols that are only $\approx 0.1\ \mu m$ in diameter. In continental aerosols, the larger particles are generally efficient condensation points for cloud droplets. The small (Aitken) particles have negligible contribution toward cloud condensation.

For compounds that do not dissociate (e.g., neutral organics) the value of n_i that appears in the equation for B_2 is straightforward to estimate. For dissociating species (e.g., inorganic salts), J_2 should include both the dissociated and undissociated species in solution. If the degree of dissolution is v_i (e.g., for HCl the maximum value is 2) and the initial mass is m_i, $n_i = v_i m_i/M_i$, where M_i is the molecular weight. Generally, dissociation leads to small values of P_w/P_w^*. If only a portion of the species i is soluble in the aqueous phase (n_i^{sol}) and a portion is insoluble (n_i^{insol}), then particle growth will be retarded somewhat since B_2 is now defined as $(3/4\pi)\left(n_i^{sol} + n_i^{insol}\right)V_w$. This makes the solute effect more pronounced than the previous case when all of i was soluble in the aqueous phase (i.e., n_i in the previous equation is equal to n_i^{sol}). Experimental verification of these different cases is reported in the literature (Warneck, 1986).

Example 3.17 Vapor Pressure above Aqueous Droplets

Determine the vapor pressure over (1) a 1 μm aqueous droplet containing 5×10^{-13} g of NaCl per particle at 298 K, and (2) a 1 μm of pure water droplet.

1. For NaCl, $v = 2$, $m_s = 0.1$, $M_s = 58$. $B_1 = 0.66/298 = 2.2 \times 10^{-3}\ \mu m$, $B_2 = 3.44 \times 10^{13}$ (2) $(5 \times 10^{-13})/58 = 0.55\ (\mu m)^3$. $\ln P_w/P_w^* = \left(2.2 \times 10^{-3}\right) - \left(0.55/1^3\right) = -0.55$. Hence, $P_w/P_w^* = 0.57$. Since $P_w^* = 223.75$, $P_w = 13.7$ mm Hg.
2. $\sigma_{a/w} = 72$ ergs cm^{-2}, $V_w = 18$ cm^3 mol^{-1}, $r = 1 \times 10^{-4}$ cm, $R = 8.314 \times 10^7$ ergs (Kmol)$^{-1}$, $T = 298$ K. Hence $\ln P_w/P_w^* = (2)(72)(18)/\left(1 \times 10^{-4} \times 8.314 \times 10^7 \times 298\right) = 1.04 \times 10^{-3}$. Hence, $P_w/P_w^* = 1.001$. Therefore, $P_w = P_w^* = 23.75$ mm Hg.

3.3.6 Air-Water Equilibrium in Waste Treatment Systems

The equations and methods of calculation of flux in the atmosphere, as described in Section 3.3.4, also apply to the estimation of air emissions from wastewater storage impoundments, settling ponds, and treatment basins (Springer et al., 1986). Wastewaters are placed temporarily in open confined aboveground facilities made of earthen material. In some cases, the storage of wastewater in such impoundments is for the purpose of settling of solids by gravity. In other cases, wastewater treatment using mechanical agitation or submerged aerators is conducted in these impoundments for the purposes of neutralization, chemical reaction, biochemical oxidation, and evaporation. All these activities cause a significant fraction of the species from water to volatilize to the atmosphere. Natural forces in the environment can dominate the transfer process. Examples of these are wind, temperature, humidity, solar radiation, ice cover, etc., at the particular site. All of these effects have to be properly accounted for if accurate air emissions of VOCs are to be made.

Consider a surface impoundment as shown in Figure 3.7. The natural variable that has the most influence on air emissions is wind over the water body. Both k_w and k_a depend on wind velocity. An average overall mass transfer coefficient K_w over a 24-hour period can be estimated. The flux of material from water will be given by

$$N_i = K_w \cdot \left(C_{iw} - \frac{C_{ia}}{K_{AW}} \right) \tag{3.63}$$

If the two phases are at equilibrium, $C_{iw} = C_{ia}/K_{AW}$, and the net overall flux is zero. In other words, at equilibrium the rates of both volatilization and absorption are the same, but opposite in direction. The air movement over the water body due to wind induced friction on the surface is transferred to deeper water layers, thereby setting

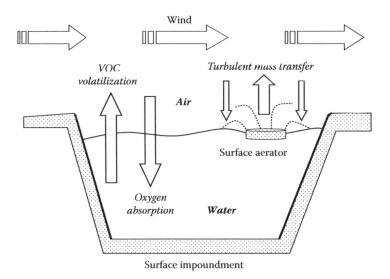

FIGURE 3.7　Transport of a volatile organic compound (VOC) from a surface impoundment. Note the two distinct areas of mass transfer, the turbulent zone where surface aerators enhance the volatilization, and the quiescent area where natural processes drive the evaporation of the chemical from water. Surface winds carry the VOC away from the source. Simultaneous to volatilization of VOCs from water, absorption of oxygen from the atmosphere also occurs.

up local circulations in bulk water and enhancing exchange between air and water. Thus, the wind-induced turbulence tends to delay approach to equilibrium between the phases. The effects of wind are far greater on k_a than on k_w, and hence depending on which one of these two controls mass transfer the overall effect on K_w will be different.

Example 3.18   Flux of Benzene from a Typical Surface Impoundment

A surface impoundment of total area 2×10^4 m^2 is 5 m deep and is well mixed by a submerged aerator. The average concentration of benzene in water was 10 μg L^{-1}, and the background air concentration was 0.1 μg L^{-1}. If the induced mass transfer is described by $k_w = 50$ m h^{-1} and $k_a = 2500$ m h^{-1}, estimate the mass of benzene emitted during an 8-hour period.

$1/K_w = 1/k_w + 1/k_a K_{AW} = 1/50 + 1/(0.23)(2500) = 1/46$ h m^{-1}. Hence $K_w = 46$ m h^{-1}. N_i 4600 $(10 \times 10^{-9} - 0.1 \times 10^{-9}/0.23) = 4.4 \times 10^{-5}$ g (cm^2 h)$^{-1}$. Amount emitted in an 8-hour period $= 4.4 \times 10^{-5}$ (10^4) (2×10^4) $(8) = 70,400$ g $= 70.4$ kg.

This calculation is for a given volume of water where there is no additional input of benzene with time. This is called a "batch" process. In practice, most surface impoundments receive a continuous recharge of material via a given volumetric inflow and outflow of wastewater. This is called a "continuous" process. We shall revisit these aspects later in Chapter 4 where applications of kinetics in environmental reactors will be discussed.

Another wastewater treatment process that involves air-water contact is flotation. Flotation is the process where tiny air bubbles are forced into a tall water column (Figure 3.2). Bubbles can be formed either by pressurizing or depressurizing the solution ("dissolved air flotation") or by introducing air under pressure through a small orifice in the form of tiny bubbles ("induced air flotation"). The purpose of bubble aeration is to the dissolved oxygen level in water. Oxygen enhancement is essential for biological treatment of wastewaters and activated sludge and also for the precipitation of iron and manganese from wastewaters. An illustrative example for the case of absorption of oxygen into water is given next.

Example 3.19   Oxygen Absorption in a Wastewater Tank

The oxygen levels in wastewater will be lower than its equilibrium value because of high oxygen demands. Improved oxygen levels in water can be achieved by bubbling air (oxygen) through fine bubble porous diffusers (see Figure 3.2). This is called "submerged aeration." Large aeration units can be assumed to be homogeneous in oxygen concentration due to the turbulent liquid mixing within the water. The oxygen absorption rate (air to water) is given by

$$N_i = (K_w a) \cdot \left(C_{iw}^{eq} - C_{iw} \right) \tag{3.64}$$

where C_{iw}^{eq} is the equilibrium saturation value of oxygen in water (mol m^{-3}). This is given by the air-water partition constant, $C_{iw}^{eq} = P_i/K_{AW}'$. C_{iw} is the concentration of oxygen in bulk water $(mol\, m^{-3})$. To account for other wastewater constituents (e.g., DOCs, surfactants, cosolvents), C_{iw} in wastewater can be related to the C_{iw} in tap water by a factor β for which a value of 0.95 is recommended. K_w is the overall liquid phase mass transfer coefficient for oxygen $(m\, s^{-1})$. a is called the specific air-water interfacial area provided by the air bubbles. It has units of $m^2\, m^{-3}$ and is given by $(6/d)\varepsilon_g$, where d is the average bubble diameter at the sparger and ε_g is the void fraction. The void fraction is defined as the ratio of air volume to the total volume (air + water) at any instant in the aeration vessel. It is a difficult quantity to measure. Hence $(K_w a)$ is lumped together and values reported. $K_{O_2}a$ is temperature dependent. Eckenfelder (1989) gives $(K_w a) = (K_w a)_{20^\circ C}\theta^{(t-20)}$ where θ is a parameter between 1.015 and 1.04 for most wastewater systems. $K_w a$ is characteristic of both the specific physical and chemical characteristics of the wastewater. It is generally estimated in the laboratory by performing experiments on the specific wastewater sample (see Eckenfelder, 1989).

In a specific bubble aeration device, $K_w a$ was determined to be 0.8 h^{-1} for a wastewater sample at 298 K. At 298 K P_i in air is 0.2 atm. The air-water partition constant for oxygen at 298 K is 0.774 $atm\, m^{-3}\, mol^{-1}$. The latter two parameters give us equilibrium concentration of oxygen in water, $C_{iw}^{eq} = 0.26\ mol\, m^{-3}$ $(= 2.6 \times 10^{-4}\ mol\, L^{-1})$. If the wastewater is completely mixed (homogeneous) and has zero dissolved oxygen concentration, then the maximum rate of oxygen absorption is $N_i^{max} = 0.208\ mol\, m^3\, h^{-1}$. If the total volume of the wastewater is 28,000 L $(= 1000\ ft^3)$, the oxygen absorbed will be 5.8 $mol\, h^{-1}$ $(= 93\ g\, h^{-1})$.

3.4 SOIL-WATER AND SEDIMENT-WATER EQUILIBRIUM

Contaminated sediments in the marine or freshwater environment exist in the United States and other parts of the world. Examples in the United States are the Great Lakes and New Bedford harbor contaminated with PCBs, Indiana harbor contaminated with PAHs, and the Hudson river sediment contaminated with metals and organic compounds, to name a few. Developing countries also suffer extensive contamination of major rivers and tributaries. There are many areas in the United States and elsewhere, where extensive sediment contamination has led to outright ban on fishing and recreation. Some well-known examples of sediment contamination are the oil spill in Prince William Sound near Alaska from the Exxon Valdez oil tanker disaster and the Kuwait oil fires after the first Persian Gulf War.

Sediments generally accumulate contaminants over a period of time. Over time, as inputs cease, the previously contaminated sediments become sources of pollutants to the water column. Contaminants enter the marine food chain and further bioaccumulate in birds and mammals. The soil-water environment is another area of significance. In the United States and other parts of the world, groundwater is a major source of clean water. Improperly buried waste (e.g., old landfills, leaking underground petroleum storage tanks) releases contaminants that leak into the groundwater. Inadvertent solvent spills and intentional dumping also create major threats to the limited clean water supplies in the world. In the groundwater environment, the equilibrium between the soil and pore water is the issue of concern.

3.4.1 PARTITIONING INTO SOILS AND SEDIMENTS FROM WATER

The soil and sediment environments can be characterized by a solid phase composed mainly of mineral matter (kaolinite, illite, montmorillonite), organic macromolecules (humic and fulvic acids of both plant and animal origin), and pore water. The mineral matter is composed of oxides of Al, Si, and Fe. Typical Louisiana soil and sediment have compositions given in Table 3.12. The sand, silt, and clay fractions predominate, whereas the total organic carbon fraction is only a small fraction of the total.

A soil or sediment has micro-, meso-, and macropores that contain pore water. Hydrogen and hydroxyl ions in pore water control the pH. Different metal ions control the ionic strength of the pore water. Both pH and ionic strength are important in controlling the dissolution and precipitation reactions at the mineral-water interfaces. The most important process as far as transport of organic and metal ions in the sedimentary and subsurface soil environment is concerned is "sorption." By "sorption" we mean both "adsorption" on the surface as well as "absorption" (physical encapsulation) within the solids. Metal ions and organic molecules in pore water will establish sorption equilibria with the solid phase. The mineral matter in the soil develops coatings of organic matter, thereby providing both an inorganic mineral surface as well as an organic coated surface for sorption. Figure 3.8 is a schematic of the different sorption sites.

The mineral surfaces (oxides of Al, Si, Fe) are electrically charged and have electrical double layers arising from the ions present in pore water. Water strongly interacts with the charged surfaces and adopts a three-dimensional hydrogen-bonded network near the surface in which some water molecules are in direct contact with the metal oxide and others attached via H-bonds to other molecules (Thiel, 1991). The strongly oriented water molecules at the surface have properties different from that of bulk water. The near-surface water molecules are sometimes referred to as "vicinal water." Drost-Hansen (1965) has summarized the properties of vicinal water.

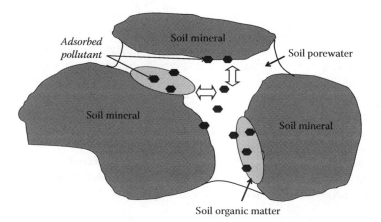

FIGURE 3.8 Partitioning of a solute between soil and pore water. Various niches are shown where pollutants reside in the soil pore.

Typically, a metal oxide such as SiO_2 is strongly hydrated and possesses Si-OH (silanol) groups. The surface charge density in this case is dependent on the pH of the adjacent pore water. Metal ions can enter into exchange reactions with the silanol groups forming surface complexes (Stumm, 1993).

The adsorption of compounds on soil or sediment surfaces occurs through a variety of mechanisms, each of which can be assumed to act independently. The overall adsorption energy is given by

$$\Delta G_{ads} = \sum_n \Delta G_n \qquad (3.65)$$

where each ΔG_n represents the contribution from the specific nth mechanism. These mechanisms include (1) interactions with the electrical double layer, (2) ion exchange including protonation followed by ion exchange, (3) coordination with surface metal cations, (4) ion-dipole interactions, (5) hydrogen bonding, and (6) hydrophobic interactions. The first two mechanisms are of relevance only for the adsorption of ionizable species. Coordination with surface cations is important only for compounds (e.g., amines) that are excellent electron donors compared to water. Ion-dipole interactions between an uncharged molecule and a charged surface are negligible in aqueous solution. Only compounds with greater hydrogen bonding potential than water can have significant hydrogen bonding mechanisms. Most neutral nonpolar compounds of hydrophobic character will interact through the last mechanism, namely hydrophobic interactions.

3.4.2 Adsorption of Metal Ions on Soils and Sediments

Figure 3.9 represents a mineral surface (silica) where an electron-deficient silicon atom and groups of electron-rich oxygen atoms are presented to the aqueous phase. The surface hydroxyl groups are similar to water molecules in that both can form hydrogen bonds. The adsorption energy for water on a silica surface is $-500\,mJ\,m^{-2}$. The energy decreases as the surface water molecules are progressively replaced. Water is therefore more favored by most mineral surfaces. There are two possible orientations of water molecules on a mineral surface, as depicted in Figure 3.9. An interesting consequence of these orientations is that water molecules in direct contact with the metal oxide can adopt specific favorable conformations and can simultaneously participate in the tetrahedral three-dimensional network that gives it the special features noted earlier.

The strongly polar mineral surface possesses a surface charge density and a double layer near it. The surface charge distribution can be obtained from the application of the Poisson-Boltzmann equation. It is given by

$$\sigma_e = \left(\frac{2\varepsilon RTI}{\pi F^2} \right)^{1/2} \sinh\left(\frac{zF\psi}{2RT} \right) \qquad (3.66)$$

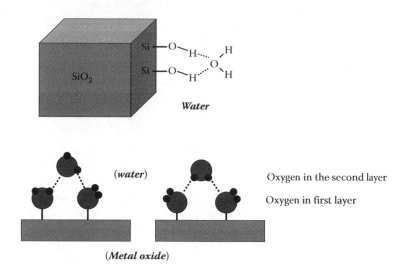

FIGURE 3.9 Typical mineral surface (silica) where an electron-deficient silicon atom and groups of electron-rich oxygen atoms are presented to the aqueous phase.

where

z is the valence of ions in the electrolyte (e.g., for KCl, z = 1)

ε is the dielectric constant of water (= 7.2×10^{-10} C V^{-1} m^{-1} at 298 K)

I is the ionic strength (mol m^{-3})

R is the gas constant (8.3 J K^{-1} mol^{-1})

F is the Faraday constant (96,485 C mol^{-1})

ψ is the surface potential (J C^{-1} V^{-1})

The surface charge density is determined by the pH of the adjacent solution. Therefore both H$^+$ and OH$^-$ are called "potential determining ions." Based on the reactivities of these ions with the surface hydroxyl groups, the following two acid-base surface exchange reactions can be written:

$$\equiv M - OH_2^+ \rightleftharpoons \equiv M - OH + H^+$$
$$\equiv M - OH \rightleftharpoons \equiv M - O^- + H^+$$

(3.67)

with equilibrium constants given by

$$K_1 = \frac{\left[\equiv M - OH\right]\left[H^+\right]}{\left[\equiv M - OH_2^+\right]}; \quad K_2 = \frac{\left[\equiv M - O^-\right]\left[H^+\right]}{\left[\equiv M - OH\right]}$$

(3.68)

The OH bonds in bulk water have an intrinsic activity derived solely from the dissociation of water molecules, whereas the activity of the surface OH groups involved in the above surface acid equilibrium constants will introduce the additional work (energy) required to bring an H^+ ion from the bulk to the surface, which is given by the electrostatic potential of the surface. Thus,

$$K_1 = K_1^{intr} \cdot \exp\left(\frac{F\psi}{kT}\right)$$

$$K_2 = K_2^{intr} \cdot \exp\left(\frac{F\psi}{kT}\right)$$

(3.69)

The intrinsic terms quantify the extent of H^+ exchange when the surface is uncharged. With increasing pH, the surface dissociation increases, since more of the surface H^+ ions are consumed, and hence increasingly it becomes difficult to remove hydrogen ions from the surface. These equations (3.69) represent this effect mathematically.

Example 3.20   Intrinsic and Conditional Equilibrium Constants

Derive the equations for the conditional equilibrium constants in terms of the intrinsic constants given above.

For the two ionization reactions of concern the conditional equilibrium constants are K_1 and K_2 as given by the equations given earlier. The activity coefficients of surface adsorbed species are assumed to be equal. The equilibrium constants are conditional since they depend on the surface ionization, which depends on the pH. The total free energy of sorption is composed of two terms: $\Delta G_{tot}^0 = \Delta G_{intr}^0 + \Delta z F \psi_0$, where the second term is the "intrinsic" free energy term. The third term represents the "electrostatic" term that gives the electrical work required to move ions through an interfacial potential gradient. Δz is the change in charge of the surface species upon sorption. Since $\Delta G^0 = -RT \ln K$, we can obtain

$$K_{intr} = K_{cond} \cdot \exp\left(\frac{\Delta F z \psi_0}{RT}\right)$$

(3.70)

Hence we derive the equation for K_1 and K_2.

It is clear from this discussion that the surface charge density is the difference in surface concentration ($mol \, m^{-2}$) of $\equiv M - OH_2^+$ and of $\equiv M - O^-$ moieties on the surface. Thus,

$$\sigma_e = \left[\equiv M - OH_2^+\right] - \left[\equiv M - O^-\right]$$

(3.71)

This equation gives the overall alkalinity or acidity of the surface. If the species concentrations are expressed in $mol \, kg^{-1}$, then the overall charge balance can be written as

$$C_A - C_B + \left[OH^-\right] - \left[H^+\right] = \left\{\equiv M - OH_2^-\right\} - \left\{\equiv M - O^-\right\}$$

(3.72)

where
  C_A and C_B are the added aqueous concentrations of a strong acid and a strong base, respectively
  $[OH^-]$ and $[H^+]$ are the aqueous concentrations of base and acid added

Note that {} denotes $mol\,kg^{-1}$ whereas [] represents $mol\,m^{-2}$ for $M-OH_2^+$ and $M-O^-$ species.

When σ_c is zero, there are equal surface concentrations of both species. The pH at which this happens is called pH_{pzc}, the point of zero charge. At the pH_{pzc}, the surface potential $\Psi \to 0$. Now one can relate the pH_{pzc} to the intrinsic surface activities by equating the concentrations of the surface species $\equiv M-OH_2^+$ and $\equiv M-O^-$.

$$pH_{pzc} = \frac{1}{2}\cdot\left(pK_1^{intr} + pK_2^{intr}\right) \tag{3.73}$$

If $pH < pH_{pzc}$, the surface has a net positive charge, whereas for $pH > pH_{pzc}$, the surface is negatively charged.

Getting back to the equation for σ_c and utilizing the equations for intrinsic equilibrium constants, one obtains

$$\sigma_c = \left[\equiv M-OH\right]\cdot\left[\frac{\left[H^+\right]}{K_1^{intr}}\cdot\exp\left(-\frac{F\psi}{RT}\right) - \frac{K_2^{intr}}{\left[H^+\right]}\cdot\exp\left(\frac{F\psi}{RT}\right)\right] \tag{3.74}$$

We can now substitute in this expression the equation for Ψ in terms of σ_c and get an explicit equation relating the surface hydroxyl concentration to σ_c through ionic strength, I. A listing of the characteristic values of K_1^{intr}, K_2^{intr}, and pH_{pzc} for several common oxide minerals is given in Table 3.13. In the presence of other ions in aqueous solution (in addition to H^+ and OH^-), the influence of the specifically sorbed ions on the surface introduces some complexity into this analysis. These concepts have

---

TABLE 3.13
Surface Properties of Important Oxide Minerals Present in the Environment

Surface	Composition	Surf. Area	pK_{s1}^{intr}	pK_{s2}^{intr}	pH_{pzc}
Quartz	SiO_2	0.14	−3	7	2
Geothite	$FeO(OH)$	46	6	9	7.5
Alumina	Al_2O_3	15	7	10	8.5
Iron oxide	$Fe(OH)_3$	600	7	9	8
Gibbsite	$Al(OH)_3$	120	5	8	6.5
Na-montmorillonite	$Na_3Al_7Si_{11}O_{30}(OH)_6$	600–800	—	—	2.5
Kaolinite	$Al_2Si_2O_5(OH)_4$	7–30	—	—	4.6
Illite	$KAl_3Si_3O_{10}(OH)_2$	65–100	—	—	—
Vermiculite		600–800	—	—	—
Muscovite		60–100	—	—	—

Sources: Schwarzenbach, R.P. et al., *Environmental Organic Chemistry*, John Wiley & Sons, Inc., New York, 1993; Sparks, D., *Environmental Soil Chemistry*, Academic Press, New York, 1999.

been discussed by Stumm (1993) and should be consulted for an in-depth discussion. The mathematical treatment of surface charges and surface ionization come under the so-called "surface complexation models" (Stumm and Morgan, 1996).

The metal cation adsorption data over a wide range of pH values can be represented by an equation of the form

$$\ln D = a + b \cdot pH \tag{3.75}$$

where $D = K_{sw}C_s = \phi_{sorb}/\phi_{soln}$. K_{sw} is the distribution (partition) constant between the solid and water, C_s is the mass of adsorbent per unit volume of aqueous solution, ϕ_{sorb} is the ratio of moles of metal adsorbed to the fixed initial moles added. $\phi_{soln} = 1 - \phi_{sorb}$ is the fraction of metal unadsorbed. A plot of $\ln D$ versus pH is called the "Kurbatov plot." The values of a and b have physical significance. If half of the added metal is adsorbed, then $D = 1$ and pH at this point is designated $pH_{50} = -a/b$. Then we have

$$\left(\frac{d\phi_{sorb}}{dpH} \right)_{pH_{50}} = \frac{b}{4} \tag{3.76}$$

or

$$\ln D = b \cdot \left(pH - pH_{50} \right) \tag{3.77}$$

Hence,

$$\phi_{sorb} = \frac{1}{1 + \exp\left[-b \cdot \left(pH - pH_{50} \right) \right]} \tag{3.78}$$

Thus evaluation of a and b from the Kurbatov plot gives us the fraction of metal adsorbed at any given pH.

There are other aspects that should also be considered in understanding the exchange of metal ions with clay minerals, since with surface exchange reactions, the forces of bonding within the clay structure will also physically distort their size and shape. In other words, extensive changes in the activity coefficients of the adsorbed ions are possible. Expansion or contraction of clays will complicate the simple activity coefficient relationships. Such general relationships are, heretofore, lacking. General trends have been noted and verified and hence qualitative relationships are possible. The simple models described here should serve to exemplify the general approach to the study of exchangeable metal ions on soils and sediments.

3.4.3 Adsorption of Organic Molecules on Soils and Sediments

As is evident from Table 3.12, a typical sediment or soil will have an overwhelming fraction of sand, silt, or clay, and only a small fraction of organic material. The total organic fraction is characterized mainly by humic and fulvic acids. They exist bound to the mineral matter through Coulombic ligand exchange and hydrophobic interactions. Some of the species are weakly bound and some are strongly bound. The weakly bound fraction enters into dynamic exchange equilibrium between the solid (mineral) surfaces and the adjacent solution phase. In some cases, this relationship can be characterized by a Langmuir-type isotherm (Murphy et al., 1990; Thoma, 1992). The adsorbed concentration rapidly reaches a maximum. The differences in the maxima for different surfaces are striking. The sorption of humic substances, for example, will depend not only on the site concentrations but also on its relative affinity with iron and aluminol hydroxyl groups. It is known that adsorption of DOCs on hydrous alumina is a Ph-dependent process with maximum adsorption between pH of five to six, and decreasing adsorption at higher pH (Davis and Gloor, 1981). As a consequence of this change, alumina coagulates more rapidly at pH values close to the adsorption maximum. It is also known that increasing molecular weight of a DOC leads to increased adsorption on alumina. In other words, adsorption of DOCs on alumina is a fractionation process. An adsorbed organic film such as DOCs (humic or fulvic compounds) can alter the properties of the underlying solid and can present a surface conducive for chemical adsorption of other organic compounds (Figure 3.10).

As we discussed in the previous section, mineral matter has a large propensity for water molecules because of their polar character. Nonpolar organic molecules (e.g., alkanes, polychlorinated biphenyls, pesticides, aromatic hydrocarbons) have to displace the existing water molecules before adsorption can occur on mineral surfaces. For example, on a Ru(001) metal surface, Thiel (1991) found that the adsorption of water from the gas phase involved a bond strength of $46 \, kJ \, mol^{-1}$ for water and that for cyclohexane was $37 \, kJ \, mol^{-1}$. The adsorption bond strengths are comparable. However, the area occupied by one cyclohexane molecule is equivalent to about seven water molecules. Hence, a meaningful comparison should be based on the "energy change per unit area." On this basis, the replacement of cyclohexane on the surface by water involves an enthalpy decrease of about $21–23 \, kJ \, mol^{-1}$.

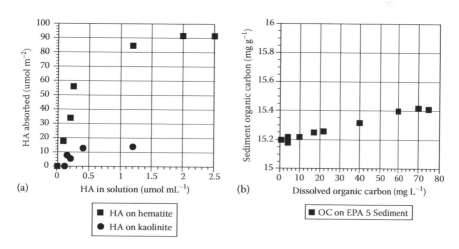

FIGURE 3.10 (a) Sorption of humic acid on hematite and kaolinite, (b) sorption of organic carbon on EPA5 sediment.

The entropy change was relatively small for both molecules. Thus, the enthalpy change per unit area is the major driving force for the displacement of cyclohexane from the metal surface. Such large enthalpy changes are possible for water molecules because of their special tetrahedral coordination and H-bonding tendency that are lacking for cyclohexane molecules.

The adsorption of several organic molecules on solids from their aqueous solutions is reported in the literature. The data is invariably reported as mass of solute (sorbate) adsorbed per unit mass of solid sorbent:

$$K_{sw}\left(L\,kg^{-1}\right) = \frac{W_i\left(mg\ sorbate/kg\ sorbent\right)}{C_{iw}\left(mg\ sorbate/L\ solution\right)} = \frac{\left(W_i^{OC} + W_i^{min}\right)}{C_{iw}} \quad (3.79)$$

The overall partition constant K_{sw} has two contributions—one from the mineral matter (min) and the other from the organic matter (oc). Thus,

$$K_{sw} = K_{sw}^{oc} + K_{sw}^{min} \quad (3.80)$$

If f_{oc} is the fraction of organic carbon on the solid ($g\,g^{-1}$) and the mineral surface area is given by S_a ($m^2\,kg^{-1}$), K_{oc} ($L\,kg^{-1}$) and K_{min} ($mg\,m^{-2}$) represent the partition constants for a compound normalized, respectively, to oc and min.

$$K_{oc} = \frac{K_{sw}^{oc}}{f_{oc}}; \quad K_{min} = \frac{K_{sw}^{min}}{S_a} \quad (3.81)$$

Schwarzenbach et al. (1993) suggest that for hydrophobic compounds, which are driven to the mineral surface via exclusion from the water structure, a linear free energy relationship (LFER) should exist between K_{min} and γ_i^*. The form of the LFER is

$$Log\ K_{min} = a \cdot log\ \gamma_i^* + b \quad (3.82)$$

Based on a limited set of data, this relationship appears to hold for a particular sorbent adsorbing different nonpolar compounds under identical conditions of pH and ionic strength (Mader et al., 1997).

The amount adsorbed on mineral surfaces is indeed very small. However, if the mineral surface is coated with even a small percent of organic macromolecules, the adsorption capacity is enhanced by several orders of magnitude. Mineral matter develops these coatings rapidly and hence natural soil and sediment can be considered to provide dual sorbent sites. The bare mineral surfaces are characterized by low K_{min}. The organic matter coating the mineral provides a highly compatible medium with which a neutral hydrophobic compound can associate and thereby limit its interaction with water.

The organic matter in sediments and soils is composed of approximately one half of the total carbon that is directly measured as organic carbon. Hence, sorption is often keyed to organic carbon rather than organic matter through the approximate relationship, $\varphi_{om} \cdot 2\varphi_{oc}$. Karickhoff et al. (1979) demonstrated that the linear isotherm described here was valid for several organic compounds over a wide range of aqueous concentrations. The linear isotherm was found to be valid up to approximately 50% of the aqueous solubility of the compound. The isotherms were reversible and showed only a 15% decrease in K_{sw} at an ion (NaCl) concentration of 20 mg mL^{-1}. They also demonstrated the linear relationship between K_{sw} and f_{oc} for a polyaromatic hydrocarbon (pyrene). Further, Means et al. (1980) extended this relationship to several PAHs on sediments and soils. The slope of the plot of K_{sw} (kg sorbate per kg sorbent) versus f_{oc} is a constant for a given compound on various soils and sediments. Thus we have

$$K_{sw} = K_{oc} f_{oc} \tag{3.83}$$

The slope K_{oc} is therefore a convenient way of characterizing the sorption of a particular hydrophobic compound.

Example 3.21   Obtaining K_{oc} from Experimental Data

1 g of soil in 500 mL of aqueous solution was spiked with 10 mg L^{-1} of an organic compound. After equilibration for 48 h, the aqueous concentration of the compound was 1 mg L^{-1}. If the soil organic carbon content was 0.02, obtain K_{sw} and K_{oc} for the compound.

$w_s = 1$ g, $V_w = 500$ mL, $C_i^{w,0} = 10$ mg mL^{-1}, $C_{iw} = 1$ mg L^{-1}, $f_{oc} = 0.02$. $W_i/w_s = (0.01 - 0.001)\,0.5/1 = 0.0045$ g g^{-1}. $K_{sw} = (W_i/w_s)/C_{iw} = 4.5$ L kg^{-1}. $K_{oc} = K_{sw}/f_{oc} = 225$ L g$^{-1} = 2.25 \times 10^5$ L kg^{-1}.

The partitioning of a solute from water to soil organic carbon is similar to the partitioning to octanol. This leads to LFER between K_{oc} and K_{ow}. As was discussed earlier, the octanol-water partition constant is given by

$$K_{ow} = \left(\frac{\gamma_{iw}}{\gamma_{io}} \right) \cdot \left(\frac{V_w^*}{V_o^*} \right) \tag{3.84}$$

Since the partitioning into the organic fraction of the mineral matter can be described similarly, we can write

$$K_{oc} = \left(\frac{\gamma_{iwc}}{\gamma_{icw}} \right) \cdot \left(\frac{V_w^*}{V_c^*} \right) \cdot \frac{1}{\rho_c} \tag{3.85}$$

where γ_{iwc} and γ_{icw} are the respective activity coefficients of solute i in water saturated with organic matter (humus) and humic saturated with water. V_w^* and V_c^* are the respective molar volumes. ρ_c is the density of the organic matter so that K_{oc} is expressed in L kg^{-1}. Using the equation for K_{ow} to substitute for V_w^* we get

$$K_{oc} = C_1 K_{ow} \left(\frac{\gamma_{io}}{\gamma_{icw}} \right) \left(\frac{\gamma_{iwc}}{\gamma_{iw}} \right) \tag{3.86}$$

Further, if $\gamma_{iwc} = \gamma_{iw}$, we have

$$K_{oc} = C_1 K_{ow} \frac{\gamma_{io}}{\gamma_{icw}} \qquad (3.87)$$

It has been observed that most solutes behave nonideally in the octanol phase to the extent that γ_i^o is given by (Curtis et al., 1986)

$$\gamma_{io} = 1.2 \cdot K_{ow}^{0.16} \qquad (3.88)$$

Therefore,

$$Log\ K_{oc} = 1.16 \cdot Log\ K_{ow} + Log\ C_2 - Log\ \gamma_{icw} \qquad (3.89)$$

where $C_2 = 1.2C_1$. Hence linear correlations between log K_{oc} and log K_{ow} are predicted. For example, Curtis et al. (1986) obtained the following linear relationship for adsorption on the natural organic matter of soils:

$$Log\ K_{oc} = 0.92 \cdot Log\ K_{ow} - 0.23 \qquad (3.90)$$

Table 3.14 displays the relationship between log K_{oc} and log K_{ow}. Table 3.15 lists the available correlations between log K_{min} and log γ_i^*. Note that γ_i^* is related directly to log K_{ow} (Section 2.2.3.4).

TABLE 3.14
Correlations between Log K_{oc} and Log K_{ow} for Various Compounds of Environmental Significance

	Log K_{oc} = a + b Log K_{ow}			
Compound Class	**b**	**a**	**r²**	**Reference**
Pesticides	0.544	1.377	0.74	Kenaga and Goring (1980)
Aromatics (PAHs)	0.937	−0.006	0.95	Lyman et al. (1982)
Aromatics (PAHs)	1.00	−0.21	1.00	Karickhoff et al. (1979)
Herbicides	0.94	0.02	—	Lyman et al. (1982)
Insecticides, fungicides	1.029	−0.18	0.91	Rao and Davidson (1980)
Phenyl ureas and carbamates	0.524	0.855	0.84	Briggs (1973)
Chlorinated phenols	0.82	0.02	0.98	Schellenberg et al. (1984)
Chlorobenzenes (PCBs)	0.904	−0.779	0.989	Chiou et al. (1983)
PAHs	1.00	−0.317	0.98	Means et al. (1980)
PCBs	0.72	0.49	0.96	Schwarzenbach and Westall (1981)

Note: K_{oc} is in L kg⁻¹ or cm³ g⁻¹.

TABLE 3.15
Correlations between Log K_{min} and Log γ_i^*

		$\text{Log } K_{min} = a + b \text{ Log } \gamma_i^*$			
Compound Class	Solid	a	b	r^2	Reference[a]
Chlorobenzenes					
PAHs, biphenyls	α-Al_2O_3	−10.68	0.70	0.94	1
Chlorobenzenes, PAHs					
Biphenyls	α-Fe_2O_3	−11.39	0.98	0.92	1
PAHs	Kaolin, glass				
	Alumina	−14.8	1.74	0.97	2
Various HOCs	Kaolinite	−12.0	1.37		3
Various HOCs	Silica	−12.5	1.37		3

Note: K_{min} is in mL m^{-2}.
[a] (1) Mader et al. (1997); (2) Backhus (1990); (3) Schwarzenbach et al. (1993).

Example 3.22   Determining K_{sw} from K_{ow}

A soil from a Superfund site in Baton Rouge, LA was found to have the following properties: clay 30%, sand 22%, silt 47%, and organic carbon content 1.13%. Estimate the soil-water partition constant for 1,2-dichlorobenzene on this soil.

For 1,2-dichlorobenzene, log K_{ow} is 3.39. Hence, log K_{oc} = 0.92(3.39) − 0.23 = 2.89. K_{oc} = 774. $K_{sw} = K_{oc}\varphi_{oc}$ = (774) (0.0113) = 8.7 P kg^{-1}.

Example 3.23   Time of Travel of a Pollutant in Groundwater

The same approach as described above is also used to describe the movement of pollutants in the subsurface groundwater (Weber et al., 1991). Sorption retards the velocity of pollutant movement in groundwater (u_p) in relation to the velocity of the groundwater itself (u_o). This can be expressed as

$$\frac{u_0}{u_P} = 1 + \frac{\rho_b}{\varepsilon} \cdot K_{sw} = R_F \qquad (3.91)$$

where R_F is the "retardation factor." For compounds that are strongly sorbing, $u_p \ll u_o$ and the pollutant concentration front is slowed down considerably. For compounds that are nonsorbing (such as chloride ions), $u_o = u_p$ and no retardation is seen. McCarty et al. (1992) reports the results of an experiment in which the retardation of various halogenated compounds present in a reclaimed municipal wastewater was injected into an aquifer in Palo Alto, CA. The fractional break-through in an observation well downfield was obtained for three adsorbing pollutants (chloroform, bromoform, and chlorobenzene) and a nonsorbing tracer (chloride ion). The results are shown in Figure 3.11. The field-measured retardation factors were 6 for chloroform and bromoform and 33 for chlorobenzene. Clearly the greater the hydrophobicity of the pollutant, the slower is its movement

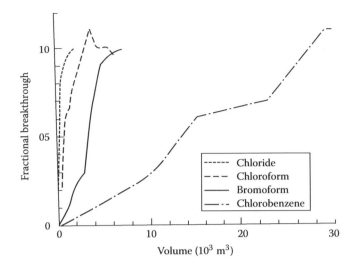

FIGURE 3.11   Sequential breakthrough of solutes at an observation well during the Palo Alto groundwater recharge study. (Reprinted from *Water Res.*, 16, Roberts, P.V., Schreiner, J., and Hopkins, G.D., Field study of organic water quality changes during groundwater recharge in the Palo Alto Baylands, 1025–1035, Copyright 1982, with permission from Elsevier.)

in the aquifer. Retardation is an important process in groundwater for two main reasons. First, if an aquifer were to become polluted with compounds of differing hydrophobicity, they would tend to appear in a down gradient well at different times in accordance with their retardation factors. This would make the concentrations and nature of water at the observation well quite distinct from the original contamination, and hence identification of the pollution source will be difficult. Second, the retardation factor will give us an idea of how much material is on the solid phase and how much is in free water, and therefore develop appropriate remediation alternatives for the restoration of both the groundwater and the aerial extent of the contaminated aquifer.

3.5   BIOTA-WATER PARTITION CONSTANT

There are a number of receptors for a pollutant released into the environment. Risk assessment is the process of understanding and minimizing the effects of pollutants on receptors. In order to illustrate this, let us consider the sediment-water environment. Transport from contaminated sediments to the overlying water column exposes the marine species to pollutants. The marine animals accumulate these chemicals, and the contaminants then make their way into the food chain of higher animals, including humans. This is the process of "bioaccumulation."

The uptake of organic chemicals in aquatic species can be modeled at different levels of complexity. The uptake depends on factors such as exposure route (dermal, ingestion by mouth, inhalation), and loss through digestion and defecation. An animal can also imbibe chemicals through its prey that has been exposed to the chemical. Although rate-based models may be better suited to describe these phenomena, thermodynamic models have been traditionally used to obtain first-order estimates of the extent of bioaccumulation (Mackay, 1982). It has been suggested that abiotic species are in near-equilibrium conditions in most circumstances. Hence, it is appropriate to discuss briefly the thermodynamic basis for modeling the bioaccumulation phenomena.

In its simplest form, a partition coefficient (also called a "bioconcentration factor," K_{bw}) is used to define the concentration level of a pollutant in an aquatic species relative to that in water. Since the major accumulation of a pollutant in an animal occurs in its lipid fraction, it is customary to express the concentration on a lipid weight basis.

The general equation for partitioning between the organism and water is given by

$$K_{bw} = \frac{C_{iB}}{C_{iw}} \tag{3.92}$$

where

C_{iB} is the animal concentration (mg kg^{-1})
C_{iw} is the aqueous concentration (mg L^{-1}) at equilibrium

If we assume that the organism is comprised of j compartments, each with a concentration of I given by C_{ij}, and with a volume fraction η_j, then we can write for the total moles of solute i in the organism as

$$\sum_j m_{ij} = \sum_j C_{ij} \eta_j V \tag{3.93}$$

where V is the total organism volume. Thus, we have

$$C_{iB} = \frac{\sum_j m_{ij}}{V} = \sum_j C_{ij} \eta_j \tag{3.94}$$

For the aqueous phase we have

$$C_{iw} = \frac{x_{iw}}{V_w} \tag{3.95}$$

At equilibrium the fugacity in all j compartments would be equal to that in the aqueous phase, $f_i^w = f_i^j$. For any compartment j

$$f_i^j = x_{ij} \gamma_{ij} f_i^0 = C_{ij} V_j \gamma_{ij} f_i^0 \tag{3.96}$$

where f_i^0 is the reference fugacity on the Raoult's law basis, $V_j = \eta_j V$. Thus,

$$C_{ij} = \left(\frac{f_i^j}{f_i^0} \right) \cdot \left(\frac{1}{V_j \gamma_{ij}} \right) \tag{3.97}$$

or

$$C_{iB} = \sum_{j} \left(\frac{f_i^j}{f_i^0} \right) \cdot \left(\frac{1}{V_j \gamma_{ij}} \right) \cdot \eta_j \tag{3.98}$$

For the aqueous phase we have

$$f_i^w = x_{iw} \gamma_{iw} f_i^0 = C_{iw} V_w \gamma_{iw} f_i^0 \tag{3.99}$$

or

$$C_{iw} = \left(\frac{f_i^w}{f_i^0} \right) \cdot \left(\frac{1}{V_w \gamma_{iw}} \right) \tag{3.100}$$

Now since $f_i^w = f_i^j$, we obtain

$$K_{Bw} = \frac{C_{iB}}{C_{iw}} = V_w \gamma_{iw} \sum_{j} \frac{\eta_j}{V_j \gamma_{ij}} \tag{3.101}$$

Since the dominant accumulation of hydrophobic solutes in an organism occurs in its lipid content, we can write

$$K_{BW} = \left(\frac{V_w}{V_L} \right) \cdot \left(\frac{\gamma_{iw}}{\gamma_{iL}} \right) \cdot \eta_L \tag{3.102}$$

where L refers to the lipid phase. Organisms with high lipid content (η_L) should have high K_{BW} values.

Since γ_{iw} is directly proportional to the octanol-water partition coefficient, K_{ow}, we can expect an LFER relationship between K_{BW} and K_{ow}. Using the definition of K_{ow} given previously along with Equation 3.102, we get

$$\frac{K_{BW}}{K_{ow}} = \eta_L \cdot \left(\frac{\gamma_{io}}{\gamma_{iL}} \right) \cdot \left(\frac{V_o}{V_L} \right) \tag{3.103}$$

For compounds that have similar volume fraction of lipids (η_L) and similar ratios of activity coefficients (γ_{io}/γ_{iw}), the ratio K_{BW}/K_{ow} should be fairly constant. This suggests that a linear one-constant correlation should suffice:

$$\text{Log } K_{BW} = a \cdot \text{Log } K_{ow} + b \tag{3.104}$$

Such a correlation was tested and confirmed by Mackay (1982). The correlation developed was for a restricted set of compounds, namely, those with log $K_{ow} < 6$, nonionizable, and those with small K_B values. The overall fit to the experimental data was

$$\text{Log } K_{BW} = \text{Log } K_{ow} - 1.32; \quad r^2 = 0.95 \tag{3.105}$$

TABLE 3.16

Log K_{BW} – Log K_{ow} Correlations

Chemical Class	Log K_{BW} = a Log K_{ow} + b			Species
	a	**b**	**r^2**	
Various	0.76	−0.23	0.823	Fathead minnow, bluegill, trout
Ether, chlorinated compounds	0.542	+0.124	0.899	Trout
Pesticides (PAHs), PCBs	0.85	−0.70	0.897	Bluegill, minnow, trout
Halogenated hydrocarbons, halobenzenes, PCBs, diphenyl oxides, P-pesticides, acids, ethers, anilines	0.935	−1.495	0.757	Various
Acridines	0.819	−1.146	0.995	Daphnia pulex

Sources: Veith, G.D. et al., An evaluation of using partition coefficients and water solubility to estimate the bioconcentration factors for organic chemicals in fish, *ASTM SPE 707*, ASTM, Philadelphia, PA, 1980, pp. 116–129; Neely, W.B. et al., *Environ. Sci. Technol.*, 8, 1113, 1974; Kenaga, E.E. and Goring, C.A.I., *ASTM SPE 707*, 78–115, 1980; Southworth, G.R. et al., *Water Res.*, 12, 973, 1978.

Thus, $K_{BW} = 0.048\ K_{ow}$. The implication is that fish is about 5% lipid or it behaves as if it is about 5% octanol by volume. This correlation can give estimates of the partitioning of hydrophobic organic compounds into biota provided (1) the equilibrium assumption is valid, and (2) the non-lipid contributions are negligible. It should be borne in mind, however, that the correlation is of dubious applicability for tiny organic species such as plankton, which have very large area to volume ratios and hence surface adsorption may be a dominant mechanism of partitioning. Other available correlations are given in Table 3.16.

Example 3.24   Bioconcentration Factor for a Pollutant

A fish that weighs 3 lb resides in water contaminated with biphenyl at a concentration of 5 mg L^{-1}. What is the equilibrium concentration in the fish?

For biphenyl, log $K_{ow} = 4.09$. Hence, log $K_{BW} = 4.09 − 1.32 = 2.77$. $K_{BW} = 589$ L kg^{-1}. Hence, $C_{iB} = 589\ (5) = 2944$ mg kg^{-1} and mass in the fish = (2944) (3) (0.45) = 3974 mg = 3.97 g.

3.6   AIR-TO-AEROSOL PARTITION CONSTANT

Particulates (aerosols) in air adsorb volatile and semivolatile compounds from the atmosphere in accordance with the expression derived by Junge (1977). If W_i ($\mu g\ m^{-3}$ air) is the amount of a solute associated with total suspended particulate concentration C_{sp} (μg m^{-3} air), and C_{ia} (ng m^{-3}) is the concentration in the adjoining air in equilibrium with it, then the partition constant between air and particulates is

$$K_{AP} = \frac{C_{ia}}{\left(W_i / C_{sp}\right)} \qquad (3.106)$$

Note that K_{AP} has units of ng m^{-3}.

Traditionally, aerosols are collected using a high-volume air sampler into which a large volume of air is pulled through a glass fiber filter that retains particulates and subsequently through a tenax bed that retains the vapors. The filter-retained material is taken to be W_i and the adsorbent retained solute is taken to be equivalent to C_{ia}. In our earlier discussion on partitioning into aerosols, we had established that the fraction adsorbed to particulates in air $\left(\phi_i^p\right)$ is determined by the subcooled liquid vapor pressure of the compound $\left(P_{s(l)}^*\right)$. Compounds with small $P_{s(l)}^*$ showed large values of ϕ_i^p. This means that the value of K_{AP} will be correspondingly large.

$$K_{AP} = \left(\frac{1 - \phi_i^p}{\phi_i^p}\right) \cdot C_{sp} \qquad (3.107)$$

The value of K_{AP} has been found to be a sensitive function of temperature. Many investigators have collected field data and developed correlations of the form

$$\text{Log } K_{AP} = \frac{m}{T} + b \qquad (3.108)$$

where m and b are constants for a particular compound. Yamasaki et al. (1982), Pankow (1987), Bidleman (1988), and Subramanyam et al. (1994) have reported correlations (Table 3.17). Pankow (1987) showed theoretically that the constants in Equation 3.101 are given by

$$m = -\frac{\Delta H_{des}}{2.303 \cdot R} + \frac{T_a}{4.606} \qquad (3.109)$$

$$b = \log\left(\frac{2.75 \times 10^5 \left(\frac{M_a}{T_a}\right)^{1/2}}{A_p t_0}\right) - \frac{1}{4.606} \qquad (3.110)$$

where
ΔH_{des} is the enthalpy of desorption from the surface (kcal mol^{-1})
T_a is the midpoint of the ambient temperature range considered (K)
A_p is the specific surface area of the aerosol (cm^2 μg^{-1})
t_0 is the characteristic molecular vibration time (10^{-13} to 10^{-12} s)

TABLE 3.17

Relationships between K_{AP} and T for Some Organics in Ambient Air

	Log K_{AP} = m/T + b			
Compound	m	b	r^2	Reference
α-Hexachlorocyclohexane	−2755	14.286	0.574	Bidleman and Foreman (1987)
Hexachlorobenzene	−3328	16.117	0.687	
Aroclor 1254	−4686	19.428	0.885	
Chlordane	−4995	21.010	0.901	
p,p'-DDE	−5114	21.048	0.881	
p,p'-DDT	−5870	22.824	0.885	
Fluoranthene	−5180	20.80	0.682	Keller and Bidleman (1984)
Fluoranthene	−4420	18.52	0.805	Yamasaki et al. (1982)
Pyrene	−4510	18.48	0.695	Keller and Bidleman (1984)
Pyrene	−4180	17.55	0.796	Yamasaki et al. (1982)
Fluoranthene	−4393	21.41		Subramanyam et al. (1994)
Fluoranthene[a]	−6040	25.68		Subramanyam et al. (1994)
Phenanthrene	−3423	19.02		Subramanyam et al. (1994)

[a] The annular denuder method was used to reduce sampling artifacts resulting from a high volume sampling procedure. All other reported data in the table are using the Hi-Vol sampling procedure. Bidleman and coworkers obtained data from Columbia, SC, Yamasaki et al. from Tokyo, Japan, and Subramanyam et al. from Baton Rouge, LA.

Since the value of b is only weakly dependent on molecular weight, M, the compound specificity on K_{AP} appears through the slope m where ΔH_{des} is characteristic of the compound. At a given T, K_{AP} decreases as ΔH_{des} increases. In general, one can write

$$K_{AP} = \frac{10^9 \cdot P_{s(l)}^*}{N_s A_p RT \cdot \exp\left(\dfrac{\Delta H_{des} - \Delta H_v}{RT}\right)} \tag{3.111}$$

where

$P_{s(l)}^*$ is the subcooled liquid vapor pressure (atm)

N_s is the number of moles of adsorption sites per cm^2 of aerosol (mol cm^{-2})

ΔH_v is the enthalpy of vaporization of the liquid (kcal mol^{-1})

R is the gas constant (= 83 cm^3 atm (mol K)$^{-1}$)

In most cases, $\Delta H_{des} - \Delta H_v$ and hence Equation 3.111 simplifies to

$$K_{AP} = \frac{1.6 \times 10^4 \cdot P_{s(l)}^*}{N_s A_p} \tag{3.112}$$

where $P_{s(l)}^*$ is expressed in mm of Hg. If K_{AP} is expressed in µg m^{-3} (a more conventional unit) denoted by the symbol K_{AP}^*, we need to multiply by 0.001.

Example 3.25   Air-to-Particulate Partitioning of a Polyaromatic Hydrocarbon

Calculate the air-particulate partition constant for a pollutant ib Baton Rouge, LA. A typical value of specific area, a_v for aerosols collected from Baton Rouge air was estimated at 5×10^{-4} $m^2 m^{-3}$. The average particulate concentration, C_{sp} in summer time in Baton Rouge is 40 µg m^{-3}. Therefore, $A_p = a_v/C_{sp} = 1.25 \times 10^{-5}$ m^2 µg^{-1}. For physical adsorption of neutral compounds, Pankow (1987) observed that N_s is generally compound independent and has an average value of 4×10^{-10} mol cm^{-2}. For phenanthrene at an ambient temperature of 298 K, the subcooled liquid vapor pressure is 5×10^{-7} atm. Hence K_{AP} at 298 K for phenanthrene in aerosols is 1.6×10^8 ng m^{-3}, or K_{AP}^* is 1.6×10^5 µg m^{-3}. Log $K_{AP}^* = 5.2$. The experimental value is 5.7 (Subramanyam et al., 1994).

Earlier in this chapter, we considered Junge's equation:

$$\phi_i^P = \frac{\rho_i a_v}{P_{s(1)}^* + \rho_i a_v} \tag{3.113}$$

Pankow (1987) showed that within a given class of compounds K_{AP} can be related to the volatility of the species, or the subcooled vapor pressure of the compound, $P_{s(1)}^*$

$$\text{Log } K_{AP} = a_1 \cdot \text{Log } P_{s(1)}^* + b_1 \tag{3.114}$$

where b_1 is temperature independent. Many investigators have shown that this correlation is useful in estimating K_{AP} (Table 3.18). The slope $a_1 \approx 1$ in most cases.

TABLE 3.18
Relationship between K_{AP} and $P_{s(1)}^*$

Compound Class	$\text{Log } K_{AP} = a' \text{ Log } P_{s(1)}^* + b'$		Location
	a'	b'	
PAHs	0.8821	5368	Portland, OR
	0.760	5100	Denver, CO
	0.694	4610	Chicago, IL
	0.631	4610	London, U.K.
	1.04	5950	Osaka, Japan
PCBs	0.610	4740	Bayreuth, Germany
	0.726	5180	Chicago, IL
	0.946	5860	Denver, CO
Chlorinated pesticides	0.740	5760	Brazzaville, Congo
	0.610	4740	Bayreuth, Germany

Source: Falconer, R.L. and Bidleman, T.F., in: Baker, J.L. (ed.), *Atmospheric Deposition of Contaminants to the Great Lakes and Coastal Waters*, SETAC Press, Pensacola, FL, 1997.

In the atmosphere, ϕ_i^p is a function of the relative humidity. Since water competes effectively with organic molecules for sorption sites on the aerosol, it reduces the fraction of adsorbed organic compound. Experimental data support the effect (Pankow et al., 1993). The modified equation for air-to-aerosol partitioning will be

$$\text{Log } K_{AP} = \frac{m}{T} + b - \text{Log } f(x_w) \qquad (3.115)$$

where m and b are the same constants described earlier.

$$f(x_w) = \frac{1}{1 + K_{Lang}x_w} \qquad (3.116)$$

where K_{Lang} is the Langmuir adsorption constant for water on aerosols. Notice that as $x_w \to 0$, $f(x_w) \to 1$. As x_w increases, $f(x_w)$ decreases, and therefore, with increasing relative humidity log K_{AP} increases.

3.7 AIR-TO-VEGETATION PARTITION CONSTANT

A major portion of the land area (80%) on earth is covered by vegetation. As such the surface area covered by vegetation cannot be overlooked as an environmental compartment. Plants also take up nutrients and organic compounds from the soil/sediment environments. Thus, plants participate in the cycling of both inorganic and organic compounds in the environment. A large number of processes, both intra- and extracellular, are identified near the root zones and leaves of plants. In the water-soil environment, a number of organic compounds are imbibed through the roots and enzymatically degraded within the plants. In the air environment, a number of studies have revealed that plant-air exchange of organic chemicals play a major role in the long-range transport and deposition of air pollutants.

Wax or lipid layers exist to prevent excessive evapotranspiration from most plant surfaces exposed to air. The combination of high surface area and the presence of wax/lipid suggests the high partitioning of hydrophobic organic compounds into vegetation. As in the case of the bioconcentration factor, we define a vegetation-atmosphere partition coefficient K_{VA}

$$K_{VA} = \frac{W_V}{f_L C_{ia}} \qquad (3.117)$$

where
  W_v is the concentration in vegetation (ng g^{-1} dry weight)
  f_L is the lipid content of vegetation (mg g^{-1} dry weight)
  C_{ia} is the atmospheric concentration (ng m^{-3})
  K_{VA} has units of m^3 air mg^{-1} lipid

A dimensionless partition coefficient K_{VA}^* can also be obtained if we use an air density of 1.19×10^6 mg m^{-3} at 298 K.

Just as K_{AP} for aerosols was related to temperature, K_{VA} has the functional relationship: ln $K_{VA} = A/T + B$. The intercept B is common for a given class of compounds.

TABLE 3.19

Correlations of K_{VA} versus 1/T for Various Compounds

PAHs: $\ln K_{VA} = (A/T) + B$

Compound	A	B
Pyrene	10,227	−35.95
Phenanthrene	9,840	−35.95
Anthracene	9,773	−35.95
Fluoranthene	10,209	−35.95
Benz[a]anthracene	10,822	−35.95
Benzo[a]pyrene	10,988	−35.95

PCBs: $\log K_{VA}^* = (A/T) + B$

Compound	A	B	r^2
2,2′,5-PCB	3688	−7.119	0.993
2,2′,5,5′-PCB	4524	−9.307	0.992
2,2′,3,5′,6-PCB	4795	−9.843	0.985
2,2′,4,4′,5,5′-PCB	6095	−13.135	0.974
2,2′,3,3′,5,5′,6,6′-PCB	5685	−11.557	0.960

Sources: Simonich, S.L. and Hites, R.A., *Environ. Sci. Technol.*, 28, 939, 1994; Komp, P. and McLachlan, M.S., *Environ. Sci. Technol.*, 31, 886, 1997.

Note: Note that K_{VA} has units of m^3 mg^{-1} of lipid and K_{VA}^* is dimensionless.

For PAHs, B was −35.95 (Simonich and Hites, 1994). Values of A for several PAHs are listed in Table 3.19.

Several investigators have shown that just as K_{AP} is correlated to $P_{s(P)}^*$, so can we relate K_{VA} to $P_{s(P)}^*$. A better correlating parameter is K_{oa}, the octanol-water partition constant. A few of these correlations are shown in Table 3.19 for PCBs.

Whereas Simonich and Hites (1994) attributed all of the seasonal variations in K_{VA} to the temperature variations, Komp and McLachlan (1997) have argued that this is due more to a combination of effects such as growth dilution, decrease in dry matter content of leaves in late autumn, and the erosion of HOCs from the surface of vegetation. These preliminary studies have shown the difficulties in understanding the complex interactions of HOCs with the plant/air system.

Example 3.26   Estimation of Uptake of Atmospheric Pollutants by Vegetation

Estimate the equilibrium concentration of pyrene on a sugar maple leaf with a lipid content of 0.016 g g^{-1}. The air concentration is 10 ng m^{-3}.

From Table 3.19, at 298 K, $\ln K_{VA} = (10,227/298) − 35.95 = −1.63$. Hence, $K_{VA} = 0.196$ m^3 mg^{-1} lipid. $W_v = K_{VA}\varphi_L C_i^a = (0.196)(16)(10) = 31$ ng g^{-1}.

3.8   ADSORPTION ON ACTIVATED CARBON
FOR WASTEWATER TREATMENT

Activated carbon has a high capacity to adsorb organic compounds from both gas and liquid streams. It is perhaps the earliest known sorbent used primarily to remove color and odor from wastewater. A large number of organic compounds have been investigated in relation to their affinity toward activated carbon (Dobbs and Cohen, 1980). The data were correlated to a Freundlich isotherm and the constant K_F and n determined. Table 3.20 is for a select number of organic compounds.

Activated carbon is available both in powdered and granular forms. Powdered form can be sieved through a 100 mesh sieve. The granular form is designated 12/20, 20/40, or 8/30. 12/20 means it will pass through a standard mesh size 12 screen but will not pass through a 20 size screen. The granular form is less expensive and is more easily regenerated and hence is the choice in most wastewater treatment plants.

Activated carbon treatment is accomplished either in a continuous mode by flowing water over a packed bed of carbon (see Figure 3.12a) or in batch (fill-and-draw) mode where a given amount of activated carbon is kept in contact with a given volume of water for a specified period of time (see Figure 3.12b). In the continuous process, the exit concentration slowly reaches the inlet concentration when the adsorption capacity of carbon is exceeded. The bed is subsequently regenerated or replaced.

TABLE 3.20
K_F and n for Selected Organic Compounds
on Granular Activated Carbon

Compound	K_F	1/n
Benzene	1.0	1.6
Toluene	26.1	0.44
Ethylbenzene	53	0.79
Chlorobenzene	91	0.99
Chloroform	2.6	0.73
Carbon tetrachloride	11.1	0.83
1,2-Dichloroethane	3.5	0.83
Trichloroethylene	28.0	0.62
Tetrachloroethylene	50.8	0.56
Aldrin	651	0.92
Hexachlorobenzene	450	0.60

Source: Data from Dobbs, R.A. and Cohen, J.M., Carbon adsorption
            isotherms for toxic organics, Report No.: EPA-600/8-80-023,
            U.S. Environmental Protection Agency, Cincinnati, OH, 1980.
Note: The carbon used was Filtrasorb 300 (Calgon Corporation).

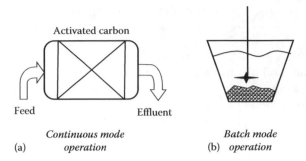

	Activated carbon	
Feed		Effluent

Continuous mode
(a)   *operation*

Batch mode
(b) *operation*

FIGURE 3.12   (a) The continuous-mode fixed-bed operation of activated carbon adsorption for wastewater treatment. Water is passed over a bed of carbon. (b) A batch mode of operation wherein a fixed weight of carbon is kept in contact with a given volume of aqueous phase. After treatment the water is replaced. This is also called the fill-and-draw or cyclic, fixed-bed batch operation. Once the carbon is exhausted, it is also replaced with a fresh batch.

Example 3.27   Activated Carbon for Treating a Wastewater Stream

At an industrial site in North Baton Rouge, Louisiana, it is proposed that the contaminated groundwater be pumped up to the surface and taken through an activated carbon unit before being discharged into a nearby lagoon. This is generally referred to as "pump-and-treat" (P & T) technology. The primary compound in the groundwater is hexachlorobutadiene (HxBD) at a concentration of 1000 μg L^{-1}. The production rate of groundwater is 200 gal min^{-1} for 10 h of operation. It is desired to achieve an effluent concentration of HxBD below the wastewater discharge limit of 27 μg L^{-1}. A batch adsorption system is contemplated to ease the load on the downstream incinerator that is planned for further destruction of the organics in the effluent air before discharge. Estimate (1) the carbon dosage required to achieve the desired level of effluent quality, (2) the amount of HxBD removed per day, and (3) the mass and volume of carbon required per day. The carbon bulk density is 20 lb ft^{-3}.

A standard batch shaker flask experiment conducted gave the following adsorption data for HxBD on a granular activated carbon (Dobbs and Cohen, 1980):

C_{iw} (mg L^{-1})	W_i (mg g^{-1} carbon)
0.098	93.8
0.027	50.6
0.013	34.2
0.007	25.8
0.002	17.3

Plot the isotherm data as log W_i versus C_{iw} to fit a Freundlich isotherm (Figure 3.13). The values of K_F and n are 245 and 0.44, respectively (correlation coefficient is 0.989).

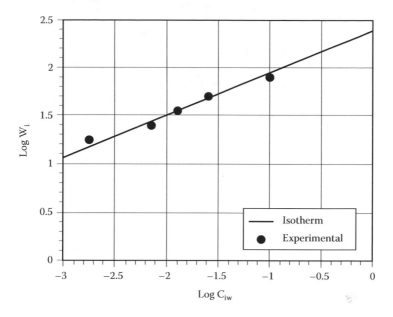

FIGURE 3.13 Freundlich isotherm fit for the adsorption of hexachlorobutadiene from water on to granular-activated carbon.

1. Using the Freundlich isotherm determine the adsorption capacity of the carbon for the required effluent concentration, viz., 27 µg L^{-1}. This gives w_i = 245 (0.027)$^{0.44}$ = 50 mg HxBD/g carbon. Notice that fortuitously this is one of the experimental points in the batch shaker flask experiments. Thus, for every 0.05 g of HxBD removed we will need 1 g carbon.
2. Amount of HxBD to be removed in one day of operation is (1 − 0.027) mg L^{-1} (200) gal min^{-1} (3.785) L gal^{-1} (1.44 × 10^3) min d^{-1} (2.2.5 × 10^{-3}) lb g^{-1} (0.001) g mg^{-1} = 2.34 lb d^{-1}. Hence, the mass of carbon required per day is 2.34/0.05 = 47 lb d^{-1}.
3. The total volume of carbon required is 47/20 = 2.3 ft^3 d^{-1}.

PROBLEMS

3.1$_2$ Consider a total evaluative environment of volume 10 × 10^{10} m^3 with volume fractions distributed as follows: air 99.88%, water 0.11%, soil 0.00075%, sediment 0.00035%, and biota 0.00005%. Consider two compounds benzene and DDT (a pesticide) with properties as follows:

Property	Benzene	DDT
K''_{AW} (Pa m^3 mol^{-1})	557	2.28
K_{sw} (L kg^{-1})	1.1	12,700
K_{bw} (L kg^{-1})	6.7	77,400

The temperature is 310 K and the densities of soil and sediment are $1500\,kg\,m^{-3}$ while that of the biota is $1000\,kg\,m^{-3}$. Determine the concentrations and mass of each compound in the various compartments using the Fugacity Level I model, assuming 1000 mol of each was released. Comment on your results.

3.2$_2$ Confined disposal facilities (CDFs) are storage facilities for contaminated sediments. Being exposed to air, the contaminated sediment is a source of volatile organic compounds to the atmosphere. Consider a two-compartment evaluative environment of volume $5 \times 10^5\,m^3$, with the following volume fractions: air 60% and sediment 40%. Consider benzene for which $K'_{AW} = 557\ Pa\,m^3\ mol^{-1}$ and $K_{sw} = 5\ L\ kg^{-1}$. If the temperature is 298 K, and the density of sediment is $1200\ kg\ m^{-3}$, determine the mass fraction of benzene in air using Fugacity Level I model.

3.3$_3$ Mercury exists in the environment in a variety of forms. Organic mercury exists predominantly as dimethyl mercury $(CH_3)_2Hg$, which is easily assimilated by the biota. Hence, fish in lakes of several states in the United States are known to contain high concentrations of Hg. Dimethyl mercury has the following properties: $K_{AW} = 0.31$ (molar concentration ratio), $K_{sw} = 7.5\ L\ kg^{-1}$, $K_{bw} = 10\ L\ kg^{-1}$. Consider an evaluative environment of $10^9\ m^3$ of air in contact with $10^5\ m^3$ of water in a lake containing $10^3\ m^3$ of sediment and $10\ m^3$ of biota. The sediment density is $1000\ kg\ m^{-3}$ and that of biota is $1500\ kg\ m^{-3}$.
1. What is the fraction accumulated by the biota in the lake? Use Fugacity Level I model.
2. If an influent feed rate of $100\ m^3\ h^{-1}$ of air with a concentration of $0.1\ mol\ m^{-3}$ is assumed, calculate the fraction assimilated by the biota using Fugacity Level II model.

3.4$_3$ Dimethylsiloxane (DMS) is a prevalent compound in various soaps, shampoos, and shaving creams. It is expected to pose problems when it gets into the natural environment through sewage and eventually distributes to all compartments. Consider a $100\ m^2$ area of sewage treatment unit that contains 10 m height of atmosphere above 1 m of a water column at 300 K. The water contains 5% by volume of solids and 10% by volume of bioorganisms. Consider the following properties: Henry's constant $= 125\ Pa\,m^3\,mol^{-1}$, $\log K_{ow} = 5.1$, $\log K_{sw} = 2.9$, $\log K_{bw} = 3.1$. The solids density is $0.015\ kg\ L^{-1}$ and the bioorganism density is $1\ kg\ L^{-1}$. Determine which compartment will have the highest amount of DMS?

3.5$_3$ During the flooding of residential areas of New Orleans by Hurricane Katrina on August 29, 2005, well-mixed floodwaters containing suspended Lake Pontchartrain sediments entered the homes and businesses of New Orleans. After the floodwaters stagnated, sediment settled out of the water column and was deposited on floors and other interiors of the homes, where it remained till the floodwaters were pumped out of the city 2 weeks later. Work by Ashley et al. (2007) provided evidence that sediment deposited inside the homes had contaminant concentrations much larger than that deposited outside. One such example was chlordane, a common organochlorine pesticide, detected in the sediment inside the homes. A concern arises about the potential for this species to volatilize from the sediment and be found inside the vapor phase of the homes. Additionally high concentrations of mold spores represent another sink for these particles from the gas phase, and returning residents and first responders can inhale both the air inside the homes and the aerosolized mold spores. Using Fugacity Level I model determine the concentrations of chlordane present in both the gas phase (μg m^{-3}) and the aerosol phase (mg kg^{-1} of mold) inside the homes. For the aerosol phase the fugacity capacity can be obtained from $Z_Q = 10^{(\log K_{oa} + \log f_{om} - 11.91)} Z_{A\rho Q} \times 10^9$. Note that log $K_{oa} = 8.872$, $f_{om} = 0.60$, $\rho_Q = 5.31 \times 10^{-18}$ kg m^{-3}. Chlordane properties are as follows: molecular weight = 409.8, $K'_{AW} = 4.91$ Pa m^3 mol^{-1}, $K_{sw} = 4265$ L kg^{-1}. Sediment density is 2.5 g cm^{-3}. Volumes of air, aerosol, and sediment are 401.6 m^3, 6.7×10^{-8} m^3, and 0.329 m^3, respectively. (*Note*: This problem was provided by Nicholas Ashley)

3.6$_2$ Siloxanes are used in the manufacture of a number of household and industrial products for everyday use. It is therefore of importance to know its distribution in the environment. Consider octamethyl cyclotetrasiloxane (D4), whose properties are as follows: $K_{AW} = 0.50$, log $K_{ow} = 5.09$, log $K_{sw} = 2.87$. For an evaluative environment of total volume 10^9 m^3, which is 50% air, 40% water, and 10% sediment, which compartment will have the highest fraction of D4? Use Fugacity Level I model.

3.7$_3$ Consider the flooding of New Orleans by Hurricane Katrina. The floodwater from Lake Pontchartrain extended over a square area of 10 miles × 10 miles to a depth of 3 m. Assume that the sediment from the lake formed a thin layer under the floodwater extending over the entire city to a depth of 1 ft. The atmosphere above the city extends to a completely mixed depth of 1 km. The temperature is 300 K. Gasoline components (benzene, toluene, ethylbenzene, and xylene) were observed in the floodwaters and originated from submerged vehicles and gasoline pumping stations. Consider the component ethylbenzene. Apply Fugacity Level I model to obtain the fraction of ethylbenzene present in the floodwater. Henry's constant for ethylbenzene is 851 Pa m^3 mol^{-1}, sediment-water partition constant is 1.4×10^4 L kg^{-1}, sediment density is 0.015 kg L^{-1}.

3.8$_3$ Wetlands can be useful in phytoremediation of contaminants through uptake of pollutants by plants. A wetland consists of soil, plants, and water in equilibrium with the air above. Consider an evaluative environment with the following volumes: air = 10^7 m^3, water = 10^5 m^3, soil = 10^3 m^3, plants = 10^3 m^3. Consider a chlorinated compound, trichloroethane, with the following properties: Henry's constant = 3090 kP L mol^{-1}, soil-water partition constant = 15 L kg^{-1}, Plant-water partition constant = 300 L kg^{-1}, soil density = 0.01 kg L^{-1}, plant density = 0.1 kg L^{-1}. Evaluate the fraction of pollutants in the plants.

3.9$_2$ Estimate Henry's constants for the following compounds using (1) the bond contribution method of Hine and Mookerjee and (2) the group contribution scheme of Meylan and Howard: ammonia, toluene, 1,2,4-trichlorobenzene, benzyl alcohol, pentachlorophenol, 1,1,1-trichloroethane.

Bond	Log K'_{AW} (Hine and Mookerjee)
C–H	−0.11
C–Cl	0.30
C_{ar}–H	−0.21
C_{ar}–Cl	−0.11
C_t–Cl	0.64
C_t–H	0.00
O–H	3.21
N–H	1.34
C_{ar}–C_{ar}	0.33
C–O	1.00
C–C	0.04

Note that C_{ar} denotes an aromatic carbon and C_t denotes a tertiary C atom.

3.10_3 The gas phase concentration of hydrogen peroxide and its Henry's constant are of critical significance in obtaining its concentration in cloud droplets. H_2O_2 is an abundant oxidant in the atmosphere capable of oxidizing S(IV) in atmospheric water and is thought to be the dominant mechanism of converting SO_2 from the atmosphere into a stable condensed phase. H_2O_2 has been observed in rainwater even in remote maritime regions.

The following data were reported by Hwang and Dasgupta (1985) for the equilibrium between air and water for hydrogen peroxide at various temperatures:

At 10°C		At 20°C		At 30°C	
P_i^* (atm)	C_i^w (mg L⁻¹)	P_i^* (atm)	C_i^w (mg L⁻¹)	P_i^* (atm)	C_i^w (mg L⁻¹)
2.8×10^{-7}	0.08	1.1×10^{-6}	0.05	8×10^{-7}	0.104
3.5×10^{-7}	0.10	2.2×10^{-6}	0.10	1×10^{-8}	1.3×10^{-3}
4.4×10^{-7}	0.12	2.8×10^{-6}	0.12	5×10^{-9}	6.5×10^{-4}
				1×10^{-9}	1.6×10^{-4}

1. Obtain the K_{AW} for hydrogen peroxide as a function of temperature.
2. Estimate the heat of volatilization of H_2O_2 from water. Clearly state any assumptions made.

3.11_2 1. The heat evolved upon dissolution of atmospheric ozone into water is determined to +5 kcal mol⁻¹ at 298 K. Given that ozone has a Henry's constant of 9.4×10^{-3} mol L⁻¹ atm⁻¹ at 298 K, determine its Henry's constant at 273 K. Note that the definition of Henry's constant is for the process air to water.

2. If the atmospheric ozone concentration in a polluted air is 100 ppbv, what will be the concentration in equilibrium with a body of water in that region at 298 K? Note that you need to assume no losses of ozone other than absorption into water.

3.12_2 The following solubility data was obtained for an organic compound (toluene) in the presence of humic acid extracted from a wastewater sample:

Humic Acid Concentration (mg of Organic C L^{-1})	Solubility of Toluene (mg L^{-1})
0	476
8	488
14	502
25	504
39	517

Estimate the K_C for toluene on the humic acid.

It was determined that the concentration of humic acid in the wastewater pond was 20 g m^{-3}. If the pure water Henry's constant for toluene is 0.0065 atm m^{-3} mol^{-1}, estimate the partial pressure of toluene in the air in equilibrium with the wastewater.

3.13_2 Consider 2 L of a 2 mM solution of NH_4OH in a closed vessel brought to equilibrium with 100 mL of nitrogen. If the final pH of the aqueous solution is 8, what is the mass of ammonia in the gas phase at equilibrium? The temperature is 298 K. $K_{AW} = 0.016$ atm mol^{-1} L^{-1}. $K_{a1} = 1.7 \times 10^{-5}$ M. Note that you have to derive the equation for the aqueous mole fraction $\alpha NH_3 \cdot H_2O$.

3.14_3 Calculate the pH of an aqueous solution that is in equilibrium with 100 ppmv of CO_2 in the gas phase. The liquid-to-gas volume is 100:1. The temperature is 298 K. The necessary parameters are given in the text.

3.15_1 100 mL of an aqueous solution containing 0.0004 mM n-nonane is placed in a closed cylindrical vessel of inner diameter 5 cm and total height 10 cm. What fraction of n-nonane will be present at the air-water interface in this system? If the total height of the vessel is reduced to 5.5 cm what will be the adsorbed fraction? Assume that n-nonane does not adsorb to the vessel wall.

3.16_2 Chlorofluorocarbons (CFCs) form an important class of atmospheric pollutants implicated in the formation of ozone hole above the polar regions. These are no longer used as a result of the worldwide agreement to ban their production and utilization. One most common CFC was CCl_3F (Freon-11) used as refrigerants. If the dry air concentration of Freon-11 is 10×10^{-6} ppmv, determine its concentration in atmospheric moisture as a function of water contents ranging from 10^{-9} to 10^{-4} in a closed system. K_{AW} for Freon-11 is 5.

3.17_2 The surface area of an average aerosol in Baton Rouge, LA, is 5×10^{-4} m^2 m^{-3}. Determine the fraction of a polyaromatic hydrocarbon (phenanthrene) adsorbed on the aerosol. Assume that the value of $\rho_i = 0.10$ in Junge's equation. The average temperature during the collection of the aerosols was 73°F on April 16, 1993. The average precipitation was 0.46 in. of water. The average particulate phase concentration of phenanthrene measured on that day was 0.21 ng m^{-3}. The air/particulate partition constant for phenanthrene was related to temperature in Kelvins through the equation log $K_{AP} = -3423/T + 19$. Determine the flux of phenanthrene in μg m^{-2} d^{-1}.

3.18_2 The concentration of a pesticide (chlorpyrifos) measured in air from a foggy atmosphere in Parlier, CA, was 17.2 $ng\,m^{-3}$ at about 25°C. Henry's constant for chlorpyrifos at 25°C is 170×10^{-6} and the air-water partition constant is 6.4×10^{-4} at 12.5°C. The average fog diameter is 1 μm. If the fog deposition rate is 0.2 $mm\,h^{-1}$, and the fog lasts for 4 h, what is the surface flux of chlorpyrifos?

3.19_2 Liss and Slater (1974) determined that over the Atlantic Ocean the mean atmospheric concentration of a chlorofluorocarbon (CCl_3F, Freon-11, molecular weight 137) is 50×10^{-6} ppmv, whereas the mean aqueous concentration is 7.76×10^{-12} $cm^3\,mL^{-1}$ water. The overall aqueous phase mass transfer coefficient was estimated as 11 $cm\,h^{-1}$. Determine the magnitude and direction of freon flux? If the ocean surface area is 3.6×10^{14} m^2, and the total annual flux of freon is 3×10^{11} g, what fraction is contributed by the Atlantic Ocean?

3.20_3 1. The aqueous solubility of hexachlorobenzene is 5 μg L^{-1}. Estimate its octanol-water partition constant. Use this information alone to calculate the aqueous concentration that will be present in a sediment suspension containing 100 mg of a sediment with an organic carbon fraction of 0.02.

     2. If the equilibrium concentration is equal to the saturation aqueous solubility, what will be the sediment concentration?

     3. If the aqueous solution in (1) above also contains 0.05 mol fraction of ethanol, what will be the value of the sediment-water partition constant?

3.21_2 The solubility of pyrene in an aqueous solution of polystyrene latex colloids is given here:

[Latex] (mg L^{-1})	[Pyrene] (mg L^{-1})
0	90
5	340
10	440
15	650
20	880

Obtain the value of the partition constant for pyrene on latex colloids. Assume that latex is 100% organic carbon and estimate the partition constant from all reasonable correlations. Compare your results and comment.

3.22_2 In a field study, a nonadsorbing (conservative) tracer (chloride) was injected into a groundwater aquifer (unconfined and relatively homogeneous) along with two adsorbing organic compounds (carbon tetrachloride and tetrachloroethylene). A test well 5 m downstream from the injection well was monitored for all three compounds. The average time of arrival (breakthrough time) at the monitoring well was 66 days for chloride ion, 120 days for carbontetrachloride, and 217 days for tetrachloroethylene.

     1. Obtain from the above data the retardation factors for the sorbing compounds.

     2. Obtain the field-determined partition constants. The soil organic carbon is 0.02.

     3. Compare with K_d obtained using the correlation of Curtis et al. (1986) given in text. Explain any difference.

3.23_3 The bioconcentration factor can be predicted solely from the aqueous compatibility factor (measured as solubility). Derive an appropriate relationship between the two factors. How does it compare with the predictions using K_{ow}? Use pyrene and benzene as examples. For a fish that weighs 1.5 lb, how much pyrene can be accumulated within its tissue?

3.24_2 It is desired to remove benzene from an aqueous waste stream at an industrial waste site, which produces approximately 20,000 gal of water per day with a pH of 5.3 and a benzene concentration of 2000 μg L^{-1}. A batch adsorption system is to be used. A preliminary bench scale analysis was conducted to get the adsorption data of benzene on two types of carbons (I and II) at a pH of 5.3. The data are given here:

Carbon Dose (mg L^{-1})	Initial Aq. Concn. (mg L^{-1})	Final Eqm. Conc. in Water (mg L^{-1})	
		I	II
96	19.8	14.2	12.5
191	19.8	10.3	12.0
573	19.8	6.2	11.0
951	19.8	5.8	9.0
1910	19.8	4.0	5.0

1. Determine which isotherm (Freundlich or Langmuir) would better represent the data?
2. What is the maximum adsorption capacity of carbon I and II?
3. If the effluent water quality for benzene is 20 μg L^{-1}, how much of carbon would be required per day? Which carbon type would you recommend?

3.25_2 The sediment in University Lake near Baton Rouge is known to be contaminated with PCBs (predominantly trichlorobiphenyl) to an average concentration of 10 μg kg^{-1}. The sediment has an f_{oc} of 0.04. Calculate the PCB loading in a one pound catfish dwelling in the lake. Assume equilibrium between the various phases in the system.

3.26_2 Inorganic Hg ($HgCl_2$, $Hg(OH)_2$) can be deposited to a water body via wet and dry deposition. Consider the Henderson Lake, a small lake east of Lafayette, LA, where a Hg advisory was issued by the Louisiana Department of Health and Hospitals in 1996. The lake is approximately 7.2 m in depth. Consider a global average atmospheric concentration of $HgCl_2 = 0.1$ ng m^{-3}. Estimate the deposition rate of $HgCl_2$ to the lake. An average annual rainfall of 0.1 cm h^{-1} can be assumed. K_{AW} for $HgCl_2$ is 2.9×10^{-8}.

3.27_2 Estimate the fractional mass of a pesticide (aldrin) that is washed out by rainfall at an intensity of 1 mm h^{-1} over an area 10^4 m^2. Appendix A gives the physicochemical data for aldrin.

3.28_3 A closed storage vessel is filled with water and ethanol (mole fraction 0.002). The vessel is only 60% full. If equilibrium is established above the solution, what is the partial pressure of ethanol in the vapor? Does the vessel stand to explode? The OSHA explosion limit for ethanol is 1000 ppmv.

3.29_3 Vapor, aerosol, and rainwater concentrations of a PCB (2,2′,4-isomer) in a urban area are given here (Duinker and Bouchertall, 1989). Vapor: 100 pg m^{-3}; aerosol: 0.3 pg m^{-3}.

Estimate the total washout ratio for this compound at ambient temperature of 25°C. Log $K_{ow} = 5.7$, log $P_i^*(atm) = -6.5$. Compare the theoretical and experimental washout ratios. Average surface area of an aerosol is 1×10^{-6} cm^2 cm^{-3} air and $\rho_i = 1.7 \times 10^{-4}$ atm cm.

3.30_2 Concentration of sulfate in the atmosphere over oceans is -0.06 μg m^{-3}. If the water content of cloud is 1 g m^{-3}, estimate the pH of the cloud water. Assume that all of the sulfate exist as sulfuric acid.

3.31_2 Based on the data in Table 4.12, estimate the percent gas phase resistance for the evaporation of the following chemicals from water: naphthalene, chloroform, PCB, ammonia.

3.32_3 Below the water table, the groundwater aquifer is a closed system with respect to exchange of CO_2 with a separate vapor phase. Without an external source or sink for dissolved CO_2 in groundwater (i.e., no carbonate), the dissolved carbon concentration should be constant. Assuming that the groundwater at the water table is in equilibrium with the atmosphere with CO_2 at $10^{-3.5}$ atm partial pressure, construct a plot showing the change in CO_2 with pH in the closed system.

3.33_2 What will be the vapor pressure of aqueous solution droplets of the following diameters: 0.01, 0.05, 0.1, 0.5, and 1 μm. The solution contains 1×10^{-13} g of NaCl per particle at 298 K. Plot the value of P_w/P_w^* as a function of droplet size.

3.34_3 PSBs are ubiquitous in the environment. The average air concentration in 1986 over Lake Superior for a PCB congener (2,3′,4,4′-tetrachlorobiphenyl) was measured as 38.7 pg m^{-3} and that in the surface water was 32.1 pg P^{-1}. Obtain the magnitude and direction of air-water exchange flux. Estimate the transfer coefficient using data in Table 4.12 and Appendix A.

3.35_2 Phthalate esters are used as plasticizers. Di-n-butyl phthalate (DBP) is found both in marine waters and the atmosphere. The mean concentrations in the Gulf of Mexico are 94 ng P^{-1} in the surface water and 0.3 ng m^{-3} in the air. Estimate the direction and magnitude of annual flux of DBP. The reported vapor pressure and aqueous solubility of DBP at 25°C are 1.4×10^{-5} mm Hg and 3.25 mg P^{-1}, respectively (Giam et al., 1978).

3.36_2 A facility operates a surface impoundment for oil field–produced water that contains benzene. The permitted annual emission is 6000 kg. The facility operator makes sure that the impoundment is operated only during those atmospheric conditions of instability when there is maximum mixing and dispersal of benzene. Estimate the mean aqueous concentration of benzene allowed in the impoundment. The area of the impoundment is 10^4 m^2.

3.37_2 Estimate the vapor pressure of the following compounds at 25°C:

Compound	Melting Point (°C)
Chlordane	103
CCl_4	−23
p,p = -DDT	108
Diethyl phthalate	−40.5
2-Methyl phenol	30.9

3.38_2 A sample of silica is coated with 0.1% of organic carbon. The silica has a surface area of $20\ m^2\ g^{-1}$. Assume that only 10% of the area is occupied by organic carbon. Estimate the adsorption constant for pyrene on modified silica. Use data from Table 4.17 for the mineral contribution to adsorption.

3.39_2 1. Estimate the sorption constant of the following compounds on a soil that has an organic carbon content of 1%: pentachlorobenzene, phenanthrene, endrin, 2,2'-dichlorobiphenyl, parathion.

     2. Estimate the effect of (a) 10 mg P^{-1} of DOC in water and (b) 0.1% methanol in water on the partition constant of phenanthrene on the above soil.

3.40_2 The following data is for the adsorption of a cation dodecylpyridinium bromide on Borden sand:

Log W_i (mol kg^{-1})	Log C_i^w (mol P^{-1})
−4.8	−7.0
−4.6	−6.5
−4.3	−6.0
−4.1	−5.5
−3.8	−5.0
−3.4	−4.5
−3.1	−4.0

Determine which isotherm (linear or Freundlich) best fits the data.

3.41_2 Contaminated sediment from a harbor is proposed to be dredged and stored in an open confined disposal facility (CDF). The sediment concentrations of Aroclor 1242 and 1254 are 687 and 446 mg kg^{-1}, respectively. The sediment suspension in water is expected to be highest immediately after loading the CDF. The highest suspended sediment concentration expected using the selected dredged head is 490 mg P^{-1}. The area of the CDF is $1.3 \times 10^5\ m^2$. What is the expected air emission of Aroclor 1242. Properties for Aroclor 1242 and 1254 should be obtained from literature.

3.42_2 On April 17, 1992, a high volume sampler was used to obtain the following atmospheric data in Baton Rouge, LA: average air temperature = 71°F, concentration of fluoranthene on particulates in air = 0.51 ng m^{-3}, particulate concentration = 100 μg m^{-3}. Estimate the dry (vapor) deposition rate of fluoranthene. A deposition velocity of 0.1 m s^{-1} can be assumed.

3.43_2 The mean concentration of formaldehyde in air is 0.4 ppb, while that in seawater is 40 nM. What is the direction and magnitude of flux? The partition constant for formaldehyde between seawater and air can be estimated from $-\log K'_{AW} = -6.7 + 3069/T$, where K'_{AW} is in mol $(L\,atm)^{-1}$.

3.44_2 What is the fraction of each of the following compounds partitioned from the atmosphere to a sugar maple leaf. Assume an ambient temperature of 25°C and a typical lipid content of 0.01 g g^{-1}. The air concentration of each compound can be assumed to be 10 ng m^{-3}: naphthalene, phenanthrene, anthracene, and pyrene. Plot your results versus log K_{oa} and comment.

3.45_3 Sodium azide (NaN_3) is used as a component of air bags in automobiles ($= 10^6$ kg in 1995). When dissolved in water it gives rise to volatile hydrazoic acid, HN_3. It is a weak acid and the reaction $HN_3 \rightleftharpoons H^+ + N_3^-$ has a pK_a of 4.65 at 25°C. The intrinsic Henry's constant K_{AW} for the neutral species is 0.0034 at 25°C (Betterton and Robinson, 1997).

1. What will be the apparent air/water partition constant at a pH of 7?
2. The enthalpy of solution is -31 kJ mol^{-1}. Calculate the air/water partition constant at 4°C.
3. In wastewater at 25°C, if the azide concentration is 0.1 mM at pH of 6.5, will the gas phase HN_3 exceed the threshold limit value (TLV) of 0.1 ppmv?

3.46_2 A spill of 1,2-dichloroethane (DCA) occurred in a chemical plant producing pesticides. Because of the delay in cleanup, some of the material seeped into the groundwater flowing at a velocity of 1.5 m d^{-1}. The soil at the site has an organic carbon content of 0.001, a porosity of 0.3, and bulk density of 1.4 g cm^{-3}. The nearest community that receives drinking water from the aquifer is located 500 m away in the direction of groundwater flow. How long will it take for the chemical to show up in the community drinking water? Assume no biodegradation of DCA.

3.47_3 K_{AW} for p,p'-DDT (a pesticide) at various temperatures were obtained for deionized water and saline water (Cetin et al., 2006). From the data calculate the following: (1) $\Delta H^0_{a \rightarrow w}$, (2) $\Delta S^0_{a \rightarrow w}$, and (3) Setschenow constant for salting out.

	K_{AW} (−)	
T (°C)	Deionized Water	Saline Water (0.5 M NaCl)
5	0.19 ± 0.04	1.17
15	0.52	1.9 ± 0.63
20	0.69 ± 0.15	—
25	0.97 ± 0.27	3.4 ± 0.64
35	2.8 ± 0.39	5.5

3.48_2 Hexabromobenzene (HBB) is found along with flame-retardant materials such as in carpets, bedding, insulation, and furniture accessories. It is found as a contaminant in soils and sediments in our waterways. It has a structural formula of C_6Br_6, molecular weight of 557.52, and a melting point of 327°C. (1) Determine the log K_{ow} of HBB using a group contribution method and a derivative method; (2) from the data in (1) determine the aqueous solubility of BHH expressed in $mol\,L^{-1}$; (3) from the above data determine the sediment-water partition constant K_{sw} using an appropriate correlation. The organic fraction of the soil is 0.05, neglect the mineral contribution to adsorption. (4) 10 g of soil with 10 $\mu g\,kg^{-1}$ of HBB is mixed with 100 mL of water and agitated for 24 h. The solution was allowed to reach equilibrium and the solid fraction separated. It was found to contain HBB at a concentration of 5 $\mu g\,kg^{-1}$. What will be the concentration of HBB in the water that was separated?

3.49_3 2,3-dichlorophenol is a contaminant resulting from wood and pulp industry operations. It has a molecular weight of 163, melting point of 45°C, and boiling point of 210°C. (1) Obtain the octanol-water partition constant of the compound; (2) from (1) obtain the aqueous solubility of the compound; (3) obtain the vapor pressure of the compound from an appropriate correlation; (4) using data from (2) and (3) obtain Henry's constant, expressed in dimensionless molar ratio. (5) A wastewater stream of 100 m^2 with a depth of 1 m is contaminated with the compound at a concentration of 1 $\mu g\,L^{-1}$. A measurement of flux to air showed 10 μg in a chemical trap obtained in 1 h of sampling. What is the concentration of the chemical in ambient air above the stream? Note that the individual water-side and air-side mass transfer coefficients for the chemical are $1 \times 10^{-5}\,m\,s^{-1}$ and $0.005\,m\,s^{-1}$, respectively.

3.50_2 Consider 100 m^3 of lake water in equilibrium with 100 m^3 of air. A 1 lb catfish resides in the lake water. The air above the lake is found to have traces of a pesticide (lindane) at a concentration of 1 $\mu g\,m^{-3}$. Determine the concentration of the pesticide in the catfish.

3.51_3 Ethylbenzene is an aromatic hydrocarbon that is commonly found in contaminated sediments and waters. Its molecular weight is 106, log K_{ow} is 3.15, vapor pressure is 1.3 kPa at 25°C, aqueous solubility is 0.0016 $mol\,L^{-1}$ at 25°C, enthalpy of vaporization is 2.1 $kJ\,mol^{-1}$, and enthalpy of solution is 46.2 $kJ\,mol^{-1}$. (1) Find its vapor pressure at 30°C; (2) find its aqueous solubility at 30°C; (3) from the above data find its Henry's constant at 30°C and express in $kPa\,L^{-1}\,mol^{-1}$. (4) If the air above a lagoon has ethylbenzene at a partial pressure of 0.1 kPa at 30°C in equilibrium with the water, what is the aqueous concentration at that temperature? (5) If the water in the lagoon is overlying the sediment in equilibrium with it at 30°C, and if the sediment organic carbon fraction is 0.02, what is the concentration of ethylbenzene in the sediment (answer in $mg\,kg^{-1}$)?

REFERENCES

Acha, R.-C. and Rebhun, M. (1992) Biding of organic solutes to dissolved humic substances and its effects on adsorption and transport in the aquatic environment. *Water Research* **26**, 1645–1654.

Ashely, N.A., Valsaraj, K.T., and Thibodeaux, L.J. (2007) Elevated in-home sediment contaminant concentrations: The consequences of a particle settling-winnowing process from Hurricane Katrina floodwaters. *Chemospehre* **70**, 833–840.

Ashworth, R.A., Howe, G.B., Mullins, M.E., and Rogers, T.N. (1988) Air-water partitioning coefficients of organics in dilute aqueous solutions. *Journal of Hazardous Materials* **18**, 25–36.

Backhus, D.A. (1990) Colloids in groundwater—Laboratory and field studies of their influence on hydrophobic organic contaminants, PhD dissertation, MIT, Cambridge, MA.

Baker, J.T. (1997) *Atmospheric Deposition of Contaminants to the Great Lakes and Coastal Waters*. Pensacola, FL: SETAC Press.

Betterton, E.A. and Robinson, J.R. (1997) Henry's law constant of hydrazoic acid. *Journal of the Air & Waste Management Association* **47**, 1216–1219.

Bidleman, T.F. (1988) Atmospheric processes: Wet and dry deposition of organic compounds controlled by their vapor-particle partitioning. *Environmental Science and Technology* **22**, 361–367.

Bidleman, T.F. and Foreman, W.T. (1987) Vapor-particle partitioning of semi-volatile organic compounds. In: Eisenreich, S.E. (ed.), *Sources and Fates of Aquatic Pollutants*, pp. 27–56. Washington, DC: ACS.

Boehm, P.D. and Quinn, J.G. (1973) Solubilization of hydrocarbons by the dissolved organic matter in sea water. *Geochimica et Cosmochimica Acta* **37**, 2459–2477.

Briggs, G.G. (1973) *Proceedings of the Seventh British Insecticide Fungicide Conference*, Vol. 11, pp. 475–478.

Broeker, W.S. and Peng, T.H. (1974) Gas exchange rates between air and the sea. *Tellus* **26**, 21–35.

Brownawell, J.W. (1986) Role of colloidal organic matter in the marine gro-chemistry of PCBs, PhD dissertation, MIT, Massachusetts, MA.

Callaway, J.Y., Gabbita, K.V., and Vilker, V.L. (1984) Reduction of low molecular weight halocarbons in the vapor phase above concentrated humic acid solutions. *Environmental Science and Technology* **18**, 890–893.

Capel, P.D., Leuenberger, C., and Giger, W. (1991) Hydrophobic organic chemicals in urban fog. *Atmospheric Environment* **25A**, 1335–1346.

Carter, C.W. and Suffet, S.J. (1982) Binding of DDT to dissolved humic materials. *Environmental Science and Technology* **16**, 735–740.

Cetin, B., Ozer, S., Sofuoglu, A., and Odabasi, M. (2006) Determination of Henry's law constants of organochlorine pesticides in deionized and saline water as a function of temperature. *Atmospheric Environment* **40**, 4538–4546.

Chiou, C.T., Kile, D.E., Brinton, T.I., Malcolm, R.L., Leenheer, J.A., and MacCarthy, P. (1987) A comparison of water solubility enhancements of organic solutes by aquatic humic materials and commercial humic acids. *Environmental Science and Technology* **21**, 1231–1234.

Chiou, C.T., Malcolm, R.L., Brinton, T.I., and Kile, D.E. (1986) Water solubility enhancement of some organic pollutants and pesticides by dissolved humica and fulvic acids. *Environmental Science and Technology* **20**, 502–508.

Chiou, C.T., Porter, P.E., and Schmedding, D.W. (1983) Partition equilibria of nonionic organic compounds between soil organic matter and water. *Environmental Science and Technology* **17**, 227–231.

Curtis, G.P., Reinhard, M., and Roberts, P.V. (1986) Sorption of hydrophobic organic compounds by sediments. *ACS Symposium Series* **323**, 191–216.

Davidson, C.I. and Wu, Y.L. (1992) Acidic *Precipitation-Sources, Deposition and Canopy Interactions*, vol. 3. Lindberg, S.E., Page, A.L., and Norton, S.A. (Eds.), pp. 103–216. New York: Springer-Verlag.

Davis, J.A. and Gloor, R. (1981) Adsorption of dissolved organics in lake water by aluminum oxide. Effect of molecular weight. *Environmental Science and Technology* **15**, 1223–1229.

Dobbs, R.A. and Cohen, J.M. (1980) Carbon adsorption isotherms for toxic organics, Report No.: EPA-600/8-80-023. Cincinnati, OH: U.S. Environmental Protection Agency.

Drost-Hansen, W. (1965) Aqueous interfaces. methods of study and structural properties. Parts I and II. In: Ross, S. (ed.), *Chemistry and Physics of Interfaces*, pp. 13–42. Washington, DC: American Chemical Society Publications.

Duinker, J.C. and Bouchertall, F. (1989) On the distribution of atmospheric polychlorinated biphenyl congeners between vapor-phase, aerosols, and rain. *Environmental Science and Technology* **23**, 57–62.

Eckenfelder, W.W. (1989) *Industrial Water Pollution Control*, 2nd edn. New York: McGraw-Hill Pub. Co.

Eisenreich, S.J., Looney, B.B., and Thornton, J.D. (1981) Airborne organic contaminants in the Great Lakes ecosystem. *Environmental Science and Technology* **5**, 30–38.

Falconer, R.L. and Bidleman, T.F. (1997) In: Baker, J.L. (ed.), *Atmospheric Deposition of Contaminants to the Great Lakes and Coastal Waters*. Pensacola, FL: SETAC Press.

Fendinger, N.J. and Glotfelty, D.E. (1988) A laboratory method for the experimental determination of air-water Henry's law constants. *Environmental Science and Technology* **22**, 1289–1293.

Fendinger, N.J., Glotfelty, D.E., and Freeman, H.P. (1989) Comparison of two experimental techniques for determining the air/water Henry's law constants. *Environmental Science & Technology* **23**, 1528–1531.

Gauthier, T.D., Shane, E.C., Guerin, W.F., Seitz, W.R., and Grant, C.L. (1986) Fluorescence quenching method for determining the equilibrium constants for PAHs binding to dissolved humic materials. *Environmental Science and Technology* **20**, 1162–1166.

Giam, C.S., Chan, H.S., Neff, G.S., and Atlas, E.L. (1978) Phthalate ester plasticizers: A new class of marine pollutant. *Science* **119**, 419–421.

Gill, P.S., Graedel, T.E., and Wechsler, C.J. (1983) Organic films on atmospheric aerosol particles, fog droplets, cloud droplets, raindrops, and snowflakes. *Reviews of Geophysics and Space Physics* **21**, 903–920.

Glotfelty, D.E., Seiber, J.N., and Liljedahl, A. (1987) Pesticides in fog. *Nature* **325**, 602 605.

Gossett, J.M. (1987) Measurements of Henry's law constants for C1 and C2 chlorinated hydrocarbons. *Environmental Science and Technology* **21**, 202–208.

Graedel, T.E. and Crutzen, P.C. (1993) *Atmospheric Change: An Earth System Perspective*. New York: W.H. Freeman Co.

Haas, C.N. and Kaplan, B.M. (1985) Toluene-humic acid association equilibria: Isopiestic measurements. *Environmental Science and Technology* **19**, 643–645.

Hine, J. and Mookerjee, P.K. (1975) The intrinsic hydrophilic character of organic compounds: Correlations in terms of structural contributions. *Journal of Organic Chemistry* **40**, 292–298.

Hunter-Smith, R.J., Balls, P.W., and Liss, P.S. (1983). Henry's Law constants and the air-sea exchange of various low molecular weight halocarbon gases. *Tellus* **35B**, 170–176.

Hwang, H. and Dasgupta, P.K. (1985) Thermodynamics of the hydrogen peroxide-water system. *Environmental Science and Technology* **19**, 255–261.

Junge, C.E. (1977) Basic considerations about trace constituents in the atmosphere as related to the fate of global pollutants. In: Suffet, I.H. (ed.), *Fate of Pollutants in the Air and Water Environments. Part I*, pp. 7–25. New York: John Wiley & Sons, Inc.

Karickhoff, S.W., Brown, D.S., and Scott, T.A. (1979) Sorption of hydrophobic pollutants on natural sediments. *Water Research* **13**, 241–248.

Keller, C.D. and Bidleman, T.F. (1984) Collection of airborne polycyclic aromatic hydrocarbons and other organics with a glass fiber filter-polyurethane foam system. *Atmospheric Environment* **18**, 837–845.

Kenaga, E.E. and Goring, C.A.I. (1980) Relationship between water solubility, soil sorption, octanol-water partitioning, and concentration of chemicals in biota. *ASTM SPE* **707**, 78–115.

Kim, J.P. and Fitzgerald, W.F. (1986) Sea-air partitioning of mercury in the equatorial Pacific ocean. *Science* **231**, 1131–1133.

Komp, P. and McLachlan, M.S. (1997) Influence of temperature on the plant/air partitioning of semivolatile organic compounds. *Environmental Science and Technology* **31**, 886–890.

Leighton, D.T.J. and Calo, J.M. (1981) Distribution coefficients of chlorinated hydrocarbons in dilute air-water systems for groundwater contamination applications. *Journal of Chemical and Engineering Data* **26**, 382–385.

Ligocki, M.P., Leuenberger, C., and Pankow, J.F. (1985) Trace organic compounds in rain II. Gas scavenging of neutral organic compounds. *Atmospheric Environment* **19**, 1609–1617.

Liss, P.S. and Slater, P.G. (1974) Flux of gases across the air-sea interface. *Nature* **247**, 181–184.

Lyman, W.J., Reehl, W.F., and Rosenblatt, D.H. (1982) *Handbook of Chemical Property Estimation Methods*. Washington, DC: ACS Press.

Mackay, D. (1979) Finding fugacity feasible. *Environmental Science and Technology* **13**, 1218–1222.

Mackay, D. (1982) Correlation of bioconcentration factors. *Environmental Science and Technology* **16**, 274–278.

Mackay, D. (1991) *Multimedia Environmental Models*. Chelsea, MI: Lewis Publishers, Inc.

Mackay, D. and Leinonen, P.J. (1975) Rate of evaporation of low solubility contaminants from water bodies to the atmosphere. *Environmental Science and Technology* **9**, 1178–1180.

Mackay, D., Paterson, S., and Shiu, W.Y. (1992) Generic models for evaluating the regional fate of chemicals. *Chemosphere* **24**, 695–717.

Mackay, D. and Shiu, W.Y. (1981) A critical review of Henry's law constants for chemicals of environmental interest. *Journal of Physical and Chemical Reference Data* **10**, 1175–1199.

Mackay, D., Shiu, W.Y., Bobra, A., Billington, J., Chau, E., Yeun, A., Ng, C., and Szeto, F. (1982) Volatilization of organic pollutants from water, EPA Report No.: 600/3-82-019. Springfield, VA: National Technical Information Service.

Mader, B.T., Goss, K.U., and Eisenreich, S.J. (1997) Sorption of nonionic, hydrophobic organic chemicals to mineral surfaces. *Environmental Science and Technology* **31**, 1079–1086.

McCarty, P.L., Roberts, P.V., Reinhard, M., and Hopkins, G. (1992) Movement and transformation of halogenated aliphatic compounds in natural systems. In: Schnoor, J.L. (ed.), *Fate of Pesticides and Chemicals in the Environment*, pp. 191–210. New York: John Wiley & Sons, Inc.

Means, J.C., Wood, S.G., Hassett, J.J., and Banwart, W.L. (1980) Sorption of polynuclear aromatic hydrocarbons by sediments and soils. *Environmental Science and Technology* **14**, 1524–1528.

Meylan, W.M. and Howard, P.H. (1992) *Henry's Law Constant Program*. Boca Raton, FL: Lewis Publishers.

Munger, J.W., Jacob, D.J., Waldman, J.M., and Hoffmann, M.R. (1983) Fogwater chemistry in an urban atmosphere. *Journal of Geophysical Research* **88**(C9), 5109–5121.

Munz, C. and Roberts, P.V. (1986) Effects of solute concentration and cosolvents on the aqueous activity coefficient of halogenated hydrocarbons. *Environmental Science and Technology* **20**, 830–836.

Munz, C. and Roberts, P.V. (1987) Air-water phase equilibria of volatile organic solutes. *Journal of American Water Works Association* **79**, 62–70.

Murphy, E.M., Zachara, J.M., and Smith, S.C. (1990) Influence of mineral-bound humic substances on the sorption of hydrophobic organic compounds. *Environmental Science and Technology* **24**, 1507–1516.

Neely, W.B., Branson, D.R., and Blau, G.E. (1974) Partition coefficient to measure bioconcentration potential of organic chemicals in fish. *Environmental Sciences and Technology* **8**, 1113–1115.

Nicholson, B.C., Maguire, B.P., and Bursill, D.B. (1984) Henry's law constants for the trihalomethanes: Effects of water composition and temperature. *Environmental Sciences and Technology* **18**, 518–521.

Pankow, J.F. (1987) Review and comparative analysis of the theories of partitioning between the gas and aerosol particulate phases in the atmosphere. *Atmospheric Environment* **21**, 2275–2283.

Pankow, J.F., Storey, J.M.E., and Yamasaki, H. (1993) Effects of relative humidity on gas/particle partitioning of semivolatile organic compounds to urban particulate matter. *Environmental Science and Technology* **27**, 2220–2226.

Perona, M. (1992) Solubility of hydrophobic organics in aqueous droplets. *Atmospheric Environment* **26A**, 2549–2553.

Rao, P.S.C. and Davidson, J.M. (1980) Estimation of pesticide retention and transformation parameters required in nonpoint source pollution models. In: Oercash, M.R. and Davidson, J.M. (eds.), *Environmental Impact of Nonpoint Source Pollution*, pp. 23–67. Ann Arbor, MI: Ann Arbor Science Publishers.

Roberts, P.V., Schreiner, J., and Hopkins, G.D. (1982) Field study of organic water quality changes during groundwater recharge in the Palo Alto Baylands. *Water Research* **16**, 1025–1035.

Schellenberg, K., Lauenberger, C., and Schwarzenbach, R.P. (1984) Sorption of chlorinated phenols by natural and aquifer materials. *Environmental Science and Technology* **18**, 652–657.

Schwarzenbach, R.P., Gschwend, P.M., and Imboden, D.M. (1993) *Environmental Organic Chemistry*. New York: John Wiley & Sons, Inc.

Schwarzenbach, R.P. and Westall, J. (1981) Transport of nonpolar organic compounds from surface water to groundwater: Laboratory sorption studies. *Environmental Science and Technology* **15**, 1360–1367.

Seinfeld, J.H. (1986) *Atmospheric Chemistry and Physics of Air Pollution*. New York: John Wiley & Sons, Inc.

Seinfeld, J.H. and Pandis, S.N. (1998) *Atmospheric Chemistry and Physics*, 2nd edn. New York: John Wiley & Sons, Inc.

Simonich, S.L. and Hites, R.A. (1994) Vegetation-atmosphere partitioning of polycyclic aromatic hydrocarbons. *Environmental Science and Technology* **28**, 939–943.

Smith, F.L. and Harvey, A.H. (2007) Avoid common pitfalls when using Henry's law. *Chemical Engineering Progress* **103**(9), 33–39.

Southworth, G.R., Beauchamp, J.J., and Schmeider, P.K. (1978) Bioaccumulation potential of PAHs in *Daphnia pulex*. *Water Research* **12**, 973–977.

Sparks, D. (1999) *Environmental Soil Chemistry*. New York: Academic Press.

Springer, C., Lunney, P.D., Valsaraj, K.T., and Thibodeaux, L.J. (1986) Emission of hazardous chemicals from surface and near surface impounds to air. Part A: Surface impoundments, Final Report to EPA on Grant No.: CR 808161-02. Cincinnati, OH: ORD-HWERL, U.S. Environmental Protection Agency.

Stumm, W. (1993) *Chemistry of the Solid-Water Interface*. New York: John Wiley & Sons, Inc.

Stumm, W.F. and Morgan, J.M. (1996) *Aquatic Chemistry*, 3rd edn. New York: John Wiley & Sons, Inc.

Subramanyam, V., Valsaraj, K.T., Thibodeaux, L.J., and Reible, D.D. (1994) Gas-to-particle partitioning of polyaromatic hydrocarbons in an urban atmosphere. *Atmospheric Environment* **28**, 3083–3091.

Suntio, L.R., Shiu, W.Y., Mackay, D., and Glotfelty, D.E. (1987) A critical review of Henry's law constants. *Reviews in Environmental Contamination and Toxicology* **103**, 1–59.

Takada, H. and Ishiwatari, R. (1987) Linear alkylbenzenes in urban riverine environments in Tokyo: Distribution, source and behavior. *Environmental Science and Technology* **21**, 875–883.

Thiel, P.A. (1991) New chemical manifestations of hydrogen bonding in water adlayers. *Accounts of Chemical Research* **24**, 31–35.

Thoma, G.J. (1992) Studies on the diffusive transport of hydrophobic organic chemicals in bed sediments, PhD dissertation, Louisiana State University, Baton Rouge, LA.

Turner, L.H., Chiew, Y.C., Ahlert, R.C., and Kossen, D.S. (1996) Measuring vapor-liquid equilibrium for aqueous systems: Review and a new technique. *AIChE Journal* **42**, 1772–1788.

Valsaraj, K.T. (1988) On the physico-chemical aspects of partitioning of hydrophobic non-polar organics at the air-water interface. *Chemosphere* **17**, 875–887.

Valsaraj, K.T. (1993) Hydrophobic compounds in the environment: Adsorption equilibrium at the air-water interface. *Water Research* **28**, 819–830.

Valsaraj, K.T., Thoma, G.J., Reible, D.D., and Thibodeaux, L.J. (1993) On the enrichment of hydrophobic organic compounds in fog droplets. *Atmospheric Environment* **27A**, 203–210.

Veith, G.D., Macek, K.J., Petrocelli, S.R., and Carroll, J. (1980) An evaluation of using partition coefficients and water solubility to estimate the bioconcentration factors for organic chemicals in fish. *ASTM SPE 707*. Philadelphia, PA: ASTM, pp. 116–129.

Warneck, P.A. (1986) *Chemistry of the Natural Atmosphere*. New York: Academic Press.

Warner, H.P., Cohen, J.M., and Ireland, J.C. (1987) Determination of Henry's law constants of selected priority pollutants, EPA-600/D-87/229. Springfield, VA: NTIS.

Weber, W.J., McGinley, P.M., and Katz, L.E. (1991) Sorption phenomena in subsurface systems: Concepts, models and effects on contaminant fate and transport. *Water Research* **25**, 499–528.

Wijayaratne, R.D. and Means, J.C. (1984) Sorption of polycyclic aromatic hydrocarbons by natural estuarine colloids. *Marine Environmental Research* **11**, 77–89.

Yamasaki, H., Kuwata, K., and Miyamoto, H. (1982) Effects of ambient temperature on aspects of airborne polycyclic aromatic hydrocarbons. *Environmental Science and Technology* **16**, 189–194.

Yurteri, C., Ryan, D.F., Callow, J.J., and Gurol, M.D. (1987) The effect of chemical composition of water on Henry's law constant. *Journal of the Water Pollution Control Federation* **59**, 950–956.

4 Applications of Chemical Kinetics in Environmental Systems

In Chapter 2, we discussed the differences between "open" and "closed" systems. Natural environmental systems are open systems. Both material and energy are exchanged across the system boundaries. However, it is convenient and useful to model them as closed systems as well if we consider a fixed composition and ignore any changes in the composition. Many waste treatment operations are conducted as closed systems, where exchange of energy and/or mass occurs among compartments within the system, but not necessarily with the surroundings.

The rate laws described in Chapter 2 are for closed systems, where the concentration of a pollutant changes with time, but there is no mass exchange with the surroundings. Rarely this represents chemical kinetics in the natural environment, where we encounter time-varying inputs and outputs. The discussion that follows is derived from the chemical engineering literature (Hill, 1977; Levenspiel, 1999; Fogler, 2006).

Applications of kinetics in environmental engineering fate and transport modeling and in waste treatment operations are numerous. The discussion will start with the conventional reactor theory. Various types of reactors (ideal and nonideal) will be discussed. This will be followed by several applications in the hydrosphere, atmosphere, and lithosphere. Finally, the chemical kinetics for the biosphere with special relevance to biodegradation, enzyme kinetics, and bioaccumulation will be discussed. Several problems are provided, which serve to illustrate the concepts presented within each section.

4.1 WATER ENVIRONMENT

In this section, we will discuss applications of chemical kinetics and reactor models in the water environment. The first part of the discussion will be illustrations of fate and transport in the natural environment. The second part will be examples in water pollution control and treatment.

4.1.1 FATE AND TRANSPORT

4.1.1.1 Chemicals in Lakes and Oceans

A number of anthropogenic chemicals have been introduced into our lakes, rivers, and oceans. In order to elucidate the impact of these chemicals on marine species, birds, mammals, and humans, we need a clear understanding of their fate and transport in

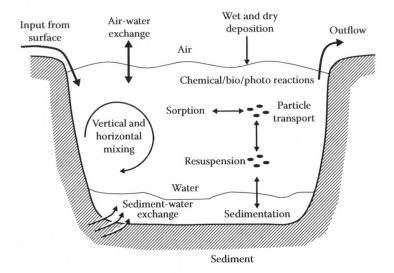

FIGURE 4.1   Schematic of various fate and transport processes for a pollutant entering a lake.

the water environment. A chemical introduced into a lake, for example, is subjected to a variety of transport, transformation, and mixing processes. There are basically two types of processes: (1) transport processes such as advection, dispersion, and diffusion within and between compartments (water, sediment, air) and (2) reaction processes that chemically transform compounds via photolysis, microbiological reactions, and chemical reactions. These processes do not occur independent of one another and in many cases are influenced by one another. Figure 4.1 is an illustration of these processes (Schwarzenbach et al., 1998).

At its simplest, a model will consist of a one-dimensional vertical transport between the sediment, water, and air. The horizontal variations in concentrations are neglected. For a deep lake, this is a good approximation, whereas for shallow water bodies the approximation fails. One can increase the spatial resolution of the model to obtain more sophisticated models.

In order to obtain the time constants for the various processes involved in the model, let us first construct a zeroth order chemodynamic model for the lake. In a lake, we have an epilimnion and a hypolimnion. To a good first approximation we shall consider the epilimnion as a well-mixed CSTR (Figure 4.2). The various processes that occur can each be assumed to be a first order loss process. A material balance for the compound A can be written as follows:

$$Input = Output + Reaction + Accumulation.$$

$$Q_0 C_{A0} = Q_0 C_A + r_{tot} V + V \cdot \frac{dC_A}{dt} \qquad (4.1)$$

where
   C_A is the total concentration of the compound in the epilimnion
   V is the volume of the epilimnion

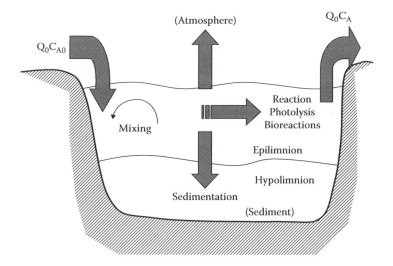

FIGURE 4.2   A CSTR one-box model for the epilimnion of a lake.

Note that $Q_0/V = k_F$, the rate constant for flushing, and that the overall rates will be composed of the several first order processes listed in Figure 4.2.

$$r_{tot} = \left(k_{voln} + k_{sed} + k_{rxn} + k_{photo} + k_{bio} \right) C_A = k_{tot} C_A \qquad (4.2)$$

Therefore,

$$\frac{dC_A}{dt} = k_F \left(C_{A0} - C_A \right) - k_{tot} C_A \qquad (4.3)$$

Expressions for individual rate constants are given in Table 4.1. Note that a similar equation was used in Example 4.1 to illustrate the loss mechanisms in a wastewater surface impoundment.

Example 4.1   Fate of Tetrachloroethene (TetCE) in Lake Griefensee, Switzerland

Schwarzenbach et al. (1998) have estimated the following parameters for Lake Griefensee: Average area = 2×10^6 m², Average epilimnion depth = 5 m, Particulate matter concentration = 2×10^{-6} kg L^{-1}, Average particulate settling velocity = 2.5 m d^{-1}, Average K_w for TetCE = 0.15 m d^{-1}, Average load of TetCE per day to the lake = 0.1 kg d^{-1}, Average flow rate of water = 2.5×10^5 m³ d^{-1}. Estimate the steady state removal of TetCE in the lake.

Firstly, note that TetCE is a refractory compound. Hence for the zeroth approximation let us assume that chemical, photochemical, and biological reactions are negligible.

TABLE 4.1

Expressions for First Order Rate Constants in a CSTR Model for the Epilimnion of a Lake

Process	Rate Constant	Expression
Flushing	k_F	Q_F/V_e
Volatilization	k_{voln}	$\left(\dfrac{K_w}{h_e}\right)\left(1-\phi_A^p\right)$
Sedimentation	k_{sed}	$\left(\dfrac{U_{set}}{h_e}\right)\phi_A^p$
Reaction loss	$k_{rxn} + k_{photo} + k_{bio}$	Specific to chemicals

Notes: Q_F = feed rate, $m^3\ d^{-1}$; V_e = epilimnion volume, m^3; K_w = overall liquid phase mass transfer coefficient of A, $m\ d^{-1}$; h_e = height of the epilimnion, m; ϕ_A^p = fraction of A bound to particulates, $\dfrac{K_{OC}C_C}{1+K_{OC}C_C}$, where K_{oc} is the organic carbon–based partition constant of A $(L\ kg^{-1})$, C_C is the particulate organic carbon concentration, $L\ kg^{-1}$; and U_{set} = average settling velocity of particulates, $m\ d^{-1}$.

K_{AW} for TetCE is 0.6 at 298 K. Log K_{ow} = 2.1. Hence $\log K_{ow} = (0.92)(2.10) - 0.23 = 1.70$. Hence $K_{oc} = K_c = 50\ L\ kg^{-1}$. $\phi_A^p = (50)(2\times10^{-6})/\left[1+(50)(2\times10^{-6})\right] = 1\times10^{-4}$. $k_{voln} = (0.15)(1)/5 = 0.03\ d^{-1}$, $k_{sed} = (2.5)(1\times10^{-4})/5 = 5\times10^{-5}\ d^{-1}$, $k_F = 2.5\times10^5/(5)(2\times10^6) = 0.025\ d^{-1}$. At steady state, $\dfrac{C_{A0}-C_A}{C_A} = \dfrac{k_{voln}+k_{sed}}{k_F} = 1.2$. Hence $1 - C_A/C_{A0} = 0.55$.

Example 4.2   Evaporation from a Well-Stirred Surface

Many compounds of low solubility (such as pesticides, PCBs, and hydrocarbons) evaporate from open waters in lakes and oceans. For a 1 m^2 of area with depth Z cm, estimate the half-life for evaporation of the following compounds: benzene, biphenyls, aldrin, and mercury.

The evaporation is assumed to occur from a volume Z cm × 1 cm^2 = Z cm^3 of surface water that is well mixed as a result of surface turbulence. A material balance over the volume Z cm^3 gives

$$Input = Output + Reaction + Accumulation.$$

$$0 = K_w\left(C_w - C_a K_{AW}\right) + 0 + Z\cdot\frac{dC_w}{dt} \tag{4.4}$$

where the rate of loss by reaction is zero on account of the refractory nature of the chemical. Rearranging and integrating with $C_w = C_0$ at t = 0, we get

$$C_w = \frac{C_a}{K_{AW}} + \left(C_0 - \frac{C_a}{K_{AW}}\right)\cdot\exp\left(-K_w\frac{t}{Z}\right) \tag{4.5}$$

If background air concentration is negligible, $C_a = 0$

$$C_w = C_0 \cdot \exp\left(-K_w \frac{t}{Z}\right) = C_0 \cdot \exp\left(-k_{voln} t\right) \qquad (4.6)$$

Hence, half-life is $t_{1/2} = 0.693/k_{voln} = 0.693\ Z/K_w$. Mackay and Leinonen (1975) give values of K_w for several compounds at 298 K:

Compound	P_i^* (mm Hg)	K_w (m h⁻¹)	$t_{1/2}$, h for Z = 1 m
Benzene	95.2	0.144	4.8
Biphenyl	0.057	0.092	7.5
Aldrin	6×10^{-6}	3.72×10^{-3}	186.3
Mercury	1.3×10^{-3}	0.092	7.5

Note that the half-life of both biphenyls and mercury are the same, although their vapor pressures vary by a factor of 2. Note also that K_w is obtained from the individual transfer coefficients k_w and k_a and requires knowledge of K_{AW} as well. These can be obtained by applying the diffusivity correction for k_w of oxygen (20 cm h⁻¹) and k_a for water (3000 cm h⁻¹), if experimental values of individual mass transfer coefficients are not available.

4.1.1.2 Chemicals in Surface Waters

The surface water in a fast-flowing stream is generally unmixed in the direction of flow, but is laterally well mixed. This suggests that a plug flow model will be applicable in these cases. The appropriate equation is the advection-dispersion equation with the axial dispersion term neglected. Figure 4.3 depicts a river stretch where we apply a material balance across the volume $WH\Delta x$:

$$\text{Input} = \text{Output} + \text{Reaction} + \text{Accumulation}.$$

$$uC_w(x)WH = uC_w(x+\Delta x)WH + rWH\Delta x + \text{Loss to air}$$

$$+ \text{Loss to sediment} + WH\Delta x \frac{dC_w}{dt}$$

At steady state, we have $dC_w/dt = 0$. The rate of loss to air $= K_w\left(C_w - \dfrac{C_a}{K_{AW}}\right)(W\Delta x)$. The rate of loss to sediment $= K_s\left(C_w - \dfrac{W_i}{K_{AW}}\right)(W\Delta x)$. If a first order rate of reaction is considered, $r = k_r C_w$. The overall material balance is

$$u \cdot \left(\frac{C_w(x+\Delta x) - C_w(x)}{\Delta x}\right) = -\frac{K_w}{H}\left(C_w - \frac{C_a}{K_{AW}}\right) - \frac{K_s}{H}\left(C_w - \frac{W}{K_{sw}}\right) - k_r C_w \qquad (4.7)$$

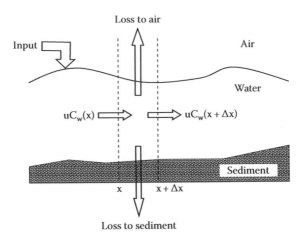

FIGURE 4.3 Material balance on a section of a stream assuming complete mixing across the width and depth of the stream.

Since time, $t = x/u$, $\Delta t = \Delta x/u$, and taking $\lim_{\Delta x \to 0}$, we obtain the following differential equation:

$$\frac{dC_w}{dt} = -\frac{K_w}{H}\left(C_w - \frac{C_a}{K_{AW}}\right) - \frac{K_s}{H}\left(C_w - \frac{W}{K_{sw}}\right) - k_r C_w \qquad (4.8)$$

If both sediment and air concentration remain constant, we can solve this equation using the initial condition $C_w = C_0$ at $t = 0$ to get (Reible, 1998)

$$C_w = C_0 e^{-\alpha t} + \frac{\beta}{\alpha}\left(1 - e^{-\alpha t}\right) \qquad (4.9)$$

where $\beta = \dfrac{K_w C_a}{HK_{aw}} + \dfrac{K_s W}{HK_{sw}}$ and $\alpha = \dfrac{K_w}{H} + \dfrac{K_s}{H} + k_r$. Note that α is a composite rate constant that characterizes the loss mechanism within the water.

Example 4.3   Loss of Chloroform from a Shallow Stream

For the Amite River near Baton Rouge, LA, the depth is 2 m, f_{oc} is 0.02 for the sediment, and has an average flow velocity of 1 m s^{-1}. For an episodic spill of chloroform in the river, determine its concentration 20 miles downstream of Baton Rouge.

Assume that the background concentration of chloroform in air is negligible. Since chloroform is highly water soluble, and its hydrophobicity is small, we can assume that its concentration in the sediment will be low. Thus β is negligible. Because of the high stream velocity and low sorption to sediment, the water to air mass transfer is likely to dominate over water to sediment mass transfer.

Hence, $\alpha = \dfrac{K_w}{h} + k_r$. The reaction in water is hydrolysis, and its rate is given by 4.2×10^{-8} h^{-1}. Clearly, this is low and chloroform is a refractory compound. From Table 4.12, $\dfrac{1}{K_w} = \dfrac{1}{4.2 \times 10^{-5}} + \dfrac{1}{(3.2 \times 10^{-3})(0.183)}$. Hence K_w is 3.9×10^{-5} m s^{-1} = 0.14 m h^{-1}. Since t = x/u = (20 miles) (1609 m mile^{-1})/(1 m s^{-1}) (3600 s h^{-1}) = 8.9 h. Hence α = 0.14/2 = 0.07 h^{-1}. C_w/C_0 = exp[–(0.07) (8.9)] = 0.53. Hence, the concentration at the monitoring station will be 53% of its concentration at the spill point.

4.1.1.3   Biochemical Oxygen Demand in Natural Streams

Bacteria and microorganisms decompose organic matter in wastewater via enzyme catalysis. The amount of oxygen that the bacteria need to decompose a given organic compound to products (CO_2 and H_2O), and in the process produce new cells, is called the "biochemical oxygen demand" (BOD). The reaction is represented as

$$\text{Organic matter} + O_2 \rightarrow CO_2 + H_2O + \text{new cells}$$

BOD is used to characterize the extent of pollution in rivers, streams, lakes, and in wastewater under aerobic conditions. BOD is measured using a 5-day test, i.e., the oxygen demand in 5 days of a batch laboratory experiment. It is represented as BOD_5. BOD_5 test is a wet oxidation experiment whereby the organisms are allowed to breakdown organic matter under aerobic conditions. It is a convenient method, but does not represent the total BOD. A 20-day test (BOD_{20}) is considered the "ultimate BOD" of a water body. BOD measurements are used to define the strength of municipal wastewater, determine the treatment efficiency by observing the oxygen demand remaining in an effluent waste stream, determine the organic pollution strength in surface waters, and to determine the reactor size and oxygen concentration required for aerobic treatment processes.

The "oxygen demand" L in water is given by a first order reaction rate:

$$-\frac{dL}{dt} = k_1 L \tag{4.10}$$

Upon integration we get

$$L = L_0 e^{-k_1 t} \tag{4.11}$$

where L_0 is called the "ultimate oxygen demand" for organic matter decomposition. (Note that there are additional oxygen demands for those compounds that contain nitrogen, which we will ignore.) Equation 4.11 shows that the oxygen demand decreases exponentially with time. The total oxygen demand of a sample is the sum of waste consumed in t days (BOD_t) and the oxygen remaining to be used after t days. Hence

$$L_0 = BOD_t + L \tag{4.12}$$

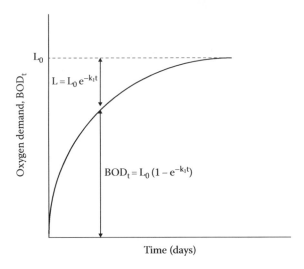

FIGURE 4.4   A typical BOD curve.

or

$$BOD_t = L_0\left(1 - e^{-k_1 t}\right) \tag{4.13}$$

Figure 4.4 represents a typical BOD curve. The value of k_1 depends on the type of system under study. Typical values range from 0.05 to 0.3 d^{-1}. Since BOD tests are carried out at 20°C, k_1 has to be corrected for other temperatures using the following equation:

$$\left(k_1 \text{ at } \theta°C\right) = \left(k_1 \text{ at } 20°C\right)\cdot\left(1.047\right)^{(\theta-20)} \tag{4.14}$$

Apart from the BOD, another important and related term used in wastewater engineering is the "oxygen deficit" Δ (expressed in mg L^{-1} or kg m^{-3}) that is defined as

$$\Delta = C_{O_2}^* - C_{O_2} \tag{4.15}$$

where
    $C_{O_2}^*$ is the saturation concentration of oxygen in water (mg L^{-1} or kg m^{-3})
    C_{O_2} is the existing oxygen concentration in water (mg L^{-1} or kg m^{-3})

The following example will clarify the relationship between L_0 and Δ for a flowing stream such as a river.

Example 4.4   Oxygen Deficit in a Polluted Natural Stream

Bio-organisms decompose organic molecules by consuming oxygen from water in the process. At the same time, dissolution of oxygen from air into water tends to restore the oxygen level in a polluted natural stream. The latter process is called "reaeration," and the former process is termed "deoxygenation." The rate constant for deoxygenation is k_d and that for reaeration is k_r, both being first order rate constants. Assume that the stream is in plug flow. The rate of increase in oxygen deficit Δ is given by

$$\frac{d\Delta}{dt} = k_d L_0 e^{-k_d t} - k_r \Delta \qquad (4.16)$$

Integrating this equation with the initial condition, $\Delta = \Delta_0$ at $t = 0$, we get

$$\Delta = \Delta_0 e^{-k_r t} + \left(\frac{k_d L_0}{k_r - k_d}\right)\left(e^{-k_d t} - e^{-k_r t}\right) \qquad (4.17)$$

This is the well-known "Streeter-Phelps oxygen sag equation," which describes the oxygen deficit in a polluted stream. t is the time of travel for a pollutant from its discharge point to the point downstream. Thus, it is related to the velocity of the stream as $t = y/U$, where y is the downstream distance from the outfall and U is the stream velocity.

Subtracting the given value of Δ from the saturated value $C_{O_2}^*$ gives the oxygen concentration C_{O_2} at any location below the discharge point.

Equation 4.17 can be used to obtain the "critical oxygen deficit" (Δ_c) at which point the rate of deoxygenation exactly balances the rate of reaeration, i.e., $d\Delta/dt = 0$. At this point we have the following equation, which gives Δ_0:

$$\Delta_c = \left(\frac{k_d}{k_r}\right) L_0 e^{-k_d t_c} \qquad (4.18)$$

This gives the "minimum dissolved oxygen concentration" in the polluted stream. In order to obtain the value of t_c at which the value of $d\Delta/dt = 0$, we can differentiate the expression obtained earlier for Δ with respect to t and set it equal to zero. This gives a relationship for t_c solely in terms of the initial oxygen deficit (Δ_0) as follows:

$$t_c = \left(\frac{1}{k_r - k_d}\right)\cdot \ln\left[\frac{k_r}{k_d}\left(1 - \frac{(k_r - k_d)}{k_d}\frac{\Delta_0}{L_0}\right)\right] \qquad (4.19)$$

Figure 4.5 is a typical profile for C_{O_2} as a function of time t (or distance y from the discharge point) in a polluted stream. In any given stream, L_0 is given by the BOD of the stream water plus that of the wastewater at the discharge point.

$$L_0 = \frac{Q_w L_w + Q_s L_s}{(Q_w + Q_s)} \qquad (4.20)$$

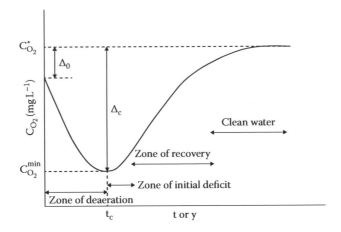

FIGURE 4.5   A typical Streeter-Phelps oxygen sag curve.

where Q_w and Q_s are volumetric flow rates of wastewater and stream water respectively. The initial oxygen deficit Δ_0 is given by

$$\Delta_0 = C_{O_2}^* - \left(\frac{Q_w C_{O_2}^w + Q_s C_{O_2}^s}{Q_w + Q_s} \right) \tag{4.21}$$

where $C_{O_2}^w$ and $C_{O_2}^s$ are respectively the oxygen concentration (mg L^{-1} or kg m^{-3}) in the wastewater and the stream water just upstream of the discharge location. The three parameters of significance obtained from this analysis are Δ, Δ_c, and t_c (or y_c). This information gives the maximum possible pollutant concentration at the discharge point. In order to utilize the Streeter-Phelps equation, one has to estimate values for k_d and k_r with precision. k_d is determined by obtaining the BOD at two known locations (a and b) in a stream and using the equation

$$k_d = \frac{y}{U} \cdot \log\left(\frac{L_a}{L_b} \right) \tag{4.22}$$

The value of k_r is obtained from an equation such as that of O'Connor and Dobbins:

$$\left(\frac{k_r}{d^{-1}} \right) = 3.9 \cdot \left(\frac{U}{H^3} \right)^{1/2} \tag{4.23}$$

where
    U is the stream velocity (m s^{-1})
    H is the mean depth (m)

The constant in the equation for k_r corrects for the bed roughness in a stream that affects the stream velocity. k_r is effected by the presence of algae (that effect C_{O_2} through diurnal variations from photosynthesis), surface active substances (oil/grease at the air-water interface), and other pollutants that effect BOD. Typical values of k_r vary from $0.1 - 0.23$ d^{-1} in small ponds to $0.7 - 1.1$ d^{-1} for swift streams. Corrections to k_r and k_d for the temperature of the stream are obtained using the following equation:

$$\left(k \text{ at } \theta°C\right) = \left(k \text{ at } 20°C\right) \cdot \left(\omega\right)^{(\theta-20)} \tag{4.24}$$

where ω is a temperature coefficient, which is 1.056 for k_d and 1.024 for k_r.

Temperature influences the oxygen sag in a manner such that with increasing temperature, the Δ_c value is reached faster. Whereas the rate of aeration decreases with temperature, the rate of deoxygenation increases with temperature. Therefore, we can substantiate the faster response of the stream at higher temperature. Diurnal variations in dissolved oxygen result from increased CO_2 due to algal respiration in the daytime and decreased dissolved oxygen levels at night time. Thus, dissolved oxygen levels are largest during the afternoon and are least at night.

4.1.2 WATER POLLUTION CONTROL

This section describes the applications of chemical kinetics and reactor models for selected wastewater pollution control operations.

4.1.2.1 Air Stripping in Aeration Basins

Wastewater is treated in lagoons into which air is introduced in either of two ways: (1) as air bubbles at the bottom of the lagoon or (2) mechanical surface aerators placed at the water surface to induce turbulence. In either case, transfer of volatile organic compounds (VOCs) from water to air with simultaneous transfer of oxygen from air to water is the objective. The analysis of oxygen uptake (absorption) and VOC desorption (stripping) are complementary to one another (Matter-Muller et al., 1981; Munz, 1985; Valsaraj and Thibodeaux, 1987). In this section, we will illustrate the kinetics of desorption of VOCs from wastewater lagoons by diffused aeration using air bubbles.

A typical VOC stripping operation using air bubbles is depicted in Figure 4.6. Let us consider first a "batch operation" ($Q_0 = 0$). If the reactor is completely mixed,

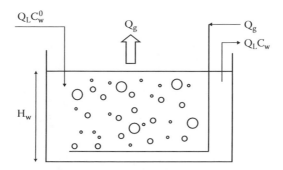

FIGURE 4.6   Schematic of submerged bubble aeration in a wastewater lagoon.

the concentration in the aqueous column C_w is the same everywhere. Consider a single air bubble as it rises through the aqueous column. It continuously picks up solute as it moves up and exits the column carrying the solute to the atmosphere. Obviously there is a trade-off between a more contaminated medium (water) and a less contaminated medium (air). The rate of transfer is controlled by diffusion across the thin boundary layer surrounding the bubble. The overall water-phase mass transfer coefficient is K_w. The rate expression for the mass transfer to a single gas bubble is given by

$$\frac{dC_w}{dt} = \frac{A_b}{V_b} \cdot K_w \left(C_w - C_w^* \right) \tag{4.25}$$

where C_w^* is the concentration in the aqueous phase that would be in equilibrium with the concentration associated with the bubble, C_g. Note that the effect of external pressure on the bubble size has been neglected in deriving Equation 4.25. The equilibrium concentration in the vapor phase of the bubble is given by air-water partitioning constant and is $C_g = K_{AW} C_w^*$.

In an aeration apparatus, it is more convenient to obtain the specific air-water interfacial area, a_v (m² per m³ of total liquid volume). Hence, $\dfrac{A_b}{V_b} = \dfrac{6}{D_b} = a_v \dfrac{V_w}{V_a}$ where D_b is the average bubble diameter, V_w is the total liquid volume, and V_a is the total air volume in the reactor at any time. The ratio V_a/V_w is called the "gas hold up," ε_g. Therefore, we can write $a_v = \dfrac{6}{D_b} \cdot \varepsilon_g$. Using these definitions we can rewrite the equation for rate of change of concentration associated with the bubble as

$$\frac{dC_g}{dt} + \left(K_w a \right) \left(\frac{V_w}{V_a} \right) \frac{1}{K_{AW}} C_g = \left(K_w a \right) \left(\frac{V_w}{V_a} \right) C_w \tag{4.26}$$

Since the rise time of a single bubble is really small ($\tau \approx$ a few seconds), a reasonable assumption is that during this time C_w is a constant. This means we can integrate this equation to get the concentration of the pollutant associated with a single bubble. Using the initial condition, $C_A^g \left(\tau = 0 \right) = 0$, we have

$$C_g \left(\tau \right) = K_{AW} C_w \left[1 - e^{ -\left(\left(K_w a \right) \left(\frac{V_w}{V_a} \right) \frac{1}{K_{AW}} \tau \right) } \right] \tag{4.27}$$

The rate of change of pollutant concentration within the aqueous phase is given by

$$\frac{dC_w}{dt} = -\left(\frac{Q_g}{V_w} \right) C_g \left(\tau \right) \tag{4.28}$$

where Q_g is the volumetric flow rate of air. The initial condition for the aqueous phase is $C_w(0) = C_w^0$ for a batch reactor. We also note that the residence time of a gas bubble is $\tau = \dfrac{V_a}{Q_g}$. Hence, upon integration we get the following:

$$\ln\left(\frac{C_w}{C_w^0}\right) = -\left(\frac{Q_g}{V_w}\right) K_{AW} \left[1 - e^{-\left[\left(K_w a\right)\left(\frac{V_w}{Q_g}\right)\frac{1}{K_{AW}}\right]}\right] t = -k_{rem} t \qquad (4.29)$$

where k_{rem} is the first order removal rate constant from the aqueous phase. One can define partial gas-phase saturation as $\phi' = 1 - \exp(-\phi H_s)$, where $\phi = \left(K_w a\right)\left(\dfrac{V_w}{Q_g}\right)\left(\dfrac{1}{K_{AW} H_s}\right)$. In this definition, H_s is the depth at which the air bubble is released in the lagoon. Thus we can rewrite the equation for k_{rem} as follows:

$$k_{rem} = \left(\frac{Q_g}{V_w}\right) K_{AW} \left(1 - e^{-\phi H_s}\right) \qquad (4.30)$$

Two special limiting cases are to be noted:

1. If the exit gas is saturated and is in equilibrium with the aqueous phase, $1 - \exp(-\phi H_s) \to 1$, and $k_{rem} = (Q_g/V_w)K_{AW}$. This condition can be satisfied if $(K_w a)$ is large.

2. The second limiting case is $\left(K_w a\right)\left(\dfrac{V_w}{Q_g}\right)\dfrac{1}{K_{AW}} \ll 1$, for which $k_{rem} = (K_w a)$. This represents the case when the exit gas is far from saturation. For large K_{AW} and Q_g values, this limiting case will apply. This is the case in most surface aeration systems, where a large volumetric flow of air is in contact with an aqueous body.

Let us now consider a continuous flow system. If the continuous flow system is in "plug flow," then at steady state the ideal residence time for the aqueous phase is $t = V_w/Q_L$, and substituting $C_w^0 = C_0$ in Equation 4.30 for a batch reactor should give us the appropriate equation for a PFR bubble column:

$$\ln\left(\frac{C_w}{C_0}\right) = -\left(\frac{Q_g}{Q_L}\right) K_{AW} \left(1 - e^{-\phi H_s}\right) \qquad (4.31)$$

The term $(Q_g/Q_L)K_{AW} = S$ is called the "separation factor" in chemical engineering, and gives the maximum separation achievable if the exit air is in equilibrium with the aqueous phase.

If the mixing in the aqueous phase is large, a CSTR approach can be used to model the process. The overall mass balance is then given by

$$\frac{dC_w}{dt} = \frac{Q_L}{V_w}\left(C_0 - C_w\right) + \frac{Q_g}{V_w}\left(C_g^0 - C_g(\tau)\right)$$ (4.32)

Since the entering gas is always clean, $C_g^0 = 0$. Assuming steady state $dC_w/dt = 0$. Therefore, we obtain

$$\frac{C_w}{C_0} = \frac{1}{1 + S\left(1 - e^{-\phi H_s}\right)}$$ (4.33)

The net of removal in a CSTR is given by

$$R_{CSTR} = 1 - \frac{C_w}{C_0} = \frac{S\left(1 - e^{-\phi H_s}\right)}{1 + S\left(1 - e^{-\phi H_s}\right)}$$ (4.34)

and for a PFR

$$R_{PFR} = 1 - \frac{C_w}{C_0} = 1 - \exp\left(-S\left(1 - e^{-\phi H_s}\right)\right)$$ (4.35)

A general nomograph can be drawn to obtain the removal efficiency as a function of both S and ϕH_s. Figure 4.7 represents the nomograph. With increasing S, R increases.

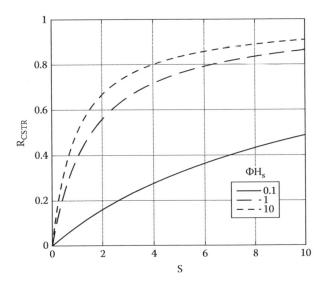

FIGURE 4.7 Nomograph for steady state removal in an aeration lagoon operated as a CSTR. Air bubbles are introduced at a depth H_s below the surface of water.

For a given S, increased ϕH_s increases the value of R. ϕH_s can be large if K_wa is large. Large values of S can be realized if either Q_g/Q_L or K_{AW} is large. Since a $(V_w/V_a) = 6/D_b$ (where D_b is the average bubble diameter or more accurately the Sauter mean bubble diameter), decreasing the bubble size could markedly improve the aeration efficiency. The nomograph is useful in designing the size (volume) of a reactor required to obtain desired removal efficiency if the values of liquid and gas flow rates are fixed for a given compound.

In some wastewater aeration ponds, one encounters a thin layer of oil or surfactant foam above the water layer as a natural consequence of the composition of waste-waters. The oil layer can act as an additional compartment for the accumulation of a pollutant from the aqueous phase. The solute activity gradient between the aqueous and oil layers helps establish transport from solvent to the aqueous phase, which competes with the unidirectional solute transport by air bubbles from the aqueous phase. The additional removal mechanism by solvent extraction has been known to effectively increase the rate of removal from the aqueous phase. This process of flota-tion and extraction using air bubbles into an organic solvent layer is termed "solvent sublation." The process of solvent sublation has been investigated in bubble column reactors (see Valsaraj and Thibodeaux, 1987; Valsaraj, 1994 for details). It has been shown to be effective in removing several organic compounds and metal ions.

Example 4.5   Reactor Sizing for a Desired Removal in Air Stripping

1,2-dichloroethane (DCA) from a contaminated groundwater is to be removed using air stripping. Obtain the size of a bubble column reactor required for 80% removal of DCA from 10,000 gal d^{-1} of groundwater using a 1 foot diameter column.

The air-water partition constant K_{AW} for DCA is 0.056. $Q_L = 10,000$ gpd $= 0.027$ m^3 min^{-1}. Consider a ratio of $Q_g/Q_L = 100$. Hence $Q_g = 2.7$ m^3 min^{-1}. Since the column radius $r_c = 0.152$ m, $A_c = \pi r_c^2 = 0.0729$ m^2. Hence the super-ficial gas velocity $u_g = Q_g/A_c = 0.617$ m s^{-1}. This parameter appears in the cor-relation for ε_g, a, and k_la. Although several correlations are available for bubble column reactors, we choose the one recommended by Shah et al. (1982):

$$\frac{\varepsilon_g}{(1-\varepsilon_g)^4} = 0.2\alpha^{1/8}\beta^{1/12}\frac{u_g}{(gD_c)^{1/2}} \text{ and } a = \frac{1}{3D_c}\alpha^{1/2}\beta^{1/10}\varepsilon_g^{1.13}, \text{ where } \alpha = \frac{gD_c^2\rho_L}{\sigma} \text{ and}$$

$\beta = \dfrac{gD_c^3}{v_L^2}$. The different parameters are g $= 9.8$ m^2 s^{-1}, $\rho_L = 1000$ kg m^{-3}, $v_L = 1 \times 10^{-6}$ m^2 s^{-1}, and $\sigma = 0.072$ N m^{-1}. Using these parameters, $\dfrac{\varepsilon_g}{(1-\varepsilon_g)^4} = 2.087$.

A trial and error solution gives $\varepsilon_g = 0.35$. Hence, we obtain a $= 525$ m^{-1}. This gives a/$\varepsilon_g = 1500 = 6/D_b$, and the average bubble diameter in the column is $D_b = 0.004$ m. Since the aqueous phase diffusivity of DCA is 9.9×10^{-10} m^2 s^{-1} (see Wilke-Chang correlation—Reid et al., 1987), we obtain $k_wa = 0.076$ s^{-1}. As before, $\dfrac{1}{K_wa} = \dfrac{1}{k_wa} + \dfrac{1}{k_gaK_{AW}}$, where k_w and k_g are individual phase mass trans-fer coefficients. For predominantly liquid phase controlled chemicals (such as DCA), $K_wa = k_wa$. Thus for the given conditions, S $= 5.6$. If the desired removal in the CSTR is $R_{CSTR} = 0.80$, we have $\phi = 1 - \exp(\phi H_s) = 0.71$, and $\phi H_s = 1.24$.

Since $\phi H_s = \dfrac{K_w a V_w}{K_{AW} Q_g}$, we obtain $V_w = 0.033 \text{ m}^3$. Therefore, the height of the reactor is $H_s = V_w/A_c = 0.45 \text{ m}$.

If back mixing is insignificant, the bubble column will behave as a plug flow reactor. In such a case, for $R_{PFR} = 0.80$, $\phi H_s = 0.34$, which gives $V_w = 0.011 \text{ m}^3$, and hence $H_s = 0.15 \text{ m}$. Thus, if axial back mixing is avoided in the bubble column, a high degree of removal can be obtained in small reactors. In reality, this is rarely achieved owing to the fact that at large air rates, the bubbles are larger (low a values) and the axial dispersion increases.

4.1.2.2 Oxidation Reactor

In this section, we will discuss the application of a redox process for wastewater treatment. Consider the oxidation of organic compounds by ozone. Ozone (O_3) is an oxidant used for disinfection and organic compound removal from water. Ozone has a large aqueous solubility (8.9 mg L^{-1}), and it has a high reactivity ($\Delta G_f^0 = +163 \text{ kJ mol}^{-1}$). Its redox potential is high, making it a powerful oxidant.

$$O_3 + 2H^+ + 2e^- \underset{}{\overset{E_H^0 = 2.1 \text{ V}}{\rightleftharpoons}} O_2 + H_2O \tag{4.36}$$

Wastewater treatment using ozone requires an ozone generator, a contactor, and off-gas treatment devices. Ozone is generated using an electric arc in an atmosphere of oxygen. The ozone contactor system is a submerged diffuser where ozone is bubbled into the water column immediately upon generation. The typical depth of the water column is 20 ft (Ferguson et al., 1991). With contact times as small as 10 min in most water treatment plants, ozone can perform microbial destruction, but can only partially oxidize organic compounds. Hence it is used in combination with H_2O_2 and UV oxidation processes (Masten and Davies, 1994). Series (cascade) reactors are required to complete the oxidation process.

The reaction of ozone in "pure" water is a radical chain reaction comprising of the formation and dissipation of the powerful hydroxyl radical (OH$^\bullet$). The sequence begins with the base-catalyzed dissociation of ozone:

$$\begin{aligned} O_3 + OH^- &\rightarrow HO_2^\bullet + O_2^{\bullet-} \\ HO_2^\bullet &\rightarrow H^+ + O_2^{\bullet-} \end{aligned} \tag{4.37}$$

This acid-base equilibrium has a pK_a of 4.8. Since, per reaction two radicals are produced, the rate constant $k_1 = 2 \times 70 \text{ L mol}^{-1} \text{ s}^{-1}$. The subsequent reactions lead to the regeneration of the catalyst OH$^-$ and the formation of O_2 and HO$^\bullet$. Staehelin and Hoigne (1985) published a concise summary of the various reactions involved in the overall scheme (Figure 4.8). The reaction involving the consumption of ozone by the superoxide anion ($O_2^{\bullet-}$) has a rate constant $k_2 = 1.5 \times 10^9 \text{ L mol}^{-1} \text{ s}^{-1}$. In the presence of an acid, the last reaction proceeds in two steps:

$$\begin{aligned} O_3^{\bullet-} + H^+ &\xrightarrow{k_3 = 5 \times 10^{10} \text{ L mol}^{-1} \text{ s}^{-1}} HO_3^\bullet \\ HO_3^\bullet &\xrightarrow{k_4 = 1.4 \times 10^5 \text{ L mol}^{-1} \text{ s}^{-1}} HO^\bullet + O_2 \end{aligned} \tag{4.38}$$

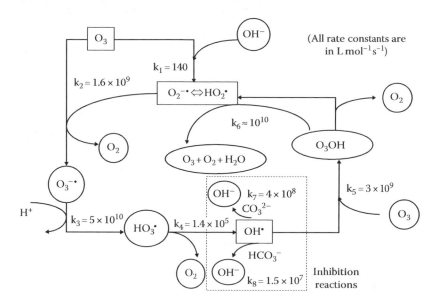

FIGURE 4.8 Radical chain reaction mechanism for ozone decomposition in pure water. Inhibition by carbonate alkalinity is also shown. (Reprinted with permission from Stehelin, J. and Hoigne, J., Decomposition of ozone in water in the presence of organic solutes acting as promoters and inhibitors of radical chain reactions, *Environ. Sci. Technol.*, 19, 1206. Copyright 1985 American Chemical Society.)

The hydroxyl radical forms an adduct with ozone, which gives rise to $HO_2^{\bullet}$ and O_2 via decomposition thus propagating the chain. The $HO_2^{\bullet}$ also reacts with the adduct forming O_3, O_2, and water with a rate constant k_6 of approximately 10^{10} L mol^{-1} s^{-1}.

Radical scavengers that react with $OH^{\bullet}$ can efficiently terminate this chain reaction. In fresh water that is slightly alkaline (due to carbonate and bicarbonate species), the hydroxyl radical is efficiently scavenged by inorganic carbonate species. Hence the decay rate of ozone is reduced. The presence of organic solutes (such as alcohols, DOC) and other species (such as phosphates) in water impacts the radical chain reaction in different ways. The following discussion is based on the scheme suggested by Staehelin and Hoigne (1985) (Figure 4.9). A solute M can interfere with the radical chain reaction involving the ozonide and hydroxyl radical in four different ways: (1) Solute M can react directly with ozone to produce the ozonide radical and an M^+ ion with a rate constant k_{13}. (2) Solute M can be oxidized to M_{ox} by direct reaction with O_3 with a rate constant k_{14}. (3) As discussed in Figure 4.8, the solute M can react with the hydroxyl radical in two ways. The reaction can lead to the scavenging of $OH^{\bullet}$ as in the case of CO_3^{2-} or HCO_3^-. This step is generally represented by the formation of the product Φ in Figure 4.9 with a rate constant k_{12}. (4) In some cases, the solute M reacts with $OH^{\bullet}$ to form the peroxo radical $ROO^{\bullet}$, which further adds O_2 to give M_{ox} and reforming the $HO_2^{\bullet}$. This is characterized by the rate constant k_{10} and k_{11} in Figure 4.9.

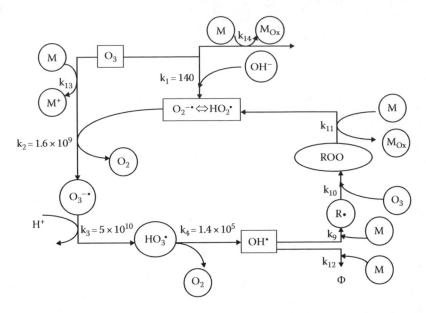

FIGURE 4.9 Radical chain reaction mechanism for ozone decomposition in impure water. (Reprinted with permission from Stehelin, J. and Hoigne, J., Decomposition of ozone in water in the presence of organic solutes acting as promoters and inhibitors of radical chain reactions, *Environ. Sci. Technol.*, 19, 1207. Copyright 1985 American Chemical Society.)

Let us now ascribe to the reaction scheme (Figure 4.9) a reaction rate law by using steady state approximations for $O_2^{-\bullet}$ and $OH^\bullet$ species, and consider the loss of ozone by all possible reaction pathways. The rate of ozone reaction with M directly is $k_{14}[M][O_3]$, and the total rate of initiation of chain reactions (i.e., the formation of $O_2^{-\bullet}$ and $O_3^{-\bullet}$) is $k_{13}[M][O_3] + 2k_1[OH^-][O_3]$. The rate at which $OH^\bullet$ is converted to $HO_2^{-\bullet}$ is $k_9[M][OH^\bullet]$, and the rate of scavenging of $OH^\bullet$ by M is $k_{12}[M][OH^\bullet]$. The rate of ozone consumption via chain reactions is given by

$$-\frac{d[O_3]}{dt} = k_2\left[O_2^{\bullet-}\right]\left[O_3\right] \tag{4.39}$$

Applying the PSSA for $O_2^{-\bullet}$ we have

$$\left[O_2^{\bullet-}\right]_{ss} = \frac{2k_1\left[OH^-\right]\left[O_3\right] + k_9\left[M\right]\left[OH^\bullet\right]_{ss}}{k_2\left[O_3\right]} \tag{4.40}$$

Assuming PSSA for $OH^\bullet$ we obtain

$$\left[OH^\bullet\right]_{ss} = \frac{k_{13}\left[M\right]\left[O_3\right] + k_2\left[O_2^{\bullet-}\right]_{ss}\left[O_3\right]}{\left(k_9 + k_{12}\right)\left[M\right]} \tag{4.41}$$

Solving these equations, the steady state concentration of the hydroxyl radical is

$$\left[OH^{\bullet}\right]_{ss} = \left(\frac{2k_1\left[OH^-\right] + k_{13}\left[M\right]}{k_{12}\left[M\right]}\right)\left[O_3\right] \tag{4.42}$$

Hence we have the following equation:

$$-\frac{1}{\left[O_3\right]} \cdot \frac{d\left[O_3\right]}{dt} = 2k_1\left[OH^-\right]\left(1 + \frac{k_9}{k_{12}}\right) + k_{13}\left[M\right]\left(\frac{k_9}{k_{12}}\right) \tag{4.43}$$

Considering all of the chain reactions for the loss of ozone, we have

$$-\frac{1}{\left[O_3\right]} \cdot \frac{d\left[O_3\right]}{dt} = k_1\left[OH^-\right] + 2k_1\left[OH^-\right]\left(1 + \frac{k_9}{k_{12}}\right) + k_{13}\left[M\right]\left(\frac{k_9}{k_{12}}\right) \tag{4.44}$$

Including the total ozone consumption via nonchain reactions as well, we can write this equation as

$$-\frac{1}{\left[O_3\right]} \cdot \frac{d\left[O_3\right]}{dt} = k_1\left[OH^-\right] + 2k_1\left[OH^-\right]\left(1 + \frac{k_9}{k_{12}}\right) + k_{13}\left[M\right]\left(\frac{k_9}{k_{12}}\right) + k_{14}\left[M\right] \tag{4.45}$$

At constant $[OH^-]$ (or pH) and constant $[M]$ we can write

$$\frac{d\left[O_3\right]}{dt} = -k_{tot}\left[O_3\right]_{tot} \tag{4.46}$$

Thus, the rate is pseudo first order with a rate constant k_{tot}. A plot of $\ln\,[O_3]$ versus t should yield a straight line with slope k_{tot} for a specified $[OH^-]$ and $[M]$. For a batch reactor, this equation will give the rate of disappearance of ozone in the presence of different substrates. The identity of solute M may be different at different points along the chain. Thus, for example, the species M undergoing oxidation and forming radicals that start the chain reaction with the rate constant k_{13} is called an "initiator" (represented I). Those that terminate the chain by reacting with $OH^{\bullet}$ are called "terminators" or "suppressors" (represented S). Those species that react with $OH^{\bullet}$ to reform $O_2^{-\bullet}$ with rate constant k_9 are called "propagators" (represented P). The "direct" reaction of M with ozone characterized by the rate constant k_{14} is designated k_d. Staehelin and Hoigne (1985) have identified different species in water that perform the mentioned functions. These are listed in Table 4.2 along with the respective rate constants. In order to differentiate these species correctly in the rate equation, they also generalized the rate constant to give

$$k_{tot} = k_1\left[OH^-\right] + \left(2k_1\left[OH^-\right] + k_1\left[M\right]\right)\left(1 + \frac{k_p\left[M\right]}{k_s\left[S\right]}\right) + k_d\left[M\right] \tag{4.47}$$

TABLE 4.2
Common Types of Initiators, Propagators, and Scavengers Found in Wastewater, and Values of k_{tot} for Ozone Depletion in Water Containing Different Species, M

Initiator	Promoter	Scavenger
Hydroxyl ion	Aromatic	Bicarbonate ion
Ferrous ion	Alcohols	Carbonate ion
DOCs (humic and fulvic acid)		DOCs (humic acid)

[M]	pH	k_{tot} (s^{-1})
0 (none)	4.0	0.15 ± 0.02
50 mM PO_4^{2-}	4.0	0.072 ± 0.006
7 µM t-BuOH	4.0	0.055 ± 0.004
50 µM t-BuOH	4.0	0.02 ± 0.002

Source: Staehelein, J. and Hoigne, J., *Environ. Sci. Technol.*, 19, 1206, 1985.

where $k_I = k_{13}$, $k_d = k_{14}$, $k_P = k_9$, and $k_S = k_{12}$. For most hard wastewaters where bicarbonate is the predominant scavenger of OH•, we have $k_I[M] \gg 2k_1[OH•]$, and hence

$$k_{tot} = k_I[M]\left(1 + \frac{k_p[M]}{k_s[S]}\right) + k_d[M] \qquad (4.48)$$

Now that we have gained an understanding of the kinetics of ozone oxidation processes, we shall next see how this information can be utilized in analyzing ozonation reactors.

Example 4.6 Wastewater Oxidation Using Ozone in a Continuous Reactor

Consider a diffuse aerator reactor operated in the continuous mode (Figure 4.10). A given influent feed rate of water (Q_L, m^3 s^{-1}) is contacted with an incoming gas stream of ozone at a concentration $[O_3]_g^{in}$ (mol m^{-3}) at a volumetric flow rate of Q_G (m^3 s^{-1}). We are interested in obtaining the extent of ozone consumption within the reactor if "pure" water is used and also in the presence of other solutes. Consider a CSTR for which a mass balance for ozone in the aqueous phase within the reactor gives

$$V_L\frac{d[O_3]}{dt} = Q_L\left([O_3]_0 - [O_3]\right) + K_w a V_L\left([O_3]^* - [O_3]\right) - k_{tot}V_L[O_3] \qquad (4.49)$$

where $[O_3]_0 = 0$, $K_w a$ is the mass transfer coefficient for ozone from gas to water, $[O_3]^*$ is the saturation concentration of ozone in water, and k_{tot} is the first

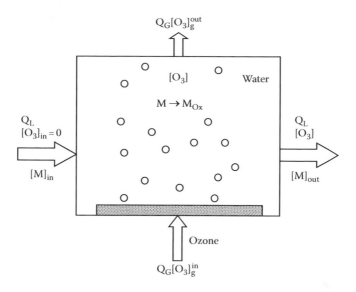

FIGURE 4.10   Schematic of an ozone reactor for wastewater oxidation.

order decomposition of ozone as described earlier. At steady state, since $d[O_3]/dt = 0$, we have

$$\frac{[O_3]}{[O_3]^*} = \frac{1}{1+\left(\dfrac{Q_L}{V_L}+k_{tot}\right)\cdot\dfrac{1}{K_w a}} \qquad (4.50)$$

With increasing k_{tot}, the ratio decreases, indicating that the exit ozone concentration is lower and more of ozone is consumed within the CSTR. Consider a typical ozone wastewater oxidation reactor with $Q_L = 2500$ m³ h⁻¹ and $V_L = 1500$ m³. $K_w a$ for ozone is typically 0.03 min⁻¹ (Roustan et al., 1992). From Staehelin and Hoigne (1985), $k_{tot} = 3$ min⁻¹ at pH 4 in the presence of 50 mM phosphate. Hence the ratio is $[O_3]^{ss}/[O_3]^* = 0.0098$. If the partial pressure of ozone P_{O_3} in the incoming gas phase is known, then $[O_3]^* = P_{O_3}/K''_{AW}$, where $K''_{AW} = 0.082$ is Henry's constant expressed in atm m³ mol⁻¹. For a typical $P_{O_3} = 0.0075$ atm, $[O_3]^* = 0.092$ mol m⁻³. Hence $[O_3]^{ss} = 9 \times 10^{-4}$ mol m⁻³. A mass balance for ozone over the entire reactor yields

$$Q_G\left([O_3]_g^{in} - [O_3]_g^{out}\right) = \left(Q_L + k_{tot} V_L\right)[O_3]^{ss} = 4 \text{ mol·min}^{-1}$$

If $Q_G = 1000$ m³ h⁻¹ (= 16 m³ min⁻¹), $[O_3]_g^{in} - [O_3]_g^{out} = 0.24$ mol m⁻³. Since $[O_3]_g^{in} = P_{O_3}/RT = 0.31$ mol m⁻³, we obtain $[O_3]_g^{out} = 0.07$ mol m⁻³. Hence the ozone transfer efficiency in the reactor is 77%.

4.1.2.3 Photochemical Reactions and Wastewater Treatment

Photochemical reactions are useful in treating wastewater streams. An application in this area is the use of semiconductors (e.g., TiO_2) in enhancing the UV-promoted oxidation of organic compounds. The reaction pathway provided by TiO_2 is complicated (Legrini et al., 1993). TiO_2 is a semiconductor. It has a structure that is composed of a valence band (filled electronic level) and a conduction band (vacant electronic level) that are separated by a band gap (see Figure 4.11). As an electron jumps from the valence band to the conduction band, a hole (positive charge) is left behind in the valence band. This e^- jump can be brought about through excitation by light (UV or visible). Organic molecules that are thereby oxidized can scavenge the hole left behind in the valence band. The photoexcited TiO_2 with the electron-hole pair $(e^- + h^+)$ is depicted as

$$TiO_2 \xrightarrow{\ h\nu\ } TiO_2\left(e^- + h^+\right)$$

Two types of oxidation reactions have been noted in aqueous suspensions of TiO_2. One is the electron transfer from the organic molecule adsorbed on the surface (RX_{ads})

$$TiO_2\left(h^+\right) + RX_{ads} \rightarrow TiO_2 + RX_{ads}^{+\cdot}$$

and the second one is the e^- transfer to the adsorbed solvent molecule (H_2O_{ads})

$$TiO_2\left(h^+\right) + H_2O_{ads} \rightarrow TiO_2 + HO_{ads}^{\cdot} + H^+$$

This reaction appears to be of greater importance in the oxidation of organic compounds. The abundance of water makes this a feasible reaction. Similarly, adsorbed hydroxide ion (OH_{ads}^-) also appears to participate in the reaction

$$TiO_2\left(h^+\right) + OH_{ads}^- \rightarrow TiO_2 + HO_{ads}^{\cdot}$$

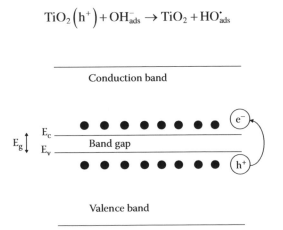

FIGURE 4.11   Valence and conduction bands in a metal oxide semiconductor.

The ever-present O_2 in water provides another avenue of reaction by acting as an electron acceptor:

$$TiO_2(e^-) + O_2 \rightarrow TiO_2 + O_2^{-\bullet}$$

The addition of hydrogen peroxide has been shown to catalyze this process, presumably via a surface dissociation of H_2O_2 to $OH^{\bullet}$ that subsequently oxidizes organic compounds.

$$TiO_2(e^-) + H_2O_2 \rightarrow TiO_2 + OH^- + OH^{\bullet}$$

This reaction is sensitive to pH changes and the electrical double layer properties of TiO_2 in aqueous solutions. Therefore, the UV-promoted dissociation of organic compounds by TiO_2 in the presence of H_2O_2 is significantly affected by pH and solution ionic strength.

The large-scale development of semiconductor-promoted reactions awaits optimal design of the photoreactor, photocatalyst, and radiation wavelengths. The critical issue in this regard is a suitable design for the treatment of a sample that can simultaneously absorb and scatter radiation. Work in this area appears to be promising and is summarized in Table 4.3.

4.1.2.4   Photochemical Reactions in Natural Waters

Natural waters are a mixture of several compounds, and, in most cases, transient radicals or species are produced by photochemical reactions. Zepp (1992) has summarized a variety of transient species. Their effects on photochemical reactions are such that the rates of reactions can vary by orders of magnitude depending on the type and concentration of the transient. Dissolved organic compounds (DOCs—humic and fulvic compounds), inorganic chemicals (nitrate, nitrite, peroxides, iron, manganese), and particulates (sediments, biota, and minerals) can produce transients in natural waters upon irradiation. The transients participate in and facilitate the redox reactions of other compounds in natural waters.

TABLE 4.3

Photodegradation of Organics by TiO_2/UV Process

Phenol, TOC reduced by 35% in 90 min

Acetic, benzoic, formic acids, ethanol, methanol, TOC reduced by >96% in 10 min

Degradation of aniline, salicylic acid, and ethanol

1,2-Dimethyl-3-nitrobenzene and nitro-*o*-xylene from industrial wastewater, >95% TOC removed in 30 min

Degradation of pentachlorophenol

Degradation of other organic compounds (trichloroethylene, chlorobenzene, nitrobenzene, chlorophenols, phenols, benzene, chloroform)

Degradation of a pesticide, atrazine

Source: Legrini, O. et al., *Chem. Rev.*, 93, 671, 1993.

TABLE 4.4
Photochemical Transient Species in Natural Waters

Transient Species	Source of Transient	Typical Steady State Concentration (mol L^{-1})	Typical Half-Life (h)
$^1O_2$	DOM	10^{-14}–10^{-12}	≈ 3.5
e_{aq}^-	DOM	$\approx 10^{-11}$	≈ 4.3
OH$^\cdot$	NO_3^-, NO_2^-, H_2O_2	10^{-17}–10^{-15}	≈ 170
$O_2^{-\cdot}$	DOM	10^{-8}–10^{-7}	≈ 2

Sources: Zepp, R.G., Sunlight-induced oxidation and reduction of organic xenobiotics in water, in: Schnoor, J.L. (ed.), *Fate of Pesticides and Chemicals in the Environment*, John Wiley & Sons, Inc., New York, 1992, pp. 127–140; Jacoby, M., *Chem. Eng. News*, 76, 47, 1998.

There are a variety of transients that have been identified in natural waters. Some of the important ones are (1) solvated electron, (2) triplet and singlet oxygen, (3) superoxide ions and hydrogen peroxide, (4) hydroxyl radicals, and (5) triplet excited state of dissolved organic matter. The level of steady state concentrations of some of these species and their typical half-lives in natural waters are given in Table 4.4. Solvated electron (e_{aq}^-) is a powerful oxidant, observed in natural waters during irradiation. It reacts rapidly with electronegative compounds (both organic and inorganic). Its reaction with O_2 is the primary pathway for the production of superoxide anions in natural waters. The major source of e_{aq}^- is aquatic humic compounds. The quantum yield for their production is approximately 10^{-5}. Dissolved organic matter in natural waters is also known to absorb photons to generate singlet and triplet excited states. These then decay by transferring energy to dissolved oxygen to produce singlet oxygen. Singlet oxygen is an effective oxidant, and the quantum yield for its formation is 0.01–0.03 in UV and blue spectra. Superoxide ions and hydrogen peroxide are found both in lakes and in atmospheric moisture (fog and rain). They are longer lived than other transients and are powerful oxidants for most organics. Algae and other biota are known to quench the action of superoxide ions.

In general, the kinetics and concentrations of photochemical transients in natural waters can be understood through the application of the steady state theory. The formation of transient (Tr) is via the reaction

$$S \xrightarrow[h\nu]{O_2} Tr \tag{4.51}$$

with the rate $r_{f,Tr} = \phi I$. The transient species can either decay by itself or by reaction with a solute molecule A:

$$\begin{aligned} Tr &\xrightarrow{k_{Tr}} P \\ Tr + A &\xrightarrow{k_r} P \end{aligned} \tag{4.52}$$

The total rate of disappearance of the transient is

$$r_{d,Tr} = k_r [A][Tr] + k_{Tr} [Tr] \tag{4.53}$$

At steady state, $r_{d,Tr} = r_{f,Tr}$, and hence

$$[Tr]^{ss} = \frac{\phi I}{k_r [A] + k_{Tr}} \tag{4.54}$$

The overall rate of the reaction of A is

$$r_A = -\frac{d[A]}{dt} = k_r \cdot \left(\frac{\phi I}{k_r [A] + k_{Tr}} \right) \cdot [A] \tag{4.55}$$

If the concentration of A is sufficiently low, as is the case in most natural systems, then $k_r[A] \ll k_{Tr}$, and hence we have

$$r_A = \frac{k_r}{k_{Tr}} \cdot \phi I [A] = k_r' [A] \tag{4.56}$$

where k_r' is a pseudo first order rate constant. Zepp (1992) has summarized how each of the terms comprising k_r' can be obtained experimentally.

4.2   AIR ENVIRONMENT

In this section, the use of concepts from chemical kinetics and mass transfer theory for modeling the fate and transport of chemicals in the air environment and in the design of air pollution control devices is addressed.

4.2.1   FATE AND TRANSPORT MODELS

4.2.1.1   Box Models

It is possible to explore environmental systems through the use of "box models." These "models" simulate the complex behavior of a natural system by applying useful simplifications. For example, box models are derived from the concept of CSTRs. Each phase is considered to be a well-mixed compartment, and pollutants exchange between the different compartments.

CSTR box models are used in understanding the transport of chemicals from mobile and stationary sources (automobiles, chemical and coal-powered power plants) to the atmosphere, transport and fate of pollutants between air and water, sediment and water, and between soil and air. The sediment-water system is pictorially represented in Figure 4.12a. Chemicals enter the water stream either with the inflow or by wet and dry deposition from the atmosphere. A third pathway is the resuspension of solids from the sediment beneath the stream. The chemicals dissipate

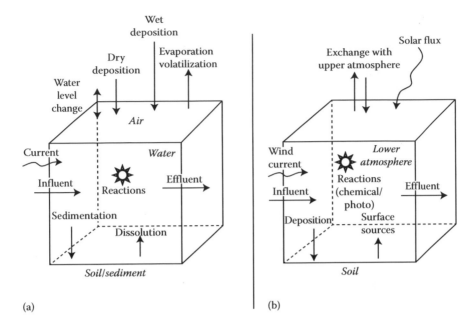

FIGURE 4.12 Schematic of typical "box" models for (a) a natural aquatic stream, and (b) atmosphere. The various transport and fate processes in each case are shown.

from the stream either by chemical reaction or biodegradation or are removed with the outflow and by evaporation from the surface. They may also subside into the underlying sediment by sedimentation with solids.

A similar CSTR box model can also be developed for atmospheric transport of chemicals (Figure 4.12b). Stationary sources such as industrial plants, waste pits, and treatment plants provide pollutant inputs to the atmosphere. Mobile sources such as automobiles also contribute to pollutant influx to the atmosphere. Deposition of aerosol-associated pollutants removes pollutants from the atmosphere. Advective flow due to winds constitutes the inflow and outflow from the CSTR. Chemical and photochemical reactions change concentrations within the box.

Consider a well-mixed box of downwind length L, crosswind width W, and vertical height H. Figure 4.13 depicts the atmosphere of volume $V = LWH$ that is homogeneously mixed (i.e., no concentration gradients). A large urban atmosphere with good mixing (turbulence) and a wide distribution of emission sources in the lower atmosphere or a small indoor volume that experiences excellent mixing and ventilation can be represented by Figure 4.13. Global atmospheric models rely on the assumption that the troposphere is a well-mixed CSTR. In some cases, the northern and southern hemispheres are considered to be two separate reactors with different degrees of mixing, but linked via chemical exchange. These models are useful in obtaining the residence times of important species in the atmosphere (Seinfeld, 1986). For the case in Figure 4.13, a mass balance on species A within the control volume will give

Rate of input = Rate of output + Rate of reaction + Rate of accumulation.

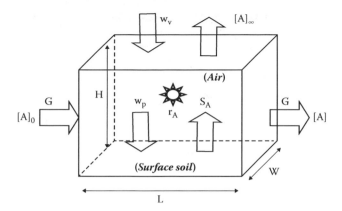

FIGURE 4.13   Material balance for a well-mixed box model for an urban atmosphere.

$$G[A]_0 + w_v A_s [A]_\infty + S_A A_s = G[A] + w_p A_s [A] + w_v A_s [A] + r_A V + V \frac{d[A]}{dt}$$
$$(4.57)$$

Dividing throughout by V and noting that Q/V = U and $A_s/V = 1/H$, we get

$$\frac{d[A]}{dt} = \frac{U}{L} \cdot ([A]_0 - [A]) + \frac{w_v}{H} \cdot ([A]_\infty - [A]) - \frac{w_p}{H} \cdot [A] + \frac{S_A}{H} - r_A \quad (4.58)$$

where the overall loss by reaction is given by $r_A = k[A]$, U is the wind velocity (m s^{-1}), w_v is the ventilation velocity away from the box (m s^{-1}), w_p is the surface deposition velocity from the atmosphere (m s^{-1}), S_A is the surface source strength of A (μg (m^2 s)$^{-1}$), $[A]_\infty$ is the concentration of A in the upper atmosphere (μg m^{-3}), and $[A]_0$ is the concentration of A in the incoming air (μg m^{-3}). A_s is the surface area (m^2). The following examples illustrate the application of this model.

Example 4.7   A CSTR Box Model for an Urban Area

Assume that the pollutant is conservative (nonreactive) and that velocities w_p and w_v are negligible. Then the overall mass balance reduces to the following equation:

$$\frac{d[A]}{dt} = \frac{U}{L} \cdot ([A]_0 - [A]) + \frac{S_A}{H} \quad (4.59)$$

At steady state, d[A]/dt = 0 and steady state concentration in the box is given by

$$[A]_{ss} = [A]_0 + \frac{S_A}{U} \cdot \frac{L}{H} \quad (4.60)$$

Thus if $[A]_0 = 0$, $[A]_{ss}$ is proportional to S_A/U, i.e., the larger the source strength, the larger is the steady state concentration in the box.

If the process is unsteady state, then we can solve for $[A]$ as a function of t with the initial condition that at $t = 0$, $[A] = [A]_{initial}$ and obtain the following equation:

$$[A] = \left(\frac{S_A}{U} \cdot \frac{L}{H} + [A]_0\right)\left(1 - e^{-\frac{U}{L} \cdot t}\right) + [A]_{initial}\, e^{-\frac{U}{L} \cdot t} \tag{4.61}$$

and for $[A]_{initial} = 0$ and $[A]_0 = 0$,

$$[A] = \left(\frac{S_A}{U} \cdot \frac{L}{H}\right)\left(1 - e^{-\frac{U}{L} \cdot t}\right) \tag{4.62}$$

which states that the concentration increases toward the steady state value as $t \to \infty$. The rate constant is U/L. The inverse L/U is called the "residence time" in the box.

Urban areas receive air pollution from stationary and mobile sources. In the United States, for example, Los Angeles, CA, experiences air pollution resulting from automobile exhaust, whereas in Baton Rouge, LA, the problem results from industrial sources. In the case of Los Angeles, a city nestled in a valley between mountains, we can consider the valley to be a CSTR where periodic strong Santa Ana winds replace the air. The prevailing winds bring with them a background level of pollutant (e.g., carbon monoxide, CO) from surrounding areas that adds to the existing background CO in the valley. The automobiles within the city emit CO as a component of its exhaust gas. Figure 4.14 schematically represents the various processes. The total source strength in moles per hour is given by S_{TOT} (mol h^{-1}) = S_A (mol (m^2 h)$^{-1}$) LW = Rate of CO in by prevailing winds + Rate of CO input by

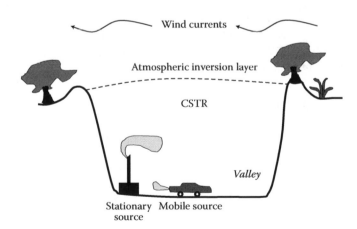

FIGURE 4.14 A simplified well-mixed "box" model for an urban area nestled between mountains (e.g., Los Angeles, CA).

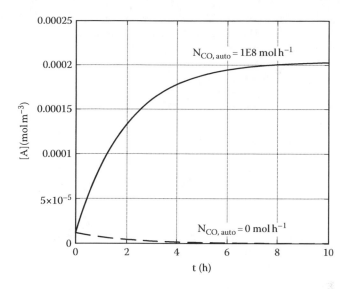

FIGURE 4.15 Change in concentration of carbon monoxide with time in an urban area as a result of automobile emissions. A comparison is made with one that is devoid of automobile contributions.

automobiles. The total volumetric flow rate of air through the valley is G ($m^3\,h^{-1}$) = U ($m\,h^{-1}$) WH (m^2) and U ($m\,h^{-1}$) = GL/V. Hence

$$[A] = \frac{S_{TOT}}{G}\left(1 - e^{-\frac{G}{V}\cdot t}\right) + [A]_0\, e^{-\frac{G}{V}\cdot t} \tag{4.63}$$

Let us take a specific example where $V = 1 \times 10^{12}\ m^3$, $G = 5 \times 10^{11}\ m^3\,h^{-1}$. CO in by wind = 2×10^6 mol h^{-1}, CO in by automobiles = 1×10^8 mol h^{-1}, and $[A]_{initial} = 1.2 \times 10^{-5}$ mol m^{-3} (background CO). The equation is $[A] = 2.04 \times 10^{-4}\ (1 - e^{-t/2}) + 1.2 \times 10^{-5}\ e^{-t/2}$. The steady state concentration $[A]_{ss}$ is 2.04×10^{-4} mol m^{-3}. Figure 4.15 depicts the change in CO with time. If the contribution from automobiles is negligible, the $[A]$ in the CSTR is continually diluted.

Example 4.8 An Indoor Air Pollution Model

In this case, we consider a pollutant that may not only react but also has a source within the indoor atmosphere. We disregard both w_v and w_p. The overall mass balance in this case is

$$\frac{d[A]}{dt} = \frac{U}{L}\cdot\left([A]_0 - [A]\right) + \frac{S_A}{H} - k[A] \tag{4.64}$$

The term $U/L = (UWH)/(LWH) = G/V$ is called the air exchange rate, I_a and $S_A/H = (S_A LW)/(LWH) = S_{TOT}/V$ where S_{TOT} is the surface source strength expressed in $\mu g \ s^{-1}$. Using the initial condition, $[A] = [A]_{initial}$ at $t = 0$, we get

$$[A] = \left(\frac{\left(\frac{S_{TOT}}{V} + [A]_0 \, I_a\right)}{I_a + k}\right) \cdot \left(1 - e^{-(I_a + k)t}\right) + [A]_{initial} \, e^{-(I_a + k)t} \qquad (4.65)$$

The steady state concentration ($t \rightarrow \infty$) is

$$[A]_{ss} = \frac{\left(\frac{S_{TOT}}{V} + [A]_0 \, I_a\right)}{(I_a + k)} \qquad (4.66)$$

For the special case where $[A]_0 = 0$ and $k = 0$ (conservative pollutant) and where unsteady state is applicable

$$[A] = \left(\frac{S_{TOT}}{I_a V}\right) \cdot \left(1 - e^{-I_a t}\right) \qquad (4.67)$$

If a room of total volume $V = 1000 \ m^3$ where a gas range emits 50 mg h^{-1} of NO$_x$ is considered for an air exchange rate $I_a = 0.2 \ h^{-1}$, and a first order rate of loss at 0.1 h^{-1}, we have at steady state $[A]_{ss} = (50/1000)/(0.2 + 0.1) = 0.16$ mg m^{-3} = 160 $\mu g \ m^{-3}$.

Example 4.9   A Global Mixing Model

Globally, the stratosphere and the atmosphere are considered to be separate compartments (boxes) with internal mixing, but with a slow exchange between the two boxes (Reible, 1998; Warneck, 1999). This is one of the reasons why nonreactive chemicals (e.g., chlorofluorocarbons, or CFCs) released to the troposphere slowly make their way into the stratosphere where they participate in other reactions (e.g., with ozone). If the slow exchange rate is represented G (see Figure 4.16), the rate of change in concentration in each CSTR is given by

$$\text{Rate of accumulation} = \text{Rate of input} - \text{Rate of output}$$

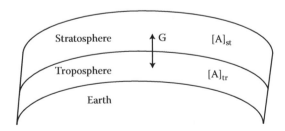

FIGURE 4.16   Schematic of the model assuming well-mixed troposphere and stratosphere with a slow exchange between the two compartments.

$$V_{tr} \frac{d[A]_{tr}}{dt} = G\left([A]_{st} - [A]_{tr}\right)$$

$$V_{st} \frac{d[A]_{st}}{dt} = G\left([A]_{tr} - [A]_{st}\right)$$

(4.68)

Consider radioactive materials released through atomic weapons testing in the early 1960s. The release occurred in the stratosphere where they reacted only slowly. Upon reaching the troposphere, however, they were quickly reacted away. In such a case, $[A]_{st} \gg [A]_{tr}$ and hence

$$\frac{d[A]_{st}}{dt} = -\frac{G}{V} \cdot [A]_{st} = -\frac{1}{\tau_{st}} \cdot [A]_{st}$$

(4.69)

where τ_{st} is the exchange (mixing) time between the stratosphere and troposphere. Upon integration with $[A]_{st} = [A]_{st}^0$ at $t = 0$, i.e., when weapons testing first began, we get

$$[A]_{st} = [A]_{st}^0 \cdot e^{-\frac{\tau}{\tau_{st}}}$$

(4.70)

The concentration in the stratosphere should decline exponentially with time. In other words, a plot of $\ln [A]_{st}$ versus t will give $1/\tau_{st}$ as the slope from which τ_{st} can be ascertained. Warneck (1988) has listed values of mixing time for several radioactive tracers. A similar approach can also be used for appropriate tracers to calculate mixing times within the troposphere and the stratosphere, if the two boxes are further subdivided into north and south hemispheres. Average values obtained are summarized in Table 4.5.

Example 4.10   Atmospheric Residence Time

Atmospheric residence time is important in atmospheric models. This tells us the duration a molecule spends in the atmosphere before it is removed either by wet or dry deposition to the surface of the earth. This is based on the perfectly mixed

---

TABLE 4.5
Exchange (Mixing) Times between and within
Compartments Using Data on ^{85}Kr

Exchange Compartments	Mixing Time (Years)
Stratosphere-troposphere	1.4
Troposphere-troposphere	1.0
Stratosphere-stratosphere	4.0

Source: Warneck, P., *Chemistry of the Natural Atmosphere,* Academic Press, New York, 1988.

box approach. Starting with 1 cubic volume of the lower atmosphere, we have the following mass balance (Warneck, 1999; Seinfeld and Pandis, 2006):

Input rate by flow – Output rate by flow + Rate of production

– Rate of removal = Accumulation

$$F_A^{in} - F_A^{out} + S_A - R_A = \frac{dm_A}{dt} \qquad (4.71)$$

where m_A is the total mass of species A in the unit volume of air considered. The average residence time is defined as $\tau_A = m_A / \left(S_A - F_A^{out}\right) = m_A / \left(S_A - F_A^{in}\right)$, since at steady state $R_A + F_A^{out} = S_A + F_A^{in}$. For the entire atmosphere at steady state $F_A^{in} = F_A^{out} = 0$ and $S_A = R_A$. Atmospheric concentrations are expressed either in mass per unit volume of air, C_A ($\mu g\ m^{-3}$), or as a "mixing ratio," $\xi_i = C_A/C_{tot}$, where C_A is the molar concentration of A in air and C_{tot} is the total molar concentration of air. Note that since ideal gas law applies $\xi_A = P_A/P$. C_A ($\mu g\ m^{-3}$) and ξ_A (ppmv) are related as follows: $\xi_A = 8.314\ (T/PM_A)\ C_A$ (see also Appendix D). For sulfur, which has a total rate of production of $2 \times 10^{14}\ g\ y^{-1}$, and an average mixing ratio of 1 ppb, we find the total mass at steady state to be $(1 \times 10^{-9}\ g\ g^{-1})\ (5 \times 10^{21}\ g)$ where 5×10^{21} g is the total mass of the atmosphere. Hence $m = 5 \times 10^{12}$ g. Thus the average residence time is $\tau_A = 5 \times 10^{12}/2 \times 10^{14} = 0.025$ y. Note that if $\tau_A = G/R_A$ and $R_A = k_A G$, we have $\tau_A = 1/k_A$.

4.2.1.2 Dispersion Models

When pollutant sources are not widely distributed across a large area, and mixing is insufficient, the CSTR box model described earlier becomes inapplicable. Examples are point source emissions from smokestacks, episodic spills on the ground, explosions, or accidental releases of air pollutants. We shall use the general case of a continuous smokestack emission to illustrate the principles of air pollution modeling from point sources.

We saw in Chapter 2 that temperature gradient in the atmosphere determines whether the atmosphere is stable or unstable. The variation in temperature with height determines the extent of buoyancy-driven mixing within the atmosphere. To do so, we consider the ideal case where a parcel of dry air is allowed to rise adiabatically through the atmosphere. The air parcel experiences a decrease in temperature as it expands in response to decreasing temperature with increasing elevation. This is called the "adiabatic lapse rate," Γ_{adia} (= dT/dz), and is a standard value of 1°C/100 m for dry air, as we saw in Chapter 2. For a saturated air, this value is slightly smaller (0.6°C/100 m). If the prevailing atmosphere has an "environmental lapse rate," $\Gamma_{env} > \Gamma_{adia}$, the atmosphere is considered unstable and rapid mixing occurs thereby diluting the air pollutant. If $\Gamma_{env} < \Gamma_{adia}$, the atmosphere is stable and little or no mixing occurs.

The second important parameter is the wind velocity. In general, the velocity of wind above the surface follows a profile given by Deacon's power law, $\dfrac{U_z}{U_{z_1}} = \left(\dfrac{z}{z_1}\right)^p$, where U_z is the wind velocity at height z, U_{z_1} is the wind velocity at height z_1, and p is a positive exponent (varies between 0 and 1 in accordance with the atmospheric stability).

These effects of temperature and wind velocity are introduced in the dispersion models through the Pasquill stability criteria. However, we are still left with one additional issue that of the terrain itself. Obviously the wind velocity near a surface is strongly influenced by the type of terrain. Rough surfaces or irregular terrain (buildings or other structures) will give rise to varying wind directions and velocities near the surface. Open wide terrains do not provide that much wind resistance.

The dispersion of gases and vapors in the atmosphere is influenced by the degree of atmospheric stability, which in turn is determined by the temperature profile in the atmosphere, the wind direction, wind velocity, and surface roughness. These factors are considered in arriving at equations to determine the dispersion of pollutants in a prevailing atmosphere. Let us consider the emission of a plume of air pollutant from a tall chimney as shown in Figure 4.17. The emission occurs from a point source and is continuous in time. The process is at steady state. The air leaving the chimney is often at a higher temperature than the ambient air. The buoyancy of the plume causes a rise in the air parcel before it takes a more or less horizontal travel path. We apply the convective-dispersion equation that was derived in Chapter 2, extended to all three Cartesian coordinates. We consider here a nonreactive (conservative) pollutant. Hence, we have the following equation:

$$U \frac{\partial [A]}{\partial x} = D_x \frac{\partial^2 [A]}{\partial x^2} + D_y \frac{\partial^2 [A]}{\partial y^2} + D_z \frac{\partial^2 [A]}{\partial z^2} \qquad (4.72)$$

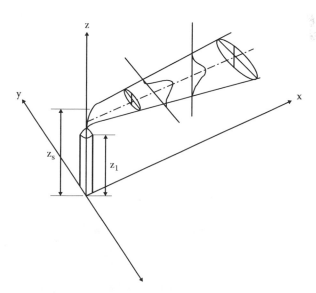

FIGURE 4.17   Emission of a plume of air pollutant from a tall chimney.

For stack diffusion problems, from experience the following observations can be made:

1. In the x-direction, the mass transfer due to bulk motion, $U\partial[A]/\partial x$, far exceeds the dispersion, $D_x\partial^2[A]/\partial x^2$.
2. The wind speed on the average remains constant, i.e., U is constant.

The simplified expression is then

$$U\frac{\partial[A]}{\partial x} = D_y\frac{\partial^2[A]}{\partial y^2} + D_z\frac{\partial^2[A]}{\partial z^2} \tag{4.73}$$

Upon integration and rearranging, the following solution results:

$$[A](x,y,z) = \frac{Q_s}{\pi U \sigma_y \sigma_z} \cdot \exp\left[-\frac{1}{2}\left(\frac{y^2}{\sigma_y^2} + \frac{(z-z_s)^2}{\sigma_z^2}\right)\right] \tag{4.74}$$

where $\sigma_y^2 = \dfrac{2D_y x}{U}$ and $\sigma_z^2 = \dfrac{2D_z x}{U}$ are the respective squares of the mean standard deviations in y and z directions for the concentrations. This equation is only applicable for the concentration in the downwind direction up to the point in the x-direction where the ground level concentration (z = 0) is significant. After this, "reflection" of pollutants will occur, since soil is not a pollutant sink and material will diffuse back to the atmosphere. This gives rise to a mathematical equivalence of having another image source at a distance—z_s (Figure 4.18). A super-position of the two solutions

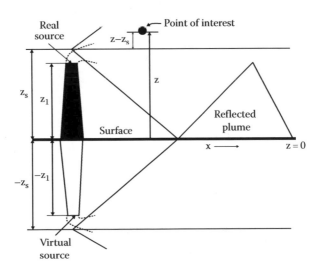

FIGURE 4.18 The reflection of a gaseous plume of a pollutant mathematically analyzed using an imaginary source.

gives the final general equation for pollutant concentration anywhere downwind of the stack. The concentration of vapor in the air any point from a plume of effective height z_s is given by

$$[A](x,y,z) = \frac{Q_s}{\pi U \sigma_y \sigma_z} \cdot \left[e^{-\frac{(z_s-z)^2}{2\sigma_z^2}} + e^{-\frac{(z_s+z)^2}{2\sigma_z^2}} \right] \cdot e^{-\frac{y^2}{2\sigma_y^2}} \qquad (4.75)$$

The concentration is given in g m^{-3} if Q_s is expressed in g s^{-1}. It can be converted to the more common units of μg m^{-3} or ppmv using appropriate conversions.

The standard deviations σ_y and σ_z have been correlated to both mechanical turbulence (wind speed at 10 m height) and buoyant turbulence (time of day). These give rise to what are called "Pasquill stability categories" given in Table 4.6. Based on the stability class, appropriate standard deviations for any distance x downwind of the source are obtained (Figure 4.19).

The effective height of the stack z_s (m) is the sum of the actual stack height z_1 (m) and the plume rise Δz (m). The plume rise can be estimated using a number of different expressions (Wark et al., 1998). One parameter that is often sought in air pollution modeling is the so-called ground level center-line concentration:

$$[A](x,0,0) = \frac{Q_s}{\pi U \sigma_y \sigma_z} \cdot e^{-\left(\frac{z_s^2}{2\sigma_z^2}\right)} \qquad (4.76)$$

The maximum concentration at a receptor along the ground-level center-line will be given by substituting $z_s^2 = 2\sigma_z^2$

$$[A]_{max} = \frac{0.1171 Q_s}{U \sigma_y \sigma_z} \qquad (4.77)$$

TABLE 4.6
Pasquill Stability Categories

	Day			Night	
	Incoming Solar Radiation			Cloud Cover	
Wind Speed at 10 m Height (m s^{-1})	**Strong**	**Moderate**	**Slight**	**Mostly Overcast**	**Mostly Clear**
<2	A	A–B	B	E	F
2–3	A–B	B	C	E	F
3–5	B	B–C	C	D	E
5–6	C	C–D	D	D	D
>6	C	D	D	D	D

Notes: Class A is the most unstable and Class F is the most stable. Class B is moderately unstable and Class E is slightly stable. Class C is slightly unstable. Class D is neutral stability and should be used for overcast conditions during day or night.

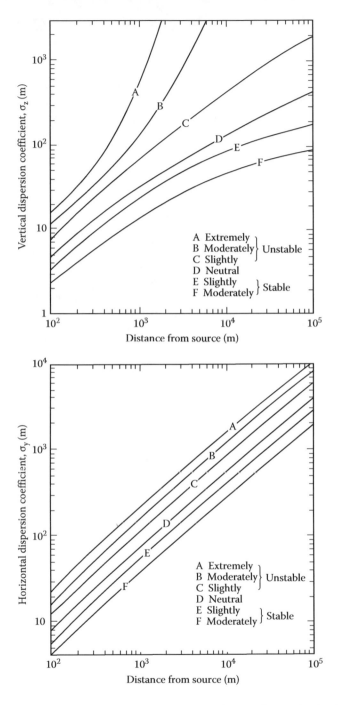

FIGURE 4.19 Vertical and horizontal dispersion coefficients for use in the Gaussian air dispersion model. (Courtesy of Turner, D.B., *Workbook of Atmospheric Dispersion Estimates*, HEW, Washington, DC, 1969.)

Example 4.11   Accidental Spill Scenario

Early morning, on a cloudy day, a tanker truck carrying benzene developed a leak spilling all of its 5000 gal on the highway forming a pool 50 m × 50 m. What will be the ground-level concentration 1 km downwind of the spill? The wind speed averaged 2 m s^{-1} and the ambient temperature was 25°C.

The spill was 1500 m^2 in area of pure benzene. Evaporation of benzene from a pure phase is gas-phase mass transfer controlled with k_g = 0.37 m s^{-1}. P_i^* = 0.125 atm. Hence Q_s = (0.37 cm s^{-1}) (3600 s h^{-1}) (0.125 atm) (1500 × 10^4 cm^2)/[(8.205 × 10^{-5} cm^3 atm K^{-1} mol^{-1}) (298 K)] = 1.7 × 10^5 mol h^{-1} = 3.7 × 10^3 g s^{-1}. At U = 2 m s^{-1}, for an overcast morning the stability class is D. At 1000 m, then σ_y = 70 m and σ_z = 30 m. Hence, [A] (1000 m, 0, 0) = (3700 g s^{-1})/(3.14) (2 m s^{-1}) (70 m) (30 m) = 0.280 g m^{-3} = 280 mg m^{-3}. This is higher than the threshold limit value (TLV) of benzene (32 mg m^{-3}).

4.2.2   AIR POLLUTION CONTROL

There are two kinds of industrial air pollutants—particulates and vapors (gases). The control of these two classes of air pollutants is dependent on their mechanism of capture in specific reactors. Processes that remove particulates rely on physical forces, i.e., gravity, impaction, electrical forces, or diffusion to bring about separation. Vapors and gases are separated using chemical processes such as adsorption, absorption, and thermal processes. In the selection and design of any method, concepts from reaction kinetics and mass transfer play important roles. We will describe selected methods with a view to illustrating the applications of kinetics and mass transfer theories.

Gases and vapors in polluted air streams are treated by chemical methods such as absorption into a liquid stream (water), adsorption onto a solid adsorbent (activated carbon, silica, alumina), or by thermal means (incineration, combustion). We shall discuss each of these with the aim of illustrating the chemical kinetics and mass transfer aspects.

4.2.2.1   Adsorption

Adsorption is a surface phenomenon as we discussed in Chapter 2. Adsorption on solids is an effective technology for pollutant removal from water or air. The design of both water treatment and air treatment are similar. Hence, our discussion in this instance for air pollution is the same as for water pollution. Adsorption from both air and water are conducted in large columns packed with powdered sorbent such as activated carbon. A feed stream containing the pollutant is brought into contact with a fresh bed of carbon. As the pollutant moves through the bed, it gets adsorbed and the effluent air stream is free of the pollutant. Once the bed is fully saturated, the pollutant breaks through and the effluent stream has the same concentration as the influent feed. The progression of the bed saturation is shown in Figure 4.20. At any instant in the bed, there are three distinctly discernible regions. At the point where the pollutant stream enters, the bed is quickly saturated. This is called the "saturated zone." This zone is no longer effective in adsorption. Ahead is a zone where adsorption is most active. This small band is called the "active (adsorption) zone." Farther from

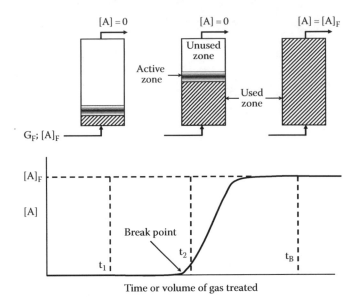

FIGURE 4.20 The movement of the active (adsorption) zone through a fixed bed of activated carbon. Breakthrough is said to occur when the adsorption zone has reached the exit of the carbon bed.

this zone is the unused bed where there is no pollutant. Provided the feed rate is constant, the adsorption zone slowly moves through the bed until it reaches the exit at which stage breakthrough of the pollutant is attained. The rate of movement of the adsorption zone gives us information on the breakthrough time.

If the wave front is ideal, it will be sharp as shown in Figure 4.21. However, because of axial dispersion (nonideal flow), the wave front will broaden (Figure 4.21).

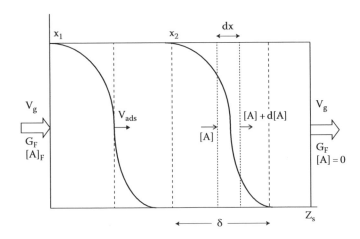

FIGURE 4.21 The analysis of an adsorption wave through a bed of activated carbon.

The velocity of wave front advance can be obtained by determining the capacity of the bed for a given feed rate G_F at a concentration $[A]_F$ in the feed stream. A mass balance over the entire column gives

Mass flow rate of pollutant into the adsorption zone = Rate of adsorption

$$G_F[A]_F = \frac{M_F}{\rho_g} \cdot [A]_F = \rho_{ads} A_C V_{ads} W_0 \tag{4.78}$$

where
M_F is the mass flow rate
ρ_g is the density of air
ρ_{ads} is the density of the adsorbent
V_{ads} is the adsorption zone velocity
A_C is the cross-sectional area
W_0 is the mass of pollutant per unit bed weight

The general Freundlich isotherm (see Chapter 2) in the following form is found to best fit the data over a wide range of concentrations for activated carbon:

$$W_0 = \left(\frac{[A]_F}{K_{Freun}}\right)^{1/n} \tag{4.79}$$

Hence,

$$V_{ads} = \frac{V_g}{\rho_{ads}} \cdot K_{Freun}^{1/n} \cdot [A]_F^{\left(1-\frac{1}{n}\right)} \tag{4.80}$$

where $V_g = M_F/\rho_g A_C$ is the superficial gas velocity.

Consider a differential volume of the bed $A_C dx$. A material balance on the pollutant gives

Input by flow = Output by flow + Mass gained on the bed by transfer from gas to solid

$$G_F[A] = G_F([A]+d[A]) + K_{mt}A_C([A]-[A]_{eq}) \cdot dx \tag{4.81}$$

where $[A]_{eq}$ is the equilibrium value of air concentration that would correspond to the actual adsorbed concentration W in the differential volume. K_{mt} is the overall mass transfer coefficient (time^{-1}) from gas to solid. Rearranging, we obtain

$$-V_g d[A] = K_{mt}([A] - K_{Freun}W^n)dx \tag{4.82}$$

Rearranging and integrating using the boundary conditions that $[A] = [A]_F$ at $x = 0$ and $[A] = 0$ at $x = \delta$,

$$\int_0^\delta dx = -\left(\frac{V_g}{K_{mt}}\right) \cdot \int_{[A]_F}^0 \frac{d[A]}{\left([A] - K_{Freun} W^n\right)} \tag{4.83}$$

An overall mass balance over the differential volume gives $[A]G_F = A_C W \rho_{ads} V_{ads}$. Utilizing the expression for v_{ad} and rearranging, we get $K_{Freun} W^n = [A]^n [A]_F^{1-n}$. Hence,

$$\delta = -\left(\frac{V_g}{K_{mt}}\right) \cdot \int_{[A]_F}^0 \frac{d[A]}{\left([A] - [A]^n [A]_F^{1/n}\right)} \tag{4.84}$$

Note that as $K_{mt} \to \infty$, $\delta \to 0$ and the process becomes equilibrium-controlled. Utilizing the dimensionless variable $\eta = [A]/[A]_F$ and $K_{mt}\delta/V_g = St$, the "Stanton number," we can write

$$St = \int_0^1 \frac{d\eta}{\eta - \eta^n} \tag{4.85}$$

Since the integral is undefined at $\eta = 0$ and 1, we take the limits as 0.01 (leading edge) and 0.99 (trailing edge) so that $[A]$ approaches 1% of the limiting value at both limits. We then have the following definite integral:

$$St = 4.505 + \left(\frac{1}{1-n}\right) \ln\left[\frac{1 - (0.01)^{n-1}}{1 - (0.99)^{n-1}}\right] \tag{4.86}$$

If we assume that the mass in the adsorption zone is much smaller than the mass in the saturated zone, the breakthrough time can be obtained:

$$t_B = \frac{H_B - \delta}{V_{ads}}$$

where H_B is the total height of the adsorbent bed.

Example 4.12   Breakthrough Time for an Activated Carbon Adsorber

Obtain the breakthrough time for a pollutant in a carbon bed at an air mass velocity of 2.0 kg s^{-1} at 1.1 bar and 30°C with an influent pollutant concentration of 0.008 kg m^{-3}. The Freundlich isotherm parameters for the pollutant are $K_F = 500$ kg m^{-3} and $n = 2$. The mass transfer coefficient is 50 s^{-1}. The bed length is 2 m with a cross-sectional area of 6 m^2 and a bulk density of 500 kg m^{-3}.

Gas density, ρ_g = P/RT = (1.1) (29)/(303) (0.083) = 1.3 kg m^{-3}. V_{ads} = 2.0 (500)$^{0.5}$ (0.008)$^{0.5}$/(1.3) (500) (6) = 1.0 × 10^{-3} m s^{-1}. Hence,

$$\delta = \frac{2.0}{(50)(6)(1.3)}\left(4.595 + \frac{1}{1.0}\ln\frac{1-(0.01)^{1.0}}{1-(0.99)^{1.0}}\right) = 0.047 \text{ m. Breakthrough time } t_B =$$

$(2.0 - 0.047)/1.0 × 10^{-3} = 1953 \text{ s} = 0.54 \text{ h}.$

4.2.2.2 Thermal Destruction

Organic compounds that form a large fraction of air pollutants in industrial effluents can be oxidized in "incinerators" at high temperatures. These devices are also called "thermal oxidizers" or "afterburners." Both direct thermal oxidation and catalytic oxidation are practiced in the industry.

A schematic of a direct thermal oxidizer for gases is shown in Figure 4.22. Thermal incineration is also used in destroying organic compounds in a liquid stream or sludge. A "rotary kiln" incinerator is used for the latter and a schematic is shown in Figure 4.23.

Cooper and Alley (1994) present the theory and design of a typical incinerator. Combustion is the chemical process of rapid reaction of oxygen with chemical compounds resulting in heat (Chapter 2). Most fuels are made of C and H, but may also include other elements (S, P, N, and Cl). Although the exact mechanism of combustion can be very complex, the basic overall process can be represented by the following two reactions:

$$C_nH_m + \left(\frac{n}{2} + \frac{m}{4}\right)O_2 \xrightarrow{k_1} nCO + \frac{m}{2}H_2O$$
$$nCO + \frac{n}{2}O_2 \xrightarrow{k_2} nCO_2$$

(4.87)

where n and m are the numbers of C and H atoms in the parent hydrocarbon. The rates of reaction are $r_{HC} = k_1[HC][O_2]$, $r_{CO} = k_2[CO][O_2] - nk_1[HC][O_2]$, and

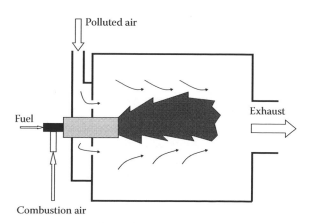

FIGURE 4.22 Schematic of a thermal incinerator (afterburner).

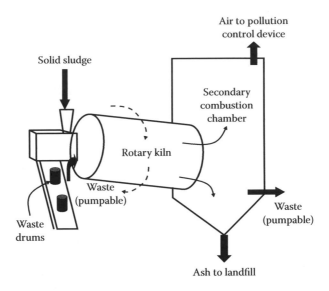

FIGURE 4.23   Schematic of a rotary kiln incinerator.

$r_{CO_2} = k_2[CO][O_2]$. Combustion is frequently carried out in the presence of excess air. The oxygen mole fraction is usually in the range 0.1–0.15 for a HC mole fraction of 0.001. Hence, $r_{HC} = k_1'[HC]$, $r_{CO} = k_2'[CO] - nk_1'[HC]$, and $r_{CO_2} = k_2'[CO]$. In Chapter 2, we saw that these reaction rates reflect the following series reaction:

$$HC \xrightarrow{k_1'} CO \xrightarrow{k_2'} CO_2$$

It is to be expected that since reaction rates are sensitive to temperature, so is the efficiency of an incinerator. Since reactions in thermal oxidizers occur at varying rates, sufficient time should be given for the reaction to go to completion. Mixing in the incinerator should be sufficient to bring the reactants together. Thus, three important time factors are to be considered in the design of an incinerator. These are the "residence time," $\tau_{res} = V_R/G = Z_s/V_g$; "chemical reaction time," $\tau_{rxn} = 1/k$; and "mixing time," $\tau_{mix} = Z_s^2/D_e$. V_R is the total reactor volume (m^3), G is the volumetric waste gas flow rate at afterburner temperature ($m^3 \ s^{-1}$), V_g is the gas velocity ($m \ s^{-1}$), Z_s is the reaction zone length (m), and D_e is the turbulent diffusivity ($m^2 \ s^{-1}$). The relative magnitudes of these times are represented in terms of two dimensionless numbers: Peclet number, $Pe = Z_s V_g/D_e$, and the Damköhler number, $Da = Z_s k/V_g$. If Pe is large and Da is small, then mixing is rate controlling in the thermal oxidizer, while for small Pe and large Da, chemical kinetics controls the rate of oxidation. At most temperatures used in afterburners, a reasonably moderate V_g is maintained so that mixing is not rate limiting.

The rate constant k for a reaction is related to its activation energy (Chapter 2), $k = A \exp(-E_a/RT)$. For a variety of hydrocarbons, values of A and E_a are available (Table 4.7) and hence k at any given temperature can be obtained.

TABLE 4.7
Thermal Oxidation Reaction Rate Parameters

Compound	A (s^{-1})	E$_a$ (kJ mol^{-1})
Acrolein	3.3×10^{10}	150
Benzene	7.4×10^{21}	401
1-Butene	3.7×10^{14}	243
Chlorobenzene	1.3×10^{17}	321
Ethane	5.6×10^{14}	266
Ethanol	5.4×10^{11}	201
Hexane	6.0×10^{8}	143
Methane	1.7×10^{11}	218
Methylchloride	7.3×10^{8}	171
Natural gas	1.6×10^{12}	206
Propane	5.2×10^{19}	356
Toluene	2.3×10^{13}	236
Vinyl chloride	3.6×10^{14}	265

Source: Buonicore, A.T. and Davis, W.T., *Air Pollution Engineering Manual*, Van Nostrand Reinhold, New York, 1992.

The reaction zone in the incinerator can be modeled as a plug flow reactor. The overall destruction efficiency is given by

$$\eta = 1 - \frac{[A]_{exit}}{[A]_{feed}} = 1 - e^{-k\tau_{res}} \qquad (4.88)$$

Example 4.13   Incinerator for Air Pollution Control

At what air velocity should a toluene/air mixture be introduced in a 20 ft incinerator to obtain 99.99% efficiency at a temperature of 1300°F?

T = 1300°F = (5/9) (1300) + 255.4 = 977 K. From Table 4.7, A = 2.3×10^{13} s^{-1} and E$_a$ = 236 kJ mol^{-1}. Hence, k = 2.3×10^{13} exp[−236/(8.31 × 10^{-3}) (977)] = 5.3 s^{-1}. Hence E = 1 − exp(−kτ_{res}) = 0.9999, kτ_{res} = 9.21, τ_{res} = 1.74 s and since τ_{res} = Z$_s$/V$_g$, we have V$_g$ = 11.5 ft s^{-1}.

4.2.3   ATMOSPHERIC PROCESSES

In this section, we shall analyze gas- and liquid phase chemical reactions in relation to various atmospheric processes. These applications involve the use of the principles of chemical kinetics and transport that we discussed in Chapter 2 and also in previous sections of this chapter.

4.2.3.1 Reactions in Aqueous Droplets

Acid rain is a problem in some of the industrialized nations (Hutterman, 1994). It has been particularly severe in parts of Western Europe and the Northeastern United States. Similar issues have become prominent in several developing nations (e.g., India, China). In India, acid rain has defaced one of the cherished monuments of the world, the Taj Mahal. Its exquisite marble sculpture is eroding due to uncontrolled SO_2 emissions from the local industries. In Germany, the beautiful Black Forest is severely affected by acid rain. In the United States, the forests of the northeast are known to be affected by acid precipitation involving pollutants from the region's industrial belt (Bricker and Rice, 1993). Table 4.8 lists specific pollutants and average concentrations in the northeast United States. pH values as low as 4 with high concentrations of SO_4^{2-} and NO_3^- have been observed in parts of the U.S. northeast.

Oxides of sulfur and nitrogen released to the atmosphere from fossil fuel combustion form strong acids by combining with the atmospheric moisture. These acids include H_2SO_4 and HNO_3. They react with strong bases (mainly NH_3 and $CaCO_3$) and also associate with atmospheric dust (aerosols). Below-cloud scavenging by rain and fog wash these particles to the surface with low pH. This is called "acid rain."

In the absence of strong acids in the atmosphere, the pH of rainwater is set by the dissolution of CO_2 to form carbonic acid (H_2CO_3). The equilibrium is

$$CO_2 + H_2O \rightleftharpoons H_2CO_3 \rightarrow H^+ + HCO_3^- \tag{4.89}$$

At a constant partial pressure of carbon dioxide, $P_{CO_2} = 350$ ppm $= 10^{-3.5}$ atm, we have $[H_2CO_3] = (1/K_{AW}'')P_{CO_2} = (10^{-1.5})(10^{-3.5}) = 10^{-5}$ M. Hence $[H^+] = K_{c1}[H_2CO_3] = (10^{-6.4})(10^{-5}) = 10^{-11.4}$ M. Thus, the pH of water is 5.7, which is 1.3 units less than neutral. Natural rainwater is therefore slightly acidic. When strong acids are present that lower the pH to below 4.3, this reaction is driven to the left and carbonic acid makes no contribution to acidity. The components of strong acids are derived from SO_4^{2-} and NO_3^-.

Gaseous SO_2 in the atmosphere dissolves in atmospheric moisture to form sulfuric acid and contributes to sulfate in atmospheric particles. SO_2 arises mainly from anthropogenic emissions (burning of fossil fuels). The wet deposition of sulfate by

TABLE 4.8
Typical Constituents of Rainwater in the Northeastern United States

Ion	Rainwater Concentration Collected in Ithaca, NY, on July 11, 1975 (µM)	Average Concentration in Northeast U.S. (µM)
Sulfate	57	28 ± 4
Nitrate	44	26 ± 5
Ammonium	29	16 ± 5
pH	3.84	4.14 ± 0.07

Sources: Park, D.H., *Science*, 208, 1143, 1980; Galloway, J.N. et al., *Science*, 194, 722, 1976.

rainfall occurs near industrial locations that emit SO_2 to the air. The emission of total S to the atmosphere from natural sources is estimated to be ~8.4×10^{13} g y^{-1}, of which ~1.5×10^{13} g y^{-1} is attributed to gaseous SO_2. Episodic events such as volcanic eruptions also contribute to the global emission of S compounds to the atmosphere; they disrupt the quasi-steady state conditions that exist in the atmosphere.

Let us estimate what sources contribute most to H^+ in the atmosphere. If only CO_2 is the contributing factor toward acidity, the pH of rainfall on earth will be 5.7. If the average annual rainfall is 70 cm y^{-1}, and the surface area of the earth is 7×10^{18} cm^2, this gives a total deposition of $[H^+] = (10^{-5.7})(10^{-3})(70)(7 \times 10^{18}) = 1.2 \times 10^{12}$ mol y^{-1}. The total acidity from all natural sources is $\approx 8 \times 10^{12}$ mol y^{-1}. In comparison, the anthropogenic sources due to fossil fuel burning, automobile, and other industrial sources contribute NO_x and SO_2, which leads to ~7.4×10^{12} mol of $[H^+]$ per year. It is, therefore, clear that human activities are contributing to the acidity of our atmosphere. Over geologic time our atmosphere has been acidic, but has increased only slightly in acidity. Natural alkalinity resulting from NH_3 tends to neutralize the acidity to some extent. The total NH_3 emission is $\approx 3 \times 10^{12}$ mol y^{-1}, and hence a reduction of $\approx 3 \times 10^{12}$ mol of $[H^+]$ per year can be attributed to reaction with NH_3. This still leaves $\approx 12 \times 10^{12}$ mol $[H^+]$ that are being annually added to our atmosphere.

Gaseous SO_2 is photochemically activated in the atmosphere and subsequently oxidized to SO_3 that reacts further with water to form H_2SO_4:

$$
\begin{aligned}
SO_2 &\xrightarrow{\;h\nu\;} SO_2^* \\
SO_2^* + O_2 &\to SO_3 + O \\
SO_2 + O &\to SO_3 \\
SO_3 + H_2O &\to H_2SO_4
\end{aligned}
\tag{4.90}
$$

There are two absorption bands for SO_2, one is a weak band at 384 nm, which gives rise to the excited triplet state of SO_2, and a strong absorption at 294 nm, which gives rise to a higher energy excited singlet state. These reactions are slow and do not account for the observed rates of SO_2 oxidation in the atmosphere ($\approx$ 0.01 to 0.05 h^{-1}). The oxidative process in the gas phase is driven by the abundantly available highly reactive free radical species OH$^{\bullet}$ in the atmosphere.

$$
SO_2 + OH^{\bullet} \to \cdots\cdots\cdots \to H_2SO_4
\tag{4.91}
$$

Oxidation of SO_2 to sulfate in atmospheric moisture is catalyzed by species such as ozone, hydrogen peroxide, metal ions (Fe^{III}, Mn^{III}), and by nitrogen (N^{III} and NO_2). The rate constants and rate expressions for several of these reactions have been studied by Hoffmann and coworkers and are given in Table 4.9. For pH values <4, the predominant oxidizer is H_2O_2, whereas for pH = 5, O_3 is an order of magnitude more powerful as an oxidizer. Only at pH $\geq$ 5 is the catalyzed oxidation by Fe and Mn of any significance. The following example is to illustrate how the change in pH of an open system can be computed as the sulfate content in the atmosphere increases. The problem can be extended to realistic atmospheric conditions, as described by Seinfeld and Pandis (2006).

TABLE 4.9
Oxidation of SO_2 to Sulfate by Different Species in the Atmosphere

Oxidizer	$r = \dfrac{d\left[S(IV)\right]}{dt}$
Ozone	$(k_0[SO_2 \cdot H_2O] + k_1[HSO_3^-] + k_2[SO_3^{2-}])[O_3]_w$
Hydrogen peroxide	$\left(\dfrac{k\left[H^+\right]\left[HSO_3^-\right]}{1 + K\left[H^+\right]}\right)\left[H_2O_2\right]_w$
Mn(II)	$k_2'\left[Mn(II)\right]\left[HSO_3^-\right]$
Fe(II)	$k = [Fe(III)][SO_3^{2-}]$

Source: Hoffmann, M.R. and Calvert, J.G., *Chemical Transformation Modules for Eulerian Acid Deposition Models*, Vol. II, The Aqueous-Phase Chemistry, National Center for Atmospheric Research, Boulder, CO, 1985.

Note: $k_0 = 2.4 \times 10^4$ L (mol s)$^{-1}$, $k_1 = 3.7 \times 10^5$ L (mol s)$^{-1}$, $k_2 = 1.5 \times 10^9$ L (mol s)$^{-1}$, $k = 7.5 \times 10^7$ L (mol s)$^{-1}$, $K = 13$ L mol^{-1}, $k_2' = 3.4 \times 10^3$ L $(mol\ s)^{-1}$, $k = 1.2 \times 10^6$ L (mol s)$^{-1}$.

Example 4.14   Effect of SO_2 Oxidation on the pH of an Open System (Aqueous Droplet) That Contains HNO_3, NH_3, H_2O_2, and O_3

Consider an aqueous droplet with HNO_3, NH_3, H_2O_2, O_3, and SO_2 as the main constituents. The following initial conditions ($t = m_0$) were chosen by Seinfeld and Pandis (2005): $[S(IV)]_{total} = 5$ ppb, $[HNO_3]_{total} = 1$ ppb, $[NH_3]_{total} = 5$ ppb, $[O_3]_{total} = 5$ ppb, $[H_2O_2]_{total} = 1$ ppb, $[S(VI)]_{t=0} = 0$, and water content $= 10^{-6}$.

For an open system, we can assume that the partial pressures of the different species remain constant. As S(IV) gets oxidized to S(VI), the new species of interest will comprise of SO_4^{2-} and HSO_4^-. We shall indicate $[S(VI)] = [SO_4^{2-}] + [HSO_4^-] + [H_2SO_4]_{aq}$. The electroneutrality equation is

$$\left[H^+\right] + \left[NH_4^+\right] = \left[OH^-\right] + \left[HSO_3^-\right] + 2\left[SO_3^{2-}\right] + 2\left[SO_4^{2-}\right] + \left[HSO_4^-\right] + \left[NO_3^-\right]$$

From the equation for [S(VI)] we can obtain (see Seinfeld, 1986)

$$\left[SO_4^{2-}\right] \approx \frac{\left[S(VI)\right]}{1 + \dfrac{\left[H^+\right]}{K_{s4}}} \quad \text{and} \quad \left[HSO_4^-\right] \approx \frac{\left[S(VI)\right]}{1 + \dfrac{K_{s4}}{\left[H^+\right]}}$$

with $K_{s4} = 0.012$ mol L^{-1}.

The rates of conversion to S(VI) by H_2O_2 and O_3 are given in Table 4.9. Hence,

$$\left[S(VI)\right]_{t=1\ min} = \left[S(VI)\right]_{t=0} + \left(\frac{d\left[S(VI)\right]}{dt}\right)_{t-\Delta t} \cdot \Delta t$$

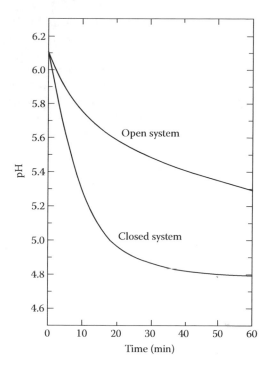

FIGURE 4.24 pH as a function of time for both open and closed systems. The conditions for the simulation are as follows: $[S(IV)]_{total} = 5$ ppb, $[NH_3]_{total} = 5$ ppb, $[HNO_3]_{total} = 1$ ppb, $[O_3]_{total} = 5$ ppb, $[H_2O_2]_{total} = 1$ ppb, $\theta_\ell = 10^{-6}$, $pH_o = 6.17$. (From Seinfeld, J.H. and Pandis, S.N.: *Atmospheric Chemistry and Physics*. 390. 1998. Copyright Wiley-VCH Verlag GmbH & Co. KGaA. Reproduced with permission.)

We then obtain a new value for $[HSO_4^-]$ and $[SO_4^{2-}]$. Substituting this in the electroneutrality equation, we get the new $[H^+]$, and hence the new pH. Continuing in this manner to obtain $[S(VI)]$ at different times, we can follow the changes in pH as more of $S(IV)$ is converted to $S(VI)$ by H_2O_2 and O_3. This is provided by Seinfeld and Pandis (2006) (Figure 4.24). In 60 min, the pH decreased to 5.3 from its initial value of 6.1.

The assumption of an open system is questionable. In a given cloud volume, it is unlikely that the partial pressures of the different species will remain constant over the duration of the reaction. In such a case, one should consider the changes in partial pressures of NH_3 and HNO_3. A detailed account of this aspect of atmospheric reaction modeling is given by Seinfeld and Pandis (1998) and is beyond the scope of this book.

Although SO_2 oxidation in the gas phase to form H_2SO_4 (aerosol) is a linear function of $OH^\bullet$ concentration, it indirectly depends on the concentration of NO_x in the atmosphere as well. As seen earlier, the oxidation in the aqueous droplet also depends on the concentration of other oxidants such as H_2O_2 and O_3. It can be shown that if the calculations in the Example 4.14 are repeated with a different concentration of

HNO$_3$, the level of NO$_3^-$ in the aqueous phase will influence [S(VI)] in the droplet. The conversion of NO$_x$ to HNO$_3$ within the droplet is primarily a function of the photochemical reactions in the gas phase, and hence responds directly to changes in NO$_x$ levels in the atmosphere. The atmospheric moisture content is obviously an important factor in deciding the fraction of S(IV) converted to S(VI). With increasing moisture content, the fraction oxidized also increases. Thus a cloud with a large moisture content will have high acidity due to a large concentration of S(VI). In conclusion, acid rain, which is mostly a regional problem, is a complex process that depends on the prevailing local conditions, particularly the levels of other species present in the atmosphere. Models exist to predict the acidity to be expected in precipitation if sources and their strengths are identified with confidence.

In our analysis thus far, we have only considered reactions within aqueous droplets in the atmosphere. However, this may not always be the controlling resistance for conversion of S(IV) to S(VI). The evolution of acidity in atmospheric precipitation (rain or fog) depends on factors such as diffusion of species toward the droplet from the air and reaction within the droplet.

Consider an aqueous droplet falling through the atmosphere. There are five processes that must be considered. These are as follows (see Schwartz and Freiberg, 1981; Seinfeld, 1986):

1. Diffusion of solute in the gas phase, characterized by the diffusion constant, D$_g$.
2. Mass transfer across the air-water interface of the droplet characterized by the mass transfer coefficient, k$_{mt}$, and the progress toward equilibrium at the interface. The latter is characterized by the air-water equilibrium constant, K$_{AW}$.
3. For a species such as SO$_2$, its dissolution in the aqueous phase is immediately followed by a dissociation reaction. The dissociation is as follows: $\left[SO_2 \cdot H_2O \right]_{aq} \overset{k_1}{\underset{k_2}{\rightleftharpoons}} H^+ + HSO_3^-$, for which the equilibrium concentration of [SO$_2$ H$_2$O]$_{aq}$ can be obtained (see Example 2.35) from
$$\frac{\left[SO_2 \cdot H_2O \right] - \left[SO_2 \cdot H_2O \right]_{eq}}{\left[SO_2 \cdot H_2O \right] - \left[SO_2 \cdot H_2O \right]_{eq}} = e^{-\alpha t},$$ where $\alpha = k_1 + k_{-1}([H^+]_{eq} + [HSO_3^-]_{eq})$.
4. If the droplet is not well mixed, then diffusion within the aqueous phase will play a role. This is characterized by the diffusion constant, D$_w$.
5. The final item to be considered is chemical reaction within the droplet. This is what we discussed earlier and is characterized by the reaction rate constant, $k = -\dfrac{1}{\left[S(VI) \right]} \cdot \dfrac{d\left[S(VI) \right]}{dt}$.

Seinfeld (1986) has analyzed each of these steps in detail. He derived an equation for the characteristic time (τ) in each case. Table 4.10 summarizes the expressions for τ. The terms, their definitions, and typical values are also given. If the characteristic time for any one step is larger than that for the chemical reaction within the

TABLE 4.10
Characteristic Times for S(IV) Oxidation in an Aqueous Droplet

Process	Expression for τ	Typical Value
Diffusion in the gas phase	$\tau_g = R^2/4D_g$	2.5×10^{-6} s
Equilibrium at the air-water interface	$\tau_i = (2\pi MRT/\alpha^2)D_w K_{AW}^2$	0.15 s
Dissociation in the aqueous droplet	$\tau_d = [k_1 + k_{-1}([H^+] + [HSO_3^-])]^{-1}$	2×10^7 s
Diffusion in the aqueous phase	$\tau_a = R^2/4D_w$	0.025 s
Reaction within the aqueous droplet	$\tau_r = -[S(IV)]/\{d[S(IV)]/dt\}$	≈ 1.5 s

Source: Seinfeld, J.H., *Atmospheric Chemistry and Physics of Air Pollution*, John Wiley & Sons, Inc., New York, 1986.

Parameters: R = 10 μm, $D_g = 0.1$ cm^2 s^{-1}, $k_1 = 3.4 \times 10^6$ s^{-1}, $k_{-1} = 2 \times 10^8$ mol (L s)$^{-1}$, $D_w = 10^{-5}$ cm^2 s^{-1}, $MK_{aw}^2/\alpha^2 = 98$, pH = 4, $P_{SO_2} = 1$ ppbv, $P_{H_2O_2} = 1$ ppbv.

Definitions of characteristic times:

τ_g—Time to achieve steady state concentration in the gas phase around the droplet.

τ_i—Time to achieve local equilibrium at the interface.

τ_d—Time to achieve equilibrium for the dissociation reaction.

τ_a—Time to achieve uniform steady state concentration in the droplet.

τ_r—Time to convert 1/e of the reactants to products.

droplet, then equilibrium will not be achieved in that step and the observed rate at which the products are formed will be smaller than the reaction rate. This has interesting consequences as far as acid rain is concerned.

From Table 4.10, one observes that for S(IV) oxidation in a 10 μm droplet, τ_r is larger than all other τ values and hence the conversion is reaction limited. In general, for all practical purposes, τ_g and τ_d are small. τ_i, τ_a, or τ_r is then rate limiting. Unless the droplets are much smaller, τ_a is large compared to τ_i and τ_r. τ_r is both pH- and species dependent.

Example 4.15   Scavenging of Conservative Pollutants by Rain or Fog

Consider rain or fog drops falling through the atmosphere. Each drop reaches terminal velocity determined by its size and scavenges pollutants as they drop to the surface. This is called "below-cloud scavenging." The rate of increase in concentration in the droplet is given by the mass transfer to the drop (Seinfeld and Pandis, 2006)

$$\pi \frac{D_P^3}{6} \cdot \frac{d[A]_{aq}}{dt} = \pi D_P^2 K_{mt} [A]_g$$

$$\frac{d[A]_{aq}}{dt} = \frac{6}{D_P} \cdot K_{mt} [A]_g$$

where we assume that the pollutant is irreversibly absorbed in to the drop and the pollutant is nonreactive. For a falling drop at its terminal velocity, U_T, we can write

$$\frac{d[A]_{aq}}{dt} = U_T \cdot \frac{d[A]_{aq}}{dz}$$

since $U_T = \dfrac{dz}{dt}$. Thus,

$$\frac{d[A]_{aq}}{dz} = \frac{6}{D_P} \cdot \frac{K_{mt}}{U_T} \cdot [A]_g$$

If we assume that $[A]_g$ is constant (homogeneous atmosphere), we can obtain the following equation upon integration:

$$[A]_{aq} = [A]_{aq}^0 + \frac{6}{D_P} \cdot \frac{K_{mt}}{U_T}[A]_g \cdot z$$

The mass scavenged (W) by a droplet of volume $\pi D_P^3/6$ will be given by $\left([A]_{aq} - [A]_{aq}^0\right)$ times the droplet volume

$$W = \frac{\pi D_P^2 K_{mt}}{U_T} \cdot [A]_g \cdot z$$

If the rainfall intensity is R_I (mm h^{-1}), the number of drops falling per unit area per hour will be

$$N_d \left(m^2 \cdot h^{-1}\right) = 0.006 R_I \frac{K_{mt}[A]_g}{U_T D_P} \cdot z$$

Hence, the rate of removal of the pollutant by rain (fog) drops from below-cloud volume of V is given by

$$-V\frac{d[A]_g}{dt} = N_d \cdot W \cdot A_c$$

Noting that $V/A_c = z$, we have

$$-\frac{d[A]_g}{dt} = 0.006 R_I \frac{K_{mt}}{U_T D_P} \cdot [A]_g$$

which upon integration gives

$$[A]_g = [A]_g^0 \cdot e^{-\lambda t}$$

where $\lambda = 0.006\dfrac{R_jK_{mt}}{U_TD_P}$ is called the "scavenging ratio." The amount of pollutant scavenged from the atmosphere by rain or fog depends on the drop diameter D_P. Note that λ is inversely proportional to D_P; the smaller the drop diameter, the better the scavenging efficiency. Small drops also have low U_T and hence have a large residence time in the atmosphere, which increases the mass transfer to the drop.

4.2.3.2 Global Warming and the Greenhouse Effect

Our atmosphere was, at one time, oxygen-rich. The highly oxidative atmosphere could not sustain early life forms. Gradually the atmosphere became oxygen-depleted (nitrogen-rich) and evolved into the one that we have today. The present atmosphere, conducive to life, is believed to be sustained by a symbiosis between the biota and the various atmospheric processes. This is the central theme of the so-called Gaia hypothesis (Lovelock, 1979). In short, our planet is a gigantic experiment, whether by design or chance.

The composition of air is 78% (v v^{-1}) nitrogen and 20% (v v^{-1}) oxygen. All other gases together constitute the remaining 2% (v v^{-1}) of our atmosphere; these are called "trace gases." Of the trace gases, a few are of special relevance. Rare gases such as argon do not vary in concentration to any measurable extent. Other gases such as CO_2, CO, NO_x, CH_4, and chlorofluorocarbons (CFCs) are variable. Even though these species are at trace concentrations, they exert profound effects on the environment.

The atmosphere does not absorb incoming energy from the sun in the visible region of the spectrum. The stratosphere absorbs only a part of the UV radiation. The cloud and the earth's surface reflect a large fraction of the radiation. A portion of the sun's energy is used to heat the surface of the earth, which reradiates heat in the infrared region of the spectrum (4–100 μm). A portion of the reradiated energy (13–100 μm) is absorbed in the atmosphere, mostly by water and CO_2. Water drops in the atmosphere also absorb energy in the region between 4 and 7 μm. It is in the window of 7–13 μm that heat escapes freely into space (Figure 4.25).

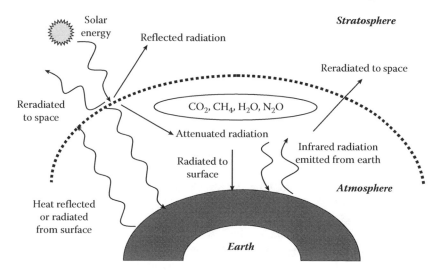

FIGURE 4.25   Solar energy absorbed and reflected in the earth's atmosphere.

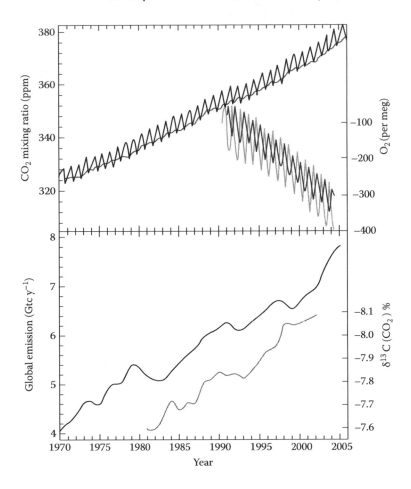

FIGURE 4.26 Monthly average CO_2 concentration observed continuously at Mauna Lao, Hawaii. (Reprinted from Solomon, S. et al. (eds.), *Climate Change 2007: The Physical Basis. Contribution of Working Group 1 to the Fourth Assessment Report of the Intergovernmental Panel on Climate Change*, Cambridge University Press, Cambridge, U.K., 2007, 996pp. With permission.)

The atmosphere acts as a greenhouse trapping moderate amounts of heat energy that is reradiated from the earth's surface, and thus maintaining a comfortable average temperature that sustains life on earth. The atmospheric CO_2 and H_2O are mainly responsible for this. During the preindustrial era, the CO_2 concentration in the atmosphere remained fairly constant. In 1850, the average concentration was ≈270 ppmv. Since then the concentration has steadily increased. In 1957, it was 315 ppmv and in 1992 it was 356 ppmv. Figure 4.26 shows the concentration of CO_2 observed at the Mauna Lao observatory in Hawaii. The oscillations in CO_2 concentration reflect the seasonal cycles of photosynthesis and respiration by the biota in the northern hemisphere. Notice the steady increase in CO_2. This increase is attributed to the burning of fossil fuels (coal and oil), and is therefore strictly anthropogenic in origin (Figure 4.27). Human intervention

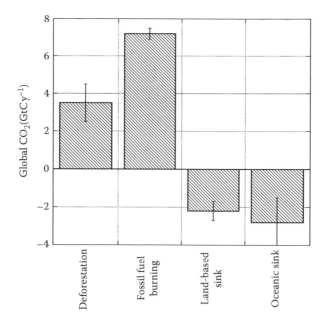

FIGURE 4.27   Percent contributions to CO_2 emissions.

through deforestation of tropical forests also plays a role by reducing photosynthesis that fix CO_2 from the atmosphere. The increase in CO_2 emissions worldwide increases the capacity of the earth's atmosphere to trap the outgoing radiation. The other trace gases such as CH_4, CO, NO_x, and CFCs also perform a similar function, since their concentrations have also increased steadily with time (see Table 4.11). Presently, CO_2 contributes the largest toward the greenhouse effect. Figure 4.28 indicates the radiative forcing effects of various greenhouse gases (GHG) as predicted by the most recent evaluation by the International Panel on Climate Change (2007).

The ability of CO_2 to initiate the greenhouse effect has been contemplated as far back as 1896 by the Swedish chemist Svänte Arrhenius (of the Arrhenius equation for rate constant in Chapter 2). But only in the past decade has the prospect of a global greenhouse effect gained attention and been recognized to have serious repercussions within the lifetime of the present generation. Most recently, however, some good news on this front has emerged. Globally the rate of input of CFCs into the atmosphere has slowed down. This can be traced to the worldwide ban on its production and the implementation of the Montreal Protocol.

Let us turn our attention toward some of the natural processes that remove greenhouse gases from the atmosphere. We will focus on CO_2 in the following discussion. The most significant pathway is the dissolution of CO_2 in the ocean water and final conversion to HCO_3^- and CO_3^{2-}. This "fixing" of carbon is rapid in the ocean surface (~500 m of the upper mixed layer), and hence can be considered to be an equilibrium process. Weathering of rocks and minerals, dissolution of $CaCO_3$ in oceans, and increased bacterial activity in some regions also contribute to the transfer of CO_2 from the atmosphere to the oceans.

TABLE 4.11

Greenhouse Gases in the Atmosphere

Parameter	CO_2 (ppmv)	CH_4 (ppbv)	CFC-12 (pptv)	N_2O (ppbv)
Preindustrial revolution (1750–1800)	280	715	0	270
2005 concentration	379 ± 0.65	1774 ± 1.8	538 ± 0.18	319 ± 0.12
Rate of annual accumulation	1.8	0.015	17	0.8
Atmospheric lifetime, years	50–200	12 ± 3	102	120
Source	Fossil fuel burning, deforestation	Rice fields, cattle, landfills, fossil fuel production	Aerosol propellants, refrigerants, foams	Fertilizers, biomass burning
Total emissions, million tons per year	5500	550	≈ 1	25

Sources: Adapted from Houghton, J.T. et al., *Climate Change: The IPCC Scientific Assessment*, New York: Cambridge University Press, 1990; IPCC, *Climate Change: The Physical Chemical Basis*, International Panel on Climate Change, Cambridge University Press, New York, 2007.

To start this discussion, we shall first consider the total carbon in solution: $[CO_2]_{tot} = [CO_2]_{aq} + [HCO_3^-] + [CO_3^{2-}]$, and the "Revelle buffer factor," R_B, as

$$\frac{\left[CO_2\right]_{tot}}{P_{CO_2}} \left(\frac{\partial P_{CO_2}}{\partial \left[CO_2\right]_{tot}}\right)_{[Alk]}$$ where [Alk] is the alkalinity of water. This is defined

(see Chapter 2) as $[Alk] = [OH^-] - [H^+] + [HCO_3^-] + 2[CO_3^{2-}]$. Since we have seen in

Chapter 2 that $\left[CO_2\right]_{tot} = \frac{P_{CO_2}}{K''_{AW}} \cdot \left(1 + \frac{K_{a1}}{\left[H^+\right]} + \frac{K_{a1}K_{a2}}{\left[H^+\right]^2}\right)$ we have

$$\left[Alk\right] = \frac{K_w}{\left[H^+\right]} - \left[H^+\right] + \frac{P_{CO_2}}{K''_{AW}} \cdot \left(\frac{K_{a1}}{\left[H^+\right]} + \frac{2K_{a1}K_{a2}}{\left[H^+\right]^2}\right) \qquad (4.92)$$

Rewriting the equation for the Revelle buffer factor as

$$R_B = \frac{\left[CO_2\right]_{tot}}{P_{CO_2}} \cdot \frac{\left(\dfrac{\partial P_{CO_2}}{\partial \left[H^+\right]}\right)_{[Alk]}}{\left(\dfrac{\partial \left[CO_2\right]_{tot}}{\partial \left[H^+\right]}\right)_{[Alk]}} \qquad (4.93)$$

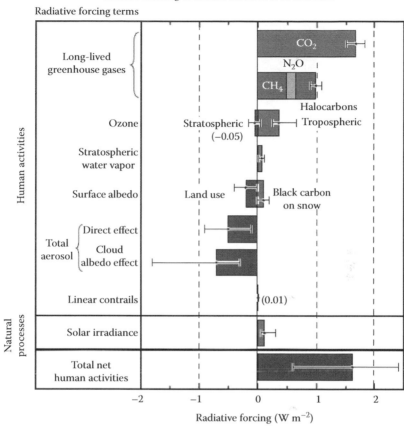

Radiative forcing of climate between 1750 and 2005

FIGURE 4.28 Radiative forcing effects of various GHG gases. (Reprinted from Solomon, S. et al. (eds.), *Climate Change 2007: The Physical Basis. Contribution of Working Group 1 to the Fourth Assessment Report of the Intergovernmental Panel on Climate Change*, Cambridge University Press, Cambridge, U.K., 2007, 996pp. With permission.)

and performing some algebraic manipulations one obtains (see Butler, 1982 for details)

$$\left(\frac{\partial P_{CO_2}}{\partial \left[H^+ \right]} \right)_{\left[Alk \right]} \approx \frac{P_{CO_2}}{\left[H^+ \right]} \tag{4.94}$$

and

$$\left(\frac{\partial \left[CO_2 \right]_{tot}}{\partial \left[H^+ \right]} \right)_{\left[Alk \right]} \approx \frac{\left[CO_2 \right]_{aq} + \left[CO_3^{2-} \right]}{\left[H^+ \right]} \tag{4.95}$$

Hence,

$$R_B \simeq \frac{[CO_2]_{tot}}{[CO_2]_{aq} + [CO_3^{2-}]}$$ (4.96)

For typical seawater samples that have an average pH of 8, we can approximate [Alk] $\approx$ [HCO$_3^-$] = $10^{-2.7}$, and [CO$_2$]$_{tot}$ $\approx$ [HCO$_3^-$] = $10^{-2.7}$. Hence,

$$[CO_2] = \frac{10^{-pH}}{K_{a1}}[HCO_3^{2-}] = 10^{-4.7}, [CO_3^{2-}] = [HCO_3^-]\frac{K_{a2}}{10^{-pH}} = 10^{-3.8}, \text{ and } R_B \approx 11,$$

compared with an experimental value of $\approx$9.5 at 298 K (Butler, 1982).

The equation for R_B gives the change in partial pressure of CO$_2$ required to produce a specified change in [CO$_2$] in seawater, if [Alk] is constant. Thus if $R_B \approx 10$, Butler (1982) estimated that about a 10% change in atmospheric concentration is required to bring about a total change of 1% in seawater CO$_2$ concentration. He also estimated that the preindustrial era (1750–1800) must have had ~145 mol CO$_2$ per m^2 of ocean surface area and the mixed layer of ocean surface water must have had ~900 mol m^{-2}. A 10% increase in atmospheric CO$_2$ (~15 mol m^{-2}) should correspond to only a 1% (~9 mol m^{-2}) increase in CO$_2$ in surface water. Extrapolating, we can conclude that in about two and one-half years, the oceans can absorb about 50% of the increased CO$_2$ in the atmosphere. Thus, there is a significant lag in the CO$_2$ absorption by the world's oceans. This feed-forward mechanism tends to increase the atmospheric CO$_2$ concentration.

Sophisticated and complex climate models have been developed to forecast the CO$_2$ increase, and how it could affect the climate and crops in different regions of the world. We shall not discuss these topics here; the student is referred to the recent document by IPCC (2007) for further details.

As already stated, the largest exchange of carbon (as CO$_2$ or carbonates and organic molecules) occurs between the ocean and the atmosphere. The exchange between the biota and atmosphere is equally important (see Figure 4.29). The average time that a CO$_2$ molecule remains free in the air is ~4 years before it is taken up by the biota or ocean. However, the adjustment time, i.e., the time taken by atmospheric CO$_2$ level to reach a new equilibrium, if either the source or sink is disturbed, is ~50 to 200 years. The net flux into or out of oceans depends both on the partial pressure of CO$_2$ in the atmosphere and the concentration of total carbon in surface waters.

One can construct a simple model to relate the changes in atmospheric CO$_2$ to increased global production of CO$_2$. The following example is an illustration of such a model.

Example 4.16   A Two-Box Model for the Effect of Increased Atmospheric CO$_2$ on the World's Oceans

MacIntyre (1978) proposed a model to relate the effects of increased CO$_2$ emissions on the world's oceans. Since the ocean is in theoretical equilibrium with the atmosphere, dissolved CO$_2$ is quickly converted to HCO$_3^-$. Organic

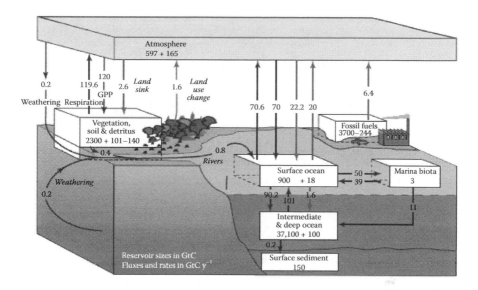

FIGURE 4.29 Global carbon reservoirs and fluxes. Numbers underlined indicate accumulation of CO_2 due to human action. Units are gigatons of carbon for reservoir sizes and gigatons of carbon per year for fluxes. (Reprinted from Solomon, S. et al. (eds.), *Climate Change 2007: The Physical Basis. Contribution of Working Group 1 to the Fourth Assessment Report of the Intergovernmental Panel on Climate Change*, Cambridge University Press, Cambridge, U.K., 2007, 996pp. With permission.)

materials that sediment in the ocean carry HCO_3^- downward. Fresh CO_2 from the atmosphere replaces the lost CO_2 in the surface ocean. Plants photosynthesize and remove CO_2 from the air, and are primarily responsible for keeping the oceans undersaturated with CO_2. As we discussed earlier, any change in CO_2 concentration is quickly buffered in the ocean. However, if the rate of change in CO_2 exceeds the rate of establishment of equilibrium, there will exist a disequilibrium between the ocean and the atmosphere. Both natural processes (photosynthesis) and anthropogenic (fossil fuel burning) will upset the equilibrium. How does the ocean then react to this change?

MacIntyre (1978) suggested that the terrestrial biosphere is in equilibrium with the atmosphere, and the marine biosphere is in equilibrium with the surface ocean (up to a depth of ~500 m the ocean is completely mixed in a short period of time). He also made the assumption that alkalinity in the ocean is only due to carbonates, and that the input of carbonates from rivers is balanced exactly by its precipitation as $CaCO_3$ in sediments. The sediment is, however, not in equilibrium with the atmosphere or the surface ocean. A two-box model such as represented in Figure 4.30 can then be envisaged.

Utilizing the expression already derived, we obtain

$$\left(\frac{\partial P_{CO_2}}{\partial [CO_2]_{tot}}\right)_{[Alk]} = R_B \cdot \left(\frac{P_{CO_2}^0}{[CO_2]_{tot}^0}\right)$$

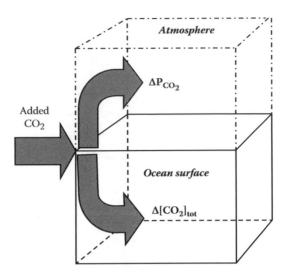

FIGURE 4.30 A two-box model for the distribution of CO_2 between the atmosphere and the surface ocean.

where $P^0_{CO_2}$ and $[CO_2]^0_{tot}$ are preindustrial values. In other words,

$$\frac{\Delta P_{CO_2}}{P^0_{CO_2}} = R_B \cdot \left(\frac{\Delta[CO_2]_{tot}}{[CO_2]^0_{tot}} \right)$$

Thus, if atmospheric P_{CO_2} increases by y%, $[CO_2]_{tot}$ increases by y/R_B%. For a given alkalinity, we can obtain the following equation: $[Alk] = [CO_2]_{tot}(\alpha_1 + 2\alpha_2) + [OH^-] - [H^+]$ and hence for a given $[H^+]$ we can obtain $[CO_2]_{tot}$. Since alkalinity in oceans is also caused by borate species, we can write the following general equation: $[Alk] = [CO_2]_{tot}(\alpha_1 + 2\alpha_2) + [OH^-] - [H^+] + B_T\alpha_B$, with α_1 and α_2 as defined earlier. Since $[CO_2]_{tot}$ is known for any given $[Alk]$, we can now obtain the changes in P with changes in $[CO_2]_{tot}$. This is shown in Figure 4.31. For a buffer factor R_B of 9.7 (at 15°C), a 10% increase in P_{CO_2} causes a 1% change in $[CO_2]_{tot}$. R_B increases with increasing P_{CO_2}. It can be observed that if P_{CO_2} increased from the present value of 330–600 ppmv, R_B changes to 17.4. Thus a doubling of P_{CO_2} from its present level leads to 5%–6% change in $[CO_2]_{tot}$ (Stumm and Morgan, 1996).

In drawing further conclusions from this analysis, one should remember that although the surface ocean is likely in equilibrium with the atmosphere, the deeper water will reach equilibrium only slowly. This is brought about by mixing due to upwelling of cold water from deeper layers and subsiding warm water toward the bottom. This thermal inertia (lag) will likely delay the overall readjustment to equilibrium for the earth.

Now that we have learned about the increased CO_2 content in the atmosphere, the next question is how this impacts the temperature of the atmosphere. "Radiative forcing" (ΔF) is a term that represents the amount of heating per m² of the surface

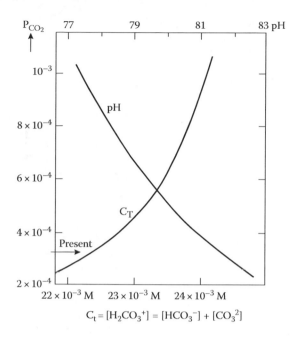

$$C_t = [H_2CO_3^+] = [HCO_3^-] + [CO_3^{2-}]$$

FIGURE 4.31   Effect of oceanic P_{CO_2} upon C_T and pH of the ocean water. The calculations have been made for the following conditions: seawater at 15°C, $P_{total} = 1$ atm, [Alk] = constant = 2.47×10^{-3} eq L^{-1}, $[B(OH)_4^-] + [H_3BO_3] = 4.1 \times 10^{-4}$ M, $1/\mathcal{H}_a = 4.8 \times 10^{-2}$ mol (L atm)$^{-1}$, $K_{c1} = 8.8 \times 10^{-7}$, $K_{c2} = 5.6 \times 10^{-10}$, and $K_{H_3BO_3} = 1.6 \times 10^{-9}$. (From Stumm, W. and Morgan, J.J.: *Aquatic Chemistry*. 3rd edn. 922. 1996. Copyright Wiley-VCH Verlag GmbH & Co. KGaA. Reproduced with permission.)

contributed by a GHG. This is related to the temperature change contributed by the GHG in the form $\Delta T = \lambda \Delta F$, where λ is a climate sensitivity parameter (K (W m^2)$^{-1}$). The NRC (1983) report stated that the ΔT associated with CO$_2$ fluctuations is given by

$$\Delta T = \eta \cdot \ln\left(\frac{P_{CO_2}}{P_{CO_2}^0}\right) \tag{4.97}$$

where

    η is a constant
    P_{CO_2} is the current partial pressure of CO$_2$
    $P_{CO_2}^0$ is the reference partial pressure of CO$_2$

Thus, the increase in surface temperature is a nonlinear function of P_{CO_2}. If we define a scaling factor, ΔT^*, which is the temperature increase due to a doubling of $P_{CO_2}^0$, we get $\Delta T^* = \eta \cdot \ln 2$. Therefore,

$$\Delta T = \left(\frac{\Delta T^*}{\ln 2}\right) \cdot \ln\left(\frac{P_{CO_2}}{P_{CO_2}^0}\right) \tag{4.98}$$

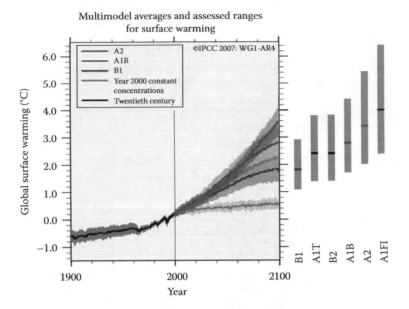

FIGURE 4.32 Projected ΔT for our planet as per the IPCC 2007 Assessment. (Reprinted from Solomon, S. et al. (eds.), *Climate Change 2007: The Physical Basis. Contribution of Working Group 1 to the Fourth Assessment Report of the Intergovernmental Panel on Climate Change*, Cambridge University Press, Cambridge, U.K., 2007, 996pp. With permission.)

Each of the greenhouse gas will contribute a certain ΔT and hence the cumulative change is $\Delta T_{overall} = \sum_i (\Delta T)_i$. The increase in temperature of the surface can have catastrophic consequences. The most recent IPCC assessment provides several scenarios for the projected ΔT for our planet (Figure 4.32). A rise in surface temperature can lead to changes in ocean levels. Thus areas could be flooded and islands and coastal plains can disappear. Regions that are presently the food baskets of the world can be hit with severe drought, and agricultural production will be curtailed in those regions. At the same time, semiarid regions of the world may become more conducive to agriculture. Attendant possibilities of international conflicts exist. If we are to mitigate these likely effects, international cooperation and strong leadership are necessary to curtail the emissions of CO_2 and other greenhouse gases.

Example 4.17   Global Temperature Change with a CO_2 Increase in the Atmosphere

The preindustrial era partial pressure of CO_2 was 280 ppmv. If a doubling of CO_2 in the atmosphere changes the atmospheric temperature by 3°C, what will be the partial pressure of CO_2 for an observed temperature change of 1°C?

$\Delta T = 1°C$, $\Delta T^* = 3°C$, $P^0 = 280$ ppmv. Hence $\dfrac{1}{3} = \dfrac{1}{0.693} \ln\left(\dfrac{P_{CO_2}}{280}\right)$, which gives $P_{CO_2} = 352$ ppmv.

4.2.3.3   Ozone in the Stratosphere and Troposphere

In this section, we will explore the chemical kinetics of atmospheric systems that involve interactions between several species. An interesting example is the chemistry of ozone, both in the upper stratosphere and the lower atmosphere (troposphere). The troposphere extends up to 16 km and the stratosphere extends up to 50 km above the surface of the earth. The pressure in the stratosphere decreases exponentially with increasing altitude. This has interesting consequences for chemical reactions in the stratosphere. Reduced pressure leads to reduced reaction rates. The reduced temperature in the upper troposphere impacts reactions with high activation energies; the rates of these reactions are lowered. Thus the lower troposphere is a far more reactive region than the upper troposphere.

The major constituents of the atmosphere are nitrogen and oxygen. Generally, these are not the dominant reactive species. Trace species such as ozone, hydroxyl radical, CO_2, SO_2, CH_4, NO_x, and CFCs are the ones that significantly impact atmospheric chemistry. We have already seen that $OH^\cdot$ is the most reactive species and appropriately termed the "atmospheric detergent." It is typically present at mixing ratios of 1 to 4×10^{-14}, despite the fact that the atmosphere contains 21% molecular oxygen by volume. The greenhouse gases (CO_2, CH_4, and CFCs) were discussed in the previous section. SO_2 and NO_x participate in the acidity of the atmosphere and were also discussed in Section 4.2.3.2.

From the point of view of exploring different reaction kinetics in the stratosphere and troposphere, ozone is a good choice. The total mass of ozone in dry air is ~3.3×10^{12} kg. It has a maximum concentration of ~500 µg m^{-3} in the upper stratosphere at about 30 km height. In the troposphere, it varies between 60 and 100 µg m^{-3} in clean air. The significance of ozone in the stratosphere lies in its ability to shield the earth from the harmful effects of the sun's UV radiation. However, in the lower troposphere it is an undesirable species since it leads to the formation of smog. U.S. federal regulations on air quality stipulates that ozone concentration greater than 235 µg m^{-3} is harmful and can cause breathing problems and eye irritations. Several cities in the United States have low air quality due to nonattainment of ozone levels.

Let us consider the formation and dissipation of ozone in the upper stratosphere. At −30 km height, ozone forms via dissociation of molecular oxygen into O atoms by solar radiation ($\lambda < 240$ nm). In the presence of a third body, Z, the oxygen atoms react with O_2 to produce ozone.

$$O_2 \xrightarrow[p_1]{h\nu} 2O$$

$$2\left(O + O_2 + Z \xrightarrow{k_2} O_3 + Z^*\right)$$

(4.99)

The overall reaction is $3O_2 \rightarrow 2O_3$. The species Z^* in this case represents vibrationally excited O_2 or N_2 molecule. UV radiation also splits up O_3 into O_2 and O.

$$O_3 \xrightarrow[p_3]{h\nu} O_2 + O$$

(4.100)

This reaction is the predominant mechanism by which ozone performs the function of shielding the earth from harmful UV light.

Ozone also reacts with O atom to give rise to O_2 as follows:

$$O_3 + O \xrightarrow{\;\;k_4\;\;} 2O_2 \qquad (4.101)$$

This set of four reactions constitutes what is called the "Chapman mechanism" for ozone formation and dissociation in the upper stratosphere.

Let us now consider the rates of formation and destruction of ozone in the stratosphere. The monoatomic O species obey the following rate expression:

$$\frac{d[O]}{dt} = 2J_1[O_2] + J_3[O_3] - [O]\left(k_2[O_2][Z] + k_4[O_3]\right) \qquad (4.102)$$

where J_i is the photolysis rate constant given by $\sum_i \phi_i(\lambda) I_i(\lambda) \sigma_i(\lambda) \Delta\lambda$ as described in Chapter 2. We prefer to distinguish this from the chemical rate constants expressed as k_2 and k_4.

For ozone the equation is

$$\frac{d[O_3]}{dt} = k_2[O][O_2][Z] - J_3[O_3] - k_4[O_3][O] \qquad (4.103)$$

At steady state we can write $\dfrac{d[O]}{dt} = \dfrac{d[O_3]}{dt} = 0$. Hence,

$$[O] = \frac{2J_1[O_2] + J_3[O_3]}{k_2[O_2][Z] + k_4[O_3]}$$

$$[O_3] = \frac{k_2[O][O_2][Z]}{J_3 + k_4[O]} \qquad (4.104)$$

The ozone concentration can be obtained by simultaneously solving these equations. A quadratic in $[O_3]$ will result. Retaining only the positive square root for the solution, we obtain after some simplification,

$$[O_3] = [O_2]\left(\frac{J_1}{2J_3}\right)\left[\left(1 + \frac{4J_3 k_2[Z]}{J_1 k_4}\right)^{\frac{1}{2}} - 1\right] \qquad (4.105)$$

In order to simplify this expression, we should have some idea about the relative magnitudes of the various rate constants. Generally, in the stratosphere $[Z] = [N_2] \sim 10^{24}$ molecules m^{-3}, $J_1 \sim 1.5 \times 10^{-10}$ min^{-1}, and $J_3 \sim 0.019$ min^{-1}. $k_2 \sim 1.2 \times 10^{-43}$ m^6 (molecule^2 min)$^{-1}$, $k_4 \sim 7.5 \times 10^{-20}$ m^3 (molecule min)$^{-1}$, $[O_2] \sim 10^{23.5}$ molecules m^{-3}. Hence, $[O_3] = [O_2]\left(\dfrac{J_1}{J_3} \cdot \dfrac{k_2}{k_4} \cdot [N_2]\right)^{\frac{1}{2}} \approx 10^{18.8}$ molecules$\cdot m^{-3}$ and

$[O] = 10^{13.4}$ molecules $\cdot m^{-3}$.

The steady state concentration of $[O_3]$ given by the equation earlier can be rearranged to obtain the expression

$$[O_3] = \frac{k_2[O_2][Z]}{k_4\left(1 + \dfrac{J_3}{k_4[O]}\right)} \tag{4.106}$$

with $\dfrac{J_3}{k_4[O]} \approx \dfrac{0.019}{7.5 \times 10^{-20} \times 10^{13.4}} = 1 \times 10^4$. Hence, $[O_3] \approx \dfrac{k_2}{J_3}[O_2][Z][O]$. Thus we have the ratio

$$\frac{[O_3]}{[O]} \approx \frac{k_2}{J_3}[O_2][Z] \tag{4.107}$$

The steady state concentration ratio $[O_3]/[O]$ should remain constant ($\sim 2.5 \times 10^5$), if $[O_2]$ remains constant. This ratio can vary if either $[O]$ or $[O_3]$ varies due to other reactions. Since the production of O atoms by photolysis of O_2 is a slow process (J_1 is very small), any competing reaction that decomposes O_3 faster will reduce the ratio rapidly. These competing reactions involve trace species such as OH·, NO, and CFCs. The general catalytic cycle follows the scheme

$$\begin{aligned} X + O_3 &\rightarrow XO + O_2 \\ XO + O &\rightarrow O_2 + X \end{aligned} \tag{4.108}$$

with the net reaction

$$O + O_3 \rightarrow 2O_2 \tag{4.109}$$

Note that the species X is neither formed nor removed from the system and thus acts as a catalyst.

The reaction with OH· is estimated to account for the decomposition of only -15% of the stratospheric ozone. Excited ozone molecules ($\lambda < 310$ nm) give rise to excited oxygen atoms which in turn decompose H_2O or CH_4 to provide OH·

$$\begin{aligned} O_3 &\xrightarrow{h\nu} O^* + O_2 \\ O^* + CH_4 &\rightarrow CH_3 + HO^\cdot \\ O^* + H_2O &\rightarrow 2HO^\cdot \end{aligned} \tag{4.110}$$

The OH radicals react with ozone in a self-propagating series of reactions to give two O_2 molecules:

$$\begin{aligned} HO^\cdot + O_3 &\rightarrow O_2 + HO_2 \\ HO_2 + O_3 &\rightarrow O_2 + HO^\cdot \end{aligned} \tag{4.111}$$

If the last two reactions are fast in comparison to the reaction of ozone with Z, then the ratio $[O_3]/[O]$ will decrease.

Although a number of species can catalyze the dissociation of ozone in the stratosphere, a major concern at present is the influence of chlorinated compounds called chlorofluorocarbons (CFCs). These have been in widespread use as refrigerants and aerosol propellants and are purely anthropogenic in origin. They comprise mainly CF_2Cl_2 (Freon 12, CFC-12) and $CFCl_3$ (Freon 11, CFC-11) and are quite inert in the troposphere. Once released to the atmosphere they are slowly transported to the stratosphere where they remain for a very long time. Prior to the signing of the Montreal Protocol in 1987, several tons of these compounds were pumped into the atmosphere. For example, the concentration of CFC-12 continually increased (see Figure 4.35) at a rate of 17 pptv y^{-1} with a mean concentration of ~484 pptv in 1990. Its atmospheric lifetime is ~130 years. Molina and Rowland (1974) first alerted the scientific community to the fact that CFCs can adversely affect the stratospheric ozone layer. Radiations with wavelengths of 180–220 nm in the stratosphere are absorbed by CFCs to produce the highly reactive Cl and ClO species (represented generally as ClO_x).

$$CCl_2F_2 \xrightarrow{\ hv\ } CF_2Cl + Cl^{\bullet}$$
$$CCl_2F_2 + O^{\bullet} \xrightarrow{\ k_1\ } CF_2Cl + ClO \tag{4.112}$$

The ClO_x can react with ozone and O atoms as follows:

$$Cl^{\bullet} + O_3 \xrightarrow{\ k_2\ } ClO^{\bullet} + O_2$$
$$ClO + O^{\bullet} \xrightarrow{\ k_3\ } Cl + O_2 \tag{4.113}$$

The net effect is $O + O_3 \rightarrow 2O_2$. The Cl atoms are subsequently trapped by hydrocarbons (e.g., CH_4) to form HCl, which appears in the troposphere and are washed down by rain. From Equations 4.112 and 4.113, we have

$$\frac{d[ClO]}{dt} = k_1[CCl_2F_2][O^{\bullet}] + k_2[O_3][Cl^{\bullet}] - k_3[ClO][O^{\bullet}] \tag{4.114}$$

Therefore, the steady state concentration of ClO is given by

$$[ClO]_{ss} = \frac{k_1}{k_3}[CCl_2F_2] + \frac{k_2}{k_3}\frac{[O_3]}{[O^{\bullet}]}[Cl^{\bullet}] \tag{4.115}$$

Thus, the steady state concentration of ClO can be maintained in the atmosphere as long as CCl_2F_2 is added to the stratosphere, even if no CCl_2F_2 is photolyzed. If photolysis of CCl_2F_2 does occur, the second term becomes significant. $Cl^{\bullet}$ will be finite and $[ClO]_{ss}$ will increase. Thus, Equations 4.112 and 4.113 will sustain high concentrations of ClO at the expense of O_3. In other words, CFCs act as catalysts to remove ozone from the stratosphere.

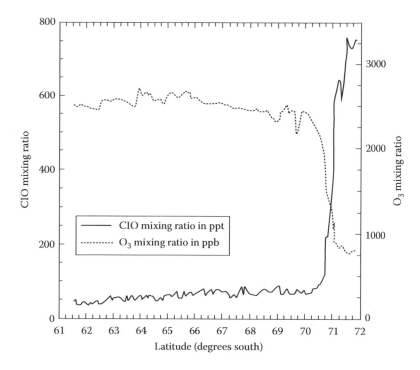

FIGURE 4.33   Simultaneously observed ClO and O_3 concentrations obtained on September 21, 1987, by the ER-2 in the Antarctic, with corrections made for variations in potential temperature. (Reprinted from Anderson, J.G. et al., *J. Geophys. Res. Atmos.*, 94(D9), 11465, 1989. American Geophysical Union. With permission.)

Depletion of the ozone layer poses a problem. It can affect the amount of UV radiation reaching earth and can increase the rate of skin cancer and other related health issues. Indeed depletions of ozone have been observed over the Antarctic regions and, to a lesser extent, in the Arctic regions; depleted ozone has been linked to the increased ClO_x in these regions (Figure 4.33).

With the signing of the Montreal Protocol, the industrialized nations of the world curtailed production of CFCs. This action has slowed the rate of increase of CFCs in the atmosphere. It is encouraging to see that industries have now replaced CFCs with alternatives that are "environmental friendly." It is possible to use a simple box model for the atmosphere to predict the effect of a voluntary reduction of CFC emissions.

Example 4.18   Effect of CFC Source Reduction on the Future Atmospheric Concentration

Figure 4.34 represents the entire atmosphere as a single box that receives CFCs at a constant production rate of S_{tot} mol $(cm^2 \ y)^{-1}$. Even though the rate of CFC production has declined since the Montreal Protocol was signed, no significant depletion of CFCs in the stratosphere has yet been noted because of its low reactivity. The long

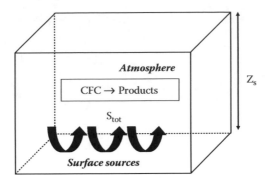

FIGURE 4.34 Box model for the production and dissipation of CFCs in the atmosphere.

lifetime of CFCs guarantees a continuous but slow increase in atmospheric concentration even after the Montreal Protocol. Let us assume that for several decades the rate of release S_{tot} can be assumed to be constant. The only mechanism by which CFCs are removed from the system involves photolysis reactions.

Let the overall rate constant for the reaction of a CFC be k (y^{-1}) and its concentration be [CFC] (mol m^{-3} of air). From Section 4.2.1.1, using $w_v = u = w_p = 0$ and $r_i = -k[CFC]$ we have the following differential equation:

$$\frac{d[CFC]}{dt} + k[CFC] = \frac{S_{tot}}{Z_s} \tag{4.116}$$

Solving this equation with the initial condition that at t = 0, [CFC] = [CFC]$_0$, we obtain

$$[CFC] = [CFC]_0 \, e^{-kt} + \frac{S_{tot}}{kZ_s}\left(1 - e^{-kt}\right) \tag{4.117}$$

Note that $\dfrac{S_{tot}}{kZ_s} = \dfrac{Q_s}{kV_{atm}}$, where Q_s is the source strength in mol y^{-1} and V_{atm} is the atmospheric volume (m^3). At some time in the future, the CFC concentration should reach a steady state, i.e., as $t \rightarrow \infty$, $[CFC]_{ss} \rightarrow S_{tot}/kZ_s$. Hence

$$\frac{[CFC]}{[CFC]_0} = \frac{[CFC]_{ss}}{[CFC]_0} + \left(1 - \frac{[CFC]_{ss}}{[CFC]_0}\right)e^{-kt} \tag{4.118}$$

Since [CFC], [CFC]$_0$, and [CFC]$_{ss}$ are concentrations in moles of CFC per m^3 of air, they are easily converted to conventional units of ratio of volume of CFC to air (either in ppmv or ppbv) by multiplying with 0.0224 m^3 CFC mol^{-1}, which is the molar volume of CFC.

Let us first calculate the values of [CFC], [CFC]$_0$, and [CFC]$_{ss}$ for the year 1987 when the Montreal Protocol took effect. In 1987, [CFC] ~ 450 pptv for CFC-12. If we choose k – 0.0065 y^{-1}, a production rate for CFC-12 of Q_s ~ 3×10^9 mol y^{-1} and $V_{atm} = 3.97 \times 10^{18}$ m^3, we obtain [CFC]$_{ss}$ = 2.8 ppbv. If, further, we assume a constant Q_s, then $\dfrac{[CFC]}{[CFC]_0} = 6.2 - 5.2e^{-0.0065t}$. Curve 1 of Figure 4.35 shows the

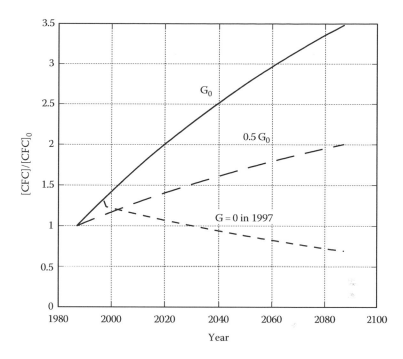

FIGURE 4.35 Atmospheric concentration of CFCs for various scenarios of the Montreal Protocol.

resulting profile. This can be considered to be the "base case," i.e., a consequence of nonimplementation of the Montreal Protocol. If the Montreal Protocol is to reach a goal of 50% reduction in net CFC emissions, then Q_s has to be replaced by $0.5Q_s$ and $[CFC]_{ss} = 1.4$ ppbv. Hence $\dfrac{[CFC]}{[CFC]_0} = 3.1 - 2.1e^{-0.0065t}$. Curve 2 of Figure 4.35 shows the expected CFC concentration in air in this case. Notice that even under this scenario, the CFC concentration continues to rise in the atmosphere, albeit at a slower rate. As a consequence, chemical reactions causing the destruction of stratospheric ozone will continue into the next century. Figure 4.35 also considers the following scenario: What if the Montreal Protocol had attempted the complete cessation of CFC production by 1997? The decay of CFC concentration will then be given by $\dfrac{[CFC]}{[CFC]_{in\,1997}} = e^{-0.0065t}$, and is shown as curve 3.

It is evident that even under these most optimistic conditions, the concentration of CFC in the year 2087 will decrease only to one-half its value in 1987. This clearly shows how long the effect of CFCs will linger in the stratosphere even if a concerted attempt is made in this century to eliminate its production.

In the lower troposphere, ozone performs a different function. It reacts with oxides of nitrogen (NO_x) and atmospheric hydrocarbons and leads to the formation of "smog." This process accounts for ~60% to 70% of the ozone destroyed in the troposphere.

Example 4.19   Kinetics of Smog Formation in Urban Areas

The chemistry of smog, as we have discussed, is tied to the chemistry of NO_x species in the atmosphere. The basic reaction is the photochemical dissociation of NO_2 ($\lambda = 435$ nm) that gives rise to NO and O. The O then rapidly reacts with O_2 in the presence of a third body Z (e.g., N_2) to produce ozone. The cycle is completed when O_3 reacts with NO to regenerate NO_2.

$$NO_2 \xrightarrow[\text{p1}]{hv} NO + O; \quad J_1 \approx 0.6 \text{ min}^{-1}$$

$$O + O_2 + Z \xrightarrow{k_2} O_3 + Z^*; \quad k_2 \approx 6.1 \times 10^{-34} \text{ cm}^6 \left(\text{molecule}^2 \cdot s\right)^{-1} \qquad (4.119)$$

$$O_3 + NO \xrightarrow{k_3} O_2 + NO_2; \quad k_3 \approx 1.8 \times 10^{-14} \text{ cm}^3 \left(\text{molecule} \cdot s\right)^{-1}$$

The rate constants given are at 298 K. As explained earlier, we distinguish photochemical rate constants from chemical rate constants by utilizing the symbol J for the former. The rates of formation of the different species are

$$\frac{d[NO_2]}{dt} = -J_1[NO_2] + k_3[O_3][NO]$$

$$\frac{d[O]}{dt} = J_1[NO_2] - k_2[O][O_2][Z] \qquad (4.120)$$

$$\frac{d[O_3]}{dt} = k_2[O][O_2][Z] - k_3[NO][O_3]$$

Applying the pseudo steady state approximation for the intermediate species [O], we obtain

$$[O]_{ss} = \frac{J_1}{k_2} \cdot \frac{[NO_2]}{[O_2][Z]} \qquad (4.121)$$

If the NO_2 formation and dissipation is at steady state, we have

$$[O_3] = \frac{J_1}{k_3} \cdot \frac{[NO_2]}{[NO]} \qquad (4.122)$$

This is an important result, since it states that the concentration of ozone is dependent not on the magnitude of NO_x alone but on the ratio $[NO_2]/[NO]$. As $J_1 \rightarrow 0$ (i.e., night time conditions) in the presence of excess of ozone the ratio becomes very large, whereas the lowest ratios are observed in bright sunshine (high $J_1 \sim 20$ h^{-1}).

   If we now start with a system with initial condition $[O_3]_0 = 0$, and $[NO]_0 = 0$, the stoichiometry states that $[O_3] = [NO_2]_0 - [NO_2] = [NO]$. Thus

$$\frac{J_1[NO_2]}{[NO]} = [NO_2]_0 - [NO_2], \; [NO_2] = \frac{k_3[NO][NO_2]}{J_1 + k_3[NO]}.$$ Now, since $[O_3] = [NO_2]_0 - [NO_2]$, we obtain the following quadratic in $[O_3]$:

$$k_3[O_3]^2 + J_1[O_3] - J_1[NO_2]_0 = 0 \tag{4.123}$$

Since $[O_3] > 0$, we keep only the positive root of this quadratic, and obtain

$$[O_3] = \frac{1}{2}\left[\left\{ \left(\frac{J_1}{k_3}\right)^2 + 4\frac{J_1}{k_3}[NO_2]_0 \right\}^{\frac{1}{2}} - \frac{J_1}{k_3} \right] \tag{4.124}$$

We know that $J_1 \sim 0.6$ min^{-1} k_3, which is in cm^3 (molecule s)$^{-1}$, can be converted into ppmv min^{-1} by multiplying with 1.47×10^{15}. Hence $k_3 \sim 26$ ppmv min^{-1} and $J_1/k_3 = 0.02$ ppmv. The variation in $[O_3]$ with $[NO_2]_0$ is shown in Figure 4.36. When $[NO_2]_0 = 0$, $[O_3] = 0$ and the ozone concentration increases as more NO is converted to NO_2.

In urban areas with smog, the concentration of ozone is higher than that predicted here. This can be interpreted using the expression $[O_3] = J_1[NO_2]/k_3[NO]$. Any reaction that increases the transformation of NO to NO_2 can bring about increased ozone concentrations. What types of materials are responsible for this effect? Table 4.12 summarizes the most common constituents of urban smog and their adverse effects.

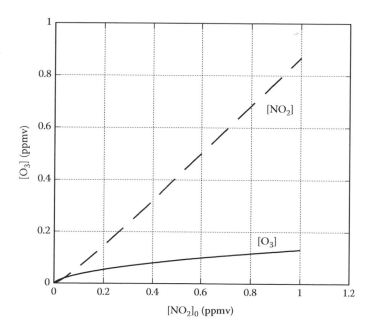

FIGURE 4.36   Ozone concentration as a function of initial nitrogen dioxide concentration.

TABLE 4.12

Common Constituents of Urban Smog That React with Ozone

Compound	Formula	Health Effects
Aldehydes	R–CHO	Eye irritation, odor
Hydrocarbons	$R–CH_3$	Eye irritation, odor
Alkyl nitrates	$R–ONO_2$	
Peroxyacyl nitrates (PAN)	R–CO–OONa	Toxic to plants, eye irritant
Aerosols	$(NH_4)_2SO_4$, NH_4NO_3, etc.	Visibility reduction

During the daytime, automobiles generate NO via combustion:

$$N_2 + O_2 \rightarrow 2NO \tag{4.125}$$

However, automobiles also generate hydrocarbons (HR) and CO via incomplete combustion, both of which react with atmospheric OH radicals to produce peroxy radicals:

$$OH + HR + O_2 \rightarrow RO_2 + H_2O \tag{4.126}$$

In general, only C_3 and higher hydrocarbons are reactive and are called "nonmethane hydrocarbons" (NMHC). The peroxy radicals are very efficient in converting NO to NO_2:

$$RO_2 + NO + O_2 \rightarrow RO + NO_2 + HO_2 \tag{4.127}$$

Thus, the ratio $[NO_2]/[NO]$ and O_3 concentration increase during the daytime. NO_2 and HR also react producing aldehydes, ozone, and peroxy acetyl nitrate (PAN). For example,

$$
\begin{aligned}
CH_3 - CH_3 + OH + O_2 &\rightarrow CH_3CH_2O_2 + H_2O \\
CH_3CH_2O_2 + NO + O_2 &\rightarrow CH_3CHO + HO_2 + NO_2 \\
OH + CH_3CHO &\rightarrow CH_3CO^{\cdot} + H_2O \\
CH_3CO^{\cdot} + O_2 &\rightarrow CH_3COOO^{\cdot} \\
CH_3COOO^{\cdot} + NO_2 &\rightarrow CH_3COOONO_2
\end{aligned}
\tag{4.128}
$$

PAN, so formed, is an eye irritant. Thus to eliminate smog in an urban area, the first line of defense is to reduce HR emissions and NO from automobiles and power plants.

In urban areas such as Los Angeles, California, where smog is a common occurrence, the concentrations of hydrocarbons (saturated and unsaturated) and aldehydes are very large. They are produced from automobile emissions. As an example, a gasoline-powered vehicle exhaust consists of approximately 78% N_2, 12% CO_2, 5% H_2O (vapor), 1% unused oxygen, ~2% each of CO and H_2, ~0.08% of hydrocarbons, and ~0.06% NO. The remaining several hundred pptv of oxidized hydrocarbons consist of aldehydes, formaldehyde being the dominant fraction. Typical smog composition in Los Angeles and the diurnal profile are shown in Figure 4.37. Notice the peak

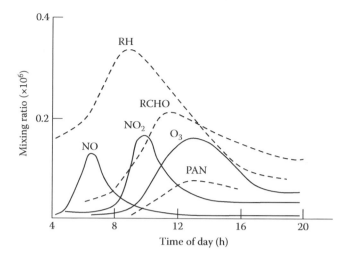

FIGURE 4.37   Diurnal variation in concentrations of various species in typical urban smog.

concentrations of ozone at noontime when smog is severe. The chemistry of smog in urban areas is incredibly complex due to the presence of a variety of hydrocarbons and aldehydes that participate in reactions with ozone and nitrogen oxides. Seinfeld (1986) has discussed the salient aspects of these reactions to which the reader is referred to for more details.

4.3   SOIL AND SEDIMENT ENVIRONMENTS

The land surface contributes about 25% of the earth's surface area. Compounds move between soil and water in the groundwater environment. Sediment is a sink for contaminants entering the water in lakes, rivers, and estuaries. Similarly, soil and air compartments exchange chemicals. We discuss three examples of transport models, one for each interface—soil-water, sediment-water, and soil-air. We also discuss soil remediation concepts that use principles from chemical kinetics.

4.3.1   FATE AND TRANSPORT MODELING

We discuss three cases here: groundwater, sediment-water, and soil-air exchange of chemicals.

4.3.1.1   Transport in Groundwater

Leaking underground storage tanks, old unlined landfills, surface impoundments, and accidental spills are the main causes for groundwater contamination. There are two regions in the subsurface that are subject to contamination—the unsaturated (vadose) zone and the saturated zone (Figure 4.38). Liquid spilled on the surface gradually migrates downward. The light nonaqueous phase liquid (LNAPL) floats on the groundwater table. A dense nonaqueous phase liquid (DNAPL) sinks and pools at

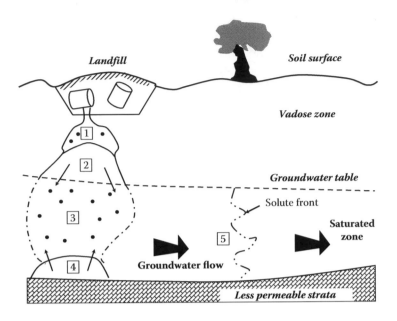

FIGURE 4.38 Groundwater contamination from a leaking source.

the bottom of the impervious layer. Contaminants slowly dissolve in the groundwater. In the zone of solubilization, globules of the organic solvent slowly dissolve with time. The transport of the contaminant plume in groundwater, therefore, consists of two processes—dissolution of globules (ganglia) of pure fluid and movement of the solubilized fraction.

4.3.1.1.1 Solubilization of Ganglia

A NAPL globule trapped in a soil pore will slowly dissolve in the groundwater. The rate of dissolution will depend on the mass transfer coefficient k_L (m s^{-1}) between the NAPL globule and water under specified flow conditions. The flux is given by $F_A = k_L([A]^* - [A])$. $[A]^*$ is the aqueous solubility of the NAPL and $[A]$ is the concentration in water. Consider the NAPL globule to be approximately spherical (diameter D_p, m) with a volume ratio θ_N (m^3 m^{-3} of the medium)—see Figure 4.39. The area per unit volume of the medium $a = 6\theta_N/D_p$. Note that both θ_N and D_p are functions of the distance x from the globule in the flow direction. $a(x,t) = \dfrac{6\theta_N(x,t)}{D_p(x,t)}$. Thus the overall concentration change along the x-direction is given by (Hunt et al., 1988)

$$\frac{d[A]}{dt} = \frac{k_L}{u_D}\left(\frac{6\theta_N}{D_p}\right)\left([A]^* - [A]\right) \qquad (4.129)$$

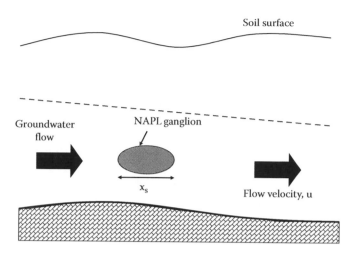

FIGURE 4.39 A NAPL ganglion dissolving in groundwater.

where [A] = f(x, t) and u_D is the Darcy velocity, defined later in this section. This equation is applicable for plug flow conditions. We assume that the reduction in NAPL volume is minimal, and hence $D_p = D_0$, and $\theta_N = \theta_0$ (the initial values). The volume fraction θ_0 is the spread over a length X_s. Integrating the resulting equation gives us the concentration of A at the edge of the plume:

$$\frac{[A]}{[A]^*} = 1 - e^{-\left(\frac{k_L}{u_D}\frac{6\theta_0}{D_0}X_s\right)} \tag{4.130}$$

Numerous correlations exist for k_L in the chemical engineering literature. Hunt et al. (1988) used the correlation $\dfrac{k_L \varepsilon_T}{1.09 u_D} = Pe^{-\frac{2}{3}}$, where Pe is the Peclet number given by $u_D D_p / D_w$, and D_w is the molecular diffusivity of the NAPL in water.

Example 4.20   NAPL Ganglion Solubilization

Let us consider an initial volume fraction θ_0 of 0.05 over a total horizontal extent of 10 m. In this case, $\theta_0 X_s = 0.5$ m. For a typical Darcy velocity of 1 m d^{-1}, and a $D_0 = 0.1$ m, and for a typical molecular diffusivity in water of 8.6×10^{-5} m^2 d^{-1}, we obtain Pe = 1157. If $\varepsilon_T = 0.4$, $k_L = 0.024$ m d^{-1}. Hence $\dfrac{[A]}{[A]^*} = 0.51$. As the globule dissolves, both its diameter and volume decreases (Powers et al., 1991). In order to obtain the actual solubilized mass in water, the change in volume should be taken into account. The ganglion lifetimes are found to be weak functions of the flow velocity. As a consequence, very large volumes of water are necessary to reduce the ganglion size. Therefore, one ends up with an expensive aboveground treatment system.

4.3.1.1.2 Transport of the Dissolved Contaminant in Groundwater

A pollutant that solubilizes in the aqueous phase is transported with the groundwater. As the solubilized pollutant moves through the heterogeneous subterranean soil pore water, it adsorbs to the soil particles. Hence, we have to consider both soluble and adsorbed contaminant mass in ascertaining the transport in the groundwater environment. Nonideal plug flow is assumed because of the soil heterogeneity. Consider a volume element in the direction of groundwater flow (Figure 4.40). Advective and dispersive terms comprise both feed and effluent rates. The advective term is U[A] and dispersion is given by $-D\partial[A]/\partial x$ evaluated at appropriate x coordinates. Accumulation of mass occurs in both solid and liquid phases. Reaction losses occur in the liquid phase as a result of chemical or microbiological processes. Note that for the porous solid the volume is $\varepsilon A_c \Delta x$. The overall material balance is

$$\text{Rate in} = \text{Rate out} + \text{Accumulation} + \text{Reaction loss}$$

$$\varepsilon A_c U[A]\Big|_x - \varepsilon A_c D \frac{\partial[A]}{\partial x}\Big|_x = \varepsilon A_c U[A]\Big|_{x+\Delta x} - \varepsilon A_c D \frac{\partial[A]}{\partial x}\Big|_{x+\Delta x}$$

$$+ \varepsilon A_c \Delta x \frac{\partial[A]_{\text{tot}}}{\partial t} + r_A \varepsilon A_c \Delta x \qquad (4.131)$$

Note that [A] is the concentration in the mobile phase (water) and $[A]_{\text{tot}}$ is the total concentration (pore water + solid). Dividing throughout by $\varepsilon A_c \Delta x$ and taking the limit as $\Delta x \to 0$, we get after rearrangement

$$D \frac{\partial^2[A]}{\partial x^2} - U \frac{\partial[A]}{\partial x} - r_A = \frac{\partial[A]_{\text{tot}}}{\partial t} \qquad (4.132)$$

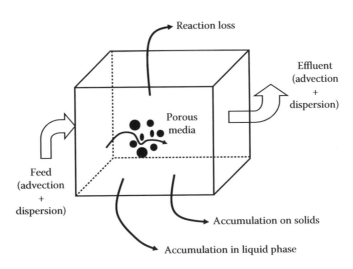

FIGURE 4.40 A volume element and transport kinetics in the groundwater environment.

This expression is the general "advective dispersive equation" in the x-coordinate. Note that in the right-hand side of the equation, $[A]_{tot} = \varepsilon[A] + (1-\varepsilon)\rho_s W_A$, the total concentration of i per m³. If transport is in all three directions, we can generalize the equation to the following:

$$\bar{\nabla} \cdot \left(D\bar{\nabla}[A] \right) - \bar{U} \cdot \left(\bar{\nabla}[A] \right) - r_A = \frac{\partial [A]_{tot}}{\partial t} \tag{4.133}$$

where $\nabla = \dfrac{\partial}{\partial x} + \dfrac{\partial}{\partial y} + \dfrac{\partial}{\partial z}$ is called the "del operator." Using the "local equilibrium assumption" (LEA) between the solid and liquid phase concentrations, $W_A = K_{sw}[A]$ and if $r_A = 0$, we get

$$D\frac{\partial^2[A]}{\partial x^2} - U\frac{\partial[A]}{\partial x} = R_F \frac{\partial[A]}{\partial t} \tag{4.134}$$

where $R_F = \varepsilon + \rho_b K_{sw}$ is the "retardation factor" as defined earlier. $\rho_b = (1-\varepsilon)\rho_s$ is the soil bulk density.

In a plug flow reactor, if the initial concentration in the fluid is $[A]_0$ and at time $t = 0$, a step increase in concentration $[A]_s$ is applied at the inlet, the following initial and boundary conditions apply to the advection-dispersion equation in the x-direction:

$$\begin{aligned}
[A](x,0) &= [A]_0 \quad \text{for } x \geq 0 \\
[A](0,t) &= [A]_0 \quad \text{for } t \geq 0 \\
[A](\infty,t) &= [A]_0 \quad \text{for } t \geq 0
\end{aligned} \tag{4.135}$$

The solution is

$$\frac{[A]-[A]_0}{[A]_s-[A_0]} = \frac{1}{2} \cdot \left[\text{erfc}\left(\frac{x - \left(\frac{U}{R_F}\right)t}{2\sqrt{\frac{Dt}{R_F}}} \right) + \exp\left(\frac{Ux}{D}\right) \cdot \text{erfc}\left(\frac{x + \left(\frac{U}{R_F}\right)t}{2\sqrt{\frac{Dt}{R_F}}} \right) \right] \tag{4.136}$$

where erfc is the complementary error function (see Appendix H).

It is instructive to consider the simplifications of the advective-dispersion equation for specific cases:

1. If dispersion (diffusion) is much larger than advection, i.e., $D\dfrac{\partial^2[A]}{\partial x^2} \gg U\dfrac{\partial[A]}{\partial x}$, and $r_A = 0$, we have

$$D\frac{\partial^2[A]}{\partial x^2} = R_F \frac{\partial[A]}{\partial t} \tag{4.137}$$

which is the well-known Fick's equation, for which, with the boundary conditions given earlier, the solution is

$$\frac{[A]-[A]_0}{[A]_s-[A]_0} = 1 - \mathrm{erf}\left(\frac{x}{\sqrt{4Dt}}\right) \tag{4.138}$$

2. If advection is larger than dispersion, i.e., $D\dfrac{\partial^2[A]}{\partial x^2} \ll U\dfrac{\partial[A]}{\partial x}$, we have,

$$-U\frac{\partial[A]}{\partial x} = R_F\frac{\partial[A]}{\partial t} \tag{4.139}$$

or, alternatively,

$$-\frac{\left(\dfrac{\partial[A]}{\partial t}\right)}{\left(\dfrac{\partial[A]}{\partial x}\right)} = \frac{U}{R_F} \tag{4.140}$$

where U is the Darcy velocity of the groundwater. The left-hand side is interpreted as the velocity of the pollutant, U'. Then, U'/U = 1/R$_F$. In other words, the retardation factor scales as the velocity of the groundwater. Large R$_F$ means U' < U and the chemical movement is retarded.

A qualitative picture of the effects of dispersion, adsorption, and reaction on the movement of pollutants is shown in Figure 4.41. Dispersion alone will make the

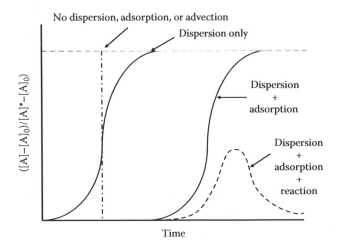

FIGURE 4.41 The effects of dispersion, adsorption, and reaction on the movement of a pollutant in the subsurface.

pollutant front be sharp. The addition of adsorption will delay the breakthrough time; however, the concentration at the sampling point will eventually reach the feed concentration at $x = 0$. The inclusion of reaction will further decrease the peak maximum and the feed concentration will never be attained at the downstream sampling point. The relative magnitudes of dispersion and advection are assessed in terms of a dimensionless Peclet number based on dispersivity, $Pe = UX_s/D$.

For soils (porous media), the dispersion coefficient has two contributions, one from molecular diffusion and another from the fluid movement through the porous media. For a porous medium, the tortuous path that a molecule has to take in traversing the pore fluid reduces the effective diffusion coefficient. $D_w\varepsilon^{4/3}$, where ε is the porosity and D_w is the molecular diffusivity of the pollutant in water, gives the effective diffusion coefficient. The dispersion due to fluid motion is related to fluid velocity through a saturated medium, which is given by Darcy's law:

$$U_D = -\kappa \cdot \frac{dh}{dx} \qquad (4.141)$$

where

U_D is the Darcy velocity (cm s^{-1})
κ is the "hydraulic conductivity" of the medium (cm s^{-1})
dh/dx is the "hydraulic gradient"

In terms of volumetric flow, $U_D = Q/A$ where Q is the volumetric flow rate of fluid through a cross-sectional area A. The negative sign for U_D indicates that the flow of fluid is in the direction of decreasing hydraulic head. Table 4.13 lists typical values of κ for representative media. It can vary over many orders of magnitude depending on the media.

The dispersion constant is given by $D = D_w \varepsilon^{4/3} + U_D\alpha_D$, where α_D is termed the "dispersivity." It is a measure of the media heterogeneity and has units of length. A proposed relationship for α_D in terms of the travel distance for a contaminant is $\alpha_D = 0.017X^{1.5}$ (Neumann, 1990), where α_D is in m and X is in m.

TABLE 4.13
Typical κ Values for Porous Media

Media	κ (cm s^{-1})
Gravel	0.03–3
Sand (coarse)	9×10^{-5}–0.6
Sand (fine)	2×10^{-5}–0.02
Silt	1×10^{-7}–0.003
Clay	8×10^{-11}–2×10^{-7}
Shale	1×10^{-11}–2×10^{-7}

Source: Bedient, P.B. et al., *Groundwater Contamination*, Prentice Hall PTR, Englewood Cliffs, NJ, 1994.

Example 4.21 Time to Breakthrough
for a Contaminant Plume in Groundwater

Estimate the concentration of chlorobenzene in the groundwater at a well 1 km from a source after 500 years. The Darcy velocity is 5 m y^{-1}. The soil has an organic carbon content of 2% and is of porosity 0.4 and a density of 1.2 g cm^{-3}.

For chlorobenzene, log K_{ow} = 2.91. Hence, log K_{oc} = (0.92) (2.91) − 0.23 = 2.45. $K_{sw} = K_{oc}f_{oc}$ = $(10^{2.45})$ (0.02) = 5.6 L kg^{-1}. $R_F = \varepsilon + (1 - \varepsilon) \rho_s K_{sw}$ = 4.4. D_w = 8.7 × 10^{-6} cm^2 s^{-1}. $\alpha_D = 0.017 X_s^{1.5}$ = 530 m = 5.3×10^4 cm. U_D = 5 m y^{-1} = 1.6 × 10^{-5} cm s^{-1}. D = 2.5 × 10^{-6} + 0.85 = 0.85 cm^2 s^{-1}. Note that Pe = 1.8 and hence both advection and dispersion are important.

Since $[A]_0$ = 0, we have $[A]/[A]_s$ for the left-hand side of the equation. Using U_D for U we have

$$\frac{[A]}{[A]_s} = \frac{1}{2}\left[\text{erfc}\left(\frac{10^5 - 1.45\times10^{-6}t}{0.51\sqrt{t}} \right) + 6.5 \cdot \text{erfc}\left(\frac{10^5 - 1.45\times10^{-6}t}{0.51\sqrt{t}} \right) \right]$$

where t is in seconds. If we use t = 500 × 3.17 × 10^7 s, we get $[A]/[A]_s$ = 0.336.

Example 4.22 Time of Travel for Advective Transport
of a Contaminant in Groundwater

Estimate the time taken for a plume of chlorobenzene to reach a groundwater well 100 m from the source if the Darcy velocity is 1 × 10^{-4} cm s^{-1} for the soil in the last example. Assume only advective transport is significant.

Since advection is dominant over dispersion, U/U_D = 1/R_F = 0.23. U = (0.23) (1 × 10^{-4}) = 2.3 × 10^{-5} cm s^{-1}. Hence t = X_s/U = 10,000/2.3 × 10^{-5} = 4.3 × 10^8 s = 13.7 y.

4.3.1.2 Sediment-Water Exchange of Chemicals

Compounds distribute between the various compartments in the environment. One of the repositories for chemicals is the sediment. Chemical exchange at the sediment-water interface is, therefore, important in delineating the fate of environmentally significant compounds. Sediment contamination arose from the uncontrolled pollutant disposal in lakes, rivers, and oceans. As environmental regulations became stricter, most pollutant discharges to our lakes and waterways became controlled. The contaminants in sediments bioaccumulate in marine species and exposure to humans become likely. Hence the risks posed by contaminated sediments have to be evaluated and sediment remediation strategies determined. The risk-based corrective action (RBCA) is predicated upon knowledge of chemical release rate from sediment and transport through air and water environments (Figure 4.42). The first step in this process requires an understanding of potential release mechanisms.

In the case of the sediment environment, a number of pathways for chemical exchange between the sediment and water can be identified. These are represented schematically in Figure 4.43. For sediments that rest in quiescent environments, diffusion (molecular) retarded by adsorption is the most ubiquitous of transport processes. Advective transport is driven by the nonuniform pressure gradients on the

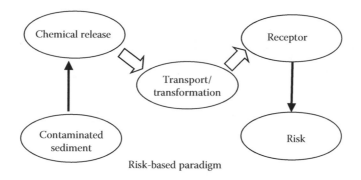

FIGURE 4.42 Risk-based corrective action (RBCA) paradigm.

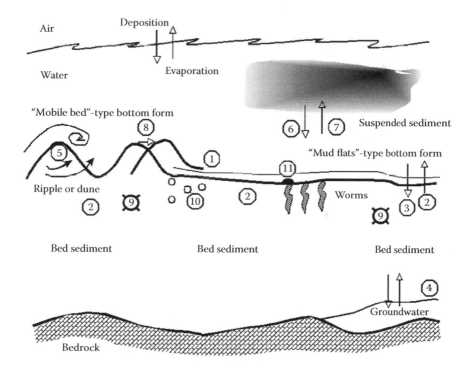

FIGURE 4.43 Schematic of the fate and transport processes in bed sediments.

rough sediment terrain. Other transport processes include active sediment particle transport. A comparison of characteristic times for a hypothetical scenario was made by Reible et al. (1991) and is shown in Table 4.14. Processes with small half-lives are likely to be the most important transport processes.

Diffusion of compounds from sediment in the absence of advection or biodegradation can be represented by the equation derived in the previous section on groundwater transport where the dispersivity is replaced by molecular diffusion. The geometry considered for modeling is shown in Figure 4.44.

TABLE 4.14

Comparison of the Characteristic Times of Sediment Transport Processes

Mechanism	Characteristic Time
Molecular diffusion (unretarded by sorption)	0.5 years (hypothetical)
Molecular diffusion (retarded by sorption)	1,900 years
Colloidal-enhanced diffusion	1,500 years
Sediment erosion (1 cm y^{-1} erosion)	10 years
Capped sediment (30 cm effective cap)	21,000 years
Bed load transport (sediment movement)	42 h
Advection (aquifer interactions)	4,000 years
Surface roughness (local advection)	69,000 years
Bioturbation	10 years

Notes: Characteristic times are order of magnitude estimates of time required to leach a typical hydrophobic organic compound (e.g., trichlorobiphenyl) from a 10 cm layer of sediment by each of various transport mechanisms. In most cases, the characteristic time is 1/e times or half-lives. For advective processes, the characteristic times represent the time for complete removal. These refer to typical sediment conditions represented in Reible et al. (1991). Although absolute values refer to a particular set of conditions, the ranked magnitudes are probably indicative of generic behavior.

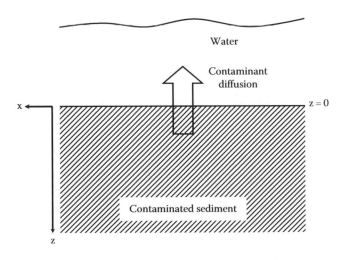

FIGURE 4.44 Schematic of the diffusive transport of a contaminant from a bed sediment to the overlying water column.

$$\frac{[A]_w}{[A]_w^0} = \text{erf}\left(\frac{z}{\sqrt{4D_s^* t}}\right)$$ (4.142)

where $D_s^* = \dfrac{D_w \varepsilon^{4/3}}{\varepsilon + \rho_p K_{sw}}$.

The release rate (flux) is

$$F_A = \left(\frac{D_s^*}{\pi t}\right)^{1/2} \cdot W_A^0 \rho_p$$ (4.143)

where W_A^0 is the initial sediment contamination.

4.3.1.3 Soil-Air Exchange of Chemicals

Figure 4.45 represents the chemical transport pathway from soil to the atmosphere. The difference from the saturated case (groundwater) is that in this unsaturated case, there exist both pore air and pore water. The mobile phase is air, while both pore water and solids are immobile with respect to the contaminant. The advective dispersion equation derived is applicable with the stipulation that [A] represents the pore air concentration and $[A]_{tot}$ is the total concentration (pore air + pore water + solid phase). D should be replaced by the effective molecular diffusivity in partially saturated air D_g and U_g is the advective gas phase velocity. R_A denotes the biochemical reaction loss of compound A in pore air and pore water.

$$D_g \frac{\partial^2 [A]_g}{\partial z^2} - U_g \frac{\partial [A]_g}{\partial z} - R_A = \frac{\partial [A]_{tot}}{\partial t}$$ (4.144)

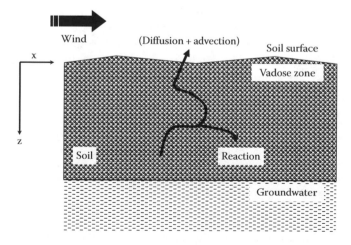

FIGURE 4.45   Transport of a contaminant from the soil to the atmosphere.

If the air-filled porosity is ε_g and ε_w is the water-filled porosity, then a total mass conservation gives $[A]_{tot}$ = mass in pore air + mass in pore water + mass on soil particles = $\varepsilon_g[A]_g + \varepsilon_w[A]_w + \rho_s(1-\varepsilon_g-\varepsilon_w)W_A$. Since local equilibrium is assumed between the three phases, $[A]_w = [A]_g/K_{AW}$, $W_A = K_{SA}[A]_g$. With these expressions, we have $[A]_{tot} = R_F[A]_g$, where $R_F = \varepsilon_g + \dfrac{\varepsilon_w}{K_{AW}} + (1-\varepsilon_g-\varepsilon_w)\rho_s K_{SA}$ is the "retardation factor." If we further consider the steady state case where $R_A = 0$ and advective velocity is negligible, we have

$$\frac{\partial[A]_g}{\partial t} = \frac{D_g}{R_F} \cdot \frac{\partial^2[A]_g}{\partial z^2} \qquad (4.145)$$

Thus molecular diffusion through pore air is the dominant mechanism in this case. The initial condition is $[A]_g(z, 0) = [A]_0$ for all z. The following two boundary conditions are applicable: (1) $[A]_g(\infty, t) = [A]_0$ for all t, and (2) $-D_g\dfrac{\partial[A]_g}{\partial z} + k_a[A]_g(z,t) = 0$, at z = 0. The second boundary condition states that at the surface, there is a reaction or mass transfer loss of chemical, thus contributing to a resistance to mass transfer at the soil surface. The solution for the flux from the surface is (Valsaraj et al., 1999)

$$F_A = k_a[A]_0 \cdot \exp\left(\frac{k_a^2 t}{D_g R_F}\right) \cdot \mathrm{erfc}\left(k_a\sqrt{\frac{t}{D_g R_F}}\right) \qquad (4.146)$$

Note that at t = 0, $F_A = k_a[A]_0$ and as $t \rightarrow \infty$, $F_A = [A]_0\sqrt{\dfrac{D_g R_F}{\pi t}}$, which is the solution to the Fick's diffusion equation. Note that F_A is a function of both D_g and R_F, both of which depend on the soil pore water content. With increasing water in the pore space, D_g and R_F will decrease. The decrease in R_F will be far more significant since K_{SA} has been observed to be a sensitive function of pore water content (Guilhem, 1999). Experimental data have verified this prediction (Figure 4.46). Thus flux will be higher from a wet soil or sediment and lower from a dry sediment. In other words, pesticides and organic compounds will be far more volatile from wet than dry soils.

Example 4.23   Pesticide Volatilization Rate from a Soil

Estimate the rate of release ($\mu g\ s^{-1}$) of dieldrin (a pesticide, molecular weight 381) applied to a soil at 100 $\mu g\ g^{-1}$ concentration. The soil has a total porosity of 0.5, f_{oc} of 1%, and a water saturation of 5%. The soil density is 2 $g\ cm^{-3}$ and area of application is 1 ha. Assume a surface mass transfer coefficient of 0.1 cm s^{-1}.

The effective diffusivity D_g in a partially saturated soil is given by $D_A\dfrac{\varepsilon_a^{10/3}}{\varepsilon_T^2}$, where D_A is the molecular diffusivity in air, ε_a is the air-filled porosity. Note that $\varepsilon_a = (1 - \theta_w)\,\varepsilon_T$, where θ_w is the water saturation. For dieldrin, $D_A = 0.028\ cm^2\ s^{-1}$. Since $\theta_w = 0.05$, $\varepsilon_a = 0.57$, $\varepsilon_w = \theta_w\varepsilon_T = 0.03$. Hence, $D_g = 0.012\ cm^2\ s^{-1}$.

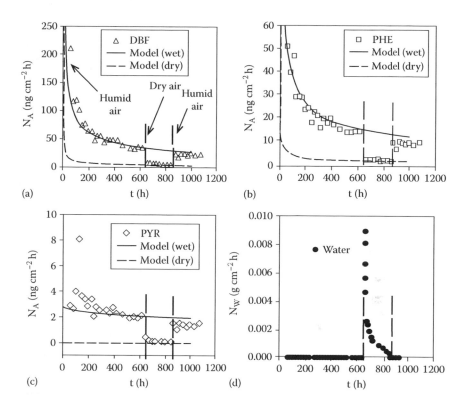

FIGURE 4.46 Effects of change in air relative humidity on (a) dibenzofuran, (b) phenanthrene, (c) pyrene, and (d) water flux from a 6.5% moisture University Lake sediment. Model curves for both humid air over "wet" sediment and dry air over "dry" sediment are shown. (Data from Valsaraj, K.T. et al., *Environ. Sci. Technol.*, 33, 142, 1999.)

$R_F = \varepsilon_g + \dfrac{\varepsilon_w}{K_{AW}} + \rho_b K_{SA}$. K_{AW} for dieldrin is 8.1×10^{-6}. From Chapter 2, we have $K_{SA} = K_{SW}/K_{AW} = f_{oc}K_{oc}/K_{AW}$. With log $K_{ow} = 5.48$, log $K_{oc} = (0.92)\,(5.48) - 0.23 = 4.81$. Hence, $K_{SA} = 7.9 \times 10^7$ P kg^{-1}, and $R_F = 6.8 \times 10^7$. $[A]_0 = \rho_b W_i/K_{SA} = 1.2 \times 10^{-6}$ µg cm^{-3}.

Hence, $F_A = (0.1)\,(1.2 \times 10^{-6})\,[\exp(1.88 \times 10^{-5}) \cdot \mathrm{erfc}(0.0137)] = 1.12 \times 10^{-7}$ µg (cm^2 s)$^{-1}$. Hence mass lost $m_i = 1.12 \times 10^{-7} \times 10^8 = 11.2$ µg s^{-1}.

4.3.2 Soil and Groundwater Transport

As depicted in Figure 4.38, surface spills of solvents can potentially contaminate two zones. The discussion in the previous section reflects the different transport mechanisms that operate in the two zones. In the vadose zone, pore air and pore water transport processes are operative, while in the saturated zone pore water transport is dominant. The remediation strategy for contaminants in the two zones also differs, the underlying phenomena being driven by the respective operative transport mechanisms.

In both zones, there are three major categories of remedial processes:

1. Containment, whereby contaminants are prevented from further spreading
2. Removal, whereby contaminants are extracted from the subsurface
3. Treatment, wherein contaminants are separated and treated by appropriate technologies.

Containment is required in cases where the movement of fluids is to be controlled before adverse effects are manifest in nearby communities that depend on drinking water supplies from the aquifer. Physical barriers such as slurry walls, grout curtains, and sheet pilings are used for containment. Hydraulic barriers, i.e., reversing the hydraulic gradient by a series of pumps and drains, are effective in containing a slow-moving contaminant plume in both zones, especially in the saturated zone.

Removal is the only option in some cases. Highly contaminated surface soils are excavated and treated before disposal. However, this is often infeasible and expensive for surface soils and groundwater in the saturated zone. A more practical solution is the so-called pump-and-treat (P&T) method. As the very name indicates, this entails bringing groundwater to the surface, separating the contaminants, and discharging the water to the subsurface or to lakes or rivers. A variation of P&T practiced in both the vadose and saturated zones is pumping air and/or applying vacuum to strip volatile materials from soil pore air or groundwater. In the vadose zone, this is called "air sparging." When water is used to flush contaminants it is called "in situ soil washing."

Treatment involves a variety of aboveground technologies for both water and air that is recovered from the subsurface. Processes that utilize intrinsic bacterial populations are useful in destroying contaminants in place. These include "natural attenuation" and "intrinsic bioremediation."

Equilibrium and chemical kinetic aspects of P&T technologies for the vadose and saturated zones in subsurface soil environments are explored in the following.

4.3.2.1 Pump-and-Treat for NAPL Removal

Figure 4.47 is a schematic for P&T groundwater remediation. This is a method for remediating contaminated groundwater. Clean water that is brought into the aquifer flushes pollutants from the contaminated region. The water brought to the surface will have to be further treated before reinjection or disposal. Although effective in containing plume migration, P&T is ineffective in removing all of the material from the subsurface. Once the free phase NAPL is removed, the residual NAPL is harder to remove by water flushing. It gives rise to the "tailing off" effect, i.e., contaminants in inaccessible regions dissolve very slowly over extended periods (Figure 4.48).

For a withdrawal rate Q from a well, the maximum removal rate of a contaminant is Q[A], where [A] is the pore water concentration. If the contaminant is present as a pure phase, the value of [A] is the same as the aqueous solubility $[A]^*$. As water is pumped through the aquifer, the equilibrium between the solids and pore water dictates the maximum mass of contaminant that can be transported through the groundwater. The process can be modeled by considering the aquifer to be a completely mixed reactor (CSTR). If we consider a total volume V_T of the aquifer,

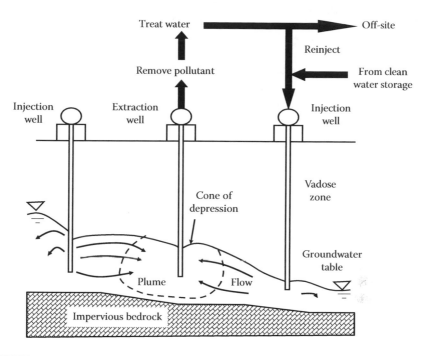

FIGURE 4.47   A basic pump-and-treat (P&T) approach for groundwater remediation.

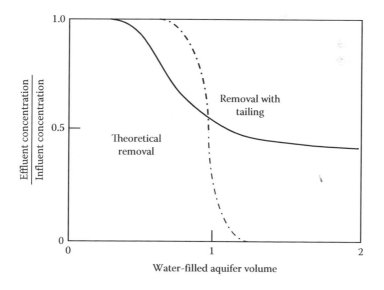

FIGURE 4.48   The phenomena of tailing-off in conventional pump-and-treat process for groundwater remediation.

the fluid volume is εV_T, where ε is the porosity of solids. A mass balance on the continuous reactor gives

Rate in with water = Rate out with water + Accumulation rate (solids + pore fluid)

$$Q[A]_{in} = Q[A] + \varepsilon V_T \frac{d[A]}{dt} + (1-\varepsilon) V_T \rho_s \frac{dW_A}{dt} \qquad (4.147)$$

If local equilibrium between solids and pore fluid is assumed, $W_A = K_{sw}[A]$. Moreover, if pure water is brought into contact from the injection well, $[A]_{in} = 0$. Hence

$$-\left(\frac{Q}{R_F}\right) \cdot \frac{[A]}{V_T} = \frac{d[A]}{dt} \qquad (4.148)$$

where Q is the volumetric flow rate, $R_F = \varepsilon + (1-\varepsilon)\rho_s K_{sw}$ is the retardation factor. Integrating this equation gives

$$\frac{[A]}{[A]_0} = \exp\left(-\frac{Qt}{R_F V_T}\right) \qquad (4.149)$$

Hence the time taken to flush 50% of the contaminant in pore water will be $t_{1/2} = 0.693 \frac{R_F V_T}{Q}$. Thus, cleanup time will increase with increasing R_F and decreasing Q. Note that $[A]_0 = \frac{M_0}{V_T R_F}$, where M_0 is the total initial mass of the contaminant. Note that $(Qt/\varepsilon V_T)$ is the number of pore volumes, N_{PV} flushed in time t. Therefore,

$$\frac{[A]}{[A]_0} = \exp\left(-\frac{\varepsilon N_{PV}}{R_F}\right) \qquad (4.150)$$

Example 4.24 Number of Pore Volumes for In Situ Flushing of Residual NAPL

Calculate the total number of pore volumes to remove 99% of the following pollutants in pore water from an aquifer with porosity 0.3 and an organic carbon content of 1%. Soil bulk density is 1.4 g cm^{-3}: (1) benzene, (2) naphthalene, and (3) pyrene.

First calculate K_{oc} from log K_{oc} = 0.937 log K_{ow} − 0.006 and $K_{sw} = f_{oc}K_{oc}$. Then calculate $R_F = \varepsilon + \rho_b K_{sw}$. For 99% removal $[A]/[A]_0 = 0.01$.

Compound	Log K_{sw}	K_{sw}	R_F	$N_{PV} = (-R_F/\varepsilon)$ ln (0.01)
Benzene	2.13	1.6	2.5	38
Naphthalene	3.36	13.9	19.8	304
Pyrene	5.13	632.1	885	13,585

Extraction from the saturated zone is incapable of removing most contaminants that have low solubility in water. These cases require excessively large volumes of water to be pumped (see Example 4.24). As a result techniques have been sought to enhance pump and treat. In Chapter 2, we noted that the use of surfactants significantly lowers the surface tension at oil-water interfaces and also increases the aqueous solubility of organic compounds through the presence of surfactant micelles. Both of these aspects have been utilized in enhancing the extent of removal using P&T methods.

Figure 4.49 identifies the different phases into which a pollutant can partition when surfactants are present in the aqueous phase in the form of micelles. If a CSTR is assumed for the aquifer, the overall mass balance with added surfactant in the aquifer gives

$$\text{Rate in} = \text{Rate out}\left(\text{water} + \text{micelles}\right) + \text{Accumulation}\left(\text{water} + \text{solids} + \text{micelles}\right).$$

$$0 = Q\left([A] - [A]_{mic}\right) + \varepsilon V_T \frac{d[A]}{dt} + \varepsilon V_T \frac{d[A]_{mic}}{dt} + \left(1 - \varepsilon\right) V_T \rho_s \frac{dW_A}{dt} \quad (4.151)$$

Since $[A]_{mic} = [Surf]K_{mic}[A]$ and $W_A = K_{sw}[A]_i$, we find

$$\frac{d[A]}{dt} = -\left(\frac{Q}{V_T}\right) \cdot \left(\frac{\left(1 + [Surf]K_{mic}\right)}{\varepsilon + \varepsilon[Surf]K_{mic} + \left(1 - \varepsilon\right)\rho_s K_{sw}}\right) \quad (4.152)$$

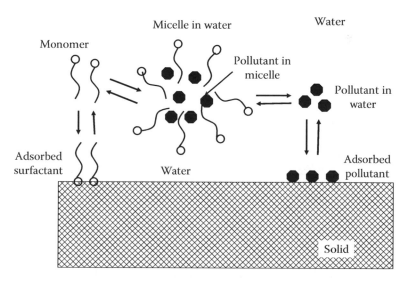

FIGURE 4.49 Partitioning of a pollutant between various compartments in the groundwater when surfactants are introduced into the subsurface.

where K_{mic} is the micelle-water partition constant for i. $[Surf] = [Surf]_{tot} - [CMC]$, where $[Surf]_{tot}$ is the total added surfactant concentration and $[CMC]$ is the critical micellar concentration. Integrating we obtain

$$\frac{[A]}{[A]_0} = \exp\left(-\left(\frac{Q}{V_T}\right) \cdot \left(\frac{1+[Surf]K_{mic}}{\varepsilon + \varepsilon[Surf]K_{mic} + \rho_b K_{sw}}\right) \cdot t\right) \qquad (4.153)$$

Example 4.25   Extraction of NAPL Residual Using a Surfactant Solution

For Example 4.24, determine the number of pore volumes required for 99% removal of benzene if a 100 mM solution of sodium dodecyl sulfate (SDS) is used for flushing. CMC of SDS is 8 mM.

From Chapter 2, $\log K_{mic} = 1.02 \log K_{ow} - 0.21 = (1.02)(2.13) - 0.21 = 1.92$. $[Surf] = 100 - 8 = 92$ mM $= 36$ g L^{-1}. $R_F' = 0.3 + (1.4)(1.6) + (0.3)(10^{1.92})(0.036) = 3.4$.

Note that $N_{PV} = (Qt)/\varepsilon V_T$. Hence, $\dfrac{[A]}{[A]_0} = \exp\left(-\dfrac{(1+[Surf]K_{mic})\varepsilon N_{PV}}{R_F'}\right)$. For

$[A]/[A]_0 = 0.01$, $N_{PV} = (4.605) \cdot \left(\dfrac{R_F'}{\varepsilon(1+[Surf]K_{mic})}\right) = 13$. Hence N_{PV} is reduced to

13 from 38 with pure water.

4.3.2.2   In Situ Soil Vapor Stripping

In the vadose zone, applying vacuum using a pump can remove contaminant (Wilson and Clarke, 1994). Figure 4.50 is a schematic of a soil vapor stripping well. Volatile contaminants desorb into the pore air that is pulled out of the subsurface. The mechanism of removal is similar to that for in situ water flushing in the saturated zone. Just as in the case of the saturated zone, the vadose zone can be considered to be a CSTR and the change in concentration in pore air is obtained from a mass balance between all the phases in the subsurface. The contaminant will be distributed between pore air, pore water, solids, and the NAPL phases. The total mass is given by

$$M_{tot} = \varepsilon_a V_{tot}[A]_g + \varepsilon_w V_{tot}\frac{[A]_g}{K_{AW}} + \rho_b K_{SA}[A]_g V_{tot} + \varepsilon_N V_{tot} K_{NA}[A]_g \qquad (4.154)$$

or concentration by

$$[A]_{tot} = \frac{M_{tot}}{V_{tot}} = \left(\varepsilon_a + \frac{\varepsilon_w}{K_{AW}} + \rho_b K_{SA} + \varepsilon_N K_{NA}\right)[A]_g \qquad (4.155)$$

where K_{SA} is the soil/air distribution constant for the chemical and K_{NA} is the NAPL/ air distribution constant. Note that $K_{SA} = \dfrac{K_{sw}}{K_{AW}}$ and $K_{NA} = \dfrac{\rho_N}{[A]^* K_{AW}}$, where ρ_N is the density of the NAPL and $[A]^*$ is the aqueous solubility of the contaminant.

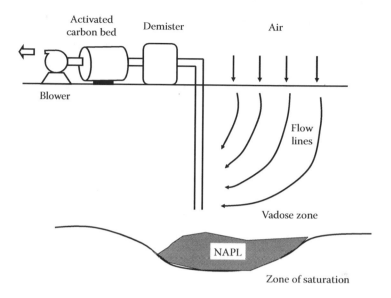

FIGURE 4.50   Schematic of a soil vapor stripping well.

Using the CSTR assumption for the vadose zone between the contaminant source and the extraction well, we have

$$\text{Rate in} = \text{Rate out} + \text{Accumulation}\,(\text{air} + \text{solid} + \text{water} + \text{NAPL}).$$

$$Q_g\left[A\right]_g^{in} = Q_g\left[A\right]_g + V_{tot}\frac{d\left[A\right]_{tot}}{dt} \qquad (4.156)$$

Since $\left[A\right]_g^{in} = 0$. The resulting differential equation can be solved with the initial condition that $[A]_g = [A]_0$ at $t = 0$ to get

$$\frac{\left[A\right]_g}{\left[A\right]_0} = \exp\left(-\frac{Q_g}{R_F V_{tot}} \cdot t\right) \qquad (4.157)$$

where $R_F = \varepsilon_a + (\varepsilon_w/K_{AW}) + \rho_b K_{SA} + (\varepsilon_N/K_{NA})$. The number of pore volumes flushed in time t is

$$N_{PV} = \frac{Q_g t}{\varepsilon_a V_{tot}} = -\left(\frac{R_F}{\varepsilon_a}\right) \cdot \ln\left(\frac{\left[A\right]_g}{\left[A\right]_0}\right) \qquad (4.158)$$

The volumetric flow rate of air is given using Darcy's law (Wilson and Clarke, 1994)

$$Q_g = \left(A_x \varepsilon_a \kappa_d\right) \cdot \frac{\left(P_0^2 - P_1^2\right)}{2X_L P_1} \tag{4.159}$$

where
   κ_d is the permeability of soil (Darcy's constant, m^2 (atm s)$^{-1}$)
   P_0 is the inlet pressure (1 atm)
   P_1 is the outlet pressure (atm)
   X_L is the distance between the inlet and exit (m)
   A_x is the area normal to the flow (m^2)

Note that N_{PV} is directly proportional to R_F and is logarithmically related to degree of removal $[A]_g/[A]_0$.

Example 4.26 Extraction of Vapors from the Vadose Zone

Estimate how much o-xylene vapors can be removed from 1 m^3 of the vadose zone of an aquifer that has a total porosity of 0.6 in 1 day. The water content is 10% and xylene content is 20%. The Darcy's constant is 0.01 m^2 (atm s)$^{-1}$. A vacuum of 0.8 atm is applied across a 10 m soil length. The soil organic fraction is 0.02 and bulk density is 1.4 g cm^{-3}. The density of xylene is 0.9 g cm^{-3} and has an aqueous solubility of 160 µg cm^{-3}.
   To calculate R_F, ε_a = (1 −0.1 − 0.2) (0.6) = 0.42, ε_w = 0.06, ε_N = 0.12. K_{AW} = 0.215, K_{oc} = $10^{2.3}$, K_{sw} = $\varphi_{oc} K_{oc}$ = 4.0 P kg. K_{NA} = (0.9)/(160) (0.215) (0.001) = 26, K_{SA} = 4/(0.215) = 18. Hence, R_F = 29. Q_g(1) (0.42) (0.01) [(1^2 − (0.2)^2)/(0.2)]

(1/20) = 0.001 m^3 s^{-1}. Hence $\dfrac{[A]_g}{[A]_0} = \exp\left(-\dfrac{(0.001)(8.64 \times 10^4)}{(29)(1)}\right) = 0.05$. Thus

removal from pore air is 95%.

As we observed in Chapter 3, air-water partition constant K_{AW} increases with temperature, while K_{SA} and K_{NA} decrease with temperature. Hence with increasing T, retardation factor decreases and the efficiency of soil vapor extraction increases. Therefore, heat treatment of soils will accelerate soil vapor stripping (Wilson and Clarke, 1994).

4.3.2.3 Incineration for Ex Situ Treatment

Thermal processes are used to destroy organic species in soils and solid waste. There are two types of incinerators employed, viz., rotary kiln and direct flame incinerators. Rotary kiln incinerators are ideal for soil and solid waste combustion. The basic design and operation of a rotary kiln incinerator is described in detail by Reynolds et al. (1991).

Example 4.27   Efficiency of a Hazardous Waste Incinerator

A chlorinated hydrocarbon waste is to be incinerated in a rotary kiln at a temperature of 1500°F. The activation energy for the reaction was observed to be 50 kcal mol^{-1} and the pre-exponential factor was 3×10^7 P s^{-1}. Laboratory work has shown that the combustion gas flow rate required was 500 SCF lb^{-1} at 70°F. If the feed rate is 1000 lb h^{-1} at a face velocity of 30 ft s^{-1} and a kiln residence time of 1 s is desired, what should be the length and diameter of the incinerator? What will be the efficiency of the reactor if it were a PFR?

From the given gas flow rate and feed rate, we get feed flow rate at 70°F. $Q_g = (500 \text{ ft}^3 \text{ lb}^{-1}) (1000 \text{ lb h}^{-1}) (1/60 \text{ h min}^{-1}) = 8333 \text{ ft}^3 \text{ min}^{-1}$. Feed flow rate at 1500°F is given by $Q_g = 8333 (T_F/T_i) = (8333) (1500 + 460/70 + 460) = 30{,}817 \text{ ft}^3$ min^{-1}. We know that $U_g = Q_g/A = X_s/\tau_R$, where X_s is the length of the incinerator and τ_R is the mean residence time. If the kiln is cylindrical, $A = \pi(D^2/4)$. Hence,

$Q_g = \pi(D^2/4) (X_s/\tau_R)$ or $D = \sqrt{\dfrac{4 Q_g \tau_R}{\pi X_s}}$. Since $U_g \tau_R = X_s = (30) (1) = 30$ ft, we have

$D = \sqrt{\dfrac{(4)(30{,}817)(1)}{(60)(3.14)(30)}} = 4.7$ ft. Thus $D = 4.7$ ft and $X_s = 30$ ft. Efficiency of a PFR

is $E = 1 - \exp(-k\tau_R)$. We need k at 1500°F. $K = (3 \times 10^7) \exp[-50{,}000/(1.98)(1960)] = 76 \text{ s}^{-1}$. Hence, $E = 1 - \exp[-(76) (2)] = 1$. Complete destruction of the material is possible.

4.4   APPLICATIONS IN ENVIRONMENTAL BIOENGINEERING

Our world is teeming with various biological species. Microorganisms lie at the base of our evolutionary chain and have played a critical role in changing the earth's environment. They possess the ability to transform complex molecules. It has long been known that microorganisms are at work in composting waste into manure. They are also critical in treating sewage and wastewater. Table 4.15 lists typical examples where microorganisms play a role in environmental engineering.

TABLE 4.15
Examples of Bioengineering in Environmental Processes

Subsurface bioremediation	Addition of microbes to waste materials (groundwater, soils, and sediments) mostly in situ
Wastewater treatment	Removal of organic compounds from industrial wastewaters, activated sludge, trickling filters
Land farming	Petroleum exploration and production waste treatment (aboveground)
Enhanced NAPL recovery	Introduction of bacterial cultures in subsurface soils for simulation of surfactant production
Oil spill cleanup	Microorganisms for oil spill biodegradation
Engineered microorganisms	To treat special waste materials (highly toxic and refractory compounds)

Soils, sediments, and groundwater contain a milieu of naturally occurring microorganisms that can breakdown toxic compounds. By providing sufficient nutrients and the necessary conditions to initiate their activity, they can act on any given type of waste.

The ultimate products of biodegradation by microorganisms are inorganic compounds (CO_2 and H_2O). This kind of a complete conversion process is called "mineralization." The following questions regarding mineralization become important: (1) Why do microorganisms transform compounds? (2) What mechanisms do they employ in breaking down complex molecules? (3) How do the microorganisms affect chemical transformation rates? The first question can be answered from thermodynamic principles. The second question needs principles from biochemistry, which we will not pursue here. The third question concerns chemical kinetics, which is the focus of this section.

Thermodynamics provides clues to the energy of microbial processes. Consider the oxidation of glucose:

$$C_6H_{12}O_6 + 8O_2 \rightarrow 6CO_2 + 12H_2O + 2870 \text{ kJ} \qquad (4.160)$$

This is called an "exergonic" reaction. For every 1 mol of glucose consumed, an organism can obtain 2870 kJ of energy. "Endergonic" reactions, on the other hand, require organisms to consume energy from the surroundings for conversion of chemicals. Microorganisms (living cells) couple exergonic and endergonic reactions to accomplish the goal of lowering the free energy. This is done using intermediate chemicals that can store energy released during an exergonic process, and transfer the stored energy to the site where an endergonic process takes place. This is the basis of living cell metabolism. Table 4.16 gives the magnitudes of ΔG^0 for some common microbial-mediated environmental processes.

In order to accomplish the types of processes discussed earlier, the living cells make use of intermediates such as adenosine triphosphate (ATP) and guanosine triphosphate (GTP). By far ATP is the most important and is the universal transfer agent of chemical energy-yielding and energy-requiring reactions. ATP formation will store energy, whereas its hydrolysis will release energy. Organisms that have adenosine diphosphate (ADP) will first convert it to ATP by the addition of a phosphate group, which requires energy to be consumed. This is called "phosphorylation." Subsequent hydrolysis of ATP to ADP releases the stored energy ($\Delta G^0 = -30$ kJ mol^{-1}). During a redox process, the electrons that are transferred between compounds by microorganisms also find home in intermediate compounds. Enzymes responsible for the redox transformations capture electrons from electron-rich compounds ("dehydrogenases") and store them in intermediate "coenzymes" such as nicotine adenine dinucleotide (NAD). A typical reaction involving the NAD/NADH pair is $NAD^+ + 2H^+ + 2e^- \rightleftharpoons NADH + H^+$.

The synthesis of pyruvic acid by the partial oxidation of glucose is called "glycolysis." This reaction is a convenient one to show how ATP mediates in a metabolic process. Shown in Figure 4.51 is the partial oxidation process. The process starts with the conversion of glucose to glucose-6-phosphate in which one ATP molecule

TABLE 4.16

A Sampling of Microbially Mediated Environmental Redox Processes

Reaction	$-\Delta G^0(w)$ (kJ mol^{-1})

Fermentation

$$\frac{1}{4}CH_2O + \frac{1}{4}H_2O \rightleftharpoons \frac{1}{4}CO_2 + \frac{1}{2}H_2 \ (g) \qquad\qquad 1.1$$

Aerobic respiration

$$\frac{1}{4}CH_2O + \frac{1}{2}O_2 \rightleftharpoons \frac{1}{4}CO_2 \ (g) + \frac{1}{4}H_2O \qquad\qquad 119$$

Nitrogen fixation

$$\frac{1}{6}N_2 \ (g) + \frac{1}{3}H^+ + \frac{1}{4}CH_2O \rightleftharpoons \frac{1}{3}NH_4^+ + \frac{1}{4}CO_2 \qquad\qquad 14.3$$

Carbon fixation

$$\frac{1}{4}CO_2 \ (g) + \frac{1}{2}H_2O \rightleftharpoons \frac{1}{4}CH_2O + \frac{1}{4}O_2 \ (g) \qquad\qquad -119$$

Notes: CH$_2$O represents 1/6 C$_6$H$_{12}$O$_6$ (glucose). $\Delta G^\ominus(w)$ was defined in Section 2.3.8 as the difference in pe^0(w) between the oxidant and reluctant.

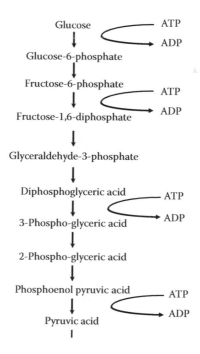

FIGURE 4.51   Partial oxidation of glucose to pyruvic acid.

is lost. Subsequent rearrangement to fructose-6-phosphate and loss of another ATP gives fructose-1,6-diphosphate. At this point, there is a net energy loss. However, during further transformations, two molecules of ATP are gained and hence overall energy is stored during the process. The conversion of glyceraldehyde-3-phosphate to pyruvic acid is the most energy-yielding reaction for anaerobic organisms. This ATP/ADP cycle is called the "fermentative mode" of ATP generation and does not involve any electron transport.

The complete oxidation of glucose should liberate 38 molecules of ATP, equivalent to the free energy available in glucose. If that much has to be accomplished the electrons generated during the process should be stored in other compounds that then undergo reduction. "Electron acceptors" generally used by microbes in our natural environment include oxygen, nitrate, Fe(III), SO_4^{2-}, and CO_2. It is to mediate the transfer of electrons from substrate to electron acceptors that the microorganisms need intermediate electron transport agents. These agents can also store some of the energy released during ATP synthesis. Some examples of these intermediates are cytochromes and iron-sulfur proteins. The same function can also be performed by compounds that act as H^+-carriers (e.g., flavoproteins). The redox potentials of some of the electron transport agents commonly encountered in nature are given in Table 4.17. Chappelle (1993) cites the example of *E. coli* that uses the NADH/NAD cycle to initiate redox reactions that eventually releases H^+ ions out of its cell. The NADH oxidation to NAD in the cytoplasm is accompanied by a reduction of the flavoprotein that releases H^+ from the cell to give an FeS protein, which further converts to flavoprotein via acquisition of $2H^+$ from the cytoplasm and in concert with coenzyme Q releases $2H^+$ out of the cell. The resulting cytochrome b transfers the electron to molecular oxygen forming water. The net result is the use of the energy from redox reactions to transport hydrogen ions out of the cell. The energy accumulated in the process is utilized to convert ADP to ATP. The entire sequence of events is pictorially summarized in Figure 4.52 and is sometimes called "chemiosmosis," the process of harnessing energy from electron transport. More details of these schemes are given in advanced textbooks such as Schlegel (1992). The main point of this discussion is the part played by electron transport intermediates and ATP synthesis in the

TABLE 4.17
Redox Potentials for Electron Transport Agents in Biochemical Systems

Component	E_H^0 (V)
NAD	+0.1
Flavoprotein	+0.34
Cytochrome c	+0.69
Oxygen	+1.23

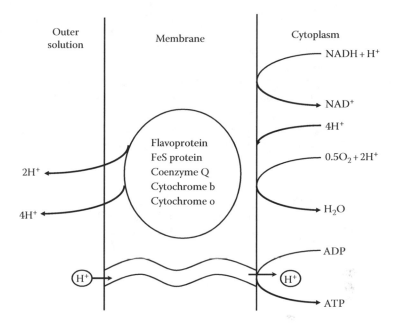

FIGURE 4.52   Electron transport and ATP synthesis in a living cell.

metabolic activities of a living cell. Thus microorganisms provide an efficient route by which complex molecules can be broken down. The entire process is driven by the energy storage and release capabilities of microorganisms that are integral parts of their metabolism.

Thermodynamic considerations of the energy of reactions mediated by microorganisms can provide us information on the feasibility of the transformations. The activity of microorganisms will be such that only those with negative overall free energies will proceed spontaneously. Those with positive free energies will need mediation by ATP- and NAD-type compounds. As we have seen, biochemical reactions are thermodynamically feasible if the overall free energy change is negative. However, even if ΔG is negative, not all of the reactions will occur at appreciable rates to be of any use in a living cell. The answer to this is catalysis caused by the "enzymes" present in all living organisms. In Chapter 2, we derived the rate laws for enzyme catalysis. In the following section, we will apply those expressions to study reactor design in environmental engineering.

4.4.1   ENZYME REACTORS

Biochemical reactors are used in environmental engineering not only for ex situ waste treatment operations but also in situ waste site remediation processes. Those that employ living cells (producing enzymes) are termed "fermentors," whereas those that involve only enzymes are called "enzyme reactors." We shall first describe the enzyme reactors, but the reader is reminded that the same principles are also applicable to fermentors (Bailey and Ollis, 1986; Lee, 1992).

There are three types of enzyme reactors similar to conventional chemical reactors such as described in Section 2.4.1. These are (1) a batch reactor, (2) a plug flow reactor (PFR), and (3) a continuous stirred tank reactor (CSTR). In addition, we will also discuss a convenient operational mode applicable to each of the reactors. This is the process of immobilizing (anchoring) the enzymes so that contact between the enzyme and substrate is facilitated.

4.4.1.1 Batch Reactor

This is the simplest mode of an enzyme reactor or fermentor. The equation representing the change in concentration of substrate with time is

$$-\frac{d[S]}{dt} = \frac{V_{max}[S]}{K_m + [S]} \tag{4.161}$$

where Michaelis-Menten kinetics is assumed to hold. Rearranging and integrating, we get

$$-\int_{[S]_0}^{[S]} \left(\frac{K_m}{[S]} + 1 \right) \cdot d[S] = V_{max} \int_0^t dt \tag{4.162}$$

$$\frac{1}{t} \cdot \ln \left(\frac{[S]_0}{[S]} \right) = \frac{V_{max}}{K_m} - \left(\frac{[S]_0 - [S]}{K_m} \right) \cdot \frac{1}{t} \tag{4.163}$$

This is the integrated rate equation for an enzymatic reaction. For a batch reactor, a plot of $(1/t) \ln([S]_0/[S])$ versus $([S]_0 - [S])/t$ should yield $-1/K_m$ as the slope and V_{max}/K_m as the intercept. From these, the Michaelis-Menten parameters can be evaluated. If we define the degree of conversion χ such that $[S] = [S]_0 (1 - \chi)$, we can write

$$\frac{1}{t} \cdot \ln \left(\frac{1}{1-\chi} \right) = \left(\frac{V_{max}}{K_m} \right) - \left(\frac{[S]_0 \chi}{K_m} \right) \cdot \frac{1}{t} \tag{4.164}$$

This equation gives us the time required for a desired conversion in a batch reactor.

Example 4.28 Pesticide Disappearance from the Soil in a Batch Reactor

Meikle et al. (1973) performed an experiment designed to estimate the biological decomposition of a pesticide (4-amino-3,5,6-trichloropicolinic acid or picloram) from a soil matrix. The experiment was conducted in a batch reactor. The following results were obtained after 423 days of treatment:

$[S]_0$ (ppmw)	$[S]$ (ppmw)	$\dfrac{[S]_0 - [S]}{(t = 423 \text{ days})}$	$\dfrac{1}{t} \ln \dfrac{[S]_0}{[S]}$
3.2	1.56	0.003877	0.001698
3.2	1.76	0.003404	0.001413
1.6	0.51	0.002577	0.002703
1.6	0.69	0.002151	0.001988
0.8	0.24	0.001324	0.002846
0.8	0.21	0.001395	0.003162
0.4	0.12	0.000662	0.002846
0.4	0.094	0.000723	0.003424
0.2	0.029	0.000404	0.004565
0.2	0.026	0.000411	0.004823
0.1	0.01	0.000213	0.005443
0.1	0.013	0.000206	0.004823
0.05	0.007	0.000102	0.004648
0.05	0.005	0.000106	0.005443

A plot of $\dfrac{1}{t} \ln \dfrac{[S]_0}{[S]}$ versus $\dfrac{[S]_0 - [S]}{t}$ was obtained as shown in Figure 4.53. The slope of the line was -0.96647, which gives $K_m = -1/\text{slope} = 1.03$ ppmw. The intercept was 0.004771, which meant $V_{max}/K_m = 0.004771$. Hence $V_{max} = 0.0049$ ppmw d^{-1}. The correlation coefficient for the straight line was 0.7852. This example should serve to illustrate how a batch reactor can be used to obtain the Michaelis-Menten constants for a biodegradation process.

4.4.1.2 Plug Flow Enzyme Reactor

The equation describing a PFR for an enzyme reaction is identical to the one described in Section 2.4.1.1.3.

$$r_S = -F_{in} \cdot \frac{d[S]}{dV} \tag{4.165}$$

where
  F_{in} in the influent feed rate
  V is the volume of the reactor

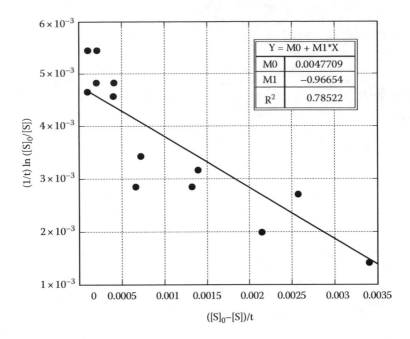

FIGURE 4.53 A linearized plot of picloram degradation rate in an enzyme batch reactor.

Thus

$$\frac{V_{max}[S]}{K_m+[S]} = -F_{in} \cdot \frac{d[S]}{dV} \tag{4.166}$$

Rearranging and integrating, we get

$$\frac{[S]_0-[S]}{\ln\left(\dfrac{[S]}{[S]_0}\right)} = V_{max} \cdot \left\{\frac{\tau}{\ln\left(\dfrac{[S]}{[S]_0}\right)}\right\} - K_m \tag{4.167}$$

where $\tau = V/F_{in}$ is the residence time for the substrate in the reactor. A plot of $\dfrac{[S]_0-[S]}{\ln\dfrac{[S]}{[S]_0}}$ versus $\dfrac{\tau}{\ln\dfrac{[S]}{[S]_0}}$ can be used to arrive at V_{max} and K_m from the slope and intercept respectively. The total volume of the reactor required for a given removal is

$$V = \left(\frac{F_{in}}{V_{max}}\right) \cdot \left[K_m \cdot \ln\left(\frac{[S]_0}{[S]}\right) + [S]_0 - [S]\right] \tag{4.168}$$

4.4.1.3   Continuous Stirred Tank Enzyme Reactor

If the influent and effluent rates are matched, then from Section 2.4.1.1.2, the following general equation should represent the behavior of a CSTR:

$$V \frac{d[S]}{dt} = F_{in} \left([S]_{in} - [S] \right) - V \cdot r_S \tag{4.169}$$

where $r_S = \dfrac{V_{max}[S]}{K_m + [S]}$. At steady state d[S]/dt is zero, and hence we have

$$[S] = -K_m + \frac{V_{max} \tau [S]}{\left([S] - [S]_{in} \right)} \tag{4.170}$$

where $\tau = V/F_{in}$. From this, one can also obtain the size (volume) of a reactor required for a specific enzymatic degradation of the substrate.

$$V = \left(\frac{F_{in}}{V_{max}} \right) \left(1 - \frac{[S]_{in}}{[S]} \right) \left(K_m + [S] \right) \tag{4.171}$$

Example 4.29   Microbial Growth and Substrate Kinetics in a CSTR

A microbial culture in a CSTR is a very efficient method for degrading substrates. The technique has been widely employed in the biotechnology field. It has also been employed in activated sludge wastewater treatment process. Indigenous bacteria are grown in a reactor that is fed with the substrate (nutrients) at a specified rate. The operating conditions (pH, dissolved oxygen, and temperature) are maintained constant within the reactor. The effluent is monitored for both the microbial population and substrate concentration. Figure 4.54 represents the CSTR.

The seed culture of microbes is placed within the reactor and nutrients provided to it in the influent stream that contains the substrate. The submerged bubble generators that are used to introduce oxygen into the reactor provide mixing. Paddle mixers are also used to gently mix the constituents. The biomass in the reactor is uniformly distributed with no clumping at the vessel wall and the yield factor y_x is assumed to be constant. A mass balance on the microbial concentration [X] should include its growth and its loss via overflow into the effluent and decay within the reactor. We have

$$V \cdot \frac{d[X]}{dt} = \mu [X] V - F[X] - k_{decay} [X] V \tag{4.172}$$

At steady state, d[X]/dt = 0, and hence

$$\mu = \frac{F}{V} + k_{decay} \tag{4.173}$$

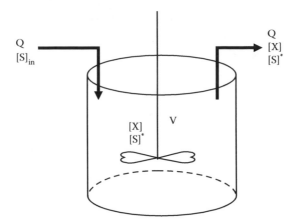

FIGURE 4.54 Schematic of a CSTR for microbial and substrate kinetics.

This is called the "dilution rate" (time^{-1}) of the microbe. The reciprocal, V/F = θ, is called the "mean cell residence time." Note that θ is the same as the specific growth rate discussed earlier. Thus in a CSTR, it is an important parameter in understanding microbial growth kinetics. μ can be large if either V is large or F is small. The maximum value of 1/θ or μ is given by μ_{max} as in the Monod equation.

$$\frac{1}{\theta} + k_{decay} = \frac{\mu_{max}[S]^*}{K_s + [S]^*} \tag{4.174}$$

$$[S]^* = \frac{K_s(1 + k_{decay}\theta)}{\mu_{max}\theta - (1 + k_{decay}\theta)} \tag{4.175}$$

We next proceed to write an overall mass balance for the substrate which gives

$$V\frac{d[S]}{dt} = F\left([S]_{in} - [S]^*\right) - V\left(\frac{\mu[X]}{y_x}\right) \tag{4.176}$$

where y_x is the yield factor as defined previously. At steady state, we have

$$[X] = y_x\left([S]_{in} - [S]^*\right) \tag{4.177}$$

Substituting for [S]* gives

$$[X] = y_x\left\{[S]_{in} - \frac{K_s(1 + k_{decay}\theta)}{\mu_{max}\theta - (1 + k_{decay}\theta)}\right\} \tag{4.178}$$

When microorganisms are grown in the reactor, and their decay proceeds simultaneously with constant k_{dec}, the observed yield factor y_x^{obs} has to be corrected to obtain the actual y_X that should be introduced into Equation 4.178. Horan (1990) derived the following equation for y_X:

$$y_X = \frac{y_X^{obs}}{1+k_{decay}\theta} \qquad (4.179)$$

As an illustration of this equation, we shall choose the special case when $k_{decay} = 0$. Other relevant parameters are chosen for a typical bacteria (*E. coli*) found in wastewater plants: $K_s = 15$ mg L^{-1}, $\mu_{max} = 25$ d^{-1}, $y_X = 0.6$, and $[S]_{in} = 15$ mg L^{-1}. Figure 4.55 displays the change in $[X]$ and $[S]^*$ as a function of the mean residence time θ in the reactor.

It should be mentioned that these equations only predict the steady state behavior. In actual operation, the unsteady state should be considered whenever the system experiences changes in influent concentrations. There will then exist a lag time before the substrate consumption and microbial growth approach a steady state. The microbial growth lags by several θ values before it adjusts to a new $[S]_{in}$. This is called the "hysteresis effect."

It is also useful to consider here the competition for a substrate S between an organism that utilizes it and other competing complexation processes within the aqueous phase. Consider Figure 4.56. While an enzymatic reaction of species S (an inorganic metal, for example) occurs via complexation with a cellular enzyme (denoted E), a competing ligand Y in the aqueous phase may bind species S. The cellular

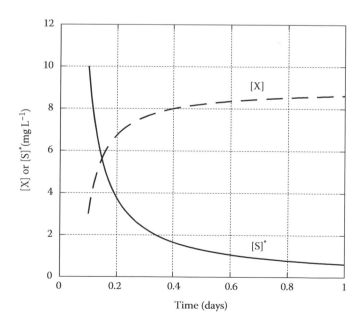

FIGURE 4.55 Microbial growth and substrate decomposition in a CSTR. $K_s = 15$ mg L^{-1}. $[S]_{in} = 15$ mg L^{-1}. Y = 0.6. $\mu_{max} = 25$ d^{-1}.

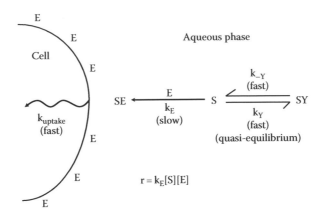

FIGURE 4.56 Kinetics of competing biological uptake and complexation in the aqueous phase. (Adapted from Morel, F.M.M. and Herring, J.G., *Principles and Applications of Aquatic Chemistry*, John Wiley & Sons, Inc., New York, 1993.)

concentration of species S is determined by a steady state between cell growth (division) and the rate of uptake of S. If $[S]_{cell}$ denotes the cellular concentration of S (mol cell^{-1}) and μ is the specific growth rate (d^{-1}), then the rate of cell growth is $r_{cell} = \mu[S]_{cell}$. From the reaction scheme in Figure 4.56, the uptake rate of species S is given by $r_{uptake} = k_E^* [S]_{tot} [E]_{tot}$, which is the rate of reaction of species S with the enzyme ligand E. This necessarily assumes that the enzyme is in excess of the concentration of species S. $[S]_{tot}$ is the total concentration of substrate S in the aqueous phase. k_E^* is the effective complex formation rate constant. For a given ionic strength and pH, k_E^* is fixed. From the steady state balance of cell growth rate and uptake rate, we get

$$\mu = k_E^* \cdot \left(\frac{[S]_{tot}}{[S]_{cell}} \right) \cdot [E]_{tot} \tag{4.180}$$

or

$$\ln [S]_{tot} = \ln \left(\frac{1}{k_E^*} \right) + \ln \left(\mu \cdot \frac{[S]_{cell}}{[E]_{tot}} \right) \tag{4.181}$$

If both $[S]_{cell}$ and $[E]_{tot}$ are constant, and k_E^* is fixed for a given pH and ionic strength, then a plot of $\ln [S]_{tot}$ versus $\ln \left(1/k_E^* \right)$ should yield a linear plot (Hudson and Morel, 1990). An interesting conclusion of that work is that "... complexation kinetics may be one of the keys to marine ecology...." (Morel and Herring, 1993).

The next two sections will describe some characteristics of bioreaction kinetics for wastewater treatment and in situ biodegradation of subsurface contaminants. It is important in these cases to facilitate the contact of pollutant with bio-organisms so that the reaction is completed in a short time. In the case of wastewater treatment, the enzymes can be isolated from the organisms and used in a completely mixed reactor. This can be made more efficient by attaching the isolated enzymes or biomass on a solid support and using it as a bed reactor. This also facilitates the separation of biomass from the solution after the reaction. For in situ biodegradation, the organisms should have easy access to the substrate in the subsurface. In soils and sediments, this is a major impediment. Moreover, nutrient and oxygen limitations in the subsurface environment will limit the growth and activity of organisms. We shall describe first the immobilized enzyme reactor for wastewater treatment. Selected aspects of in situ subsoil bioremediation will follow this.

4.4.1.4   Immobilized Enzyme of Cell Reactor

Enzymes are generally soluble in water. Hence their reuse after separation from a reactor is somewhat difficult. It is, therefore, useful to isolate and graft it on to surfaces where they can be immobilized. The surface can then act as a "fixed-bed reactor" similar to a packed column or ion-exchange column. The enzyme or cell can be easily regenerated for further use. The reaction can be carried out in a continuous mode where the substrate (pollutant) is passed through the reactor bed and the products recovered at the effluent end. Both chemical and physical methods can be used to immobilize enzymes onto solid substrates (see Table 4.18). The same methods are also useful in immobilizing living cells onto solid substrates. Both immobilized enzyme and cell reactors have been shown to have applications in wastewater treatment (Trujillo et al., 1991).

The use of an immobilized enzyme or cell reactor involves the consideration of some factors that are not necessarily addressed in a conventional CSTR. Specifically, we have to consider the different resistances to mass transfer of substrates toward the reaction site on the immobilized enzyme. The situation is very similar to that described for sorption and reaction in a natural porous medium (Section 4.2.1.2). There are three sequential steps before a substrate undergoes transformation; these are indicated in Figure 4.57. The first step, viz., the diffusion in the bulk fluid toward the liquid boundary layer and the subsequent diffusion through the boundary layer to the support surface constitute the external mass transfer resistance. The last step, viz., the diffusion to the enzyme reaction site constitutes the internal mass transfer resistance.

If the enzyme or cell is grafted on the surface of an insoluble support, only the first two steps contribute to mass transfer resistance and reaction. In other words,

---

TABLE 4.18
Immobilization Techniques for Enzymes or Cells

Chemical	Physical
Covalent bonding to inactive supports	Adsorption onto supports
Copolymerization with support structure	Entrapment in cross-linked polymers
Crosslinking multifunctional groups	Microencapsulation

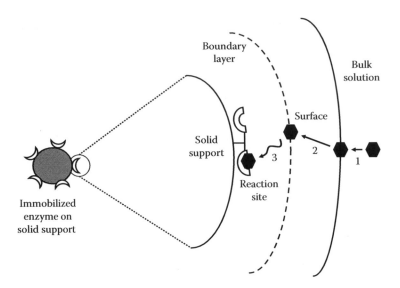

FIGURE 4.57 The steps involved in diffusion and reaction of a substrate at an immobilized enzyme support.

the external resistance controls the substrate decomposition. In such a case, the rate of transfer of solute to the surface of the support is balanced exactly by the substrate reaction at the enzyme site:

$$k_L a_v \left([S]_\infty - [S] \right) = \frac{V_{max} [S]}{K_m + [S]} \tag{4.182}$$

where
[S] is the substrate concentration at the surface
$[S]_\infty$ is the concentration in the bulk solution
k_L is the overall liquid-phase mass transfer coefficient for substrate diffusion across the boundary layer
a_v is the total surface area per unit volume of the liquid phase

In order to simplify further discussion, we define the following dimensionless variables: $[S]^* = \dfrac{[S]}{[S]_\infty}$; $\beta = \dfrac{[S]_\infty}{K_m}$; $\Theta = \dfrac{V_{max}}{k_L a_v [S]}$. The last one Θ is called the "Dämkohler number." It is defined as the ratio of the maximum rate of enzyme catalysis to the maximum rate of diffusion across the boundary layer. This equation can now be rewritten as

$$\Theta = \left(1 - [S]^* \right) \left(1 + \frac{1}{\beta [S]^*} \right) \tag{4.183}$$

If $\Theta \gg 1$, the rate is controlled entirely by the reaction at the surface and is given by $\dfrac{V_{max}[S]}{K_m + [S]}$, whereas if $\Theta \ll 1$, the rate is controlled by mass transfer across the boundary layer and is given by $k_L a_v [S]_\infty$. From this equation, one obtains the following quadratic in $[S]^*$:

$$\left([S]^*\right)^2 + \alpha[S]^* - \frac{1}{\beta} = 0 \qquad (4.184)$$

where $\alpha = \Theta - 1 + \dfrac{1}{\beta}$. The solution to this quadratic is

$$[S]^* = \frac{\alpha}{2} \cdot \left(-1 \pm \sqrt{1 + \frac{4}{\beta\alpha^2}}\right) \qquad (4.185)$$

If $\alpha > 0$, one chooses the positive root, whereas for $\alpha < 0$, one has to choose the negative root for a physically realistic value of $[S]^*$. As $\alpha \to 0$, it is easy to show that $[S]^* \to \left(\dfrac{1}{\beta}\right)^{\frac{1}{2}}$.

In order to analyze the lowering of reaction rate as a result of diffusional resistance to mass transfer, we can define an "effectiveness factor," ω. The definition is

$$\omega = \frac{\text{Actual rate of reaction}}{\text{Rate of reaction unaffected by diffusional resistance}}$$

If the reaction is unaffected by diffusional resistance to mass transfer, the rate is given by the Michaelis-Menten kinetics where the concentration at the surface is the same as the concentration in the bulk solution. In other words, no gradient in concentration exists in the boundary layer. Thus

$$\omega = \frac{\left(\dfrac{V_{max}[S]}{K_m + [S]}\right)}{\left(\dfrac{V_{max}[S]_\infty}{K_m + [S]_\infty}\right)} = \frac{(1+\beta)[S]^*}{1 + \beta[S]^*} \qquad (4.186)$$

ω varies between the limits of 0 to 1.

Let us now turn to the case where the resistances to mass transfer is within the enzyme-grafted solid surface. In this case, the diffusion of the substrate within the particle (a spherical pellet) is balanced by the rate of conversion to products. If we consider a spherical shell of thickness dr, the following mass balance for substrate [S] will hold:

$$\text{Rate of input at}\,(r + dr) - \text{Rate of output at}\,r$$

$$= \text{Rate of consumption in the volume}\,4\pi r^2 dr.$$

$$D_s \frac{1}{r^2} \frac{\partial}{\partial r} \left(r^2 \frac{\partial [S]}{\partial r} \right) = r_S \tag{4.187}$$

where
  D_s is the diffusivity of the substrate in the fluid within the pellet
  r is the radius

r_S is given by the Michaelis-Menten kinetic rate law.

$$D_s \frac{1}{r^2} \frac{\partial}{\partial r} \left(r^2 \frac{\partial [S]}{\partial r} \right) = \frac{V_{max} [S]}{K_m + [S]} \tag{4.188}$$

This second order differential equation can be cast into a nondimensional form by using the following dimensionless variables: $[S]^* = \dfrac{[S]}{[S]_\infty}$; $\beta = \dfrac{[S]_\infty}{K_m}$; $r^* = \dfrac{r}{R}$, where $[S]_\infty$ is the substrate concentration in the bulk solution, and R is the radius of the pellet. The resulting equation is

$$\frac{1}{(r^*)^2} \frac{\partial}{\partial r^*} \left((r^*)^2 \frac{\partial [S]^*}{\partial r^*} \right) = \left(\frac{R^2 V_{max}}{D_s K_m} \right) \cdot \left(\frac{[S]^*}{1 + \beta [S]^*} \right) \tag{4.189}$$

If we now define a "Thiele modulus" Φ as $\left(\dfrac{R^2 V_{max}}{9 D_s K_m} \right)^{1/2}$ so that the observed overall rate is expressed in moles per pellet volume per unit time, that is, $r_S^{obs} = \dfrac{3}{R} D_s \left. \dfrac{\partial [S]^*}{\partial r^*} \right|_{r^*=1}$, this equation can now be written as

$$\frac{1}{(r^*)^2} \frac{\partial}{\partial r^*} \left((r^*)^2 \frac{\partial [S]^*}{\partial r^*} \right) = (9 \Phi^2) \cdot \left(\frac{[S]^*}{1 + \beta [S]^*} \right) \tag{4.190}$$

This equation can be solved numerically to get $[S]^*$ as a function of r^*, and further to obtain the effectiveness factor ω as defined by

$$\omega = \frac{r_S^{obs}}{r_s} = \frac{\left(\dfrac{3}{R} D_s \left. \dfrac{\partial [S]^*}{\partial r^*} \right|_{r^*=1} \right)}{\left(\dfrac{V_{max} [S]^*}{K_m + [S]^*} \right)} \tag{4.191}$$

Solutions to ω as a function of Φ are given in sources on biochemical engineering (Bailey and Ollis, 1986; Lee, 1992) to which the reader is referred for further details.

These equations are also useful to represent the substrate uptake and utilization by a microbial cell, if we replace the Michaelis-Menten kinetics by Monod kinetics. This forms the basis of the kinetic analysis of biological floc in activated sludge processes for wastewater treatment. Powell (1967) developed a lumped parameter model for substrate utilization and kinetics. The rate of substrate utilization was given by the following equation:

$$r_s = \left(\frac{\mu_{max}}{y_s 2\sigma}\right)\left[\left(1+\sigma+\frac{[S]_0}{K_s}\right) - \left\{\left(1+\sigma-\frac{[S]_0}{K_s}\right)^2 + \frac{4[S]_0}{K_s}\right\}^{1/2}\right] \qquad (4.192)$$

where

$\sigma = \dfrac{\mu_{max}\delta R^2}{3y_s K_s D_s (R+\delta)}$ and δ is the thickness of the stagnant zone of liquid around a cell of radius R

$[S]_0$ is the substrate concentration in the bulk fluid

μ_{max} and K_s are the Monod kinetics constants

This equation is identical to that derived earlier using Michaelis-Menten kinetics.

4.4.1.5   In Situ Subsoil Bioremediation

An important application of environmental bioengineering is the use of special microbes to degrade recalcitrant and persistent chemicals in the subsurface soil and sediment environment. Primarily bacteria, algae, and fungi bring this about. Bacteria are found in large numbers in the soil; they are small in size compared to fungi. Fungi are filamentous in nature and account for a large mass of microorganisms in the soil.

Microbes in the soil can survive with or without oxygen. Accordingly, soil microbes are classified as "aerobes" (requiring oxygen) and "anaerobes" (require no oxygen). "Facultative anaerobes" are those that can grow in either environment. Aerobic microbes breakdown chemicals into CO_2 and H_2O. Anaerobes breakdown chemicals into completely oxidized species in addition to some CO_2, H_2O, and other species (H_2S, CH_4, SO_2, etc.). The biotransformation of chemicals in soil involves several steps, enzymes, and species of many types. Since enzymes are specific to a chemical, the biodegradation of complex mixtures requires that several species coexist and exert their influence on individual chemicals in the mixture. Microbes act in two ways: (1) they imbibe chemicals and the "intracellular enzymes" act to degrade the pollutant, and (2) the organism excretes enzymes ("extracellular enzymes"), which subsequently catalyze the degradation.

The general pathway of degradation of many types of pollutants by specific microbes has been elucidated over the past several decades. These aspects are described extensively in a recent book (Alexander, 1994). The specific mechanisms are not the topic of this chapter, whereas we are more interested in the equilibrium and kinetics of biodegradation in the subsoil.

In the earlier section, we noted the kinetics of enzyme catalysis and cell growth in the aqueous environment. In the subsurface soil environment, which is a two-phase system (soil + pore water), we should expect some similarities with the aqueous system. The microbes are attached to the soil surface and they act as a fixed-film (immobilized cell) reactor.

In the subsurface soil environment, the degradation of pollutants and microbial growth are most conveniently handled using a Monod-type expression (see Section 4.2.1). The substrate degradation rate per unit volume (mol m^{-3} s^{-1}) is

$$r_v = \frac{1}{y_X} \cdot \frac{\mu_{max}[S][X]}{K_s + [S]} \tag{4.193}$$

with y_X being the yield factor. The term μ_{max}/y_X is called the maximum rate of substrate utilization and is represented by a constant, k. The limits of the rate expression (4.193) are of special interest. If $[S] \gg K_s$, $r_v = k[X]$ and the rate is first order, whereas for $[S] \ll K_s$, $r_v = (k/K_s) [X] [S]$ and is second order. Further if [X] is constant, the rate is zero order in [S] at high [S] and first order at low [S].

Other rate expressions that are also based on Monod-type kinetics have been proposed to describe biotransformations in the subsurface soil environment. They are based on the basic premise that as either [X] or [S] $\to$ 0, $r_v \to 0$, and that r_v should increase linearly with both [X] and [S]. These models comprise of two types:

1. A power law relationship: $r_v = k[S]^n$
2. A hyperbolic relationship: $r_v = \dfrac{a[S]}{b+[S]}$

Note that the latter resembles Monod kinetics. These expressions do not include microbial concentration as a separate parameter and describe the general disappearance rates (biotic or abiotic) of chemicals in soils. In most instances, in the subsurface soil environment, we have low values of [S] and hence a first order expression in [S] is sufficient. Fortunately, this is the same limiting condition for both the Monod kinetics and the empirical rate laws described earlier. However, the consensus is that a theoretically based approach such as the Michaelis-Menten (or Monod) kinetics is more appropriate than an empirical rate law whenever empiricism can be avoided. A theoretical model will offer a greater degree of confidence in the testable conclusions drawn from a chemodynamic model.

For a successful biodegradation, there are several requirements. These are summarized here:

1. Availability of the chemical to the organism
2. Presence of the appropriate nutrient
3. Toxicity and/or inactivation of enzymes
4. Sufficient acclimation of the microbe to the environment

Whereas the latter three issues are specific to a microbe or a chemical, the first one is common to all organisms.

Chemicals can be held in special niches such as micropores and crevices in the soil structure that are inaccessible to the microbes. Chemicals can adsorb to solid surfaces and may not readily available in the aqueous phase where the microbes (enzymes) function. In most cases, sorption seems to decrease the rate of degradation. Sorbed molecules may not be in the right conformation to form the requisite enzyme-substrate complex. The sorption site may also be far removed from the microbial colony. We can think of two separate biodegradation processes, one that of the sorbed species and the other that of the truly dissolved species in the aqueous phase. If both are first order, $r_v^{sorbed} = \rho_b k_s [S]_{ads}$ and $r_w = k_w [S]$, where ρ_b is the soil density, $[S]_{ads}$ is the sorbed mass of S, k_s and k_w are the respective rate constants. We can write the following expressions for the fraction sorbed and truly dissolved:

$$f_{sorbed} = \frac{K_{sw}\rho_s}{1 + K_{sw}\rho_s}$$

$$f_w = \frac{1}{1 + K_{sw}\rho_s} \tag{4.194}$$

where $[S]_{ads}$ is the soil sorbent concentration. A composite rate is defined as

$$r^* = k^* [S]_{tot} \tag{4.195}$$

where $[S]_{tot}$ is the total substrate concentration and

$$k^* = f_{sorbed} k_s + f_w k_w \tag{4.196}$$

This equation satisfactorily represents the effect of sorption on the overall biodegradation if it is modeled as a first order process. Apart from sorption, biotransformations can also be affected by other environmental factors such as soil pH, temperature, soil moisture, clay content, and nutrients. Valentine and Schnoor (1986) have summarized these issues in an excellent review to which the reader should refer to for more details.

Since bioavailability is a primary limitation in bioremediation, considerable effort has been expended to facilitate desorption of chemicals from sediment and soil surfaces. Mobilization via solubilization using surfactant solutions is one of the concepts under study. Several workers have suggested that nonionic surfactant solutions used in conjunction with specific microbes can affect the mineralization of several hydrophobic organics. There are, however, several conflicting reports regarding this approach; in particular, the long-term viability of the microbes in the presence of surfactants, the presence of residual surfactants in the subsurface soil environment, and the degree of mineralization in the presence of surfactants have been disputed.

4.4.2 KINETICS OF BIOACCUMULATION IN THE AQUATIC FOOD CHAIN

In Chapter 3, we introduced the "bioconcentration factor" that defined the equilibrium partitioning of chemicals between the biota and water. This section is to model the kinetics of chemical uptake and dissipation in organisms. Connolly and Thomann (1992) summarized the state-of-the-art in this area. What follows is a concise description of their model. There are three important subareas that have to be included in such a kinetic model, namely, (1) the growth and respiration rates of each organism included in the food chain, (2) the efficiency of chemical transfer across the biological membranes (gills, gut, etc.), and (3) the rate of dissipation and excretion of chemicals by the organism.

To analyze the uptake rate, we should first consider the base organisms in the food chain, viz., phytoplankton and detrital organic material. The uptake by organisms generally depends on the ingestion, sorption, metabolism, and growth. For the base in the food chain, it is only dependent on sorption and desorption processes. If k_s (m^3 d^{-1} g^{-1} dry weight) denotes the rate of sorption from the aqueous phase and k_d (d^{-1}) the desorption rate, the rate of accumulation within the base-organisms is given by

$$\frac{d}{dt}\left(W_{org}\left[A\right]_{org}^{b}\right) = k_s W_{org}\left[A\right]_w - k_d W_{org}\left[A\right]_{org}^{b} \tag{4.197}$$

where

$\left[A\right]_{org}^{b}$ is the concentration of A in the base organism (mol kg^{-1} weight of organism)
$[A]_w$ is the concentration in the aqueous phase (mol dm^{-3})
W_{org} is the weight of the organism (kg)
k_s is a second order rate constant (dm^3 (kg d)$^{-1}$)
k_d is a first order rate constant (d^{-1})

If equilibrium is assumed between the organism and water we have

$$\frac{\left[A\right]_{org}^{b}}{\left[A\right]_w} = \frac{k_s}{k_d} = K_{BW} \tag{4.198}$$

as defined in Chapter 3.

In this case, the change in $\left[A\right]_{org}^{b}$ with t will follow the equation

$$\left[A\right]_{org}^{b} = K_{BW}\left[A\right]_w\left(1-e^{-k_d t}\right) \tag{4.199}$$

Thus $\left[A\right]_{org}^{b}$ increases in an approximate first order form to approach the equilibrium value $K_{BW}[A]_w$.

When we move up the food chain (zooplankton → fish → top predator) we have to consider not only the rate of direct uptake from water by sorption/desorption, but also by ingestion of the lower species and the portion actually assimilated into the tissues of the animal. Additionally, we also have to take into account the mass excreted by the animal (Figure 4.58). Thus we have three rate processes included in the overall

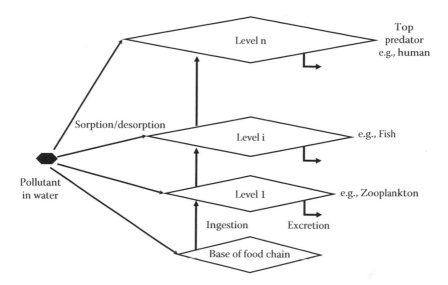

FIGURE 4.58   Food chain bioaccumulation of a pollutant from water.

uptake rate: (1) the rate of uptake by sorption, (2) the rate of uptake by ingestion, and (3) the rate of loss via desorption and excretion.

Example 4.30   Approach to Steady State Uptake in the Food Chain

The earliest attempts at modeling bioaccumulation involved exposing the first level of aqueous species to organic chemicals and obtaining the rate of uptake. We will now ascertain the approach to steady state in this system. Figure 4.59 treats the organism and water as distinct compartments (CSTRs) between which an organic chemical is exchanged. Neely (1980) described such a model; the following discussion is based on his approach.

As explained earlier let $[A]_{org}$ represent the moles of chemical A per weight of organism and, $[A]_w$ be the concentration of chemical A in water (mol dm^{-3} or mol kg^{-1}).

Expanding the left-hand side of the equation for the rate of uptake

$$\frac{d}{dt}\left(W_{org}[A]_{org}\right) = W_{org}\frac{d[A]_{org}}{dt} + [A]_{org}\frac{dW_{org}}{dt} \tag{4.200}$$

If the weight of the organism is constant, $dW_{org}/dt = 0$. However, since the organism grows as it consumes the chemical, its concentration changes in relation to the growth. The overall equation is then

$$W_{org}\frac{d[A]_{org}}{dt} = k_s W_{org}[A]_w - k_d W_{org}[A]_{org} - [A]_{org}\frac{dW_{org}}{dt} \tag{4.201}$$

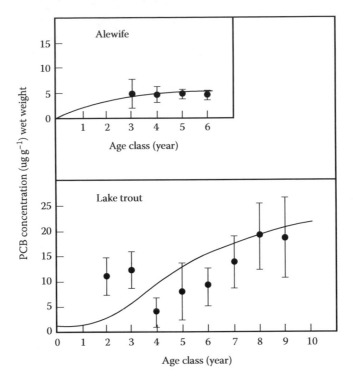

FIGURE 4.59 Comparison of observed PCB concentrations in alewife and lake trout. (Reprinted with permission from Thomann, R.V. and Connolly, J.P., *Environ. Sci. Technol.*, 18, 68. Copyright 1984 American Chemical Society.)

In order to proceed further, we denote the growth rate of the organisms as

$$\mu = \frac{1}{W_{org}} \cdot \frac{d[W]_{org}}{dt} \tag{4.202}$$

Hence

$$\frac{d[A]_{org}}{dt} = k_s [A]_w - (k_d + \mu)[A]_{org} \tag{4.203}$$

If the concentration in water is assumed to change only slightly such that $[A]_w$ is constant during a time interval, τ, we can integrate Equation 4.203 to yield

$$\frac{[A]_{org}}{[A]_w} = \left(\frac{k_s}{k_d + \mu}\right) \cdot \left(1 - e^{-(k_d + \mu)t}\right) \tag{4.204}$$

If $\mu = 0$ and τ is very large, we have $\dfrac{[A]_{org}}{[A]_w} = \dfrac{k_s}{k_d} = K_{BW}$. The approach to equilibrium will be delayed if the organism grows as it feeds on the pollutant A.

For any species i in the food chain we have the following rate equation (Connolly and Thomann, 1992)

Rate of increase in body burden = Rate of sorption

+ Rate of uptake by ingestion of lower species

− Rate of excretion.

Let $[A]^*_{org,i}$ be the whole weight body burden in the ith species (mole per organism) and is given as $[A]_{org,i} W_{org,i}$. The rate of sorption is given by the product of the uptake rate constant for species i, the organism weight, and the aqueous concentration of the chemical, i.e., $k_{si} W_{org,i} [A]_w$. This is analogous to oxygen uptake from water. k_{si} has units of dm^3 (d kg)$^{-1}$ of organism. An expression for k_{si} is $\dfrac{10^{-3} W_i E}{\varphi_i}$, where w_i is the total weight of organism i and φ_i is the fraction of lipid content (kg lipid per kg wet weight). E_i is the chemical transfer efficiency. A general expression for the rate of respiration is (Connolly and Thomann, 1992)

$$R_{ei} \left(g \cdot g^{-1} \cdot d^{-1} \right) = \beta_i W_i^\gamma e^{\rho_i T} e^{v_i u_i} \qquad (4.205)$$

where
   T is temperature (°C)
   u_i is the speed of movement of the organism in water (m s^{-1})
   β_i, γ_i, ρ_i, and v_i are constants typical to the given species

The rate of ingestion is given by the product of the chemical assimilation capacity of the organism i on another organism j, denoted as α_{ij}, the rate of consumption of organism i on organism j, denoted as C_{ij}, and the concentration of the pollutant in organism i, $[A]_{org,i}$. $C_{ij} = p_{ij} C_i$, where p_{ij} is the fraction of the consumption of i that is on j and C_i is the weight-specific consumption of i (kg of prey per kg of predator per day). The value of C_{ij} is dependent on the rate of respiration and the rate of growth: $C = (R + \mu)/\alpha$. The value of α varies from 0.3 to 0.8 and is specific to individual species. The rate of uptake via ingestion of lower species is $\sum_j \alpha_{ij} W_{org,i} C_{ij} [A]_{org,j}$. The overall rate of ingestion of a chemical by the species is given by $\sum_j^n \alpha_{ij} C_{ij} [A]_{org,j}$. The summation denotes the uptake of different species j = 1, 2, ..., n by species i.

The rate of excretion via desorption of the chemical is given by $k_{di} W_{org,i} [A]_{org,i}$, where k_{di} has units of d^{-1}.

Now, since

$$\frac{d \left(W_{org,i} [A]_{org,i} \right)}{dt} = W_{org,i} \frac{d[A]_{org,i}}{dt} + [A]_{org,i} \frac{dW_{org,i}}{dt} \qquad (4.206)$$

and denoting the growth rate, $\mu_i = \dfrac{1}{W_{org,i}}\dfrac{dW_{org,i}}{dt}$, we have

$$\frac{d\left(W_{org,i}\left[A\right]_{org,i}\right)}{dt} = W_{org,i}\frac{d\left[A\right]_{org,i}}{dt} + \left[A\right]_{org,i}W_{org,i}\mu_i \qquad (4.207)$$

The overall rate equation for bioaccumulation is, therefore, given by

$$\frac{d\left[A\right]_{org,i}}{dt} = k_{si}\left[A\right]_w + \sum_{j=1}^{n}\alpha_{ij}C_{ij}\left[A\right]_{org,i} - k_{di}\left[A\right]_{org,i} - \mu_i\left[A\right]_{org,i} \qquad (4.208)$$

This equation is applied to each age class of the organism. Within each age group the various parameters are assumed constant. Since equilibrium is rapidly attained for lower levels of the food chain, it may be appropriate to assume no significant change in concentration with time for such species. This suggests a pseudo steady state approximation for $[A]_{org,i}$. Therefore the concentration in the organism i at steady state is given by

$$\left[A\right]_{org,i}^{*} = \frac{k_{di}\left[A\right]_w + \sum_{j=1}^{n}\alpha_{ij}C_{ij}\left[A\right]_{org,i}}{\left(k_{di} + \mu_i\right)} \qquad (4.209)$$

Connolly and Thomann (1992) employed this equation to obtain the concentration of polychlorinated biphenyls (PCBs) in contaminated fish in Lake Michigan. Figure 4.59 shows the data where it is assumed that for each age class, constant values of assimilation, ingestion, and respiration rates are applicable.

If exposure of organism to pollutant A occurs only through the water route, we have at steady state

$$\left[A\right]_{org,i}^{*} = \frac{k_{si}}{\left(k_{di} + \mu_i\right)}\cdot\left[A\right]_w \qquad (4.210)$$

Thus, the bioconcentration factor is

$$K_{BW} = \frac{\left[A\right]_{org,i}^{*}}{\left[A\right]_w} = \frac{k_{si}}{\left(k_{di} + \mu_i\right)} \qquad (4.211)$$

Since $k_{si} = 10^{-3}W_i^{-\gamma}\dfrac{E_t}{p_t}$, $k_{di} = k_{si}/K_{ow}$ (Thomann, 1989), and $\mu_i = 0.01W_i^{-\gamma}$, we can write

$$K_{BW} = K_{OW}\left(\frac{1}{1+10^{-6}\left(\dfrac{K_{OW}}{E}\right)}\right) \qquad (4.212)$$

This equation indicates that K_{BW} goes through a maximum at a $K_{ow} = 10^6$. As K_{ow} increases, the value of K_{BW} decreases due to decreasing E and increasing μ_i.

The ideas presented in this section are solely to make the reader aware of the applications of chemical kinetics principles to obtain the steady state concentration of pollutants in a given food chain. Numerous parameters are needed and only average values are justifiable. These limitations are indicative of the state of knowledge in this area.

4.5   APPLICATIONS IN GREEN ENGINEERING

Manufacturing industries have traditionally relied on what is called "end-of-the-pipe" waste treatment philosophy to take care of the inevitable pollutant releases to the environment. This philosophy involves treating the waste stream (air, water, or solid waste), i.e., pollution control, at the end of the manufacturing process so as to decrease the amount discharged to the environment. In many cases, this also led to a significant portion of valuable materials at trace concentrations being returned to the environment. As processes became "greener" more emphasis was placed in recovering and recycling valuable species from the waste, thus reducing the amount released and also reducing the pollution control costs. As pollution regulations became more stringent, emphasis was placed on reducing the amount of pollutant produced at the source itself. Thus was born the concept of "environmentally conscious design," otherwise called "green engineering" (Allen and Shonnard, 2002). Figure 4.60 describes the difference between a "traditional" and "green" process.

In order to make a process design more environmentally conscious, one has to have methods and tools to properly evaluate the environmental consequences of the various chemical processes involved and products therein. This means that we have to be able to quantify environmental impacts and guide product design changes.

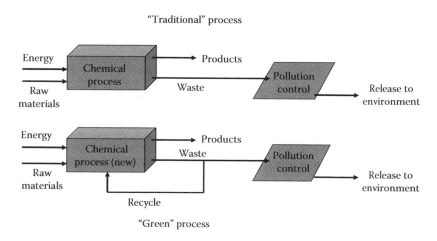

FIGURE 4.60 Differences between a "traditional" and a "green" chemical manufacturing process.

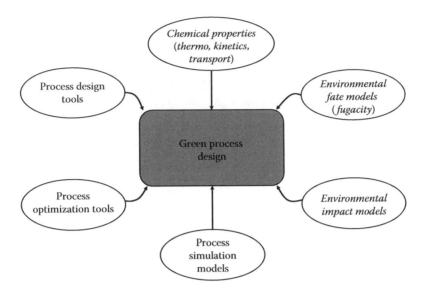

FIGURE 4.61 Various steps in an environmentally conscious design of a chemical process.

Figure 4.61 shows the various items we need to carry out an environmentally conscious design of a chemical process.

Green engineering follows a set of core principles laid down in 2003 in what is known as the "Sandestin Declaration," which are given in Table 4.19. A definition of "Green Engineering" is "Green engineering is the design, commercialization, and use of processes and products that minimize pollution, promote sustainability, and protect human health without sacrificing economic viability and efficiency" (U.S. Environmental Protection Agency, N/A). Closely aligned with the principles of green engineering are two other related concepts: (1) Sustainability defined as "meeting the needs of the current generation without impacting the needs of future generations to meet their own needs" and (2) life cycle assessment (LCA).

4.5.1 Environmental Impact Analysis (EIA)

In order to implement green processes in chemical or environmental engineering processes, we use a hierarchical (3-tiered) approach for Environmental Impact Analysis (EIA) (Allen and Shonnard, 2002):

Tier 1: Identify toxicity potential and costs when the problem or process is defined.

Tier 2: Identify material/energy intensity, emissions, and costs when the recycle/separation system is being considered.

Tier 3: Identify emissions, environmental fate, and risk when the overall system design is being considered.

TABLE 4.19
The Sandestin Declaration of Green Engineering Principles

Green engineering transforms existing engineering disciplines and practices to those that promote sustainability. Green engineering incorporates the development and implementation of technologically and economically viable products, processes, and systems to promote human welfare while protecting human health and elevating the protection of the biosphere as a criterion in engineering solutions.

To fully implement green engineering solutions, engineers use the following principles:

1. Engineer processes and products holistically, use systems analysis, and integrate environmental impact assessment tools.
2. Conserve and improve natural ecosystems while protecting human health and well-being.
3. Use life cycle thinking in all engineering activities.
4. Ensure that material and energy inputs and outputs are as inherently safe and benign as possible.
5. Minimize depletion of natural resources.
6. Strive to prevent waste.
7. Develop and apply engineering solutions, being cognizant of local geography, aspirations, and cultures.
8. Create engineering solutions beyond current or dominant technologies; improve, innovate, and invent (technologies) to achieve sustainability.
9. Actively engage communities and stakeholders in the development of engineering solutions.

Source: Adapted from Gonzalez, M.A. and Smith, R.L., *Environ. Progr.*, 22, 269, 2003. With permission.

In Tier 1, we make extensive use of the concepts derived from chemical thermodynamics and kinetics. This includes estimating thermodynamics properties of reactants and products, fugacity models for fate and transport, kinetic rate constants and mass transfer coefficients for species. These are then used in Tier 3 relative risk assessments. This approach is a stepwise, hierarchical one and involves the following tasks:

Step 1: Identify reactions and processes.

Step 2: Identify inputs/outputs rates of various species—Use EPA or other emission factors, use process flow sheet data.

Step 3: Estimate the physicochemical parameters for each species—Use thermodynamic correlations, use databases (e.g., EPIWIN).

Step 4: Obtain concentrations in various compartments using a multimedia environmental model (e.g., Fugacity model).

Step 5: Obtain relative risk index (RRI) for various categories such as global warming, ozone depletion, smog formation, acid rain, and carcinogenicity.

Step 6: Adjust parameters—(1) Go to Step 2 if risks are large and repeat calculations, or (2) Go to Step 1 if reactant substitutions are possible and repeat calculations.

The RRI for a particular impact category is the sum of contributions for all chemicals from a process weighted by their emission rates (Allen and Shonnard, 2002):

$$I_{category} = \sum_{i=1}^{N} \left(\text{dimensionless risk index for the category} \right)_i \times m_i \quad (4.213)$$

where $i = 1, \ldots, N$ is the number of compounds (species) involved. The risk indexes are generally calculated for the following five categories: global warming potential (GWP), smog formation potential (SFP), acid rain potential (ARP), inhalation toxicology potential (INHTP), and ingestion toxicology potential (INGTP). The first three potentials are available from various resources for specific chemicals and are summarized in Appendix D in Allen and Shonnard (2002). The last two (noncarcinogenicity toxicity indices) are obtained from the reference dose (RfD) and reference concentration (RfC) for ingestion (oral) and inhalation respectively. Similarly, the carcinogenicity indices for ingestion (oral) or inhalation pathways are also defined using the slope factors instead of the reference doses (see also Chapter 1 for descriptions of these terms).

To obtain the INHTP or INGTP, one has to first estimate the concentration of a chemical in the air or water environments using the fugacity model. For example, let us assume we know the concentration of the chemical in water and air. Once this is obtained, we can obtain the risk index as follows:

$$INHTP_i = \frac{\left(C_{ia} / RfC_i \right)}{\left(C_{bm,a} / RfC_{bm} \right)} \quad (4.214)$$

where bm stands for a benchmark chemical.

$$INGTP_i = \frac{\left(C_{iw} / RfD_i \right)}{\left(C_{bm,w} / RfD_{bm} \right)} \quad (4.215)$$

The carcinogenic potential (both inhalation and ingestion) are determined by substituting RfC and RfD with cancer slope factors, SF.

In order to illustrate this methodology, an example is given next.

Example 4.31   Illustration of Environmental Risk Evaluation for a Chemical Manufacturing Process

Let us consider a plastics manufacturing process that has to treat and recover materials from a gaseous stream. The gaseous stream consists of benzene, toluene, and byproducts (CO_2 and NO_x). Figure 4.62 is the overall process flow sheet. The stream that consists of 200 kg h^{-1} of benzene and toluene is introduced first into an absorption column where nonvolatile absorption oil is used to strip the benzene and toluene from the stream. The oil is distilled to separate the benzene and toluene for reuse within the manufacturing process. The oil is then recycled into the absorption column. The primary emissions from the process occur through

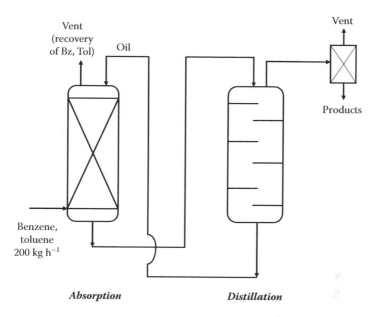

FIGURE 4.62 Overall process flow sheet for a typical chemical manufacturing process.

the vents, storage tanks, and other fugitive sources and consist of traces of benzene, toluene, and utility gases (CO_2, NO_x). These emissions are dependent on the oil flow rate through the absorption column. We need to determine the overall environmental impacts from this process and how they can be minimized as we change the process parameter, viz., oil flow rate. The methodology described here follows closely that given by Allen and Shonnard (2002).

Table 4.20 lists the various emission rates resulting from the process schematic given in Figure 4.64. Ideally, one would use a commercial process simulator such as HYSIS, ASPEN, or SimSci to carry out mass and energy balances on the various system components and calculate the emission rates of different species generated in the process. In doing so, one will have to use the standard EPA emission

TABLE 4.20

Typical Air Emission Rates from the Absorption/Distillation Process Described in Figure 4.62

Absorption Oil Flow Rate (kg mol h⁻¹)	Air Emission Rate (kg h⁻¹)			
	Benzene	Toluene	CO_2	NO_x
0	200	200	0	0
50	1	150	150	0.2
100	0.05	120	300	0.5
250	0.05	10	700	1.0
500	0.05	0.2	1500	2.0

factors that are listed for specific pollutants. We have circumvented that process to generate Table 4.20 using hypothetical values for the process as an illustration of the calculation of environmental impact indices. The emission rate for benzene rapidly decreases with increasing oil flow rate, whereas that for toluene decreases much more slowly. With increasing oil flow rates, the utility systems emit more of CO_2 and NO_x. Typical physicochemical properties required for the calculation will have to be obtained from thermodynamic arguments presented in Chapter 2 or from standard tables such as given in Appendix A.

Table 4.21 lists the various relative risk indices for individual chemicals needed for the calculation of overall environmental risk index for each category. Note that the values of GWP, SFP, and ARP for chemicals are listed in various sources (Allen and Shonnard, 2002). Similarly, the RfD and RfC values for benzene and toluene are listed by the Agency for Toxic Substance Disease Registry (http://www.atsdr.cdc.gov/). We can now utilize Equation 4.215 for the calculation of I_{GWP}, I_{SFP}, and I_{ARP} using the values from Table 4.21 for various values of the oil flow rate given in Table 4.20. For the calculation of I_{INHTP} and I_{INGTP} using Equations 4.214 and 4.215, the benchmark chemical used is benzene in Table 4.21. In order to do so, we have to calculate the concentrations in air and water compartments for benzene and toluene in the general resulting from releases from the process. This is done using a fugacity model such as described in Chapter 3. Typically a Level III model is required; however, we use a Level II model here as an illustration. We consider a three-compartment model (air, water, soil) as given in Example 3.2. A total emission rate of 0.5 mol h^{-1} each of benzene and toluene to air is assumed so that $E = 1$ mol h^{-1}. All other values are as given in Example 3.2. Note that the value of E will not affect the calculation of environmental indices in Equations 4.214 and 4.215 since the concentration ratios of benzene and toluene are used. The outputs from Level II fugacity calculation are given in Table 4.22 along with the calculated values of I_{INGTPi} and I_{INHTPi} for benzene and toluene.

Figure 4.63 shows the results of the calculations of risks for various categories as a function of the process variable, the absorption oil flow rate. I_{GWP} decreases considerably and reaches a minimum at about 50 kg mol h^{-1} of oil flow rate. The optimum therefore with respect to global warming is to work around this value. Initially, when there is no oil flow, all of the VOCs are released into the air and gradually converted to CO_2 elevating the global warming impact. However, as the oil flow rate increases, more of the VOCs are recovered and recycled, reducing the global warming impact. However, as the oil flow rate increases above the 50 kg mol h^{-1},

TABLE 4.21

Relative Risk Index for Each Chemical in the Process Described in Table 4.20

Category	Benzene	Toluene	CO_2	NO_x
Global warming potential, I_{GWPi}	3.4	3.3	1.0	40
Smog forming potential, I_{SFPi}	0.42	2.70	—	—
Acid rain potential, I_{ARPi}	—	—	—	1.07
RfD for ingestion (mg kg^{-1} d^{-1})	0.004	0.2	—	—
RfC for inhalation (mg m^{-3})	0.03	0.4	—	—

TABLE 4.22

Results from Fugacity Level II Calculations for Air and Water Concentrations

Chemical	K'_{AW} (Pa m³ mol⁻¹)	K_{sw} (L kg⁻¹)	C_{ia} (mol m⁻³)	C_{iw} (mol m⁻³)	I_{INGTPi}	I_{INHTPi}
Benzene	557	5	0.11	0.50	1.0	1.0
Toluene	676	20	0.10	0.42	42	12

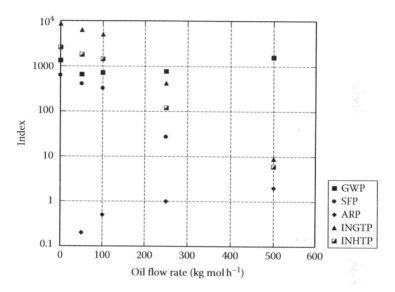

FIGURE 4.63 Calculated risks for different categories as a function of the absorption oil flow rate.

the process utilities increase the emission of CO_2 and therefore negate any advantages due to increased recovery of VOCs. The smog formation potential I_{SFP} appears to decrease considerably above 100 kg mol h⁻¹ of oil flow rate. Hence, the maximum reduction in the smog formation potential can be realized at the highest oil flow rate achievable. This also coincides with the highest recovery of toluene from the gaseous stream. The acid rain potential I_{ARP} shows a slow steady increase with oil flow rate and relates only to the increased process utilities emission of NO_x at high oil flow rates. The optimum value appears to be about 100 kg mol h⁻¹ for this case. The noncarcinogenicity indices I_{INGTP} and I_{INHTP} show similar behavior and decrease considerably above 100 kg mol h⁻¹ oil flow rate. The reductions in inhalation and ingestion toxicity are both related to the removal of benzene and toluene from the process waste stream that would otherwise enter the general environment. Thus, a single value of oil flow rate, which minimizes the overall environmental index, appears to be not realizable. However, a value between 50 and 100 kg mol h⁻¹ appears to be a good compromise for the process.

Note that in Example 4.31 of green engineering principles, we have utilized concepts from Chapters 2 and 3 in the estimation of physicochemical properties of chemicals and the fugacity Level II model to obtain the environmental fate and transport of chemicals. Additionally, in the more general case of fugacity Level III model, we will also have to utilize rate constants in air, water, and soil that were the topics of Chapter 3. This should serve as another application of the basic concepts of chemical thermodynamics and kinetics in environmental engineering that we have touched upon in this book.

4.5.2 LIFE CYCLE ASSESSMENT

Another important aspect of green engineering is the so-called LCA. This process involves following the so-called cradle-to-grave cycle of any component within a complete manufacturing, usage, and disposal of a product. This is best illustrated in Figure 4.64. A material such as, for example, a metal is used in several manufacturing operations. One follows the metal from its raw material supply chain to the production process, and further to the usage of the product in several forms. Ultimately, the manufactured products are disposed after use. During the various stages, we have the opportunity to reuse the product several times if recovered from the disposal process. From the disposed material some fraction of the metal can be recovered in its original form and used to remanufacture the product. Thirdly, the recovered metal may also be used as raw material in the production chain. During any of the product life cycle process, some fraction of the metal (in any oxidized, reduced, or native form) may be released to the environment (air, soil, or water). A material flow analysis during the entire life cycle of the metal can be done to ascertain the overall fate of the metal in any manufacturing process. The overall cycle and material flow analysis for tungsten in the United States during the years 1975–2000 is given in Figure 4.65. Once the rate of input into the environment is obtained, a fugacity model can be used to ascertain the overall environmental fate of the metal in various compartments.

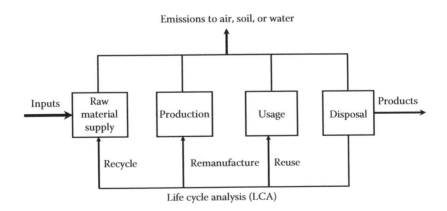

FIGURE 4.64 Steps in the LCA for a metal.

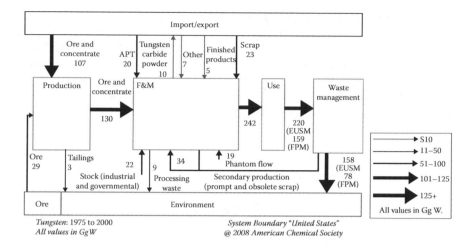

FIGURE 4.65 Overall cycle and material flows for tungsten. (From Harper, E.M. and Graedel, T.E., *Environ. Sci. Technol.*, 42(10), 3835, 2008. With permission.)

PROBLEMS

4.1₂ Consider a lake with a total capacity of 10^{10} m³ with an average volumetric water flow rate of 10^3 m³ d⁻¹. The lake receives a wastewater pulse of pesticide A inadvertently released into it by a local industry due to an explosion in a manufacturing unit. The input pulse has a concentration of 1000 µg dm⁻³. The pesticide is degradable in water to an innocuous product B with a rate constant of 0.005 d⁻¹. Determine how long it will take for 90% of the pollutant to disappear from the lake. Repeat the calculation if in addition to the reaction A → B, the pesticide A also adsorbs to suspended particles (~1000 mg L⁻¹) in the lake and settles out of the water into the sediment. K_{sw} for A is 1000 L kg⁻¹.

4.2₃ Consider a barge containing a tank full of benzene that develops a leak while in the Mississippi River near Baton Rouge, LA, and spilled 10,000 gal of its contents into the river. The stretch of the Mississippi River between Baton Rouge and New Orleans is 100 miles long. Assume that the pollutant is in plug flow and is completely solubilized in water. Estimate how long it will take to detect benzene in New Orleans, the final destination of the river. Assume that benzene does not transform via biodegradation or chemical reactions. Benzene may adsorb to bottom sediments. The following parameters are relevant to the problem: Volumetric flow rate of the river is 7.6×10^3 m s⁻¹, river velocity is 0.6 m s⁻¹, mean depth of the river is 16 m, and its mean width is 1.8×10^3 m. The average organic carbon content of the sediment is 0.2% and the average solids concentration in the river water is 20 mg L⁻¹.

4.3₃ On November 1, 1986, a fire in a chemical warehouse near Basel, Switzerland, caused the release of 7000 kg of an organo-phosphate ester (pesticide) into the river Rhine. Assuming that the river is in plug flow, determine the concentration of the pesticide detected near Strasbourg, which is 150 km from Basel.

The relevant chemical properties of the pesticide are as follows: First order rate constant for reaction in water $= 0.2$ d^{-1}, first order rate constant for reaction on sediment $= 0.2$ d^{-1}, the sediment-water partition constant $= 200$ L kg^{-1}. The river Rhine has an average volumetric flow rate of 1400 m^3 s^{-1}, a solids concentration of 10 mg L^{-1}, a resuspension rate of 5×10^{-5} d^{-1}, a total cross-sectional area of 1500 m^2, and an average water depth of 5 m.

4.4_3 A CSTR box model can be useful in estimating pollutant deposition onto indoor surfaces.

1. Consider a living room of total volume V and total internal surface area A onto which ozone deposition occurs. Let V_d represent the deposition velocity; it is defined as the ratio of the flux to the surface (μg $(m^2$ $s)^{-1}$) to the average concentration in air (μg m^{-3}). Outside air with ozone concentration C_o exchanges with the indoor air through vents at a known rate R_{exc}. At steady state, what will be the value of V_d for 75% deposition of ozone in a room that is 5 m $\times$ 5 m $\times$ 2.5 m, if $R_{exc} = 5$ L min^{-1} and $C_o = 75$ ppb? Experimental values range from 0.001 to 0.2 cm s^{-1}. What are the possible sources of such wide distributions in experimental values?

2. A home near a chemical plant has a total internal volume of 300 m^3. The normal air exchange between the indoor and outdoor occurs at a rate of 0.1 h^{-1}. An episodic release of a chlorinated compound (1,1,1-trichloroethane, TCA) occurred from the plant. The concentration of TCA was 50 ppbv in the released air. If the normal background concentration of TCA is 10 ppbv, how long will it take before the concentration inside the room reaches a steady state value?

4.5_2 A chemical spill near Morganza, Louisiana, occurred at about 1.45 p.m. on Monday 19, 1994 from a barge on the Mississippi River. The barge was carrying 2000 gal of a highly flammable product called naphtha. The U.S. Coast Guard closed the river traffic for about 4 miles from the spill to about 22 miles downriver, ostensibly, so that naphtha could dissipate. First determine from the literature the major constituents of naphtha. Using this as the key, perform calculations (considering a pulse input) to check whether the coast guard's decision was appropriate and prudent. Flow characteristics of the Mississippi River are given in Problem 4.2.

4.6_2 Tropospheric CO is an important pollutant. It has a background mixing ratio of 45–250 ppbv. In the period between 1950 and 1980, it has been shown that P_{CO} increased at ~1% per year in the Northern Hemisphere. More recent data (1990–1993) showed a significant decrease in CO in the atmosphere. Consider the fact that the largest fraction of CO in air is lost via reaction with the hydroxyl radical: $CO + OH^{\cdot} \xrightarrow{k_{OH}} CO_2 + H^{\cdot}$. If the CO emission rate from all sources is considered to be represented by a single term ΣS_i, the rate of change of P_{CO} is given by (Novelli et al., 1994): $\dfrac{dP_{CO}}{dt} = \sum_i S_i - k_{OH}P_{CO}$, what will be the concentration of CO at steady state? Utilize the expression to derive the differential expression dP_{CO}/P_{CO}. Assume that CO is mainly emitted from automobile exhaust. In the mid-1970s, stringent regulations have reduced the automobile emissions. Hence assume that no change in CO sources have occurred during 1990–1993. If P_{CO} presently is 120 ppb and the [OH] concentration in the atmosphere has increased $1\% \pm 0.8\%$ per year, what is the percent change in P_{CO} during the period 1990–1993?

4.7₂ Using the following data, establish the contribution to global warming for each of the following greenhouse gas. Assume a predicted equilibrium temperature change of 2°C for a doubling of the partial pressure of each gas.

1. Calculate the total temperature increase in year 3000 due to the combined emissions of all five gases, if the rate of accumulation is unchecked.

Compound	Preindustrial Atm. Concn. (1750–1800)	Current (1990) Concentration	Current Rate of Accumulation in Air (% per year)
CO_2	280 ppmv	353 ppmv	0.5
CH_4	0.8 ppmv	1.72 ppmv	0.9
N_2O	288 ppbv	310 ppbv	0.25
CFC-11	0	280 pptv	4
CFC-12	0	484 pptv	4

2. If the rate of accumulation of CFC-11 and CFC-12 are reduced to <0.1% through voluntary limits on CFC production, and that of N_2O is reduced to <0.1% via stricter automobile emission checks, what temperature increase can be anticipated in year 3000?

4.8₂ Free radicals formed in the atmosphere from photochemical species such as aldehydes can increase the concentration of ozone in urban smog. The following mechanism is suggested for the effect of formaldehyde upon the reaction of ozone in the troposphere:

$$HCHO \xrightarrow[J_2]{hv} 2HO_2^{\cdot} + CO_2; \quad J_2 = 0.015 \text{ min}^{-1}$$

$$HCHO + OH^{\cdot} \xrightarrow{k_4} HO_2^{\cdot} + CO + H_2O; \quad k_4 = 1.1 \times 10^{-11} \text{ cm}^3 \text{ molecule}^{-1} \cdot s$$

$$HO_2^{\cdot} + NO \xrightarrow{k_5} NO_2 + OH^{\cdot}; \quad k_5 = 8.3 \times 10^{-12} \text{ cm}^3 \text{ molecule}^{-1} \cdot s$$

$$OH^{\cdot} + NO_2 \xrightarrow{k_6} HNO_3; \quad k_6 = 1.1 \times 10^{-11} \text{cm}^3 \text{ molecule}^{-1} \cdot s$$

Along with the Chapman mechanisms, these reactions can be used to obtain $[O_3]$ as a function of t. Derive expressions for the rate of change in concentrations of [NO], $[NO_2]$, and [HCHO]. The equations should involve only terms in $[NO_2]$ and [HCHO]. From the resulting equations, how do you propose to obtain $[O_3]$ given the initial conditions $[NO]_0$, $[NO_2]_0$, and $[HCHO]_0$.

4.9₃ Consider the long-term trends of PCBs in Lake Superior. It is an oligotrophic lake with an average depth of 145 m, surface area of 8.2×10^{10} m², volume of 1.21×10^{13} m³, and a flushing time of 177 years. The DOC content is ~1 mg L⁻¹. $K_{SW} = 10^5$ L kg⁻¹. It was reported that the PCB content in Lake Superior has been steadily decreasing between 1978 and 1992 due to a combination of flushing, sedimentation, biodegradation, and volatilization. A pseudo first order rate

constant has been suggested. The following equation was said to be applicable:
$\dfrac{d[W]}{dt} = I - k'[W]$ where [W] is the mass of PCBs in the lake (kg), I is the input

of PCBs (both direct and by gas absorption, kg y^{-1}), and k' is the first order rate constant for disappearance of PCBs (y^{-1}). It is related to the *overall* first order rate constant as $k = k' + I(t)/W(t)$.

1. Justify the equation for $d[W]/dt$. State all necessary assumptions.
2. Derive the equation for the overall rate constant k.
3. k' is given as a sum $k' = k_{flushing} + k_{sedimentation} + k_{biodegradation} + k_{volatilization}$. With $k_{biodegradation} = 0$, $k_{flushing} = 0.006 \ y^{-1}$, $k_{sedimentation} = 0.004 \ y^{-1}$, and $k_{volatilization} = 0.24 \ y^{-1}$ for a combination of 82 PCB congeners, calculate the input of PCBs in year 1986 if the 1986 PCB burden is 10,100 kg.
4. How much PCB (in kg) is lost through volatilization in the year 1986?
5. Using an average Henry's constant of 1×10^{-4} atm m^3 mol^{-1} and a net 1986 PCB flux of 1900 kg y^{-1}, determine the air concentration of PCB in 1986. The 1986 water concentration is 2 ng L^{-1} (*Note*: You have to relate $k_{volatilization}$ to the mass transfer coefficient K_w through the equation $k_{volatilization} = (K_w/h)f_w$, where h is the average depth of the lake. $f_w - 0.87$ is the dissolved aqueous phase PCB fraction).
6. How much PCB will remain in the lake in year 2020?

4.10$_2$ The following data pertain to a local stream in Baton Rouge, LA:
Average stream speed: 0.36 m s^{-1}
Average stream depth: 3 m
Average water temperature: 13°C
Flow rate of water in the stream: 1 m^3 s^{-1}
BOD of the stream: 5 mg L^{-1}
$C_{O_2,w}$ in the stream: 8 mg L^{-1}.
At a point along the stream, a treatment plant discharges wastewater through a pipe approximately 2 m above the water level with the following characteristics:
Average flow rate: 0.1 m^3 s^{-1}
BOD of the wastewater: 20 mg L^{-1}
$C_{O_2,w}$ in the wastewater: 2 mg L^{-1}.
Obtain the following: (1) the dissolved oxygen profile in the stream as a function of distance downstream from the discharge point, (2) the maximum oxygen deficit and its location, (3) the concentration of the waste (BOD) 15 miles downstream of the discharge point (outfall). Deoxygenation coefficient k_d is 0.2 d^{-1}.

4.11$_2$ Consider the University Lake system in Baton Rouge, LA, which has a potential water quality problem resulting from waste discharged into the nearby Crest Lake. From the following data, determine what nutrient level (phosphorous) is to be expected at steady state in the Crest Lake, which is a completely mixed unstratified system.
Mean depth: 1.5 m
Surface area: 3.4×10^4 m^2
Mean settling velocity: 10 m y^{-1}
Mean detention time: 561 d
Phosphorous concentration in the waste: 10 mg L^{-1}.

Assume no losses due to volatilization or other reactions of P in the lake. What reduction in discharge level should be accomplished so that the P concentration is reduced to <0.2 mg L^{-1} at which no fish kills are observed in the lake?

4.12$_3$ Consider a two level system comprising of the base food chain phytoplankton, which form the food for a fish. They are both exposed to a pollutant (a PCB) in the aqueous phase. Consider the fish population to be in two age classes: 0–3 years and 4–6 years. Using the data given, obtain a plot of PCB concentration in the fish at steady state as a function of the age (year).

	Age Class	
Parameter	0–3 years	4–6 years
Growth rate, μ_i (d^{-1})	0.002	0.0005
Respiration rate: $R = \beta W^\gamma e^{\rho t} e^{vu}$ (g g^{-1} d)	$\beta = 0.05$	
	$\gamma = -0.2$	
	$\rho = 0°C^{-1}$	
	$v = 0.01$ s cm^{-1}	
Swim speed, u (cm s^{-1}): $\omega W \delta e^{\varphi T}$	$\omega = 1$	
	$\delta = 0.1$	
	$\varphi = 0 \,°C^{-1}$	
Food assimilation efficiency	0.8	
PCB assimilation efficiency	0.35	
K_B (μg kg^{-1} μg^{-1} L^{-1})	4×10^5	
Dissolved PCB concentration (ng L^{-1})	5	
Weight of fish (g)	0.5	3
Lipid fraction (kg lipid per kg weight)	0.1	0.1

4.13$_2$ Determine the half-life and 90% removal time for the following compounds from a well-mixed lake surface water: (1) pyrene, (2) 1,2-dichloroethane, (3) $p,p = $ −DDT, (4) chlorpyrifos, (5) 1,2-dichlorobenzene. The depth of mixing is 1 m.

4.14$_3$ The following data was obtained on the oxygen deficit ($\Delta = C_{O_2}^* - C_{O_2}$) in a pond with a surface aerator. Determine the mass transfer coefficient for oxygen from the data.

t (min)	Oxygen Deficit (% of Saturation)
0	78
2	62
4	52
6	44
8	37
10	31
12	27
14	23

Note $\dfrac{d\Lambda}{dt} = -k_r\Lambda$, and $k_r = k_L a$ for a surface aerator.

4.15$_2$ The absorption cross section and actinide flux for the photolysis $ClNO \xrightarrow{\text{hv}} Cl + NO$ in air are given here:

λ_i (nm)	$\sigma_{\lambda i}$ (cm^2 molecule^{-1})	$I_{\lambda i}$ (photons (cm^2 s)$^{-1}$)
280	10.3×10^{-20}	0×10^{14}
300	9.5	0.325
320	12.1	5.08
340	13.7	8.33
360	12.2	9.65
380	8.3	8.45
400	5.1	11.8

Determine the atmospheric half-life for ClNO.

4.16$_2$ In the surface waters of a natural lake, iron (Fe^{2+}) reacts with hydroxide in an oxidation precipitation reaction with a rate $-r_{oxdn} = k_{ox}[Fe^{2+}][O_2(aq)][OH^-]^2$ and photochemically dissolves by reduction with a rate $-r_{photo} = k_{photo}\{Fe^{III}L\}$ where $\{Fe^{III}L\}$ is the concentration of ligand-bound Fe^{III} on surfaces. If the lake volume is V and has a volumetric flow rate of Q, obtain the steady state concentration of Fe^{2+} in the lake.

4.17$_2$ Using a 16 W low-pressure Hg lamp photoreactor emitting light at 254 nm, a series of pesticides were subjected to photodegradation in distilled water. The concentrations, absorbances, and quantum yields are given here:

Compound	φ	A_{abs}	C (mol L^{-1})
Atrazine	0.037	0.08	2.3×10^{-5}
Simazine	0.038	0.059	1.8×10^{-5}
Metolachlor	0.34	0.005	1.0×10^{-5}

The emitted light intensity of the lamp was $I_0 = 7.1 \times 10^{-8}$ einstein (L s)$^{-1}$. Estimate the half-lives of the pesticides in water.

4.18$_2$ The air above the city of Baton Rouge, LA, can be considered to be well mixed to a depth of 1 km over an area 10 km × 10 km. CO is being emitted by both stationary and mobile sources within the city at a rate of 10 kg s^{-1}. Assume CO is a conservative pollutant. (1) For a wind velocity of 5 m s^{-1} in the city, what will be the steady state concentration of CO in the city?, (2) If after attaining steady state, the wind velocity decreases to 1.5 m s^{-1}, what will be the new steady state concentration?

4.19$_2$ Peroxyacyl nitrate (PAN) is a component of photochemical smog. It is formed from aldehydes by reaction with OH as follows:

$$CH_3CHO + OH^{\bullet} \rightarrow CH_3CO + H_2O$$

$$CH_3CO^{\bullet} + O_2 \rightarrow CH_3CO(O_2)^{\bullet}$$

$$CH_3CO(O_2)^{\bullet} + NO_2 + M \rightleftharpoons CH_3C(O)O_2NO_2 + M$$

Derive an expression for the production of PAN.

4.20$_1$ The annual production rate of CH_3Cl is 0.3 Tg y^{-1} and has an average mixing ratio of 650 pptv. What is its average residence time in the atmosphere?

4.21$_2$ Consider a lake 10^8 m^2 of surface area for which the only source of phosphorous is the effluent from a wastewater treatment plant. The effluent flow rate is 0.4 m^3 s^{-1} and has a phosphorous concentration of 10 g m^{-3}. The lake is also fed by a stream of 20 m^3 s^{-1} flow with no phosphorous. If the phosphorous settling rate is 10 m y^{-1}, estimate the average steady state concentration of P in the lake. The depth of the lake is 1 m.

4.22$_2$ An accident on a highway involving a tanker truck spilled 5000 gal of 1,2-dichloroethene (DCA) forming a 2500 m^2 pool. DCA has a vapor pressure of 0.1 atm at 298 K. The air temperature is 25°C and the wind speed averaged 4 m s^{-1}. The accident occurred on a sunny afternoon. Estimate the distance downwind of the spill that would exceed the worker standard exposure limit of 1 ppmv for DCA.

4.23$_2$ In a poorly ventilated hut in a third world country, logs are burnt to supply heat for cooking purposes. The total volume of the hut is 1000 m^3. Log burning releases CO at the rate of 2 mg h^{-1}. The air exchange rate is 0.1 h^{-1}. (1) What will be the CO concentration after 2 h of cooking inside the hut? (2) Compare the steady state indoor CO concentration with the U.S. ambient air quality criteria of 0.05 ppmv.

4.24$_2$ In 1976, a tragic release of one of the most toxic compounds known to man (dioxin, i.e., 2,3,7,8-tetrachlorodibenzo-p-dioxin) occurred in Seveso, Italy. The plant was manufacturing 2,4,5-trichlorophenol from 1,2,4,5-tetrachlorobenzene when the reaction ran away. The reactor over pressurized and the relief system opened for 5 min during which time 2 kg of dioxin escaped into the atmosphere through the plant's roof. The wind speed was 2 m s^{-1} and leak occurred on an overcast day. What is the likely concentration 10 km from the plant site?

4.25$_2$ Consider the city of Baton Rouge, LA, with a population of 400,000. The morning peak hour traffic is approximately 100,000 vehicles in an area of 300 km^2 with an average travel distance of 5 km from 7 a.m. to 10 a.m. daily. Assume each vehicle emits 2 g of CO for every 1 km traveled. Determine the CO concentration in the atmosphere at 9 a.m. The initial background concentration in the air prior to rush hour traffic is 5 mg m^{-3} and the average wind velocity is 3 m s^{-1}.

4.26$_3$ A vent from a hydrocarbon process (process gas) contains appreciable amounts of pentane (C5), hexane (C6), and heptane (C7), which are to be recovered using a brine knockout condenser and a carbon adsorption unit. The knock-out condenser already exists and produces a condensate having the composition: 10, 40, and 50 mol% C5, C6, and C7, respectively. The temperature and pressure of the condensate and tail gas are 10°C and 35 psia. Nitrogen is noncondensable under these conditions. As the project engineer, your task is to recover the alkanes in the tail gas. Assume ideal behavior. The Antoine constants for compounds are given here:

	Pentane	Hexane	Heptane
A	6.88	6.91	6.89
B	1076	1190	1264
C	233	226	217

1. First calculate the composition of the tail gas that is emitted from the knockout condenser. Express as mole fractions of C5, C6, C7, and N$_2$.
2. Express the total concentration of hexane (C6) in kg m^{-3} and define it as C$_F$, the inlet feed concentration to the adsorption bed. (Hint: use ideal gas law, 14.696 psia = 101,352 Pa). Based on laboratory data for activated carbon, the mass transfer zone length and Freundlich parameters have been calculated for the tail stream: $\delta = 0.1$ m. Freundlich isotherm is $C = K_f W^n$, where K_f is 10 kg m^{-3} and n = 2.25. C is the concentration (kg m^{-3}) and W is the adsorbate concentration (kg kg^{-1}). If the adsorbent density is 384 kg m^{-3}, the actual volumetric flow rate through the bed is 1 m^3 s^{-1}, what should be the length? The adsorber bed diameter is 2 m and breakthrough time is 1 h. Note that it is not necessary to determine or use the % recovery for the absorber as the given mass transfer zone length corresponds to 99% recovery.

4.27$_2$ What is the desired gas flow rate to obtain a 1 h breakthrough in treating an air stream using a carbon bed of density 0.35 g cm^{-3}. The inlet concentration of the pollutant is 0.008 kg m^{-3}. The bed length is 2 m and has a cross-sectional area of 6 m^2. The Freundlich isotherm parameters are K$_f$ = 500 kg m^{-3} and n = 2. The mass transfer coefficient is 50 s^{-1}.

4.28$_1$ Answer whether the following statements are true or false. Give explanations:
1. The mass transfer of a contaminant from a gas stream that is very soluble in a liquid absorbing phase is likely to be gas phase controlled.
2. Particulates formed by condensation of gases (i.e., smoke) are likely to be easily removed from a gas stream by gravity settling.
3. High ozone pollutant levels were responsible for the deaths of as many as 4000 people in London in early 1950s.
4. Atmospheric conditions during high winds and overcast skies would be expected to be stable, with relatively little crosswind or vertical pollutant dispersion.

5. Specification of absorption column diameter is usually done on the basis of the mechanical design of the column internals and is essentially independent of the compound to be stripped.

6. A crossflow scrubber is more efficient than countercurrent or cocurrent scrubbers for the removal of particulates from an air stream.

7. A doubling of carbon dioxide partial pressure in the atmosphere effectively increases the global average surface temperature by the same amount.

8. The reaction of ozone with oxides of nitrogen accounts for 60%–70% of ozone destruction in the lower troposphere.

4.29$_2$ A local industry produces a process air stream that contains vinyl chloride (molecular weight 62.5) at a concentration of 1000 ppmv. The mass flow rate of air is 1.9 kg s^{-1} at a temperature of 25°C and a total pressure of 1 atm. The environmental group has proposed using an activated carbon bed to remove vinyl chloride. Density of air is 1.2 kg m^{-3}. Given that the adsorption bed is 5 m^2 in area and 0.35 m in thickness, determine the following: (1) The lifetime of the bed before regeneration. Answer in hours. The carbon used is Ambersorb XE-347 for which the Freundlich isotherm parameters for vinyl chloride are $K_f = 4.16$ and $n = 2.95$, i.e., $C_e = K_f W_{sat}^n$, where W_{sat} is the saturation adsorbed concentration (g g^{-1}). The adsorption bed density is 700 kg m^{-3}. The overall mass transfer coefficient is 30 s^{-1}. (2) What is the overall flux of vinyl chloride through the bed? Answer in kg (m^2 s)$^{-1}$. What is the mass rate through the bed in kg h^{-1}?

4.30$_2$ Construct a two-box model for a lake that consists of an epilimnion (a well-stirred upper region) and a hypolimnion (the cold portion of the lake). r_p is the rate of removal (precipitation, sedimentation), r_b is the rate of burial, r_d is the rate of addition to hypolimnion by dissolution, Q_{exc} is the rate of exchange of water between epilimnion and hypolimnion. Write a mass balance equation for the steady state concentration of a pollutant in the hypolimnion. Note that at steady state $r_b = r_p - r_d$. Consider a lake where the average river water flow is 100 m^3 s^{-1}, an average rainfall deposition of 20 m^3 s^{-1}, and a mean burial rate of 200 mg s^{-1} for phosphorous. If the influent to the lake has a P loading of 5 μg L^{-1} from agricultural pesticide run off into the river, is the lake prone to eutrophication (i.e., biological productivity)? Mean P concentration for biological productivity in lakes is 5 μg L^{-1}.

4.31$_3$ Estimate the time to breakthrough for 1,2-dichlorobenzene to reach a well 1 km from a source in a groundwater that has a Darcy velocity of 8 m y^{-1}. The soil has a 3% organic carbon, porosity of 0.4, and a density of 1.2 g cm^{-3}. Both advection and dispersion are important in this case. Breakthrough is defined as the time at which the concentration reaches 1% of the feed (source).

4.32$_2$ The following two contaminants are present in a gasoline mixture: hexane and octane. If a gasoline spill has occurred in the subsurface due to an underground storage tank rupture and contaminated the groundwater, how quickly will these chemicals be detected in a monitoring well 100 m away? The arrival time of a conservative tracer (chloride ion) was observed to take only 50 days. The soil had an organic carbon fraction of 1% and a porosity of 0.45. The soil density was 1.3 g cm^{-3}.

4.33$_2$ During a routine navigational dredging of a river, a 55 gal drum containing pure benzene was excavated. A small leak was observed through a hole of 10 cm^2 area on the top of the drum. How much benzene has been lost to the river water by diffusion during the one year the drum has been on the bottom sediment surface? The drum has a height of 100 cm.

4.34$_2$ Estimate the total volume of water required to remove a 3% residual trichloroethylene (TCE) spill from 1 m^3 of an aquifer of porosity 0.3 and organic carbon content of 1%. The density of TCE is 1.47 g cm^{-3} and its aqueous solubility is 1100 mg L^{-1}. Assume no retention of TCE on aquifer solids.

4.35$_2$ Estimate the number of pore volumes required to reduce 99% of the pore water concentration of the following contaminants from an aquifer with 2% organic carbon content and a porosity of 0.4: 1,2-dichloroethane, 1,4-dichlorobenzene, biphenyls, Aroclor-1242.

4.36$_3$ Compare the rates of emission to air of the following pesticides applied to a surface soil at an average concentration of 10 μg g^{-1} each. The soil porosity is 0.6 with a 3% water saturation and organic carbon content of 3%. The soil density is 2 g cm^{-3} and the area of application is 1 hectare. Assume a surface mass transfer coefficient of 1 cm s^{-1}. Repeat your calculation for a saturated surface soil of 10% water saturation: (1) lindane, (2) chloropyrifos, (3) $p,p = -$DDT.

4.37$_3$ The groundwater source in Baton Rouge is high in Fe(II). The Central Treatment Facility uses aeration to oxidize Fe(II) and precipitate as iron oxyhydroxide. The reaction rate is given in Section 2.3.8.1. The observed rate constant is k = 1.6 × 10^{-7} L (mol min atm)$^{-1}$ at the water pH of 6.7. The oxygen concentration in water is to be kept at its saturation value. If the oxidation reactor is a CSTR, what residence time is required to reduce Fe(II) by 90% in the groundwater?

4.38$_3$ Emissions from wastewaters in petroleum refineries are categorized into primary and secondary. Secondary sources of emissions are generally from wastewater ponds. Before the water reaches a pond, streams from different unit operations are mixed in a large vessel where air emissions can be a problem. There is little or no chemical degradation in these vessels. The only loss mechanism is mass transfer to the vapor space. Although no aerators are placed, there is enough turbulence to mix the water. Use a CSTR model to assess the rate of air emissions from a vessel in which the inlet concentration of ethyl benzene is 1 × 10^{-5} g mL^{-1} at a flow rate of 2 L s^{-1}. The vessel cross-sectional area is 1 m^2.

4.39$_2$ Determine the number of pore volumes of air required for 99% removal of chlorobenzene from the vadose zone of an aquifer. The total volume of the aquifer is 1 m^3. It has a porosity of 0.6, an organic carbon fraction of 0.02, and bulk density of 1.4 g cm^{-3}. The zone is 5% water saturated and 5% saturated with chlorobenzene.

4.40₂ On a clear, calm Monday evening (January 18, 1999), a tanker truck carrying 22 tons of liquid anhydrous ammonia rolled over as it was making a sharp turn outside the Farmland Industries chemical plant in Pollock, LA. The liquid ammonia escaped through a valve. Assume that it formed a 2500 m^2 pool and turned into gas as it went into the air. Estimate the concentration of ammonia vapor 500 m from the accident. Is the concentration dangerous to the workers in the plant? The wind speed during the period was nominal (3 m s⁻¹).

4.41₃ Ed Garvey determined that as water flows over hot spots in the Thomson Island pool of the Hudson River, its PCB concentration increases rapidly. Data taken from 1993 to 1995 showed that the "hot-spot" sediments were responsible for about 1.5 lb of PCBs (3 lb during the summer) moving into the water column. Because the flesh of most Hudson River fish exceeds the FDA limit of 2 ppm for PCBs, various fishing bans and advisories are in effect along the river, with children and women of child-bearing age told to "eat none" (Rivlin, 1998). Use a compartmental box model to verify whether or not fish advisories or bans are warranted. Assume: (1) no PCBs in the air, (2) no PCB up-stream of Thomson Island pool, and (3) fish in equilibrium with the PCBs in the water. The PCB data are as follows: evaporation mass transfer coefficient = 5.7 cm h⁻¹, log K_{oc} = 4.5. River data: surface area for evaporation = 2.3×10^5 m², river flow = 4×10^5 L s⁻¹.

4.42₂ Benzene (=1), chlorobenzene (=2), methyl chloride (=3), and methane (=4) are combusted in a new incinerator designed for a residence time of 1 s and a burner temperature of 1200°F. However, during the startup, the conversions were all less than 20%, which is unacceptable. As the contact engineer, you are presented with two options: (1) extend the burner chamber 10 fold, thus increasing the residence time to 10 s, or (2) increase the temperature from 1200°F to 1400°F. Which option gives the highest conversions? Which would be cheapest to implement? What are the consequences or concerns of each?

4.43₂ Heterogeneous reactions on polar stratospheric clouds (PSCs) are responsible for the abnormal destruction of ozone in the Antarctic regions in the springtime. The following series of reactions are shown to be occurring on a surface, M:

$$2ClO + M \xrightarrow{k_1} Cl_2O_2 + M$$

$$Cl_2O_2 \xrightarrow[hv]{J_2} ClOO + Cl$$

$$ClOO + M \xrightarrow{k_3} Cl + O_2 + M$$

$$2\left(Cl + O_3 \xrightarrow{k_4} ClO + O_2\right)$$

Note that the overall scheme is $2O_3 \xrightarrow{hv} 3O_2$. The first step in this scheme is the slowest reaction. Justify that $\dfrac{d[O_3]}{dt} = -k_1[ClO]^2[M]$, where $k_1 = 8 \times 10^{-44}$ m^6 s^{-1} at 190 K (stratospheric temperature). Assume an average $[M] = 3 \times 10^{24}$ m^{-3} and determine the mean concentration of ClO responsible for the ozone destruction.

4.44₂ The $SO_2(g)$ from power plant smoke stacks dissolves in a surface cloud (e.g., fog) to produce S(IV). The resulting S(IV) then reacts with a dissolved species A(aq) to give S(VI) in a fog droplet. The rate of the reaction is given by $r = -\dfrac{d[S(IV)]}{dt} = k[A]_{aq}[S(IV)]_{aq}$. Rewrite this equation to give the rate in terms of the liquid water content of the fog, θ_L (see Section 3.3.1) to given $r = 3.6 \times 10^6$ $\theta_L rRT$, where the units of r are ppbv SO_2 h^{-1} and R is the universal gas constant.

4.45₂ Hurricane Katrina flooded several New Orleans homes in 2005. Contaminated sediments from Lake Ponchartrain remained as a cake and finally dried out in abandoned homes once the floodwater receded. These sediments are likely to be sources of vapors of organic compounds to the indoor air. Consider a 20 m long × 20 m wide × 5 m high room where the air flow velocity is only 10 m h^{-1}. Assume the emission rate of the organic vapor from the sediment to air is 100 ng m^{-2} h^{-1}. Assume also that the organic vapor degrades in the indoor air with a first order rate constant of 10 h^{-1} and also that the fresh air entering the room has only negligible organic vapor in it.

1.   Perform a mass balance on the organic vapor in the room and write the unsteady mass balance differential equation.
2.   Apply the steady state approximation to obtain the concentration of the organic vapor in the air.
3.   For the data given here calculate the steady state concentration. Express your answer in ng m^{-3}.

4.46₃ A pool of pure benzene of area 0.4 m^2 exists at the bottom of a lake of volume 1000 m^3. The volumetric flow rate of water in the lake is 50 L s^{-1}. If the average mass transfer coefficient (k_{mt}) for the dissolution of pure benzene into water is 0.1 m s^{-1}, determine the concentration in the outflow from the lake after 12 h of the spill. The solubility of benzene in water is 780 mg L^{-1}. Assume that the lake is well mixed and initial concentration of benzene is zero.

4.47₂ Estimate the volume of a bioreactor required to achieve a yield of 0.5 of *E. coli* bacteria using 20 mg L^{-1} of an activated sludge in the influent stream of a continuous reactor. The desired flow rate of the sludge is 10 m^3 d^{-1}. The bacteria follows Monod kinetics with $K_s = 10$ mg L^{-1} and $\mu_{max} = 10$ d^{-1}. The bacterial decay constant is 2 d^{-1}. It is desired to maintain the bacterial population at 5 mg L^{-1} through the day. Give the answer in m^3.

4.48   Air enters a room (5 m × 5 m × 8 m) at a velocity of 1 m h^{-1} and carries a gaseous pollutant at a concentration of 10 μg m^{-3}. An open-hearth furnace in the room generates the same pollutant at a rate of 24 μg m^{-2} h^{-1}. The pollutant, however, undergoes a zero order reaction in the room with a rate constant of $1 \, \mu g \, m^{-3} \, h^{-1}$.

1. Write down the appropriate unsteady state mass balance equation for the pollutant in the room.
2. Solve that equation to obtain concentration as a function of time.
3. What is the pollutant concentration leaving the room 1 h after the start of the open-hearth furnace?

4.49₁ State whether the following statements are true or false. Give brief explanations.

1. In an enzyme reactor we have to continuously provide enzymes externally for the reaction to proceed.
2. An organic pollutant will travel faster in a water-saturated column filled with soil than in a column filled with sand.
3. In the lower atmosphere smog appears when the sunlight is at its highest intensity and VOC concentration is lowest.
4. For the same degree of conversion, a plug flow reactor requires a larger volume than a CSTR.

4.50₃ A tanker collided with a tugboat and spilled 9000 gal of diesel fuel into the Mississippi River at 1:30 a.m. on July 23, 2008 near the mouth of the river in New Orleans, LA. As a result, residents of communities such as Algiers, Gretna, and the Plaquemines parish living downstream of the river were asked to conserve water, as water intakes to these communities were closed to prevent contamination of the drinking water supply. Consider the most soluble component of diesel (conduct a literature survey) and determine how fast it will move with the water to the nearest community down river. You will have to obtain the relevant flow parameters from the USGS website.

4.51₃ Heavy duty vehicles (hdv) using diesel fuel emit carcinogenic hydrocarbon, Benzo[a]pyrene (BaP), directly to the air. A detailed investigation of BaP emissions between 1961 and 2004 was recently reported by Beyea et al. (2008). Consider the Baltimore Harbor tunnel that is 2.31 km long, 6.7 m wide, and 4.1 m high. The maximum speed of vehicles within the tunnel is 80 km h⁻¹. Wind parallel to the tunnel blows in at 4 m s⁻¹. It is estimated that 2500 vehicles per hour with 12% hdv mix moving at an average speed of 46 km h⁻¹ transits through the tunnel. If the estimated emission factor for BaP was 40 μg km⁻¹ in 1961 and 5 μg km⁻¹ in 2004, what are the corresponding changes in BaP air concentrations during this period?

REFERENCES

Alexander, M. (1994) *Biodegradation and Bioremediation.* New York: Academic Press.

Allen, D. and Shonnard, D.R. (2002) *Green Engineering: Environmentally Conscious Design of Chemical Processes.* New York: Prentice-Hall.

Anderson, J.G., Brune, W.H., and Proffitt, M.H. (1989) Ozone destruction by chlorine radicals within the Antarctic vortex: The spatial and temporal evolution of ClO-O₃ anticorrelation based on in situ ER-2 data. *Journal of Geophysical Research: Atmospheres* **94**(D9), 11465–11479.

Bailey, J.E. and Ollis, D.F. (1986) *Biochemical Engineering Fundamentals*, 2nd edn. New York: McGraw-Hill, Inc.

Bedient, P.B., Rifai, H.S., and Newell, C.L. (1994) *Groundwater Contamination*. Englewood Cliffs, NJ: Prentice Hall PTR.

Beyea, J., Stellman, S.D., Hatch, M., and Gammon, M.D. (2008) Airborne emissions from 1961 to 2004 of benzo[a]pyrene from U.S. vehicles per km of travel based on tunnel studies. *Environmental Science and Technology* **42**, 7315–7320.

Bricker, O.P. and Rice, K.C. (1993) Acid rain. *Annual Review of Earth and Planetary Sciences* **21**, 151–174.

Buonicore, A.T. and Davis, W.T. (eds.) (1992) *Air Pollution Engineering Manual*. New York: Van Nostrand Reinhold.

Butler, J.N. (1982) *Carbondioxide Equilibria and Its Applications*, 2nd edn. New York: Addison-Wesley Publishing Co.

Chappelle, F. (1993) *Groundwater Geochemistry and Microbiology*. New York: John Wiley & Sons, Inc.

Connolly, J.P. and Thomann, R.V. (1992) Modelling the accumulation of organic chemicals in aquatic food chains. In: Schnoor, J.L. (ed.), *Fate of Pesticides and Chemicals in the Environment*, pp. 385–406. New York: John Wiley & Sons, Inc.

Cooper, C.D. and Alley, F.C. (1994) *Air Pollution Control—A Design Approach*, 2nd edn. Prospect Heights, IL: Waverly Press.

Ferguson, D.W., Gramith, J.T., and McGuire, M.J. (1991) Applying ozone for organics control and disinfection: A utility perspective. *Journal of the American Water Works Association* **83**, 32–39.

Fogler, H.S. (2006) *Elements of Chemical Reaction Engineering*, 4th edn. Englewood Cliffs, NJ: Prentice-Hall Inc.

Galloway, J.N., Likens, G.E., and Edgerton, E.S. (1976) Acid precipitation in the northeastern United States: pH and acidity. *Science* **194**, 722–724.

Gonzalez, M.A. and Smith, R.L. (2003) A methodology to evaluate process sustainability. *Environmental Progress* **22**, 269–276.

Guilhem, D. (1999) Sediment/air partitioning of hydrophobic organic contaminants, PhD dissertation, Louisiana State University, Baton Rouge, LA.

Harper, E.M. and Graedel, T.E. (2008) Illuminating tungsten's life cycle in the United States: 1975–2000. *Environmental Science & Technology* **42**(10), 3835–3842.

Hoffmann, M.R. and Calvert, J.G. (1985) *Chemical Transformation Modules for Eulerian Acid Deposition Models*, Vol. II. The Aqueous-Phase Chemistry. Boulder, CO: National Center for Atmospheric Research.

Horan, N.J. (1990) *Biological Wastewater Treatment Systems: Theory and Operation*. Chichester, England: John Wiley & Sons Ltd.

Houghton, J.T., Jenkins, G.J., and Ephraums, J.J. (1990) *Climate Change: The IPCC Scientific Assessment*. New York: Cambridge University Press.

Hudson, R.J. and Morel, F.M.M. (1990) Iron transport in marine phytoplankton: Kinetics of cellular and medium co-ordinatin reactions. *Limnology and Oceanography* **35**, 1002–1020.

Hunt, J.R., Sitar, N., and Udell, K.S. (1988) Nonaqueous phase liquid transport and cleanup: 1. Analysis of mechanisms. *Water Resources Research* **24**, 1247–1258.

Hutterman, A. (1994) *Effects of Acid Rain on Forest Processes*. New York: John Wiley & Sons, Inc.

IPCC. (2007) *Climate Change: The Physical Chemical Basis*, International Panel on Climate Change. New York: Cambridge University Press.

Jacoby, M. (1998) At sea with photochemistry. *Chemical and Engineering News* **76**, 47–49.

Lee, J.M. (1992) *Biochemical Engineering*. Englewood-Cliffs, NJ: Prentice-Hall Publishers Inc.

Legrini, O., Oliveros, E., and Braun, A.M. (1993) Photochemical processes for water treatment. *Chemical Reviews* **93**, 671–698.

Levenspiel, O. (1999) *Chemical Reaction Engineering*, 3rd edn. New York: John Wiley & Sons, Inc.

Lovelock, J. (1979) *Gaia: A New Look at Life on Earth*. New York: Oxford University Press.

MacIntyre, F. (1978) Toward a minimal model of the world CO_2 system. I. Carbonate-alkalinity version. *Thalassia Jugoslavica* **14**, 63–98.

Mackay, D. and Leinonen, P.E. (1975) Rate of evaporation of low solubility contaminants from water bodies to atmosphere. *Environmental Science and Technology* **9**, 1178–1180.

Masten, S.J. and Davies, S.H.R. (1994) The use of ozonation to degrade organic contaminants in wastewaters. *Environmental Science and Technology* **28**, 180A–185A.

Matter-Muller, C., Gujer, W., and Giger, W. (1981) Transfer of volatile substances from water to the atmosphere. *Water Research* **15**, 1271–1279.

Meikle, R.W., Youngson, C.R., Hedlund, R.T., Goring, C.A.I., Hamaker, J.W., and Addington, W.W. (1973) Measurement and prediction of picloram disappearance rates from soil. *Weed Science* **21**, 549–555.

Molina, M.J. and Rowland, F.S. (1974) Stratospheric sink for chlorofluoromethanes: Chlorine-atom catalysed destruction of ozone. *Nature* **249**, 810–812.

Morel, F.M.M. and Herring, J.G. (1993) *Principles and Applications of Aquatic Chemistry*. New York: John Wiley & Sons, Inc.

Munz, C. (1985) Air-water phase equilibria and mass transfer of volatile organic solutes, PhD dissertation, Department Civil Engineering, Stanford University, Stanford, CA.

Neely, W.B. (1980) *Chemicals in the Environment: Distribution, Fate and Transport*. New York: Marcel Dekker Inc.

Neumann, S.P. (1990) Universal scaling of hydraulic conductivities and dispersivities in geologic media. *Water Resources Research* **26**, 1749–1758.

Novelli, P.C., Masarie, K.A., Tans, P.O., and Lang, P.M. (1994) Recent changes in atmospheric carbon monoxide. *Science* **263**, 1587–1590.

NRC. (1983) *Climate Change: Carbon Dioxide Assessment Committee*. Washington, DC: National Research Council, National Academies Press.

Park, D.H. (1980) Precipitation chemistry patterns: A two-network data set. *Science* **208**, 1143–1145.

Powell, E.O. (1967) The growth rate of microorganisms as a function of substrate concentration. In: Anonymous (ed.), *Proceedings of the Third International Symposium on Microbiology and Physiology of Continuous Cultures*, pp. 34–39. London, U.K.: HMSO.

Powers, S.E., Loureiro, C.O., Abriola, L.M., and Weber, W.J. (1991) Theoretical study of the significance of nonequilibrium dissolution of nonaqueous phase liquids in subsurface systems. *Water Resources Research* **27**, 463–477.

Reible, D.D. (1998) *Fundamentals of Environmental Engineering*. Boca Raton, FL: Lewis Publishers.

Reible, D.D., Valsaraj, K.T., and Thibodeaux, L.J. (1991) Chemodynamic models for transport of contaminants from sediment beds. In: *The Handbook of Environmental Chemistry*, Vol. 2F. Berlin, Germany: Springer-Verlag GmbH.

Reid, R.C., Prausnitz, J.M., and Poling, B.E. (1987) *The Properties of Gases and Liquids*, 4th edn. New York: McGraw Hill Book Co.

Reynolds, J.P., Dupont, R.R., and Theodore, L. (1991) *Hazardous Waste Incineration Calculations*. New York: John Wiley & Sons, Inc.

Rivlin, M.A. (1998) Muddy waters. *The Amicus Journal* **19**(4), 30–37.

Roustan, M., Brodard, E., Duguet, J.P., and Mallevialle, J. (1992) Basic concepts for the choice and design of ozone contactors. In: Mallevialle, J., Suffet, I.H., and Chan, U.S. (eds.), *Influence and Removal of Organics in Drinking Water*, pp. 195–206. Boca Raton, FL: Lewis Publishers.

Schlegel, H.G. (1992) *General Microbiology*, 2nd edn. New York: Cambridge University Press.

Schwartz, S.E. and Freiberg, J.E. (1981) Mass transport limitations to the rate of reaction of gases and liquid droplets: Application to oxidation of SO_2 and aqueous solution. *Atmospheric Environment* **15**, 1129–1144.

Schwarzenbach, R.P., Haderlein, S.B., Muller, S.R., and Ulrich, M.M. (1998) Assessing the dynamic behavior of organic contaminants in natural waters. In: Macalady, D.L. (ed.), *Perspectives in Environmental Chemistry*. New York: Oxford University Press.

Seinfeld, J.H. (1986) *Atmospheric Chemistry and Physics of Air Pollution*. New York: John Wiley & Sons, Inc.

Seinfeld, J.H. and Pandis, S.N. (1998) *Atmospheric Chemistry and Physics*, p. 390. New York: John Wiley & Sons, Inc.

Seinfeld, J.H. and Pandis, S.N. (2005) *Atmospheric Chemistry and Physics*, 3rd edn. New York: John Wiley & Sons, Inc.

Seinfeld, J.H. and Pandis, S.N. (2006) *Atmospheric Chemistry and Physics: From Air Pollution to Climate Change*, 2nd edn. New York: John Wiley & Sons, Inc.

Shah, Y.T., Kelkar, B.G., Godbole, S.P., and Deckwer, W.D. (1982) Design parameter estimations for bubble column reactors. *AIChE Journal* **28**, 353–359.

Solomon, S., Qin, D., Chen, Z., Marquis, M., Averyt, K.B., Tignor, M., and Miller, H.L. (eds.) (2007) *Climate Change 2007: The Physical Basis. Contribution of Working Group 1 to the Fourth Assessment Report of the Intergovernmental Panel on Climate Change*. Cambridge, U.K.: Cambridge University Press, p. 996.

Staehelin, J. and Hoigne, J. (1985) Decomposition of ozone in water in the presence of organic solutes acting as promoters and inhibitors of radical chain reactions. *Environmental Science and Technology* **19**, 1206–1213.

Stumm, W. and Morgan, J.J. (1996) *Aquatic Chemistry*, 4th edn. New York: John Wiley & Sons, Inc.

Thomann, R.V. (1989) Bioaccumulation model of organic chemical distribution in aquatic food chains. *Environmental Science and Technology* **23**, 699–707.

Thomann, R.V. and Connolly, J.P. (1984) Model of PCB in the Lake Michigan lake trout food chain. *Environmental Science and Technology* **18**, 68.

Trujillo, E.M., Jeffers, T.H., Ferguson, C., and Stevenson, H.Q. (1991) Mathematically modelling the removal of heavy metals from a wastewater using immobilized biomass. *Environmental Science and Technology* **25**, 1559–1565.

Turner, D.B. (1969) *Workbook of Atmospheric Dispersion Estimates*. Washington, DC: HEW.

U.S. Environmental Protection Agency (N/A) Green engineering. https://www.epa.gov/green-engineering. Accessed January 15, 2018.

Valentine, R.L. and Schnoor, J.L. (1986) Biotransformation. In: Hern, S.C. and Melancon, S.M. (eds.), *Vadose Zone Modelling of Organic Pollutants*, pp. 191–222. Chelsea, MI: Lewis Publishers, Inc.

Valsaraj, K.T. (1994) Hydrophobic compounds in the environment: Adsorption equilibrium at the air-water interface. *Water Research* **28**, 819–830.

Valsaraj, K.T., Ravikrishna, R., Choy, B., Reible, D.D., Thibodeaux, L.J., Price, C., Yost, S., Brannon, J.M., and Myers, T.E. (1999) Air emissions from exposed, contaminated sediment and dredged material. *Environmental Science and Technology* **33**, 142–148.

Valsaraj, K.T. and Thibodeaux, L.J. (1987) Diffused aeration and solvent sublation in bubble columns for the removal of volatile hydrophobics from aqueous solutions. *ACS Preprint Abstracts (Div. Env. Chem.)* **27**(2), 220–222 abstract.

Wark, K., Warner, C.F., and Davis, W.T. (1998) *Air Pollution—Its Origin and Control*, 3rd edn. New York: Addison Wesley Publishers.

Warneck, P. (1988) *Chemistry of the Natural Atmosphere*. New York: Academic Press.

Warneck, P. (1999) *Chemistry of the Natural Atmosphere*, 2nd edn. New York: Academic Press.

Wilson, D.J. and Clarke, A.N. (eds.) (1994) *Hazardous Waste Site Soil Remediation*. New York: Marcel Dekker.

Zepp, R.G. (1992) Sunlight-induced oxidation and reduction of organic xenobiotics in water. In: Schnoor, J.L. (ed.), *Fate of Pesticides and Chemicals in the Environment*, pp. 127–140. New York: John Wiley & Sons, Inc.

Appendix A: Properties of Selected Chemicals of Environmental Significance

1. Inorganic Compounds[a,b,c]					
Compound	Mol. Wt.	Mixing Ratio, χ_i (−)	Aq. Sol., C_i^* (mol dm⁻³)	Log K_{AW}^* (kPa dm³ mol⁻¹)	Log K_{ow} (−)
CO	28	$6\text{--}12 \times 10^{-6}$	9.3×10^{-4}	+4.99	
CO_2	44	0.035	3.3×10^{-2}	+3.47	0.83
N_2	28	79	6.2×10^{-4}	+5.19	0.67
NO	30	$\leq 5 \times 10^{-8}$	1.8×10^{-3}	+4.73	
NH_3	17	$1\text{--}10 \times 10^{-8}$	28.4	+0.22	−1.37
NO_2	46	$\leq 5 \times 10^{-7}$		+4.00	
H_2	2	6×10^{-5}	7.7×10^{-4}	+5.10	0.45
O_2	32	21	2.6×10^{-4}	+4.90	0.65
O_3	48	$1\text{--}10 \times 10^{-6}$	4.4×10^{-4}	+3.91	
SO_2	64	$1\text{--}10 \times 10^{-9}$	1.5	+1.92	
H_2S	34	$\leq 2 \times 10^{-8}$	0.1	+3.01	0.96
H_2O	18	3.2	—	—	−1.15
H_2O_2	34	1×10^{-7}		−2.85	−1.08
N_2O	42	3.0×10^{-5}	2.4×10^{-2}	+3.60	0.43
HNO_3 (g)	63	$1\text{--}10 \times 10^{-8}$		−3.32	
HCl (g)	36.5			−1.39	

2. Organic Compounds

Compound	Mol. Wt.	Vapor Pressure, χ_i (–)	Aq. Sol. P_i^* (kPa)	Log K''_{AW} (kPa dm³ mol⁻¹)	Log K_{ow} (–)
(i) Hydrocarbons[d]					
Methane	16	2.8×10^4	4.1×10^{-1}	4.85	1.12
Ethane	30	4.0×10^3	8.1×10^{-2}	4.90	1.78
Propane	44	9.4×10^2	1.3×10^{-2}	4.86	2.36
n-Butane	58	2.5×10^2	2.6×10^{-3}	4.98	2.89
n-Pentane	72	7.0×10^1	5.6×10^{-4}	5.10	3.62
n-Hexane	86	2.0×10^1	1.5×10^{-4}	5.26	4.11
n-Heptane	100	6.2×10^0	3.1×10^{-5}	5.31	4.66
n-Octane	114	1.9×10^0	6.3×10^{-6}	5.51	5.18
n-Decane	142	1.7×10^{-1}	2.7×10^{-7}	5.85	6.70
Benzene	78	1.2×10^1	2.3×10^{-2}	2.75	2.13
Toluene	92	3.8×10^0	5.6×10^{-3}	2.83	2.69
Ethylbenzene	106	1.3×10^0	1.6×10^{-3}	2.93	3.15
Naphthalene	128	3.7×10^{-2}	8.7×10^{-4}	1.69	3.36
Phenanthrene	178	9.0×10^{-5}	3.4×10^{-5}	0.55	4.57
Anthracene	178	7.8×10^{-5}	3.3×10^{-5}	0.36	4.54
Pyrene	202	4.0×10^{-6}	4.4×10^{-6}	−0.04	5.13
Cyclopentane	70	4.3×10^1	2.3×10^{-3}	4.27	3.00
Cyclohexane	84	1.2×10^1	7.1×10^{-4}	4.25	3.44
D-Limonene	136	0.26	1.0×10^{-4}	1.88	4.57
β-Pinene	136		1.7×10^{-4}	−1.67	4.37
(ii) Acids and bases[d,e]					
Acetic acid	60	1.6×10^0	1.0×10^2	−1.99	−0.17
n-Propionic acid	74	4.5×10^{-1}		0.33	
n-Butanoic	88	1.0×10^{-1}	3.2×10^{-2}		0.79
n-Pentanoic	102	2.0×10^{-2}	8.1×10^{-3}		0.99
n-Hexanoic	116	6.0×10^{-3}	9.4×10^{-2}		1.90
n-Heptanoic	130	2.0×10^{-3}	1.8×10^{-2} (at 293 K)		2.72
n-Octanoic	144	1.0×10^{-3}	1.7×10^{-2} (at 373 K)		3.22
Benzoic acid	122		2.7×10^{-2}	−2.15	1.87
(iii) Alcohols and phenols[d,f,g]					
Butanol	74	9.3×10^{-1}	1.0×10^0	−0.24	0.88
Pentanol	88	3.4×10^{-1}	3.0×10^{-1}	0.17	1.16
Hexanol	102	1.4×10^{-1}	1.3×10^{-1}	0.25	2.03
Heptanol	116		1.5×10^{-2}	0.34	2.41
Octanol	130	6.8×10^{-3}	4.4×10^{-3}	0.45	2.84
Phenol	94	6.9×10^{-2}	1.0×10^0	−1.34	1.48
2-Chlorophenol	128	1.9×10^{-1}	2.2×10^{-1}	−1.24	2.17
2,4-Dichlorophenol	163		2.7×10^{-2}		2.75

(Continued)

2. Organic Compounds

Compound	Mol. Wt.	Vapor Pressure, χ_i (–)	Aq. Sol. P_i^* (kPa)	Log $K_{AW}^{\prime\prime}$ (kPa dm^3 mol^{-1})	Log K_{ow} (–)
2,4,6-Trichlorophenol	197	1.1×10^{-3}	4.0×10^{-3}	−2.20	3.38
Pentachlorophenol	266	1.5×10^{-5} (at 293 K)	5.2×10^{-5}	−1.56	5.04
2-Nitrophenol	139	1.8×10^{-2}	1.3×10^{-2}	0.13	1.89
4-Me-2-nitrophenol	153	6.2×10^{-3}	3.8×10^{-3}	0.21	2.37
4-Cl-2-nitrophenol	173	4.9×10^{-3}	3.9×10^{-3}	0.10	2.46
2,4-Dinitrophenol	230	6.8×10^{-4}	1.5×10^{-3}	−1.55	1.67
(iv) *Halocarbons*[h]					
Methylene chloride	85	5.9×10^{1}	2.3×10^{-1}	2.43	1.15
Chloroform	119	2.6×10^{1}	6.4×10^{-2}	2.59	1.93
Carbon tetrachloride	154	1.5×10^{1}	6.3×10^{-3}	3.33	2.73
1,2-Dichloroethane	99	9.1×10^{0}	8.5×10^{-2}	2.00	1.47
1,1,1-Trichloroethane	133	1.6×10^{1}	8.5×10^{-3}	3.49	2.48
Hexachloroethane	285	2.4×10^{-2}	1.7×10^{-4}	2.40	
Hexachlorobutadiene	261	3.4×10^{-2}	1.2×10^{-5}	3.44	4.90
Chlorobenzene	112	1.6×10^{0}	4.5×10^{-3}	2.54	2.91
1,2-Dichlorobenzene	147	1.9×10^{-1}	9.8×10^{-4}	2.28	3.38
1,3,5-Trichloro-benzene	181	7.8×10^{-2}	7.1×10^{-5}	3.04	4.02
Hexachlorobenzene	285	3.5×10^{-4}	2.3×10^{-6}	2.18	5.50
Endrin	381	4.0×10^{-7}	6.6×10^{-10}	−0.12	4.56
Aldrin	365	8.0×10^{-7}	5.5×10^{-7}	0.17	6.50
Lindane	291	6.4×10^{-5}	1.9×10^{-4}	−0.49	3.78
p,p′-DDT	355	9.5×10^{-8}	9.8×10^{-8}	−0.02	6.37
2,3,7,8-Tetrachloro-dibenzo-*p*-dioxin	322	1.6×10^{-7}	3.2×10^{-8}	0.70	6.64
Trichlorofluoro-methane	137	1.0×10^{2}	8.3×10^{-3}	4.10	2.16
Dichlorodifluoro-methane	121	1.5×10^{4}	1.6×10^{-2}	4.60	2.53
Biphenyl	154	1.0×10^{-3}	1.3×10^{-4}	2.19	4.09
4-Chlorophenyl	188	2.5×10^{-3}	3.1×10^{-5}	1.91	4.53
4,4′-Dichlorobiphenyl	223	8.3×10^{-5}	5.1×10^{-6}	1.21	5.33
2,3′,4,4′-Tetrachloro-biphenyl	292	6.7×10^{-6}	2.1×10^{-7}	1.49	6.31
Chlorocyclohexane	118	1.08	1.8×10^{-3}	2.78	
Bromocyclohexane	163	0.41	3.9×10^{-4}	2.97	3.20
(v) *Thio compounds and esters*[h]					
Dimethyl sulfide	94	3.8×10^{0}	3.6×10^{-2}	2.02	1.77
Thiophene	84	1.0×10^{1}	4.7×10^{-2}	2.35	1.81

(Continued)

2. Organic Compounds

Compound	Mol. Wt.	Vapor Pressure, χ_i (–)	Aq. Sol. P_i^* (kPa)	Log K_{AW}'' (kPa dm^3 mol^{-1})	Log K_{ow} (–)
Diethyl phthalate	222	8.3×10^{-4}	4.1×10^{-3}	-0.70	2.35
Di-*n*-butyl phthalate	278	9.5×10^{-6}	3.4×10^{-5}	-0.89	4.57
(vi) *Mercury and mercury compounds*[i]					
Hg	201	1.6×10^{-4}	2.8×10^{-7}	$+2.86$	0.61
HgCl$_2$	272	1.3×10^{-5}	2.5×10^{-1}	-4.14	0.52
Hg(OH)$_2$	235		5.1×10^{-4}	-2.10	-1.30
(CH$_3$)$_2$Hg	230	1.13×10^{-3}	1.3×10^{-2}	$+2.88$	2.26
(vii) *Ketones and aldehydes*[b]					
Acetone	58	2.8×10^{1}	1.3×10^{1}	0.47	-0.24
Formaldehyde	30	5.2×10^{2}	4	-1.53	0.35
Acetaldehyde	44	1.2×10^{2}	4	0.83	0.52
Benzaldehyde	106	1.3×10^{-1}	3.1×10^{-2}	0.61	1.48
Acetophenone				-5.53	
(viii) *Siloxanes*[j]					
Octamethyl-cyclotetrasiloxane	297	3.0×10^{-2}	1.8×10^{-7}	4.39	5.09
(ix) *Alkenes and haloalkenes*[b,h]					
1-Butene	56	3.6×10^{2}	4.0×10^{-3}	4.40	2.40
Styrene	104	6.3×10^{-1}	2.4×10^{-3}	2.42	3.05
Vinyl chloride	62	3.9×10^{2}	1.7×10^{-1}	3.35	0.60
Tetrachloroethene	166	2.5×10^{0}	9.1×10^{-4}	3.44	2.88
(x) *Amines*					
Aniline	93	1.3×10^{1}	3.9×10^{-1}	-0.47	0.90
Ethylamine	45	1.3×10^{2}	1.1×10^{2}	5.03	-0.30
(xi) *Perfluoroalkanoic acids*[k,l]					
Perfluorooctane-sulfonic acid (PFOS)	500	3.3×10^{-7}	1.4×10^{-3}	-3.48	4.13
Perfluorooctanoic acid (PFOA)	383	1.3×10	8.9×10^{-3}	0.39	4.40
(xii) *Polybrominated diphenyl ethers (PBDE)*[m]					
PeBDE-tetra	485	4.7×10^{-8}	2.7×10^{-8}	1.04	6.57
PeBDE-penta	564				
OBDE-hexa	643	6.6×10^{-9}	7.7×10^{-10}	1.02	6.29
OBDE-hepta	722				
OBDE-octa	801				
DBDE-deca	959	4.6×10^{-9}	$<1 \times 10^{-10}$	>1.6	6.27

(Continued)

2. Organic Compounds

Compound	Mol. Wt.	Vapor Pressure, χ_i (-)	Aq. Sol. P_i^* (kPa)	Log $K_{AW}^{''}$ (kPa dm³ mol⁻¹)	Log K_{ow} (-)
(xiii) *Alkylnitriles*[n]					
Acetonitrile	41	13.4	0.04	0.79 ± 0.04	-0.34
Propionitrile	55	5.2	1.08	1.17 ± 0.05	0.16
Butyronitrile	69	2.7	0.48	1.48 ± 0.05	0.53
Valeronitrile	83	1.5	0.28	1.65 ± 0.07	1.12
Isovaleronitrile	83			2.55 ± 0.09	1.07

Note: Excellent compilations of physicochemical properties for a variety of chemicals can be found at the following sites on the worldwide web: (1) http://www.webbook.nist.gov and (2) http://www.chemfinder.camsoft.com.

[a] In order to obtain values in conventional units, use the following conversion factors: Pressure in kPa should be multiplied by (1/101.325) to convert to atm. Aqueous solubility expressed in mol dm⁻³ is identical to mol L⁻¹. Henry's constant $K_{AW}^{''}$ given in units of kPa dm³ mol⁻¹ should be multiplied by 4.04×10^{-4} to obtain K_{AW}, which is a dimensionless molar ratio. All values are at 298 K unless otherwise indicated. Log K_{ow} values are mostly from Hansch and Leo (1995).

[b] Stumm and Morgan (1981), Liss and Slater (1974), Mackay and Leinonen (1975), Thibodeaux (1979), Morel and Herring (1993), Leo et al. (1971), Zhou and Mopper (1990), Montgomery (1996), Ji and Evans (2007).

[c] For gases, the partial pressures are typical of the atmosphere. Aqueous solubility is that at a total pressure of 101.325 kPa and at 298 K.

[d] Vapor pressure and aqueous solubility at 298 K except where noted. For solids and gases, the values are for the subcooled liquid state. Ambrose and Ghiasse (1987), Helburn et al. (2008), Cal (2006).

[e] Verschueren (1983), Montgomery (1996), Schwarzenbach et al. (1993).

[f] Values for pentachlorophenol are for the solid species.

[g] For nitrophenols, the values are for subcooled liquid species at 293 K, Schwarzenbach et al. (1988). Values of log K_{ow} are for the neutral species.

[h] Values for solids and gases are for subcooled liquid state at 298 K. Schwarzenbach et al. (1993), Mackay (1991), Sarraute et al. (2008).

[i] Values from Mason and Fitzgerald (1996), Stein et al. (1996), Iverfeldt and Lindqvist (1984).

[j] Values from Mazzoni et al. (1997).

[k] Organization for Economic Co-operation and Development: Joint meeting of the chemical committee and the working party on chemicals, pesticides and biotechnology: Hazard assessment of PFOS, ENV/JM/RD(2002)17/Final (2002).

[l] Li et al. (2007).

[m] PBDEs are commercial mixtures and consist of PeBDE (penta-, tetra-, and hexa-BDE), OBDE (hepta-, octa-, and hexa-BDE), and DBDE (deca- and nona-BDE). Data from Environment Canada (2006).

[n] Ji et al. (2008), Lee et al. (1996), Hansch and Leo (1995), Yaffe et al. (2001), Ruelle (2000).

REFERENCES

Ambrose, D. and Ghiasse, N.B. (1987) Vapour pressures and critical temperatures and critical pressures of some alkanoic acids: C1 to C10. *Journal Chemical Thermodynamics* **19**, 505.

Cal, K. (2006) Aqueous solubility of liquid monoterpenes at 293 K and relationship with calculated Log P value. *Yakugaku Zasshi* **126**, 307–309.

Environment Canada. (2006) Ecological screening assessment report on PBDEs, June 2006.

Hansch, C. and Leo, A. (1995) *Exploring QSAR. [1] Fundamentals and Applications in Chemistrh and Biology*, ACS Professional Reference Book. Washington, DC: American Chemical Society.

Helburn, R. et al. (2008) Henry's law constants for fragrance and organic solvent compounds in aqueous industrial surfactants. *Journal of Chemical & Engineering Data* **53**, 1071.

Iverfeldt, A. and Lindqvist, O. (1984) The transfer of mercury at the air–water interface. In: Brutsaert, W. and Jirka, G.M. (Eds.), *Proceedings of the International Symposium on Gas Transfer at Water Surfaces*, pp. 533–538. Dordrecht, Holland: D. Reidel Publ. Co.

Ji, C. and Evans, E.M. (2007) Using an internal standard method to determine Henry's law constants. *Environmental Toxicology and Chemistry* **26**, 231.

Ji, F. et al. (2008) Measurement of Henry's law constants using internal standards. A quantitative GC experiment for the instrumental analysis or environmental chemistry laboratory. *Journal of Chemical Education* **85**, 969.

Lee, Y.H. et al. (1996) Aqueous functional group activity cocfficients (AQUAFAC) 4: Applications to complex organic compounds. *Chemosphere* **33**, 2129.

Leo, A., Hansch, C., and Elkins, D. (1971) Partition coefficients and their uses. *Chemical Reviews* **71**, 525–613.

Li, H. et al. (2007) Measurement of low air–water partition coefficients of organic acids by evaporation from a water surface. *Journal of Chemical & Engineering Data* **52**, 1580.

Liss, P.S. and Slater, P.G. (1974) Flux of gases across the air-sea interface. *Nature* **247**, 181–184.

Mackay, D. (1991) *Multimedia Environmental Models*. Chelsea, MI: Lewis Publishers, Inc.

Mackay, D. and Leinonen, P.J. (1975) Rate of evaporation of low solubility contaminants from water bodies to the atmosphere. *Environment Science and Technology* **9**, 1178–1180.

Mason, R. and Fitzgerald, W. (1996) Sources, sinks and biogeochemical cycling of mercury in the ocean. In: Baeyens, W., Ebinghaus, R., and Vasiliev, O. (eds.), *Global and Regional Mercury Cycles*. Dordrecht, the Netherlands: Kluwer Academic Publishers.

Mazzoni, S.M., Roy, S., and Grigoras, S. (1997) Eco-relevant properties of selected organosilicon materials. *Handbook of Environmental Chemistry* **3**, 53–57.

Montgomery, J.H. (1996) *Groundwater Chemicals Desk Reference*, 4th edn. Boca Raton, FL: CRC Press.

Morel, F.M.M. and Herring, J.G. (1993) *Principles and Applications of Aquatic Chemistry*. New York: John Wiley & Sons, Inc.

Ruelle, P. (2000) The n-octanol and n-hexane/water partition coefficient of environmentally relevant chemicals predicted from the mobile order and disorder (MOD) thermodynamics. *Chemosphere* **40**, 457.

Sarraute, S. et al. (2008) Atmosphere/water partition of halocyclohexanes from vapour pressure and solubility data. *Atmospheric Environment* **42**, 4724.

Schwarzenbach, R.P., Gschwend, P.M., and Imboden, D.M. (1993) *Environmental Organic Chemistry*. New York: John Wiley & Sons, Inc.

Schwarzenbach, R.P., Stierli, R., Folsom, B.R., and Zeyer, J. (1988) Compound properties relevant for assessing the environmental partitioning of nitrophenols. *Environmental Science and Technology* **22**, 83–92.

Stein, E.D., Cohen, Y., and Winer, A.M. (1996) Environmental distribution and transformation of mercury compounds. *Critical Reviews in Environmental Science and Technology* **26**, 1–43.

Stumm, W. and Morgan, J.M. (1981) *Aquatic Chemistry*, 2nd edn. New York: John Wiley & Sons, Inc.

Thibodeaux, L.J. (1979) *Environmental Chemodynamics*, 1st edn. New York: John Wiley & Sons, Inc.

Verschueren, K. (1983) *Handbook of Environmental Data on Organic Chemicals*, 2nd edn. New York: Van Nostrand Reinhold Co.

Yaffe, D. et al. (2001) A fuzzy ARTMAP based on quantitative structure–property relationships (QSPRs) for predicting aqueous solubility of organic compounds. *Journal of Chemical Information and Computer Science* **41**, 1177.

Zhou, X. and Mopper, K. (1990) Apparent partition coefficients of 15 carbonyl compounds between air and seawater and between air and freshwater; implications for air-sea exchange. *Environmental Science and Technology* **24**, 1864–1869.

Appendix B: Standard Free Energy, Enthalpy, and Entropy of Formation for Compounds of Environmental Significance

Compound (kJ mol⁻¹)	ΔG_f^0 (kJ mol⁻¹)	ΔH_f^0 (J K⁻¹ mol⁻¹)	S_f^0
H^+ (aq)	0	0	0
H_2O (l)	−237.2	−285.8	69.9
H_2O (g)	−228.6	−241.8	188.7
OH^- (aq)	−157.5	−230.3	−10.7
CO_3^{2-} (aq)	−527.9	−677.1	−56.9
HCO_3^- (aq)	−586.8	−692.0	91.2
H_2CO_3 (aq)	−623.2	−699.7	187.0
CO_2 (g)	−394.4	−393.5	213.6
$HOCl$ (aq)	−80.0	−121.1	142.5
H_2S (g)	−33.6	−20.6	205.7
H_2S (aq)	· −27.9	−39.8	121.3
HS^- (aq)	12.0	−17.6	62.8
S^{2-} (aq)	85.8	33.0	−14.6
SO_4^{2-} (aq)	−744.6	−909.2	20.1
Cl^- (aq)	−131.6	−167.2	56.5
Cl_2 (aq)	6.9	−23.4	121.0
Ca^{2+} (aq)	−553.5	−542.8	−53
$CaCO_3$ (s)	−1127.8	−1207.4	88.0
Fe (metal)	0	0	27.3
Fe^{2+} (aq)	−78.8	−89.1	−138.0
FeS (s)	−100.6	−100.1	60.4
FeS_2 (s)	−167.2	−178.5	53.0
$Fe(OH)_3$ (s)	−697.6	−824.2	106.8
Na^+ (aq)	−262.2	−240.5	59.1
NH_3 (g)	−16.5	−46.1	192.0
NH_3 (aq)	−26.6	−80.3	111.0
NH_4^+ (aq)	−79.4	−132.5	113.4
CH_4 (g)	−50.8	−74.8	186.0

Source: Obtained from Wagman, D.D. et al., Selected values of chemical thermodynamics properties, U.S. National Bureau of Standards, Technical Notes 270-3, 1968; 270-4, 1969; 270-5, 1971.

Note: All values are at 298 K.

REFERENCE

Wagman, D.D. et al. Selected values of chemical thermodynamics properties, U.S. National
     Bureau of Standards, Technical Notes 270-3, 1968; 270-4, 1969; 270-5, 1971.

Appendix C: Selected Fragment (b_j) and Structural Factors (B_k) for Octanol–Water Partition Constant Estimation

	Fragment Constant[a]			
Fragment	b_j	b_j^{ϕ}	$b_j^{\phi\phi}$	b_j, Other Types
(a) *Hydrocarbon increments*				
–H	0.23	0.23		
>C<	0.20	0.20		
=C<aromatic	0.13[b]			
=CH-aromatic	0.355			
–CH₃	0.89	0.89		
–C₆H₅	1.90			
(b) *Oxygen increments*				
O	−1.82	−0.61	0.53	Vinyl[c] −1.21
–O-aromatic	−0.08			
–OH	−1.64	−0.44		Benzyl[c] −1.34
(c) *Carbonyl increments*				
–C(O)–	−1.90	−1.19	−0.50	
–C(O)-aromatic	−0.59			
–C(O)H	−1.10	−0.42		
–C(O)O–	−1.49	−0.56	−0.09	Vinyl −1.18, Benzyl −1.38
–C(O)O-aromatic	−1.40			
–C(O)OH	−1.11	−0.03		Benzyl −1.03
–C(O)NH₂	−2.18	−1.26		Benzyl −1.99
(d) *Nitrogen increments*				
–N<	−2.18	−0.93	−0.50	
–N< aromatic	−1.12			
–NH–	−2.15	−1.03	−0.09	
–NH-aromatic	−0.65			
–NH₂–	−1.54	−1.00		Benzyl[c] −1.35
–NO₂	−1.16	−0.03		
–CN	−1.27	−0.34		Benzyl −0.88
(e) *Halogen increments*				
–Cl	0.06	0.94		Vinyl 0.50
–Br	0.20	1.09		Vinyl 0.64

(Continued)

Structural Factors

Feature	B_k	
(a) *Geometric features*		
Multiple bond (unsaturation)		
Double bond	-0.09^d	
Triple bond	-0.50^d	
Skeletal flexing		
Hydrocarbon chains	$-(N-1)(0.12)^e$	
Alicyclic rings	$-(N-1)(0.19)$	
Chain branching		
Nonpolar chain	-0.13	
Polar chain	-0.22	
(b) *Electronic features*		
Polyhalogenation		
2 on the same C	0.60	
3 on the same C	1.59	
4 on the same C	2.88	
2 on adjacent sp^3 C	0.28	
3 on adjacent sp^3 C	0.56	
4 on adjacent sp^3 C	0.84	

Polar Fragments	In Chain	In Aromatic Ring
On same C	$-0.42(b_1 + b_2)$	
On adjacent C	$-0.26(b_1 + b_2)$	$-0.16(b_1 + b_2)$
On C separated by one C	$-0.10(b_1 + b_2)$	$-0.08(b_1 + b_2)$
Intramolecular hydrogen bonding		
With $-OH$	1.0	
With $-NH$	0.6	

[a] The superscript ϕ denotes attachment to an aromatic ring (e.g., C_6H_5-O-). $\phi\phi$ denotes fragment to two aromatic rings (e.g., $C_6H_5-CO-C_6H_5$). Values are adapted from Lyman et al. (1990) and Baum (1998).

[b] Contribution of C shared by aromatic rings is 0.225, for C shared by aromatic rings and bonded to a hetero atom or a nonisolating C is 0.44.

[c] Vinyl means an isolated double bond $>C=C<$, benzyl means $C_6H_5CH_2-$ group.

[d] Value includes deductions for removing hydrogen atoms to produce unsaturation.

[e] N denotes the number of bonds in chain or ring.

Appendix D: Concentration Units for Compartments in Environmental Engineering

Concentrations of pollutants can be expressed in several units. The following is a summary of some of the common ones in air, water, and soil matrices that also appear in the text.

D.1 COMPARTMENT—AIR

The following are the common units to represent air concentration of a pollutant:

1. A volumetric ratio (parts per million by volume)

$$\{ppm_v\} = \left(\frac{V_i}{V_T} \right) \cdot 10^6$$

   where V_i is the volume of pollutant i and V_T is the total volume (air + pollutant). The advantage of the volume unit is that gaseous concentrations reported in these units do not change upon gas compression or expansion.
2. A mixed unit C_{ia} (μg of i/m³ of air), which is related to ppm_v (at temperature T degrees K and P_T total pressure)

$$\{ppm_v\} = \left(\frac{RT}{P_T} \right) \cdot \left(\frac{1}{M_i} \right) \cdot \frac{\{\mu g \ m^{-3}\}}{10^3}$$

   where M_i is the molecular weight of pollutant i, R is the gas constant 0.08205 L atm K⁻¹ mol⁻¹, and 1000 is a conversion factor (1000 L = 1 m³). Note that at 1 atm total pressure and standard ambient temperature of 298 K, the conversion factor $RT/1000 \ P_T$ is 0.0245.
3. Atmospheric "mixing ratio"
   Note that ppm_v is also equal to $(P_i/P_T) \ 10^6$ since volumes are related to partial pressures through the ideal gas law. P_i is the partial pressure of the pollutant in the gas. Analogously, ppm_v is also equal to $(n_i/n_T) \ 10^6$, where n represents the number of moles. Note that $(P_i/P_T) \ 10^6$ is also called the "mixing ratio, χ_i."

D.2   COMPARTMENT—WATER

The four most common quantifications of a pollutant concentration in water are as follows:

1. Molarity, C_{iw} = moles of i per liter (or per dm^3) of solution.
2. Molality, m_i = moles of i per kilogram of water.
3. Mole fraction of i, $x_i = \dfrac{n_i}{\left(n_i + n_T\right)}$. Note that for most "dilute solutions," mole fraction and molarity are related through $x_i = C_{iw} V_w$, where V_w is the molar volume of water (0.018 L mol^{-1}).

   A common, but less precise, unit of pollutant concentration in water is
4. Parts per million by mass, ppm_m = (mg of i per kg of solution). Note that for water, since the density is 1 kg L^{-1}, ppm_v is equivalent to mg of i per liter of solution. $ppm_v = mg\,L^{-1}$.

For acids and bases, the preferred unit of concentration is "normality," which is the number of equivalents per liter of solution. The number of equivalents per mole of acid equals the number of moles of H$^+$ the acid can potentially produce.

D.3   COMPARTMENT—SOIL/SEDIMENT

Concentration of a pollutant on soil/sediment is expressed as

1. Parts per million by mass $(ppm_m) = (W_i/W_{soil})\,10^6$, where W_i is the grams of pollutant i and W_{soil} is the total amount of soil in grams. Note that this is equivalent to milligrams of i per kilogram of soil or sediment. Hence $ppm_m = mg\,kg^{-1}$.
2. Moles of i per square meter of surface area of soil/sediment = (mg of i per kilogram of soil) $(10^{-3}\,S_a/M_i)$, where S_a is the specific area of the soil (m^2 per kilogram of soil) and M_i is the molecular weight of i.
3. Percent by weight of i = (g of i/g of soil) 100.

Appendix E: Dissociation Constants for Environmentally Significant Acids and Bases

Compound	K_{a1}	K_{a2}	K_{a3}
Acids: $HA + H_2O \rightleftharpoons H_3O^+ + A^-$; $\quad K_a = \dfrac{\left[H_3O^+\right]\left[A^-\right]}{\left[HA\right]}$			
Formic acid	2.1×10^{-4}		
Acetic acid	1.7×10^{-5}		
Propionic acid	1.4×10^{-5}		
Butyric acid	1.5×10^{-5}		
Valeric acid	1.6×10^{-5}		
Phenol	1.0×10^{-10}		
2-Nitrophenol	6.0×10^{-8}		
2-Chlorophenol	3.0×10^{-9}		
2,4-Dinitrophenol	6.0×10^{-6}		
2,4-Dichlorophenol	1.4×10^{-8}		
2,4,6-Trichlorophenol	1.0×10^{-6}		
Pentachlorophenol	1.8×10^{-5}		
H_2CO_3	4.3×10^{-7}	4.7×10^{-11}	
H_2SO_3	1.3×10^{-2}	6.0×10^{-8}	
H_2S	9.1×10^{-8}	1.3×10^{-13}	
H_3PO_4	7.5×10^{-3}	6.2×10^{-8}	4.8×10^{-13}
H_3BO_3	7.3×10^{-10}	1.8×10^{-13}	1.6×10^{-14}
H_2SO_4		1.2×10^{-2}	
HF	3.5×10^{-4}		

Compound	K_{b1}	K_{b2}	K_{b3}

Bases: $B + H_2O \rightleftharpoons BH^+ + OH^-$; $K_b = \dfrac{\left[BH^+\right]\left[OH^-\right]}{\left[B\right]}$

Compound	K_{b1}	K_{b2}	K_{b3}
Acetate	5.5×10^{-10}		
Ammonia	1.8×10^{-5}		
N_2H_4	8.9×10^{-5}		
Aniline	2.5×10^{-5}		
Methylamine	2.2×10^{-11}		
Tributylamine	1.3×10^{-11}		
Glycine	4.4×10^{-3}	1.7×10^{-10}	
Urea	7.9×10^{-1}		
$Fe(OH)_3$	3.1×10^{-12}	5.0×10^{-12}	
$Al(OH)_3$	5.0×10^{-9}	2.0×10^{-10}	
$CaOH^+$		3.5×10^{-2}	
$MgOH^+$		2.5×10^{-3}	

Appendix F: Bond Contributions to Log K_{AW} for the Meylan and Howard Model

(A) Bond Contributions			
Bond	**Contribution, q_i**	**Bond**	**Contribution, q_i**
C–H	−0.1197	C_{ar}–H	−0.1543
C–C	0.1153	C_{ar}–C_{ar}	0.2638 (intra-ring C to C)
C–C_{ar}	0.1619		0.1490 (external C to C)
C–C_o	0.0635	C_{ar}–Cl	−0.0241
C–C_t	0.5375	C_{ar}–O	0.3473
C–N	1.3001	C_{ar}–OH	0.5967
C–O	1.0855	C_{ar}–N	0.7304
C–Cl	0.3335	CO–H	1.2101
C–Br	0.8187	CO–O	0.0714
C_{ol}–H	−0.1005	CO–CO	2.4000
C_{ol}=C_o	0.0000	O–H	3.2318
C_{ol}–C_{ol}	0.0997	O–O	−0.4036
C_{ol}–Cl	0.0426	N–N	1.0956
C_{ol}–O	0.2051	N–H	1.2835
C_{tr}–H	0.0040		
C_{tr}–C_{tr}	0.0000		

(B) Correction Factors	
Feature	**Correction Factor, Q_j**
Linear or branched alkane	−0.75
Cyclic alkane	−0.28
Monoolefin	−0.20
Linear or branched aliphatic alcohol	−0.20
Additional alcohol functional group above one	−3.00
A chloroalkane with only one chlorine	+0.50
A totally halogenated halofluoroalkane	−0.90

Source: Adapted from Baum, E.J., *Chemical Property Estimation Theory and Application*, Lewis Publishers, New York, 1997.

Note: C_{ol}, olefinic C; C_{ar}, aromatic C; C_{tr}, bonded to a triple bond.

REFERENCE

Baum, E.J. (1997) *Chemical Property Estimation Theory and Application*, New York: Lewis
    Publishers.

Appendix G: Regression Analysis (The Linear Least Squares Methodology)

Regression analysis is a powerful tool for establishing a relationship between two or more variables. There are a number of cases in environmental engineering where a dependent variable y and an independent variable x are measured to give an array of data points (x_i, y_i). Examples are linear free energy relationships and numerous examples where adsorption data, partitioning data, reaction kinetic data, etc., are fitted to a straight-line equation. In each case, a plot of x_i versus y_i is made and the best-fit curve to the data is determined. The method is called linear least squares. Consider the figure given here:

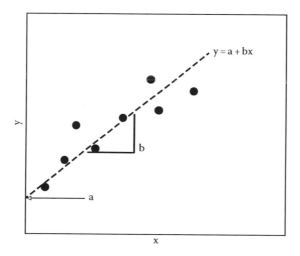

The best fit through the data points is a straight line with a slope of b and an intercept of a. The objective is to estimate (1) the slope b and intercept a from the data points, and (2) the degree of fit, i.e., how well does the data fit the straight line?

We make the primary assumption that there is negligible error in the independent variable x. For any given value of x_i, there is an observed value y_i and a corresponding calculated value $y_{i,\text{calcd}} = a + bx_i$. Thus, the residual sum of errors in the estimation is given by

$$W = \sum_i \left(y_i - y_{i,\text{calcd}}\right)^2 = \sum_i \left(y_i - a - bx_i\right)^2$$

In order to obtain the value of a and b, one minimizes the function W with respect to both a and b, i.e., differentiate the function with respect to a and b and set them equal to zero. Thus,

$$\frac{\partial W}{\partial a} = -2\sum_i \left(y_i - a - bx_i \right) = 0$$

and

$$\frac{\partial W}{\partial b} = -2\sum_i \left(y_i - a - bx_i \right) x_i = 0$$

These equations give us the following two equations:

$$na + b\sum_i x_i = \sum_i y_i$$

$$a\sum_i x_i + b\sum_i x_i^2 = \sum_i x_i y_i$$

where n is the total number of paired (x, y) data points.
Solving these equations gives us the values of a and b as follows:

$$a = \frac{\left(\sum_i x_i^2\right) \cdot \left(\sum_i y_i\right) - \left(\sum_i x_i\right)\left(\sum_i x_i y_i\right)}{\left(n\sum_i x_i^2\right) - \left(\sum_i x_i\right)^2}$$

$$b = \frac{\left(n\sum_i x_i y_i\right) - \left(\sum_i x_i \cdot \sum_i y_i\right)}{\left(n\sum_i x_i^2\right) - \left(\sum_i x_i\right)^2}$$

The respective *standard deviations* in the slope and intercept are then given by the following equations:

$$\sigma_a = \sigma_y \cdot \left\{ \frac{\sum_i x_i^2}{n\sum_i x_i^2 - \left(\sum_i x_i\right)^2} \right\}^{1/2}$$

$$\sigma_b = \sigma_y \cdot \left\{ \frac{n}{n\sum_i x_i^2 - \left(\sum_i x_i\right)^2} \right\}^{1/2}$$

where

$$\sigma_y^2 = \frac{\sum_i \left(y_i - a - bx_i\right)^2}{\left(n-2\right)}$$

The denominator in this equation $(n - 2)$ represents the total number of degrees of freedom excluding the two degrees of freedom already used up, namely the slope and intercept. σ_y is called the *standard error of estimate*.

The goodness of fit is ascertained in terms of the *regression coefficient*, which is given by the following equation:

$$r^2 = \frac{\sum_i \left(y_{i,calcd} - \bar{y}\right)^2}{\sum_i \left(y_i - \bar{y}\right)^2}$$

where
$y_{i,\,calcd}$ is the value estimated from the regression equation
y_i is the measured value
$\bar{y}$ is the mean of the measured values

In the case of an exact fit, r^2 will be unity and r^2 will decrease in magnitude as the quality of the fit of the model to the data diminishes. An r^2 of near zero indicates that the best estimate of the dependent variable is not any better than the overall mean of the independent variable estimated from the data. It is possible that a poor model can give a high r^2 and, alternatively, a low r^2 can be displayed for a good model. Hence, the correlation coefficient should only be taken as a general indicator of the goodness of the fit.

Appendix H: Error Function and Complementary Error Function Definitions

The "error function" is defined as

$$\text{erf}\left(x\right) = \left(\frac{2}{\sqrt{\pi}}\right) \cdot \int_{0}^{x} e^{-y^2} dy$$

The "complementary error function" is defined as

$$\text{erfc}\left(x\right) = 1 - \text{erf}\left(x\right)$$

Some characteristics of the error functions are given here:

erf(0) = 0	erfc(0) = 1
erf(∞) = 1	erfc(∞) = 0
erf($-\infty$) = -1	
erfc($-x$) = $-$erf(x)	erfc($-x$) = 1 $-$ erf($-x$) = 1 + erf(x) = 2 $-$ erfc(x)

The following is a partial table of error function values. For a complete tabulation, see Abramovitz and Stegun (1965).

x	erf(x)
0.00	0.00000 00000
0.02	0.02256 45747
0.04	0.04511 11061
0.06	0.06762 15944
0.08	0.09007 81258
0.10	0.11246 29160
0.20	0.22270 25892
0.30	0.32862 67595
0.40	0.42839 23550
0.50	0.52049 98778

(Continued)

x	erf(x)
0.60	0.60385 60908
0.70	0.67780 11938
0.80	0.74210 09647
0.90	0.79690 82124
1.00	0.84270 07929
2.00	0.99532 22650
3.00	1.00000

REFERENCE

Abramovitz, M. and Stegun, I.A. (1965) *Handbook of Mathematical Functions*. New York: Dover Publications.

Appendix I: Cancer Slope Factor and Inhalation Unit Risk for Selected Carcinogens

Chemical	Cancer Oral Slope Factor[a] ($mg\,kg^{-1}\,d^{-1}$)	Oral Reference Dose, RfD[b] ($mg\,kg^{-1}\,d^{-1}$)	Inhalation Unit Risk[c] (per $\mu g\,m^{-3}$)	Reference Inhalation Concentration, RfC[d] ($mg\,m^{-3}$)
Arsenic	1.5	3.4×10^{-4}	4.3×10^{-3}	—
Benzene	4.0×10^{-3}	4.0×10^{-3}	2.2×10^{-6}	3×10^{-2}
Cadmium	—	5×10^{-4} (water)	1.8×10^{-3}	—
		1×10^{-3} (food)		
Carbon tetrachloride	0.13	7.4×10^{-4}	1.5×10^{-5}	—
Chloroform	—	1×10^{-2}	2.3×10^{-5}	—
Methyl chloride	0.0075	—	—	9×10^{-2}
p,p'-DDT	0.34	5×10^{-4}	—	—
Dieldrin	0.16	5×10^{-5}	—	—
Hexachloroethane	0.014	1×10^{-3}	4.0×10^{-6}	—
PCBs	0.04 – 2.0	—	1.0×10^{-4}	—
Naphthalene	—	2.0×10^{-2}	—	3×10^{-3}
Vinyl chloride	0.72 – 1.5	3.0×10^{-3}	$(4.4 - 8.8) \times 10^{-6}$	1.0×10^{-1}

Source: U.S. EPA, IRIS database, 1989, http://www.epa.gov/iriswebp/iris/subst/index.html.

[a] *Oral slope factor*: An upper bound, approximating a 95% confidence limit, on the increased cancer risk from a lifetime oral exposure to an agent. This estimate, usually expressed in units of a population affected per mg $(kg\,d)^{-1}$, is generally for use in the low-dose region of the dose–exposure relationship, i.e., for exposures corresponding to risks less than 1 in 100.

[b] *RfD*: An estimate of a daily oral exposure to human population that is likely to be without an appreciable risk of deleterious effects during a lifetime.

[c] *Inhalation unit risk*: The upper bound excess lifetime cancer risk estimated to result from continuous exposure to an agent at a concentration of 1 $\mu g\,L^{-1}$ in water or 1 $\mu g\,m^{-3}$ in air. The interpretation of inhalation unit risk would be as follows: If unit risk = $2 \times 10^{-6}\,\mu g\,L^{-1}$, 2 excess cancer cases are expected to develop per 1,000,000 people if exposed daily for a lifetime to 1 μg of the chemical in 1 L of drinking water.

[d] *RfC*: An estimate of a continuous inhalation exposure to the human population that is likely to be without an appreciable risk of deleterious effects during a lifetime.

REFERENCE

U.S. EPA. (1989) IRIS database. http://www.epa.gov/iriswebp/iris/subst/index.html.

Appendix J: U.S. National Ambient Air Quality Standards

Pollutant	Primary Standards		Secondary Standards	
	Level	Averaging Time	Level	Averaging Time
Carbon monoxide	9 ppm (10 mg m^{-3}) 35 ppm (40 mg m^{-3})	8 h 1 h	None	
Lead	1.5 µg m^{-3}	Quarterly average	Same as primary	
Nitrogen dioxide	0.053 ppm (100 µg m^{-3})	Annual (arithmetic mean)	Same as primary	
Particulate matter (PM$_{10}$)	150 µg m^{-3}	24 h	Same as primary	
Particulate matter (PM$_{2.5}$)	15.0 µg m^{-3}	Annual (arithmetic mean)	Same as primary	
	35 µg m^{-3}	24 h	Same as primary	
Ozone	0.075 ppm (2008 std)	8 h	Same as primary	
	0.08 ppm (1997 std)	8 h	Same as primary	
	0.12 ppm	1 h (applies only in limited areas)	Same as primary	
Sulfur dioxide	0.03 ppm	Annual (arithmetic mean)	0.5 ppm (1300 µg m^{-3})	3 h
	0.14 ppm	24 h		

Source: http://www.epa.gov/air/criteria.html#1.

Index

A

C

Calo direct measurement technique, 233
CalTOX, 6
Canadian Environmental Protection Act (CEPA), 7
Cancer dose-response models, 6
Capillary condensation, 61
Capillary phenomenon, 40
Carbon dioxide (CO_2)
    alkalinity of water, 352, 354
    dissolution of, 351
    emissions, percent contributions, 350–351
    global temperature change, 358
    monthly average concentration, 350
    Revelle buffer factor, 352–354
    two-box model, oceans, 354–358
Carnot's heat engine, 19–20
Cell reactor, 401–405
Celsius thermometers, 17
Centimeter-gram-second (CGS) system, 9, 11–12
CFCs, *see* Chlorofluorocarbons
Chain reactions, 129–132
Chapman mechanism, 360
Characteristic time, S(IV) oxidation, 346–347
ChemCAN, 6
Chemical kinetics
    continuous flow reactor, 166
    definition, 4
    dispersion and reaction, 179–180
    environmental systems applications, *see*
        Environmental systems, chemical
        kinetics applications
    heterogeneous medium
        diffusive flux, 181
        first-order surface reaction rate, 181
        kinetics and transport, fluid/fluid
          interfaces, 182–185
        reaction and diffusion, 182–189
    ideal reactors
        batch reactor, 166, 168
        CSTR, 168–175
        mass balance/mass conservation, 167
        PFR, 170–175
        plug flow/tubular reactor, 169–170
        steady state and equilibrium, 175
        time/reactor volume, 167
    nonideal reactors
        dispersion model, 176–177
        tanks-in-series model, 177–179
Chemical manufacturing process
    environmental risk evaluation
        absorption oil flow rate, 418–419
        air emission rates, 416–418
        fugacity level II calculations, 418–420
        overall process flow sheet, 416–417
        relative risk index, 418
    traditional *vs.* green, 413

Chemical potentials, 48
    equilibrium adsorption isotherms,
        interfaces, 87
    fugacity model, 45
    Gibbs free energy and
        Gibbs-Duhem relationship, single
          phase, 35–36
        homogeneous single phase, 34
        partial molar Gibbs free energy, 34
        significance of, 35
        standard states for, 35–36
Chemical process industry (CPI), 84
Chemical reaction kinetics
    activation energy, 119–128
    definition, 101
    environmental catalysis, 133–147
    environmental engineering, 101
    environmental photochemical reactions,
        154–159
    enzyme catalysis, 159–166
    equilibrium, 102–104
    kinetic rate laws, 106–119
    natural systems, 101
    quasi-steady state behavior, 101
    reaction in solutions, 132–133
    reaction mechanisms, 101, 128–132
    reaction rate, order and rate constant, 104–106
    redox reactions, environmental systems,
        147–154
    time-dependent behavior, environmental
        systems, 101
    unsteady state behavior, 102
    waste treatment, 100
Chemical thermodynamics
    definition, 4
    environmental processes, applications in, 9–10
    equilibrium partitioning, 5
    role of, 9, 11
Chemiosmosis, 392
Chlordane, GAC, 92
Chlorobenzene, 376
Chlorofluorocarbons (CFCs), 362–365
Chloroform, 232, 304–305
Chlorpyrifos, 250
Chromatography, 91
Chronic daily intake (CDI), 7
Clausius-Clapeyron equation
    Antoine equation, 56
    critical point, 54
    dynamic equilibrium, 54
    entropy change, 55
    estimations, 58
    intermolecular forces, 54, 57
    molar changes, 55
    phase diagram, 54
    subcooled liquid state, 55, 57
    sublimation, 56